"Carl Zimmer lifts off the lid, dumps out the contents, and sorts through the pieces of one of history's most problematic ideas: heredity. Deftly touching on psychology, genetics, race, and politics, *She Has Her Mother's Laugh* is a superb guide to a subject that is only becoming more important. Along the way, it explains some remarkably complicated science with equally remarkable clarity—a totally impressive job all around."
—Charles C. Mann, author of *1491: New Revelations of the Americas Before Columbus*

"Carl Zimmer is not only among my favorite science writers—he's also now responsible for making me wonder why there is more Neanderthal DNA on earth right now than when Neanderthals were here, and why humanity is getting taller and smarter in the past few generations. *She Has Her Mother's Laugh* explains how our emerging understanding of genetics is touching almost every part of society and will increasingly touch our lives."
—Charles Duhigg, author of *Smarter Faster Better* and *The Power of Habit*

"With this book, Carl Zimmer rises from being our best biological science writer to being one of our very best nonfiction writers in any field, period."
—Kevin Padian, professor of integrative biology, UC Berkeley

"How every characteristic—from genes to personality—is passed down from one generation to the next is one of the most fundamental, complex, misunderstood, and misused enigmas of biology. In this beautifully written, heartfelt, and enjoyable masterpiece, Zimmer weaves together history, autobiography, and science to elucidate the mysteries of heredity and why we should care. I couldn't put this book down and can't recommend it too highly."
—Daniel E. Lieberman, Harvard University, author of *The Story of the Human Body*

"*She Has Her Mother's Laugh* is at once enlightening and utterly compelling. Carl Zimmer weaves spellbinding narrative with luminous science writing to give us the story of heredity, the story of us all. Anyone interested in where we came from and where we are going—which is to say, everyone—will want to read it."
—Jennifer Ackerman, author of *The Genius of Birds* and *Chance in the House of Fate*

"Traversing time and societies, the personal and the political, the moral and the scientific, *She Has Her Mother's Laugh* takes readers on an endlessly mesmerizing journey of what it means to be human. Carl Zimmer has created a brilliant canvas of life that is at times hopeful, at times horrifying, and always beautifully rendered. I could hope for no better guide into the complexities, perils, and, ultimately, potential of what the science of heredity has in store for the world."
—Maria Konnikova, author of *The Confidence Game*

"With his latest work, Zimmer has assured his place as one of the greatest science writers of our time. *She Has Her Mother's Laugh* is an extraordinary exploration of a topic that is at once familiar and foreign, and touches every one of us. With the eloquence of a poet and the expertise of a scientist, Zimmer has created a nonfiction thriller that will change the way you think about your family, those you love, and the past and future."
—Brian Hare, Duke University, coauthor of *The Genius of Dogs*

BY CARL ZIMMER

At the Water's Edge

Parasite Rex

Evolution: The Triumph of an Idea

Soul Made Flesh

The Descent of Man: The Concise Edition

Microcosm

The Tangled Bank

Brain Cuttings

Science Ink

Evolution: Making Sense of Life

A Planet of Viruses

SHE
HAS HER
MOTHER'S
LAUGH

The Powers,
Perversions,
and Potential of Heredity

CARL ZIMMER

DUTTON

DUTTON

An imprint of Penguin Random House LLC
375 Hudson Street
New York, New York 10014

LIBRARY OF CONGRESS CATALOGING-IN-PUBLICATION DATA
Names: Zimmer, Carl, 1966- author.
Title: She has her mother's laugh : the powers, perversions, and
potential of heredity / Carl Zimmer.
Other titles: What heredity is, is not, and may become
Description: New York, New York : Dutton,
an imprint of Penguin Random House
LLC, [2018] | Includes bibliographical references and index.
Identifiers: LCCN 2017046101| ISBN 9781101984598 (hardcover) |
ISBN 9781101984604 (ebook)
Subjects: | MESH: Heredity—genetics
Classification: LCC QH431 | NLM QU 500 | DDC 576.5—dc23
LC record available at https://lccn.loc.gov/2017046101

Printed in the United States of America
5   7   9   10   8   6

BOOK DESIGN BY AMY HILL

*To Grace, for spending this juncture between
the past and the future with me*

The whole subject of inheritance is wonderful.

—Charles Darwin

CONTENTS

SHE HAS HER
MOTHER'S LAUGH

PROLOGUE

THE WORST SCARES of my life have usually come in unfamiliar places. I still panic a bit when I remember traveling into a Sumatran jungle only to discover my brother, Ben, had dengue fever. I lose a bit of breath any time I think about a night in Bujumbura when a friend and I got mugged. My fingers still curl when I recall a fossil-mad paleontologist leading me to the slick mossy edge of a Newfoundland cliff in search of Precambrian life. But the greatest scare of all, the one that made the world suddenly unfamiliar, swept over me while I was sitting with my wife, Grace, in the comfort of an obstetrician's office.

Grace was pregnant with our first child, and our obstetrician had insisted we meet with a genetics counselor. We didn't see the point. We felt untroubled in being carried along into the future, wherever we might end up. We knew Grace had a second heartbeat inside her, a healthy one, and that seemed enough to know. We didn't even want to find out if the baby was a girl or a boy. We would just debate names in two columns: Liam or Henry, Charlotte or Catherine.

Still, our doctor insisted. And so one afternoon we went to an office in lower Manhattan, where we sat down with a middle-aged woman, perhaps a decade older than us. She was cheerful and clear, talking about our child's health beyond what the thrum of a heartbeat could tell us. We were politely cool, wanting to end this appointment as soon as possible.

We had already talked about the risks we faced starting a family in our thirties, the climbing odds that our children might have Down syndrome.

We agreed that we'd deal with whatever challenges our child faced. I felt proud of my commitment. But now, when I look back at my younger self, I'm not so impressed. I didn't know anything at the time about what it's actually like raising a child with Down syndrome. A few years later, I would get to know some parents who were doing just that. Through them, I would get a glimpse of that life: of round after round of heart surgeries, of the struggle to teach children how to behave with outsiders, of the worries about a child's future after one's own death.

But as we sat that day with our genetics counselor, I was still blithe, still confident. The counselor could tell we didn't want to be there, but she managed to keep the conversation alive. Down syndrome was not the only thing expectant parents should think about, she said. It was possible that the two of us carried genetic variations that we could pass down to our child, causing other disorders. The counselor took out a piece of paper and drew a family tree, to show us how genes were inherited.

"You don't have to explain all that to us," I assured her. After all, I wrote about things like genes for a living. I didn't need a high school lecture.

"Well, let me ask you a little about your family," she replied.

It was 2001. A few months beforehand, two geneticists had come to the White House to stand next to President Bill Clinton for an announcement. "We are here to celebrate the completion of the first survey of the entire human genome," Clinton said. "Without a doubt, this is the most important, most wondrous map ever produced by humankind."

The "entire human genome" that Clinton was hailing didn't come from any single person on Earth. It was an error-ridden draft, a collage of genetic material pieced together from a mix of people. And it had cost $3 billion. Rough as it was, however, its completion was a milestone in the history of science. A rough map is far better than no map at all. Scientists began to compare the human genome to the genomes of other species, in order to learn on a molecular level how we evolved from common ancestors. They could examine the twenty thousand–odd genes that encode human proteins, one at a time, to learn about how they helped make a human and how they helped make us sick.

In 2001, Grace and I couldn't expect to see the genome of our child, to examine in fine detail how our DNA combined into a new person. We might as well have imagined buying a nuclear submarine. Instead, our genetics counselor performed a kind of verbal genome sequencing. She asked us about our families. The stories we told her gave her hints about whether mutations lurked in our chromosomes that might mix into dangerous possibilities in our child.

Grace's story was quick: Irish, through and through. Her ancestors had arrived in the United States in the early twentieth century, from Galway on one side, Kerry and Derry on the other. My story, as far as I understood it, was a muddle. My father was Jewish, and his family had come from eastern Europe in the late 1800s. Since Zimmer was German, I assumed he must have some German ancestry, too. My mother's family was mostly English with some German mixed in, and possibly some Irish—although a bizarre family story clattered down through the generations that our ancestor who claimed to be Irish was actually Welsh, because no one would want to admit to being Welsh. Oh, I added, someone on my mother's side of the family had definitely come over on the *Mayflower*. I was under the impression that he fell off the ship and had to get fished out of the Atlantic.

As I spoke, I could sense my smugness dissolving at its margins. What did I really know about the people who had come before me? I could barely remember their names. How could I know anything about what I had inherited from them?

Our counselor explained that my Jewish ancestry might raise the possibility of Tay-Sachs disease, a nerve-destroying disorder caused by inheriting two mutant copies of a gene called HEXA. The fact that my mother wasn't Jewish lowered the odds that I had the mutation. And even if I did, Grace's Irish ancestry probably meant we had nothing to worry about.

The more we talked about our genes, the more alien they felt to me. My mutations seemed to flicker in my DNA like red warning lights. Maybe one of the lights was on a copy of my HEXA gene. Maybe I had others in genes that scientists had yet to name, but could still wreak havoc on our child. I had willingly become a conduit for heredity, allowing the biological past to make its way into the future. And yet I had no idea of what I was passing on.

Our counselor kept trying to flush out clues. Did any relatives die of cancer? What kind? How old were they? Anyone have a stroke? I tried to build a medical pedigree for her, but all I could recall were secondhand stories. I recalled William Zimmer, my father's father, who died in his forties from a heart attack—I think a heart attack? But didn't an old cousin once tell me about rumors of overwork and despair? His wife, my grandmother, died of some kind of cancer, I knew. Was it her ovaries, or her lymph nodes? She had died years before I was born, and no one had wanted to burden me as a child with the oncological particulars.

How, I wondered, could someone like me, with so little grasp of his own heredity, be permitted to have a child? It was then, in a panic, that I recalled an uncle I had never met. I didn't even know he existed until I was a teenager. One day my mother told me about her brother, Harry, how she would visit Harry's crib every morning to say hello. One morning, the crib was empty.

The story left me flummoxed, outraged. It wouldn't be until I was much older that I'd appreciate how doctors in the 1950s ordered parents to put children like Harry in a home and move on with their lives. I had no grasp of the awkward shame that would make those children all the more invisible.

I tried to describe Uncle Harry to our genetics counselor, but I might as well have tried sketching a ghost. As I blathered on, I convinced myself that our child was in danger. Whatever Harry had inherited from our ancestors had traveled silently into me. And from me it had traveled to my child, in whom it would cause some sort of disaster.

The counselor didn't look worried as I spoke. That irritated me. She asked me if I knew anything about Harry's condition. Was it fragile X? What did his hands and feet look like?

I had no answers. I had never met him. I had never even tried to track him down. I suppose I had been frightened of him gazing at me as he would at any stranger. We might share some DNA, but did we share anything that really mattered?

"Well," the counselor said calmly, "fragile X is carried on the X chromosome. So we don't have to worry about that."

Her calmness now looked to me like sheer incompetence. "How can you be so sure?" I asked.

"We would know," she assured me.

"How would we know?" I demanded.

The counselor smiled with the steadiness of a diplomat meeting a dictator. "You'd be severely retarded," she said.

She started to draw again, just to make sure I understood what she was saying. Women have two X chromosomes, she explained, and men have one X and one Y. A woman with a fragile X mutation on one copy of her X chromosome will be healthy, because her other X chromosome can compensate. Men have no backup. If I carried the mutation, it would have been obvious from when I was a baby.

I listened to the rest of her lesson without interrupting.

A few months later, Grace gave birth to our child, a girl as it turned out. We named her Charlotte. When I carried her out of the hospital in a baby seat, I couldn't believe that we were being entrusted with this life. She didn't display any sign of a hereditary disease. She grew and thrived. I looked for heredity's prints on Charlotte's clay. I inspected her face, aligning photos of her with snapshots of Grace as a baby. Sometimes I thought I could hear heredity. To my ear, at least, she has her mother's laugh.

As I write this, Charlotte is now fifteen. She has a thirteen-year-old sister named Veronica. Watching them grow up, I have pondered heredity even more. I wondered about the source of their different shades of skin color, the tint of their irises, Charlotte's obsession with the dark matter of the universe, or Veronica's gift for singing. ("She didn't get that from me." "Well, she certainly didn't get it from *me*.")

Those thoughts led me to wonder about heredity itself. It is a word that we all know. Nobody needs an introduction to it, the way we might to *meiosis* or *allele*. We all feel like we're on a first-name basis with heredity. We use it to make sense of some of the most important parts of our lives. Yet it means many different things to us, which often don't line up with each other. Heredity is why we're like our ancestors. Heredity is the inheritance of a gift, or of a curse. Heredity defines us through our biological past. It also gives us a chance at immortality by extending heredity into the future.

I began to dig into heredity's history, and ended up in an underground palace. For millennia, humans have told stories about how the past gave rise to the present, how people resemble their parents—or, for some reason, do not. And yet no one used the word *heredity* as we do today before the 1700s. The modern concept of heredity, as a matter worthy of scientific investigation, didn't gel for another century after that. Charles Darwin helped turn it into a scientific question, a question he did his best to answer. He failed spectacularly. In the early 1900s, the birth of genetics seemed to offer an answer at last. Gradually, people translated their old notions and values about heredity into a language of genes. As the technology for studying genes grew cheaper and faster, people became comfortable with examining their own DNA. They began to order genetic tests to link themselves to missing parents, to distant ancestors, to racial identities. Genes became the blessing and the curse that our ancestors bestowed on us.

But very often genes cannot give us what we really want from heredity. Each of us carries an amalgam of fragments of DNA, stitched together from some of our many ancestors. Each piece has its own ancestry, traveling a different path back through human history. A particular fragment may sometimes be cause for worry, but most of our DNA influences who were are—our appearance, our height, our penchants—in inconceivably subtle ways.

While we may expect too much from our inherited genes, we also don't give heredity the full credit it's due. We've come to define heredity purely as the genes that parents pass down to their children. But heredity continues within us, as a single cell gives rise to a pedigree of trillions of cells that make up our entire bodies. And if we want to say we inherit genes from our ancestors—using a word that once referred to kingdoms and estates—then we should consider the possibility that we inherit other things that matter greatly to our existence, from the microbes that swarm our bodies to the technology we use to make life more comfortable for ourselves. We should try to redefine the word *heredity*, to create a more generous definition that's closer to nature than to our demands and fears.

I woke up one bright September morning and hoisted Charlotte, now two months old, from her crib. As Grace caught up on her sleep, I carried Charlotte to the living room, trying to keep her quiet. She was irascible, and the only way I could calm her was to bounce her in my arms. To fill the morning hours, I kept the television on: the chatter of local news and celebrity trivia, the pleasant weather forecast, a passing report of a small fire in an office at the World Trade Center.

Having been a father for all of two months had made me keenly aware of the ocean of words that surrounded my family. They flowed from our television and from the mouths of friends; they looked up from newspapers and leaped down from billboards. For now, Charlotte could not make sense of these words, but they were washing over her anyway, molding her developing brain to take on the capacity for language. She would inherit English from us, along with the genes in her cells.

She would inherit a world as well, a human-shaped environment that would help determine the opportunities and limits of her life. Before that morning, I felt familiar with that world. It would boast brain surgery and probes headed for Saturn. It would also be a world of spreading asphalt and shrinking forests. But the fire grew that morning, and the television hosts mentioned reports that a plane had crashed into it. I rocked Charlotte as the television wove between ads and cooking tips and a second plane crashing into the second tower. The day mushroomed into catastrophe.

Charlotte's fussing faded into sleepy comfort. She looked up at me and I down at her. I realized how consumed I had become with wondering what versions of DNA she might have inherited from me. I kept my arms folded tightly around her, wondering now what sort of world she was inheriting.

PART I

A Stroke on the Cheek

CHAPTER 1

The Light Trifle of His Substance

THE EMPEROR, clad in black, hobbled into the great hall. An audience of powerful men had assembled in the Palace of Brussels on October 25, 1555, to listen to a speech by the Holy Roman emperor Charles V. At the time, he ruled over much of Europe as well as wide swaths of the New World. A few years before, Titian had painted his portrait, astride a war horse, encased in armor, brandishing a lance. But now, at fifty-five, he had become toothless and blank-stared. As he made his way to the front of the hall, he had to lean on both a cane and Prince William of Orange. Trailing Charles was his twenty-eight-year-old son, Philip. There was no question that they were related. Father and son alike had lower jaws that jutted far forward, leaving their mouths to hang open. Their shared look was so distinctive that anatomists later named it after their dynasty: the Habsburg jaw.

Father and son climbed together up a few steps onto a dais, where they turned and sat before the assembly. They listened to the president of the Council of Flanders announce that Charles had summoned the audience to witness his abdication. They would now have to transfer their allegiance from Charles V to Philip II, his rightful heir.

Charles then rose from the throne and put on a pair of spectacles. He read from a page of notes, reflecting on his forty-year reign. Over those decades, he had expanded his power across much of the world. In addition to Spain, he ruled the Holy Roman Empire, the Low Countries, and much of Italy. His power extended from Mexico to Peru, where his armies had recently crushed the Inca Empire. Waves of ships rolled back east across the Atlantic, arriving at Spanish ports to unload gold and silver.

Starting in the 1540s, however, Charles had begun to flag. He developed gout and hemorrhoids. His battles now ended in fewer victories and more stalemates. Charles grew depressed, sometimes so despondent he would lock himself away in a chamber. His chief consolation was his son. Charles had put Philip in charge of Spain when he was still just a teenager, and Philip had amply proven himself fit to inherit Charles's power.

Now, in 1555, Charles was content to make him a king. As he finished his speech, he turned to Philip. "May the Almighty bless you with a son," he said, "to whom, when old and stricken with disease, you may be able to resign your kingdom with the same good-will with which I now resign mine to you."

It took a couple of years for all the formalities to get squared away, for Charles to retire to a monastery that he filled with clocks, and for his son to be crowned. But during all that time, the transition rolled along smoothly. No one objected to transferring their allegiance. What could be more natural, after all, than a prince succeeding his father? For anyone else to take control of the empire would have been to defy the laws of heredity.

Heredity—herencia in Spanish, hérédité in French, eredità in Italian—originally came from the Latin word hereditas. The Romans did not use their word as we typically use ours today, to describe the process by which we inherit genes and other parts of our biology. They used hereditas as a legal term, referring to the state of being an heir. "If we become heirs to a certain person," the jurist Gaius wrote, "that person's assets pass to us."

It sounded simple enough, but Romans fought bitterly over heredity. Their conflicts accounted for two-thirds of all the lawsuits in Roman courts. If a wealthy man died without a will, his children would be first in line to inherit his fortune—except any daughters who had married into other families. Next in line would be the father's brothers and their children, then more distantly related kin.

Rome's system was one among many. Among the Iroquois, a child might have many mothers. In many South American societies, a child could have many fathers; any man who had sex with a pregnant woman was considered a parent to her unborn child. In some societies, kinship had meaning only through the father's line, others only through the mother's. The Apinayé of Brazil had it both ways: The men trace their ancestry back through their father's line, while the Apinayé women trace theirs back through their mother's. The words people used for their kin reflected how they organized relatives into a constellation of heredity; Hawaiians, for example, could use the same term for both sisters and female cousins.

Medieval Europe inherited some of Rome's hereditary customs, but over the centuries new rules emerged. In some countries, the sons split their father's land. In others, only the eldest inherited it. In others still, it went to the youngest son. In the early Middle Ages, daughters sometimes became heirs, too, but as the centuries passed, they were mostly shut out.

As Europe grew wealthier, new hereditary rules took hold to keep the fortunes intact. The most powerful families of all took on titles and crowns, which were handed down through hereditary succession, to a son, preferably; if not, then a daughter or perhaps a grandnephew. Sometimes the branches of a dead monarch's family would fight for the crown, justifying their claim on heredity. But these claims became hard to judge when the memories of ancestors faded.

Noble families fought this forgetting by putting their genealogies in writing. In the Middle Ages, Venice's Great Council created the Golden Book, which every son from the prominent old families of the republic signed on his eighteenth birthday. Only those whose names were recorded in the book could become members of the council. As unbroken lines of descent from noble ancestors became more important, leading families paid artists for visual propaganda. At first they represented heredity as vertical lines, but later they started painting simple trees. They might paint the founder of a noble lineage at the base of the tree, and his descendants perched on branches. The French gave these pictures a name in honor of their forking shape: *pé de grue*, meaning "crane's foot." In English, the word became *pedigree*.

By the 1400s, pedigrees had become instantly recognizable, as evidenced by a pageant that was put on in 1432 to honor Henry VI of England. The king, only ten years old at the time, had been crowned king of France. On his return to London, the city came out in force to celebrate his expanded power. Giant tableaux lined his path. He passed towers and tabernacles; Londoners dressed up as Grace, Fortune, and Wisdom, as well as a multitude of angels. The centerpiece of the citywide display was a castle constructed from green jasper, displaying a pair of trees.

One tree traced Henry's ancestry back to the early kings of both England and France. The other was a tree that traced Jesus's ancestry all the way back to King David and beyond. These trees were a blend of fact and fiction, of display and concealment. They represented only those supposed ancestors whose kinship bolstered Henry's claim to power. The trees lacked siblings and cousins, bastards and wives. The most important omission of all was the House of York, Henry's rivals to the throne. But erasing them from Henry's tree did not erase them from history. Henry VI would be murdered at forty-nine, after which the House of York seized control of England.

When Charles V abdicated in 1555, he created a pageant of his own. Father and son stood on stage, side by side. The noblemen who sat before the emperor and his prince silently endorsed the hereditary transfer of power. Perhaps, as they listened to Charles deliver his speech, they turned their gaze from father to son and back. If they settled their gaze on the royal jaws, they would not have said that Charles had *inherited* his jaw from his father. They could recognize a family resemblance, but they did not explain it with the language of thrones and estates.

To account for why Charles and Philip looked alike, sixteenth-century Europeans relied largely on the teachings of ancient Greeks and Romans. The Greek physician Hippocrates argued that men and women both produced semen, and that new life formed when the two were mixed. That blending accounted for how children ended up with a mix of their parents' characteristics. Aristotle disagreed, believing that only men produced the seeds of life. Their seeds grew on menstrual blood inside women's bodies, developing into embryos. Aristotle and his followers believed a woman could influence the traits of her children, but only in the way the soil can influence how an acorn grows into an oak tree. "The mother is not the true parent of the child which is called hers," the Greek playwright Aeschylus wrote. "She is a nurse who tends the growth of young seed planted by the true parent, the male."

The classical world had less to say about why different parents passed down different traits—why some people were tall and others short, why some were dark and others pale. One widespread notion was that new differences arose through experiences—in other words, people could pass down a trait they acquired during their lives. In ancient Rome, for example, there was a prominent family called the Ahenobarbi. Their name means "red beard," a trait that set them apart in bright contrast to Rome's dark-haired majority. The Ahenobarbi themselves had started out dark-haired as well, according to legend. But one day, a member of the Ahenobarbi clan, a man named Lucius Domitius, was traveling home to Rome when he encountered the demigods Castor and Pollux (otherwise known as the Gemini twins). They told Domitius to deliver news to Rome that they had won a great battle. And then Castor and Pollux stroked his cheek. With that divine touch, the beard of Domitius turned the color of bronze, and he then passed down his red beard to all his male descendants.

Hippocrates provided his medical authority to another story of acquired traits, about a tribe known as the Longheads. A long head was a sign of nobility for the tribe, prompting parents to squeeze the skulls of newborns and wrap them in bandages. "Custom originally so acted that through force such a nature came into being," Hippocrates said. Eventually, Longhead babies came into the world with their heads already stretched out.

Other Greeks told similar stories—of men who lost fingers, for example, and then fathered fingerless children. "For the seed," Hippocrates wrote, "comes from all parts of the body, healthy seed from healthy parts, diseased seed from diseased parts." If those parts changed during a person's life, his or her seeds changed accordingly.

The place where people lived could also shape them, the Greeks believed, and even give them some of their national character. "The people of cold countries generally, and particularly those of Europe, are full of spirit, but deficient in skill and intelligence," Aristotle declared. They were therefore unfit to govern themselves or others. Asians had skill and intelligence, but lacked spirit, which was why they lived under the rule of despots. "The Greeks, intermediate in geographical position, unite the qualities of both sets of peoples," Aristotle wrote.

The theories of Aristotle and other ancient writers were preserved by Arab scholars, from whom Europeans learned of them in the Middle Ages. In the 1200s, the philosopher Albertus Magnus declared that the temperature and humidity of people's birthplace determined the color of their skin. Indians were especially good at math, Albertus thought, because the influence of the stars was especially strong in India.

But over the next three centuries, Europeans developed a new explanation for the link between one generation and the next: They were joined by blood. Even today, Westerners still use the word *blood* to talk about kinship, as if the two were equivalent in some obvious way. But other cultures thought of kinship in terms of other substances. On the Malaysian island of Langkawi, to pick just one counterexample, people traditionally believed that children gained kinship through what they ate. They consumed the same milk as their siblings, and they later ate the same rice grown from the same soil. These beliefs are so strong among the Langkawi that if children from two families nurse from the same woman, a marriage between them would be considered incest.

The European concept of blood gave ancestry a different form. It sealed off kinship from the outside world. A child was born with the blood of its parents coursing through its veins, and inherited all that went with it. Philip II was fit to inherit his father's crown because he had royal blood, which came from his father, and his grandfather before that. Genealogies became bloodlines, serving as proof that noble families were not tainted with lower-class blood. The Habsburgs were especially protective of their royal blood, only marrying other members of their extended family. Charles V married Isabella of Portugal, for example; they were both grandchildren of King Ferdinand and Queen Isabella of Spain.

Before long, Europeans even began to sort animals according to their blood. Of all birds, falcons had the noblest blood, and falconry was thus suitable to be the sport of kings. If a falcon mated with a less noble bird, the chicks were called bastards. Noblemen also became connoisseurs of dogs and horses, paying fortunes for pure-blooded breeds. For animals no less than people, inheriting noble blood meant inheriting noble traits like bravery and strength.

No experience could hide the virtue carried in the blood of man or beast. In a medieval romance called *Octavian,* the Roman emperor of the same name unknowingly fathers a child named Florentine, who ends up being raised by a butcher. Even in that lowly household, Florentine's noble blood cannot be masked. His adoptive father sends him to the market to sell two oxen, and Florentine trades them instead for a sparrow hawk.

In the 1400s, people began to use a new word to define a group of animals that shared the same blood: a *race*. A Spanish manual from around 1430 offered breeders tips for providing a "good race" of horse. Their stallion must "be good and beautiful and of good coat and the mare that she be large and well formed and of good coat." Before long, people were assigned to races as well. A priest named Alfonso Martínez de Toledo declared in 1438 that it's easy to tell the difference between men belonging to good and bad races. It doesn't matter how they're raised, Martínez de Toledo said. Imagine that the son of a laborer and the son of a knight are reared together on some isolated mountain away from their parents. The laborer's son would end up enjoying working in a farm field, Martínez de Toledo promised, while the knight's son would take pleasure only in riding horses and sword fighting.

"The good man of good race always returns to his origins," he wrote, "whereas the miserable man, of bad race or lineage, no matter how powerful or how rich, will always return to the villainy from which he descends."

In the late 1400s, Jews in Spain found themselves defined as a race of their own. For centuries, Jews across Europe had been tormented for all sorts of concocted crimes against Christians. In fifteenth-century Spain, thousands of Jews tried to escape this persecution by converting to Christianity, becoming so-called *conversos*. The self-proclaimed "Old Christians" remained hostile, rejecting the idea that Jews could escape their sinful inheritance with a mere oath. Nor could their children, for that matter, because Jewish immorality was carried in their blood and embedded in their seed, passed down from one generation to the next. "From the days of Alexander up till now, there has never been a treasonous act that did not involve a Jew or his descendants," the Spanish historian Gutierre Díaz de Games declared in 1435.

Spanish writers began referring to unconverted Jews and *conversos* alike as the Jewish "race." Christian men were warned not to have children with a woman of the Jewish race, in the same way that a fine stallion shouldn't be bred with a mare from a lower caste. In 1449 the Spanish city of Toledo began turning this hostility into law, decreeing that even a trace of Jewish blood disqualified a subject from holding office or marrying a true Christian.

The ban spread across Spain, expanding its scope along the way. Jewish blood now barred people from getting university degrees, inheriting estates, or even entering some parts of the country. In order to define Jews as a separate race, the majority of Spain had to define itself as a race of its own. Noble families now claimed that their genealogies extended back to the Visigoths. They boasted of the cleanliness of their blood, known as *limpieza de sangre*. They extolled the pale skin of Old Christians, which revealed the *sangre azul*—blue blood—coursing in the vessels beneath. The phrase would survive for centuries and cross the Atlantic, becoming a label for upper-class New Englanders.

Official certificates of purity were required for marriages between powerful Spanish families and for lucrative government posts. The Spanish Inquisition would follow up with their own detective work, getting testimony from relatives and neighbors. The inquisitors would investigate any rumor of Jewish ancestry—a report that an ancestor worked as a clothes merchant or a moneylender could be enough to arouse suspicion. The discovery of even a single Jew in one's ancestry could spell doom. Wealthy families would hire special race researchers, called *linajudos,* to marshal proof of their *limpieza de sangre*. Of course, just about every noble family actually did have some Jewish ancestry. The *linajudos* grew rich by inventing chronicles that left it out.

The label of race emerged around the time that Europeans began colonizing other parts of the world. They discovered more people to whom they could attach the label.

"I have found no monsters," Christopher Columbus wrote in a letter from the Caribbean in 1493. Instead of Cyclopses or Amazons, he encountered people, whom he named Indians. Columbus was not sure what to make of them at first. They seemed to flout Aristotle's rule about skin color: Even though they lived under a fierce sun, their skin was not black like that of Africans. They lacked clothes, steel, or weapons. Yet Columbus was impressed by their skill in building and piloting canoes. "A galley could not compete with them by rowing, because they travel incredibly fast," he said. "They are of subtle intelligence and can find their way around those seas."

While Columbus may have found some things to admire in the Native Americans he encountered, he didn't hesitate to force them into slavery. He dispatched some to work on farms or in mines; he sent hundreds more to Spain to be sold, although most died during the voyage across the Atlantic. Conquistadors and settlers followed Columbus's example. While some theologians pleaded that they treat Native Americans more humanely, others justified slavery by race. They declared Native Americans to be natural slaves, incapable of reason and designed by God to serve European masters.

"For them there is no tomorrow and they are content that they have enough to eat and drink for a week," wrote the Spanish jurist Juan Matienzo. "Nature proportioned their bodies so that they should have strength for personal service," said another scholar. "The Spaniards, on the other hand, are delicately proportioned, and were made prudent and clever, so that they should be able to lead a political and civil life."

Yet Native Americans suffered so badly from new diseases and hard labor that their population collapsed. In response, Charles V outlawed their slavery, although many ended up as impoverished peasants toiling on haciendas.

Now a new supply of workers had to be imported to take their place: African slaves.

For centuries, a vigorous slave trade had moved people out of sub-Saharan Africa into Europe, the Near East, and South Asia. The enslavers justified the practice by dehumanizing the enslaved. In 1377, the Tunisian scholar Ibn Khaldun, declared that Africans—as well as Slavs, another enslaved population—"possess attributes that are quite similar to those of dumb animals." But Khaldun still subscribed to a Hippocratic view of heredity. The black Africans who moved north into the cold climate of Europe, Khaldun claimed, were "found to produce descendants whose colour gradually turned white."

Muslims first brought African slaves into Spain in the eighth century, and their numbers grew as Portuguese traders captured Africans and brought them back to Europe. And yet the social boundaries between slavery and freedom remained loose. Some slaves of African ancestry gained their freedom and spent the rest of their lives alongside Europeans. Some joined the crews that sailed with Columbus to the New World.

As slave traders began shipping their cargo straight to Brazil, Peru, and Mexico, Europeans developed more enduring justifications of slavery. Some declared it a curse that Africans inherited from their biblical ancestors. Theologians had long claimed that Africans were the descendants of Ham, one of Noah's sons. After Ham saw his father naked, Noah cursed him, declaring that Ham's own son, Canaan, would never know freedom. "A slave of slaves shall he be to his brothers," Noah said.

In the 1400s, European scholars revived the story of Ham, casting it as the foundation of a distinct race, its cursed essence marked by dark skin. In 1448, the Portuguese scholar Gomes Eanes de Azurara wrote that because of Ham's sin, "his race should be subject to all the other races of the world. And from this race these blacks are descended."

———

Of all the powerful families in Europe, none worked as hard to keep themselves free of hereditary taint as the Habsburgs. Their blood ran blue, as their detailed genealogies could attest. To maintain their purity—and to keep the world's greatest empire intact—the Habsburgs only married among themselves. Cousins married cousins. Uncles married nieces. And yet the more time passed, the more the Habsburgs of Spain became burdened with hereditary suffering. The Habsburg jaw was the most prominent of their afflictions. Scientists have examined the paintings of Philip II and the other Habsburg kings to make a diagnosis, and they now suspect that the Habsburgs did not actually have an enlarged lower jaw so much as a small upper jaw that failed to develop to its full size. Philip II also suffered from other troubles familiar to the Habsburg family, including asthma, epilepsy, and melancholy.

To protect the family's power, Philip II married Maria Manuela, his first cousin. Genetically speaking, though, she was even more closely related than that. Philip's parents, Charles and Isabella, were also first cousins, while Maria Manuela's own parents were Charles and Isabella's siblings. Her father was Isabella's brother, and her mother was Charles's sister. The result of this close union was a sickly son born in 1545, Don Carlos. The right side of his body was less developed than the left, causing him to walk with a limp. He was born with a hunchback and a kind of deformed rib cage called pigeon chest.

Don Carlos was ten when his father became king. The boy wailed inconsolably and often refused to eat. But his many troubles didn't stop Philip from naming Don Carlos his "universal heir" at age twelve, destined to inherit all the kingdoms Philip had inherited from his own father, Charles.

By the time Don Carlos was nineteen, however, it was obvious to everyone, his father included, that something was very wrong. One visitor to the Spanish court wrote, "He is still like a child of seven years." Philip himself agreed. "Although other children develop late," the king wrote, "God wishes that mine lags far behind all others."

In his early twenties, Don Carlos grew violent. He once hurled a servant out of a window for displeasing him. He wasted hundreds of thousands of ducats. He tried to kill a nobleman. Philip decided that his son's "natural and unique temperament" would never change, and that he could not be allowed to rule. The king put on a suit of chain mail, assembled a group of armed courtiers, and stormed his son's room. They nailed Don Carlos's windows shut, removed all the weapons, papers, and treasure from the prince's room, and turned it into a prison cell. Don Carlos died there a few weeks later, on July 24, 1568, at age twenty-three.

Philip II remarried—this time choosing his own niece, Anna of Austria. In 1578, they had a son, Philip III, who succeeded his father twenty years later. Philip III married a cousin of his own and ruled till 1621, whereupon his own son, Philip IV, took over. It was during Philip IV's reign that the Spanish Empire—long the greatest power on Earth—went into decline. The Spanish army grew weak, and Portugal slipped from Philip IV's grasp. Gold and silver continued to arrive from the New World, but it headed straight to bankers elsewhere in Europe rather than enriching the people of Spain, who suffered from plagues and famines.

Philip IV was insulated from the chaos within the confines of his huge palace. He hung masterpieces by Rubens on the walls and listened to poets sing his praises. They called him the Planet King. The endless pageantry was disturbed only by the king's worry that his planetary throne might slip out of Habsburg hands if he didn't produce a son and heir.

Along with the Habsburg jaw and other ailments, the dynasty began to suffer an increasing number of miscarriages and infant deaths. Although they were among the most pampered people on Earth at the time, they suffered a higher rate of infant mortality than Spanish peasant families. Philip IV's first wife, Elisabeth of France, had a long string of miscarriages and babies who died young before her death in 1644. Their son, Balthasar Charles, managed to survive to age seventeen before dying of smallpox in 1646. The Habsburg dynasty now faced a crisis: It had no heir to succeed Philip IV after his death.

After Balthasar's death, Philip IV married his son's fiancée—and his own niece—Mariana. In 1651 she bore the king a daughter, Margaret Theresa, who would survive for twenty-two years. But over the following years, she had two more children who died young. In 1661, when their son, Philip Prospero, died at age four, Philip IV blamed their deaths on his lust for actresses.

———

When we look back to the seventeenth century, it can be hard to understand why Philip IV didn't recognize that the heredity of his family was to blame. But hardly anyone at the time thought about heredity this way during the years of the Habsburg dynasty. One of the few exceptions was the writer Michel de Montaigne, who published an essay in 1580 called "Of the Resemblance of Children to Their Fathers."

Montaigne was a French courtier who retired from political life in 1571 to sit in a castle tower and reflect on vanity and happiness, on liars and friendship. While he found comfort in this solitude, pain intruded on his contemplation from time to time, thanks to his kidney stones. One day, Montaigne transformed the stones into grist for an essay.

"It is likely I inherited the gravel from my father," Montaigne guessed, "for he died sadly afflicted by a large stone in the bladder." Yet Montaigne had no idea how one could inherit a disease, as opposed to a crown or a farm. His father had been in perfect health when Montaigne was born, and remained so for another twenty-five years. Only in his late sixties did his kidney stones first appear, and they then tormented him for the last seven years of his life.

"While he was still so remote from the disease, how could the light trifle of his substance out of which he built me convey so deep an impress?" Montaigne wondered. "Where could the propensity have been brooding all this while?"

Simply musing in this way was a visionary act. No one in Montaigne's day thought of traits as being distinct things that could travel down through generations. People did not reproduce; they were engendered. Life unfolded as reliably as the rising of bread or the fermenting of wine. Montaigne's doctors did not picture a propensity lurking in parents and then being reproduced in their children. A trait could not disappear and be rediscovered, like a hidden letter. Doctors did sometimes observe certain diseases that were common in certain families. But they didn't think very much about why that was so. Many simply turned to the Bible for guidance, citing the passage telling of God "visiting the iniquity of the fathers upon the children unto the third and fourth generation."

Whatever Montaigne's doctors might have said about his father's kidney stones, he probably would have dismissed it. He hated doctors, like his father and grandfather before him. "My antipathy against their art is hereditary," he said.

Montaigne wondered if such an inclination could be inherited, along with diseases and physical traits. But how all of that could be carried from one generation to another in a seed, Montaigne could not begin to imagine. "The doctor who can satisfy me on this point I'll believe as many miracles of as he pleases," he promised, "provided he does not give me—as doctors usually do—a theory more intricate and fantastic than the thing itself."

Montaigne lived for another dozen years, apparently never meeting a doctor who could satisfy him about heredity. In the year of his death, an elderly Spanish doctor named Luis Mercado was appointed by Philip II to be Physician of the King's Chamber. Mercado might have met Montaigne's high standards, because he was one of the few doctors in Europe to recognize that people inherit diseases and to ask why.

For decades before his appointment to the court, Mercado had taught medicine at the University of Valladolid. A colleague there called him "modest in dress, sparing in diet, humble in character, simple in matter." At the university, Mercado had given lectures steeped in Aristotle's ideas. But his dedication to the ancients didn't prevent him from making observations of his own and publishing books with new ideas about fevers and plagues. And in 1605, at age eighty, Mercado published his masterwork: *De morbis hereditariis—On Hereditary Diseases*. It was the first book dedicated to the subject.

Mercado sought an explanation for why diseases ran in families. He dismissed the possibility that they were divine punishments. Instead, to understand hereditary diseases, Mercado believed it was necessary to understand how new lives develop. He argued that each part of the body—a hand, the heart, an eye—had its own distinctive shape, its own balance of humors, and its own particular function. In the bloodstream, Mercado claimed, the humors from each part of the body mixed together, and a mysterious formative power shaped them into seeds. Unlike Aristotle, Mercado believed that both men and women produced seeds, which were combined through sex. The same formative power acted on those joined seeds, producing from them a new supply of humors that gave rise to a new human being that developed the same parts as its parents.

Mercado believed that this cycle of generation, combination, and development was well shielded from the outside world. The willy-nilly waves of chance could not reach the hidden seeds of human life and alter their hereditary traits. He dismissed popular notions about the power of the environment—that a mother's imagination could alter her baby, or that dogs taught new tricks could pass them down to their puppies. A hereditary disease was like a stamp that marked a seed. The same stamp appeared on each new generation's seeds and gave rise to the same disease—"the bringing forth of individuals similar to oneself and deformed by the same defect," Mercado wrote.

In his experience with patients—royal and common—Mercado had seen many different kinds of hereditary diseases. Some could strike immediately—a child could be born deaf, for example—while others were slower to emerge, like the kidney stones that afflicted Montaigne as they had his father. In many cases, Mercado came to believe, parents impressed only a tendency toward a disease on their children. A child's humors might be able to weaken that impression. Or a healthy parent's seed could sometimes counter a diseased one. The defect still lurked in the child, who could then pass it down to its own children. If they didn't inherit a countervailing seed from their other parents, the disease could flare up out of hiding.

Some hereditary diseases could be treated, Mercado argued, but only slowly and incompletely. "Let us in some secluded spot teach the deaf and dumb to speak by forming and articulating the voice," he wrote. "By long practice many with hereditary affliction have regained their speech and hearing."

But for the most part, a doctor could do little, because the stamp of heredity was sealed away from a physician's reach. Mercado urged instead that people with the same defect not marry, because their children would be at greatest risk of developing the same hereditary disease. All people should seek out a spouse as different from themselves in as many individual characteristics as possible.

Mercado went remarkably far toward answering Montaigne's questions about heredity. But the world was not ready to investigate his ideas. The Scientific Revolution was decades away, and it would take two centuries more before heredity itself would come to be seen as a scientific question. No one—not even Mercado himself, it seems—could recognize that his own royal patients were in the midst of staging their own hereditary disaster. By preserving their noble blood, they were increasing the number of disease-causing mutations in their lines. They were lowering their odds of having children, and the children who beat those odds were then at grave risk of inheriting mutations that would give them a host of diseases.

———

By 1660, Philip IV had been trying to produce a male heir for forty years. In that time he fathered a dozen children. Ten died young, and the surviving two were girls. As Philip grew older, the survival of the entire Habsburg dynasty fell into jeopardy. The following year, at last, the empire celebrated the birth of a son who would become king.

Charles, the new prince, was "most beautiful in features, large head, dark skin, and somewhat overplump," according to the official gazette. Spain's royal astrologers declared the stars at Charles's birth to be well arranged, "all of which promised a happy and fortunate life and reign." When Charles was only three, his father died. On his deathbed, gazing at the crucifix on the wall before him, Philip IV could console himself that he had forged a new link in heredity's chain, leaving behind a boy king.

King Charles II of Spain proved to be the sickest Habsburg monarch of them all. "He seems extremely weak," an ambassador wrote back to France, "with pale cheeks and very open mouth." The ambassador observed a nurse usually carried him from place to place so that he would not have to walk. "The doctors do not foretell a long life," the ambassador reported.

Charles II, born six decades after Mercado published *On Hereditary Diseases,* managed to survive to manhood, although his health remained poor and his mind weak. Famines and wars unfolded around him, but he preferred to distract himself with bullfights. The only national matter with which he concerned himself was producing an heir of his own. And even in that task, he failed.

As the years passed without his queen becoming pregnant, Charles grew more ill. "He has a ravenous stomach, and swallows all he eats whole, for his nether jaw stands so much out, that his two rows of teeth cannot meet," a British ambassador reported, "to compensate which, he has a prodigious wide throat, so that a gizzard or liver of a hen passes down whole, and his weak stomach not being able to digest it, he voids in the same manner."

The Spanish Inquisition blamed the lack of an heir on witches, but their trials did nothing to help the king. It became clear he would die soon. Yet Charles managed to dither for months over whom to name as his heir. Finally, in October 1700, he selected the Duke of Anjou, the grandson of the king of France. Charles worried that the empire might collapse after his death, and so he issued a demand that his heir rule "without allowing the least dismemberment nor diminishing of the Monarchy founded with such glory by my ancestors."

But his monarchy soon began disintegrating anyway. The prospect of France and Spain forming an alliance prompted England to form an alliance of its own with many of the other great powers in Europe. Skirmishes began breaking out, both in Europe and in the New World. Eventually, the fighting would escalate into the War of Spanish Succession. The conflict would change the planet's political landscape, leaving England ascendant and Spain broken.

Yet Charles still dreamed that his empire would remain whole. He even added a codicil to his will stating his wishes that the Duke of Anjou would marry one of his Habsburg cousins in Austria. Not long afterward, he grew so ill that he could no longer hear or speak. Charles died on November 1, 1700. He was only thirty-five. There was no child left to inherit his empire, because of invisible things Charles had inherited from his ancestors. When doctors examined the king's cadaver, they found that his liver contained three stones. His kidneys were awash in water. His heart, they reported, was the size of a small nut.

CHAPTER 2

Traveling Across the Face of Time

I N 1904, a fifty-five-year-old Dutchman, heavily built and sporting a graying beard, boarded a ship bound for New York. Hugo de Vries was a university professor in Amsterdam, but he was not a hothouse inhabitant of lecture halls. He spent much of his time wandering the Dutch countryside, scanning meadows for exceptional wildflowers. An English colleague once complained that his clothes were foul and that he changed his shirt once a week.

When de Vries's ship docked in New York, he boarded a train that pushed its way across the country to California. The official reason for the journey was to visit scientists at Stanford University and the University of California at Berkeley. De Vries dutifully gave his lectures and went to the required evening banquets. But as soon as he could manage, he escaped north.

Fifty miles from San Francisco, de Vries arrived in a small farming town called Santa Rosa. With four fellow scientists in tow, he made his way from the train station to a four-acre plot ringed by low picket fences and crammed with gardens. A modest vine-covered house sat in the middle of the property, flanked by a glass-roofed greenhouse and a barn. A boxwood-lined path led from the street to the front porch of the house. Next to the path stood a blue-and-white sign informing visitors that all interviews were limited to five minutes unless they were by appointment.

Fortunately, de Vries had one. A small, stooped man about his own age, outfitted in a rough brown suit, came out to greet the visiting party. His name was Luther Burbank.

Burbank shared the house in the middle of the garden with his sister and mother. He had been expecting de Vries's arrival for months and set aside an evening and a day for the visit. He showed off his garden to the scientists and then took them to an eighteen-acre farm he tended in the Sonoma foothills. Those two plots of land, and the plants that sprouted from their soil, had made Burbank both rich and famous.

"His results are so stupendous," de Vries later wrote, "that they receive the admiration of the whole world."

This was no exaggeration. Each year, Burbank's postman brought him thirty thousand letters. Henry Ford and Thomas Edison traveled to Santa Rosa to meet him. Newspapers regularly praised Burbank, calling him "the wizard of horticulture." The Burbank potato, which he produced at age twenty-four, was already the standard breed for farmers across much of the United States. The Shasta daisy sprang into existence under Burbank's care, and quickly became a mainstay in middle-class flower beds. In his gardens, Burbank created thousands of different kinds of plants—the white blackberry, the Paradox walnut, the spineless cactus.

"Such a knowledge of Nature and such ability to handle plant life would only be possible to an innately high genius," de Vries had declared to a group of Stanford scientists on the eve of his trip to Santa Rosa. Before his meeting with Burbank, de Vries wondered how much of what was written about him was true. The *San Francisco Call* said that Burbank's flowers "thrive upon a scale so extensive as to suggest magic rather than the sober work of science." Sometimes Burbank's catalogs read like fairy tales. In one edition, de Vries saw that Burbank was now offering a stoneless plum. He simply couldn't believe such a thing could be created. When de Vries finally reached Santa Rosa, he asked for proof. Burbank led de Vries and his other visitors to a plum tree bowed down with blue fruit. He gave each man a plum, and when they bit down, their teeth met only soft sweetness. "Although we knew there was no stone in the plum, we experienced a feeling of wonder and astonishment," de Vries wrote.

De Vries was not one for much wonder. He was a scientist to his marrow, and before his trip to California, he had spent the previous two decades running experiments that helped establish the first genuine science of heredity. Not long before his visit to Burbank, it had been given a proper name: genetics.

But genetics in 1904 was like a barely started house, more footings than walls. It still left fundamental questions about heredity unanswered. De Vries knew that he and his fellow geneticists were really just newcomers to heredity's mysteries, that other people had been plumbing them for thousands of years. He respected the wisdom of animal and plant breeders, although he also recognized much of their ancient wisdom had disappeared into unrecorded oblivion. Over the course of the 1700s and 1800s, some breeders became rich. Nations looked to them to work miracles on heredity to deliver economic salvation. And at the debut of the twentieth century, there was no greater breeder than Luther Burbank. He had dedicated decades to understanding what he called "the inherent constitutional life force, with all its acquired habits, the sum of which is heredity." De Vries came to Santa Rosa to learn what Burbank had learned about heredity in order to push genetics out of its infancy.

———

Pottery shards, ancient seeds, and the bones of livestock all indicate that the first breeders started their work in earnest around eleven thousand years ago. Plants and animals, once wild, came under the control of humans, grown for their benefit. The agricultural revolution let the population of our species explode, but it also made us precariously dependent on the heredity of what we raised. When farmers planted a new field of barley seeds, or goatherds delivered a new batch of kids, they needed each new generation of plants and animals to end up like the previous one. If corn kernels randomly became as hard as glass, or if cows were born unable to produce milk, people would starve. Learning how to steer heredity could also make farmers more prosperous. If they could raise pigs that reliably grew more pork on their bodies, they gained more wealth. And once farmers could supply their goods to markets and trade networks, they could attract more customers for their particular breeds—their sweeter oranges or their more durable cowhides.

It's hard to know exactly how much early farmers understood about breeding as they carried it out. The historical record of their ideas is practically a void, but the results of their efforts were impossible to ignore. The wealth of the Habsburg kings of Spain, in fact, came in part from the mysterious art of animal breeding. The first sheep to graze the meadows of Spain were unexceptional creatures with rough wool coats. When the Moors arrived, they brought sheep with them from northern Africa, which they interbred with the resident flocks. The new cross came to be called the Merino. For centuries, Spanish shepherds bred Merinos by the millions, every year leading them on a journey across the country. The Merinos spent each summer grazing in the Pyrenees and then traveled narrow paths for hundreds of miles to the southern lowlands to pass the winter. Over many generations of breeding, Merino wool became extraordinarily soft, lush, and silky.

Merino wool turned into a precious commodity. On their journeys, Spanish shepherds would stop to shear their sheep and sell their wool at fairs to merchants from across Europe. Henry VIII of England said he would accept nothing but Merino for his royal garments. Merino wool became so valuable to Spain that smuggling a single Merino sheep out of the country was made a crime punishable by death.

In the seventeenth century, the magnificence of Merino wool was as mysterious as the suffering of the Habsburg kings. No one at the time would have guessed they shared anything in common. Some speculated that the environment in which the Merinos lived was responsible for their wool. The cold of the mountains and the heat of the tablelands influenced their seed, in the same unknowable way the terroir of a grapevine determined the taste of its wine. More evidence for this influence came from the few cases when sheep were smuggled out of Spain. In other countries, they failed to thrive. After a few generations of crossbreeding with native flocks, the sheep no longer grew good wool.

Across Europe, the growing population was clamoring for more wool—as well as for more beef and leather from cows, for more eggs from chickens. Wheat, barley, and corn were in greater demand as well. Anyone who could steer heredity in a more profitable direction stood to make a good living. A particularly successful breeder could even become a celebrity. And no breeder in the 1700s was more famous than a portly Englishman named Robert Bakewell. A duchess once referred to him as "the Mr. Bakewell who invented sheep."

Mr. Bakewell was born in 1725 on Dishley Grange, a 450-acre property that his father worked as a tenant farmer. His father encouraged him to learn new techniques by traveling to other farms around England, Ireland, and the Netherlands. He helped his father improve the farm, digging a labyrinth of channels and hatches to deliver water across the property, tripling the amount of grass that grew on it. Robert Bakewell took over Dishley Grange by the time he was thirty. A decade later the first hint of his breeding skill emerged when he won first prize at the Ashby Horse Show.

But it was with sheep that Bakewell would become famous. He and his neighbors reared a humdrum local breed known as Old Leicester. The animals were heavy, long, and flat-sided. They grew rough wool, and their mutton, a coarse-grained meat with little flavor, brought no excitement to the dinner table. But when Bakewell looked at an Old Leicester sheep, he saw a New Leicester sheep waiting to emerge. The generating powers inside the animals could, with the proper guidance, produce a breed that could make sideboards groan with huge cuts of delicious mutton—while requiring relatively little feed. Bakewell was a man of his mechanical age, engineering woolen meat-making machines.

Unlike an engineer, however, Bakewell did not understand the natural processes he was trying to manipulate. He could only guess, picking out ewes from his flock that approached his vision. Bakewell believed that the traits he could see on the outside of a sheep were linked to qualities on the inside, ones that could be passed down to offspring.

"He asserts," a visitor to Dishley Grange wrote, "the smaller the bones, the truer will be the make of the beast—the quicker she will fat—and her weight, we may easily conceive, will have a larger proportion of valuable meat."

Bakewell traveled England inspecting rams and brought home a select few to breed with his ewes. When he crossed these sheep, they did not instantly produce a uniform supply of New Leicester lambs. Instead, their litters were a hodgepodge, made up of lambs of different sizes and shapes. But Bakewell did not lose faith in his vision. He turned his exacting eye to his lambs. He picked out ones to mate with one another, or with other sheep he bought from other farms. These cycles of inspection and selection went on for years, during which time Bakewell turned his farm into a primitive laboratory. He herded his sheep into houses and sheds kept as clean as horse stables so that he could experiment on their heredity in secret. He measured his sheep and weighed them every week until slaughter. He chalked his data on slates and then transferred them to ledgers, which sadly were later lost.

In time, the sheep began to accord with the animal that gamboled in Bakewell's mind. He stopped touring England to buy rams. Instead, he employed a strategy known as in-and-in breeding. Bakewell mated cousin to cousin, brother to sister, father to daughter. Other farmers thought him mad because they believed inbreeding invariably led to disaster. That might be true for other farmers, but not for Bakewell. He was able to make sure that all the qualities he wanted in his sheep became fixed in his flock, but none of the deformities that might ruin his new breed.

After fifteen years, Old Leicester had at last become New Leicester. People found Bakewell's new breed—with its broad, barrel-shaped body; its straight, short, flat back; its small head; and its short, small-boned legs—peculiarly pleasing to the eye. New Leicester mutton might not have the fine flavor that aristocrats clamored for. One critic even declared it "only fit to glide down the throat of a Newcastle coal-heaver." But Bakewell didn't care about epicurean snobs. "My people want fat mutton and I gave it to them," he declared.

He was fibbing a bit. With a flock of just a few hundred New Leicester, Bakewell couldn't feed the millions of hungry English. Instead, he sold his sheep to other breeders, who started their own New Leicester flocks. They paid him dearly. They were even willing to do something that had previously been unheard-of: They would rent his rams for their services. Bakewell sent the rams to their appointments in two-wheel sprung carriages, suspended inside from slings. He claimed the right to take the best lambs produced by his rented rams, improving his own flock even more.

Dishley Grange itself became a destination for travelers, who came from as far as Russia to see Bakewell's work and learn about the astonishing methods of "this prince of breeders." Bakewell welcomed visits. He turned his house into a museum of heredity, filling it with sheep skeletons and brine-pickled joints, demonstrating the transformation he had brought about in his animals. It was great public relations. Bakewell's visitors wrote letters and books about his experiments. One French nobleman declared that Bakewell "had been making observations, and studying how to bring into being his fine breed of animals with as much care as one would put into the study of mathematics or any of the sciences."

In fact, Bakewell didn't leave behind a single measurement of a sheep. He published no law of heredity to explain his success. Bakewell lived at the turning point in the history of heredity, when people recognized it as something to be understood and manipulated, while still relying on the intuitions of their farming ancestors to steer it. Looking back at Bakewell's work, we can't help but turn our attention to what it lacked—the data and statistics that are essential to studying heredity today. But in his own time, Bakewell had an enormous impact, showing the world how much heredity could be stretched and sculpted. As one of his visitors wrote, "He has convinced the unbelievers of the truth of his sheepish doctrine."

———

Among Bakewell's international admirers was Frederick Augustus III, the Elector of Saxony. In 1765, Frederick received an extraordinary gift from the king of Spain: 210 Merino sheep. Frederick wanted to use the Merinos to build a thriving sheep industry in Saxony, but he worried that the livestock might not thrive outside of Spain. He consulted with Bakewell about his plan.

Bakewell assured Frederick that the traits carried in a sheep's blood would endure through generations no matter where they were bred, as long as they were properly raised. Frederick discovered Bakewell was right, and soon Germany was producing so much fine Merino wool that it could satisfy much of the demand from English factories and had enough left over to support a textile industry of its own. Around Moravia, at the heart of this new industry, a new generation of sheep breeders were inspired to achieve even more. They believed that if they could exploit the laws of heredity, they'd be able to breed even better sheep. But first they'd have to discover those laws.

In 1814 the breeders founded an organization, the title of which was—deep breath—"The Association of Friends, Experts and Supporters of Sheep Breeding for the achievement of a more rapid and more thoroughgoing advancement of this branch of the economy and the manufacturing and commercial aspects of the wool industry that is based upon it." Those who didn't want to lose too much oxygen uttering the full name simply called it the Sheep Breeders' Society.

The Sheep Breeders' Society was based in the city of Brno in Moravia (now part of the Czech Republic). They held regular meetings drawing members from as far as Hungary and Silesia. The city also hosted the Brno Pomological Society, a group of plant breeders who hoped to bring similar improvements to crops. The plant breeders had a Bakewell of their own to emulate, an English gentleman named Thomas Andrew Knight.

In the late 1700s, Knight applied Bakewell's sheepish doctrine to the flocks on his English estate and was pleased with the results. He then set out to apply the same principles to plants. His plan was to hand-fertilize plants with pollen grains. The pollen—the botanical equivalent of sperm—would make their way inside of flowers to their ovules—the equivalent of eggs. Knight would use different varieties for his experiments in order to make hybrids. And he would then use Bakewell's in-and-in breeding methods until their heredity became stable.

At first, Knight crossed apple trees. They grew so slowly that he couldn't tell if his procedure was actually working. Around 1790, Knight searched for another species that could return him faster results.

"None appeared so well calculated to answer my purpose," he later wrote, "as the common pea."

Knight was delighted to discover that his hybrid peas flowered, producing seeds that could develop into plants of their own, quickly growing high in his garden. He was also intrigued by the way traits of the parents reappeared in descendants. When he fertilized white peas with pollen of a gray-seeded variety, for example, the hybrid plant bore gray seeds.

"By this process, it is evident, that any number of new varieties may be obtained," Knight declared. If breeding was carried out scientifically, he was convinced, England need never go hungry. "A single bushel of improved wheat or peas may in ten years be made to afford seed enough to supply the whole island," he declared.

No one in England was able to make Knight's hope come true. But in Brno, plant breeders kept trying, collaborating with sheep breeders to uncover biology's mysteries. In 1816, the Sheep Breeders' Society organized a series of public debates about the nature of heredity. Some members argued that the environment impressed traits on offspring. A Hungarian count named Imre Festetics took the opposing view. Based on years of sheep breeding, he argued that healthy animals pass on their characteristics to their offspring. He observed a pattern much like what Knight had seen in peas: The traits of grandparents could disappear from their lambs, only to reappear in the following generation.

Festetics even argued that freaks of nature could leap back into a pedigree after many generations of healthy sheep. He warned against using those freaks for breeding. Inbreeding could improve flocks of sheep, Festetics declared, but only if breeders first carefully selected the stock they used. In an 1819 manifesto, Festetics urged that his fellow breeders determine the nature of these patterns scientifically, uncovering what he called "genetic rules of nature."

In later years, the Moravian breeders followed Festetics's advice. They designed breeding experiments, guided by the latest discoveries coming out of Germany's universities. One of the busiest research centers was a local Augustinian priory, led by the abbott Cyrill Franz Napp. Napp and his friars got into the breeding business to pay off the priory's massive debts, and they came to enjoy great success with sheep and crops. Yet Napp complained that breeding was "a lengthy, troublesome and random affair." The trouble would not go away until breeders changed their ways. "What we should have been dealing with is not the theory and process of breeding," Napp declared at an 1836 meeting of the Sheep Breeders' Society, "but the question should be: what is inherited, and how?"

His scientific frame of mind led Napp to set his friars loose on scientific questions. They studied how to forecast the weather, maintained a large collection of minerals, and built a massive scientific library. Napp set aside part of the grounds solely to grow rare species of plants. A monk named Matthew Klácel ran experiments in another garden—at least until his radical philosophy on nature forced him to flee to the United States. When young men entered the Augustinian order, Napp encouraged them to immerse themselves in the latest scientific advances. One of the young men in whom Napp took a special interest was a poor farmer's son named Gregor Mendel.

Mendel's first job at the priory was to teach languages, math, and science in a local school. He proved so good at it that Napp sent him to the University of Vienna for more training. Mendel took a course in physics there in which he learned how to design careful experiments, and another in botany, where he learned about the long-running debate over hybrid plants and whether two species could cross to produce a new species. When Mendel returned to the priory in 1853, he continued to teach, but his time at the university inspired him to take up scientific research. He ran the friary's weather station and investigated the possibility of communicating weather reports with semaphore flags or telegraph messages. He raised honeybees, studied sunspots, and invented chess problems. And he carried on Napp's own research by breeding plants. Mendel cross-pollinated fruit trees, raised prizewinning fuchsias, and bred varieties of beans and peas.

In 1854, Napp gave Mendel permission to run a large-scale experiment that Mendel hoped would make some sense of hybridization. The randomness that bedeviled the breeding societies might be hiding some hidden order. Mendel followed Knight's example, and planted his garden with peas.

For his experiment, Mendel grew twenty-two varieties of peas, each with a set of distinctive traits reliably passed from ancestors to descendants. He raised the plants in a greenhouse, where they couldn't be randomly pollinated by visiting bees. Mendel patiently crossed the varieties, moving pollen from one line to another. His experiment was gigantic, involving more than ten thousand plants, because he had learned in his physics classes that big samples are statistically more likely to reveal important patterns.

In one of his first experiments, Mendel crossed yellow and green plants. When he opened the pods, he got a result similar to what Knight had found sixty years before. All the peas inside were yellow. Mendel then transferred pollen between these hybrids and produced a second generation. Now only some of the peas were yellow. A fraction of the plants displayed the green color that had disappeared from sight in the previous generation.

When Mendel counted the peas, he found about three yellow plants for every green one. He then selected plants from the second generation that produced yellow peas and crossed them with the original line of yellow plants. Some of their offspring produced green peas once more. Mendel got similar results when he compared wrinkly peas to smooth ones, or tall plants to short ones.

In 1865, Mendel talked about his experiment at a meeting of Brno's Natural History Society. To make sense of the three-to-one ratio he found so often in peas, he proposed that every plant contained a pair of "antagonistic elements." When a plant produced pollen or ovules, each one received only one of those elements. And when a pollen grain fertilized an ovule, the new plant inherited its own pair of the elements. Each element could give rise to a particular trait in a plant. One might produce a green color, while another produced yellow. But Mendel argued that some elements were stronger than others. As a result, a hybrid plant with one yellow element and a green one would be yellow, because yellow is dominant over green.

This scheme could account for the three-to-one ratio, thanks to the way the elements were passed down from parents to offspring. When Mendel mated two yellow hybrids together, each plant contributed one of its two elements to each offspring. Which element a particular offspring inherited was a matter of chance. There were thus four combinations: yellow/yellow, yellow/green, green/yellow, and green/green. Working through these figures, Mendel calculated that a quarter of the plants would inherit the yellow element from both parents. Half would inherit one yellow and one green— and also end up looking yellow. Meanwhile, the remaining quarter would inherit two green elements.

Mendel's talks did not set his audience's hair on fire. None of them were so inspired by his experiments to repeat them. In hindsight, it's easy to recognize the importance of his results, but at the time they didn't stand out from the many other studies of hybrids that were also being carried out. A mentor of Mendel, the Swiss botanist Carl Nägeli, encouraged him to see if the same patterns would emerge in another species, suggesting hawkweed.

It turned out to be a bad suggestion, thanks to hawkweed's peculiar biology. When Mendel crossed hawkweed plants, he didn't produce the three-to-one ratio again. Instead, the hawkweed often reverted back to one of the ancestral forms Mendel had started with, and he was unable to alter their descendants any further. The experiment didn't make Mendel abandon his ideas about antagonistic elements, however. He added a new speculation: In hawkweed, the elements didn't get separated as pollen and ovules developed.

"Evidently we are here dealing only with individual phenomena," Mendel wrote to Nägeli, "which are the manifestation of a higher, more fundamental, law."

That law would eventually bear Mendel's name. But in the years after Mendel published his experiments, only a few other researchers cited them. One day, when Mendel was standing in his hawkweed garden with a friend, he predicted he would be proven right eventually. "My time is yet to come," he said.

When Napp died in 1868, his protégé succeeded him, and before long, the newly appointed Abbot Mendel got so ensnared in tax battles with the government that he had to abandon his experimental garden. When he died sixteen years later, in 1884, his funeral was attended by throngs of peasants and the poor. But no scientists turned up to mourn his passing.

B reeders in the United States took a different path. The American colonies produced no Bakewell of their own. No scientific breeding society emerged in the early republic to debate how precisely sheep inherited fatty mutton. American plant breeders did not set up experimental gardens to test the boundaries of species. Instead, the United States became an arena for capitalist competition as farmers battled one another with breeds they hoped would make them a fortune.

Many of those breeds were imported from Europe to the New World. In the early 1800s, thousands of Merino sheep were illegally smuggled from Spain to Vermont. The legends about the Merino prompted New England sheep farmers to abandon their flocks for the new imports. By 1837, there were a million Merinos in Vermont alone.

The American booms typically went bust. Merino speculators became convinced that textile mills would develop a bottomless appetite for wool, and the price for a single lamb climbed beyond a thousand dollars. When the Merino bubble popped, Americans promptly turned for salvation to exotic chickens—Black Polands, White Dorkings, Yellow Shanghae—until the hen fever broke, too.

Along with new animal breeds, American farmers searched for new crop varieties. They typically didn't make crosses like Knight or Mendel did. Instead, they would simply stumble across a peculiar plant. Some farmers would keep their discoveries to themselves, so as to attract more customers when they sold their goods at local markets. Others sent off their discoveries to the new seed catalog companies, hoping to get rich on orders. In Iowa, a Quaker farmer named Jesse Hiatt noticed a little apple tree growing between the rows of his orchard. He chopped the seedling down, but the following year it had returned. He cut it down again, and it returned once more. "If thee must grow, thee may," Hiatt reportedly told the tree. After ten years, the tree finally bore fruit: handsome, red-and-yellow-streaked apples with a crisp, sweet flavor. He shipped some to Missouri, to enter a contest run by Stark Bro's company. His apples won the contest, and Stark Bro dubbed his variety Delicious. It became one of their most successful varieties, and it remains so today.

Luther Burbank was born into this land of breeding in 1849. His first memory of his mother, he later recalled, was of her setting him down in a meadow at their Massachusetts farm while she gathered strawberries. Within a few years, Luther had farm work of his own to do: "the wood to bring, weeds to pull, chickens to feed, the cows to drive to pasture," he later wrote. Yet Luther still had time left over to build waterwheels and bark canoes. He inspected the apple trees in the family's orchard, learning how to spot the difference between the Baldwins and Greenings. He observed the swelling buds as they cast off brown coats and opened their pink-and-white petals. When Luther became a teenager, he planted his own garden, writing to his older brother, who had moved to California, to send him the seeds of exotic Western breeds.

The Burbanks hoped Luther would become a doctor, but at school he showed little proficiency in Latin or Greek. He was more interested in the books about natural history that his cousin, an amateur naturalist, gave him. They took walks around the countryside, where his cousin instructed him on the landscape, from the rocks to the plants that grew over them. Luther developed a fierce desire, he later said, "to know, not second-hand, but first-hand, from Nature herself, what the rules of this exciting game of Life were."

In 1868, when Luther Burbank was nineteen, all daydreams about nature and medicine were cut short. His father suddenly died, forcing his family to sell off their farm and move away. Burbank had to support his mother and sisters by farming rented fields. "Nature was calling me to the land, and when there came to me my share of my father's modest estate I could no longer resist the call," Burbank later remembered.

He decided that he had to do more than stick seeds in the ground. He needed to change the seeds themselves. When Burbank sold produce in town markets, he could see how some farmers made more money because they used better breeds. Their customers preferred bigger fruits, tastier vegetables. Farmers who planted early-growing breeds could start selling produce earlier in the year. Burbank got a grand ambition: to use the rules of the game of Life to create entirely new breeds.

In the 1860s, the concept of heredity had not penetrated the United States very far. The textbooks Burbank read in school didn't even use the word. Instead, they offered a jumble of folk explanations for why people resemble their ancestors. Burbank's physiology textbook informed him that if a woman "has a small, taper waist, either hereditary or acquired, this form may be impressed on her offspring;—thus illustrating the truthfulness of scripture, 'that the sins of the parents shall be visited upon the children unto the third and fourth generation.'"

One day Burbank spotted a new two-volume book at the Lancaster town library on animal and plant breeds. Desperate for help with his experiments, he dipped into it, and before long he had devoured the entire work. After he finished, Burbank felt as if he had been given the keys to heredity's locks. He was ready to create new kinds of crops the world had never seen. "I think it is impossible for most people to realize the thrills of joy I had in reading this most wonderful work," Burbank later said.

The book, *The Variation of Animals and Plants Under Domestication*, was written by a British naturalist named Charles Darwin. In it, Darwin cast heredity as a scientific question in urgent need of an answer. But the answer he offered would turn out to be spectacularly wrong.

———

*T*he Variation of Animals and Plants Under Domestication served as a sequel to its far better-known predecessor, *The Origin of Species*. In that earlier book, Darwin had presented the outlines of his theory of evolution. In every species and strain, Darwin observed, individuals varied from one another. Some of those variations may help some individuals survive and reproduce. The next generation will inherit those successful variations, and will pass them down in turn. Darwin called this process natural selection, and he argued that, over many generations, it could turn varieties into separate species. Over even longer periods, it could produce radically new forms of life.

The Origin of Species became one of the most influential books ever written, opening up millions of minds to the fact that life has been evolving into new species for billions of years and is continuing to evolve today. Yet Darwin knew that in the book he had glossed over some of the most important parts of evolution. While its logic was straightforward enough, Darwin couldn't explain the biology that made it possible. Yes, individuals varied, but why? Yes, offspring resembled their parents, but why? Anyone who would answer those questions would first have to explain what heredity really is.

"The laws governing inheritance," Darwin conceded, "are quite unknown."

Three decades earlier, when Darwin was twenty-eight, he began jotting down notes and questions in a series of notebooks. In their pages, we can see the slow metamorphosis of his ideas about the diversity of life. From the beginning, he already recognized the importance and mystery of heredity. When two breeds were crossed, he wondered, why did the offspring sometimes look more like one breed than the other? Why did they sometimes look like neither parent?

In search of answers, Darwin read everything he could find about heredity. Dissatisfied by what naturalists had to say, he turned to breeders for help. He read Bakewell's famous rules for producing better sheep and cows. He printed up a short pamphlet entitled *Questions About the Breeding of Animals* and sent it out in 1839 to England's leading breeders. He asked them what happened when they crossed different species or varieties— whether hybrids were produced and, if so, whether their offspring were sterile. He asked how reliably traits were passed down from generation to generation, whether animals inherited the behaviors of their parents, whether the disuse of some body part might lead it to dwindle away.

The information Darwin got back from the breeders still wasn't enough.

So he became a breeder himself. Filling his greenhouse with plants, Darwin became expert at crossing orchids. He bought rabbits so that he could compare their dimensions to wild hares. He built a pigeon house at the end of his yard and stocked it with rare breeds. He went to club meetings of pigeon breeders, and even attended the annual poultry show in Birmingham, known as "the Olympic Game of the Poultry World." Darwin marveled at the way the breeders could spot tiny variations from one pigeon to the next, and how they used those differences to produce extravagant new breeds. Pigeons, Darwin declared to his friend Charles Lyell in 1855, were "the greatest treat, in my opinion, which can be offered to human being."

Darwin turned to humans for clues to heredity as well, but he mainly studied how they went mad. Doctors had long puzzled over the causes of insanity. Some blamed alcohol, others sorrow, others sin, others masturbation. But some considered insanity to be a hereditary disease. In eighteenth-century France, a fierce debate broke out about whether hereditary diseases even existed, and the French doctors of the mind—alienists, as they were then known—started gathering data to prove they did. They filled out entrance forms when people were admitted to asylums, and they studied national censuses. Madness, the alienists decided, clearly ran in families. "Of all illnesses," the French alienist Étienne Esquirol said in 1838, "mental alienation is the most eminently hereditary."

The French alienists investigated how madness could be hereditary—what it had in common with other hereditary diseases like gout or scrofula. They contemplated the underlying mystery: the process by which traits—both illnesses and ordinary traits—were passed down through the generations. Along the way, their language experienced a subtle yet profound shift. At first French alienists only used the adjective *héréditaire* in order to describe diseases inherited from ancestors. But in the early 1800s, they began using the noun *hérédité*. Heredity was becoming a thing unto itself.

In his research into madness, Darwin plowed through a two-volume tome called *Treatise on Natural Inheritance*, published in 1850 by the French alienist Prosper Lucas. Darwin framed its pages with notes in the margins.

In English, he began to follow Lucas's example. Again and again, he wrote down the word *heredity*.

Darwin was not drawn to heredity purely out of intellectual curiosity. Marrying his first cousin Emma led him to worry what fate they might deliver to their children. He read reports from alienists about how the children of first-cousin marriages were prone to madness. His anxiety only grew as his own health failed. In his twenties, he had been fit enough to take a voyage around the planet, but after his return he developed a constellation of disorders. He vomited violently, he suffered from boils and eczema, his fingers went numb, and his heart often raced. He described himself in 1857 as a "wretched contemptible invalid." Three of Darwin's ten children died young, and the others suffered from bouts of poor health.

"It is the great drawback to my happiness, that they are not very robust," he wrote to a friend in 1858. "Some of them seem to have inherited my detestable constitution."

———

Darwin put only a little of his research on heredity in *The Origin of Species*. Instead, he saved that profound matter for a book of its own. When he began focusing his thoughts on heredity, however, he decided that all the details he had been collecting about pigeons and insanity would not be enough. He would also have to figure out the physical process that accounted for all the strange ways in which animals and plants reproduced.

At the time, Mendel was raising pea plants and hawkweed, but Darwin—like most scientists of his day—didn't even know who Mendel was. Instead, Darwin drew his inspiration from other biologists who had made a profound discovery of their own: that all of life is made of cells.

To Darwin, the central question of inheritance was what sort of substances the cells of parents transmitted to an embryo so that its cells came to resemble theirs. Whatever made muscles strong was stored in muscle cells. Whatever made brains wise or defective must be stored in brain cells.

Perhaps, Darwin thought, the cells throughout the body cast off "minute granules or atoms." He dubbed these imaginary specks gemmules.

Once released by cells, gemmules coursed through the body, gradually piling up in the sexual organs. When the gemmules from both parents combined in a fertilized egg, they enabled it to develop into a blend of cells from both parents.

Darwin wanted a catchy name for this imaginary process. Maybe something that combined *cells* with *genesis*. Darwin asked his son George, then a student at the University of Cambridge, to ask classics professors there for a name. George came back with outlandish suggestions like *atomo-genesis* and *cyttarogenesis*. Darwin settled on *pangenesis*.

Pangenesis set Darwin apart from most naturalists of his day. They explained heredity as a blending of traits—akin to mixing blue and yellow paint to produce green. Darwin looked at heredity instead as the result of distinct particles. They never fused and never lost their separate identities. Darwin readily admitted that pangenesis was "merely a provisional hypothesis or speculation." Yet it offered Darwin great powers of explanation. "It has thrown a flood of light on my mind in regard to a great series of complex phenomena," he said.

Darwin could explain why children sometimes resembled one parent more than the other with pangenesis: Some gemmules were stronger than others. The gemmules that gave rise to newborn babies were a mixture of particles that had accumulated over generations, from parents, grandparents, and on back through time. A gemmule might be overshadowed by stronger ones for thousands of years, only to leap forward and revive some ancient feature. And as experiences altered cells, they would also alter their gemmules. As a result, a trait acquired in life could be passed down to future generations.

On that last point, Darwin was simply following in a tradition that reached back over two thousand years to the writings of Hippocrates. Earlier in the nineteenth century, Darwin's predecessor, the French naturalist Jean-Baptiste Lamarck, had offered the first detailed theory of evolution, and he had made the inheritance of acquired characteristics a crucial part. A giraffe striving to reach leaves on a high branch would force a vital fluid into its neck, stretching it. Its offspring would then be born with that longer neck, and over many generations this stretching produced the neck of the giraffes we see today.

Darwin saw gemmules as acting like Lamarck's vital fluid. In his research, Darwin discovered that improved breeds of cattle grew small lungs and livers compared to free-ranging breeds. He saw this as the result of pangenesis. Farmers fed these breeds better food and expected less work from them. As a result, they didn't need to work their lungs or livers, and the organs produced different gemmules as a result.

To Darwin, cattle and other domesticated animals put pangenesis on an impressive display. In just a few thousand years, humans had altered the heredity of animals and plants in endless ways, producing greyhounds and corgis and Saint Bernards, racehorses and draft horses, apples, wheat, and corn. Breeders such as Bakewell selected individuals to breed, unknowingly choosing which animals could pass their gemmules to future generations. They crossed different strains to combine gemmules in new combinations. Breeders had exploited the same laws of inheritance that had made the evolution of all species—even ourselves—possible.

"Man, therefore, may be said to have been trying an experiment on a gigantic scale," Darwin wrote, "and it is an experiment which nature during the long lapse of time has incessantly tried."

———

Sitting in Massachusetts, young Luther Burbank read *The Variation of Animals and Plants Under Domestication* with a rush of marvel and relief. He might be a novice farmer, but now he felt he was part of something far bigger. The same biology that gave rise to all living things—their variation, their selection, and their heredity—now felt like clay he could shape with his hands. Darwin declared that variation emerged from crossbreeding, which mixed gemmules of different origins together in new combinations. By selecting which plants to breed again, Burbank could eventually produce a new variety that reliably passed down its traits to future generations.

"While I had been struggling along with my experiments, blundering on half-truths and truths," Burbank later wrote, "the great master had been reasoning out causes and effects for me and setting them down in orderly fashion, easy to understand."

In 1871, Burbank bought a seventeen-acre farm where he could carry out Darwin's causes and effects. He cross-pollinated beans. His cabbage seeds and sorghum won prizes at the local agricultural fair. And then, at the tender age of twenty-three, Burbank spotted an odd potato that would bring him agricultural immortality.

One day, as he tended a patch of Early Rose potatoes, Burbank noticed a tiny, tomato-shaped mass dangling from one of the vines. It was, he realized, something wonderfully rare: a seed ball. Farmers typically propagate potatoes by cutting up their tubers and planting the pieces, which can grow into entire new potato plants. Potatoes can also reproduce by having sex. They grow flowers, and once the ovules in the flowers are fertilized by pollen, they develop into seeds. The seeds cling together in a ball-shaped clump.

Over thousands of years of breeding, domesticated potatoes have mostly lost the ability to make seed balls. If farmers noticed one in a potato field, they usually ignored it. But Burbank had Darwin on his mind, and so, to him, finding a seed ball was like stumbling across a jewel. "Stored in every cherished seed was all the heredity of the variety," he later said.

When Burbank spotted the seed ball, it was still immature and thus not yet ready to use for breeding. To make sure he could find it again, Burbank tore a strip of cloth from his shirt and tied it around the plant. When he checked back later, however, the seed ball had dropped to the ground and disappeared from sight. For three straight days, Burbank searched for it. When he finally found it again, he opened it up and found twenty-three potato seeds inside. Burbank carefully stored them away for the winter and then planted them in the spring of 1872.

From that single seed ball grew a riot of variation. Burbank ended up with potatoes of different colors, shapes, and sizes. When he tasted the tubers, he found that two were unusually good. They were also smooth, large, and white; they stored well over the following winter. Burbank brought them to the 1874 Lunenburg town fair, where people were stunned at what he had created. The following year, Burbank sold the potato to James Gregory, a seed merchant, for $150.

The "Burbank Seedling," as Gregory generously named it, quickly became one of the best-known crops in the United States. A descendant of that variety, the Russet Burbank, carpets much of the state of Idaho. They are the only potatoes that McDonald's, the biggest purchaser of potatoes in the United States, will accept for its french fries.

Burbank's success with his potatoes convinced him that Darwin could guide him to riches. He sold his farm inventory, paid off his small mortgage, and left the stony soils of Massachusetts for California. Later, Burbank would look back in surprise at his rash move. He put it down to some impulsive streak in his ancestry. "In short I was a product of all my heredity," he wrote.

Perhaps it was likewise "an inherited sensitiveness about money," as Burbank liked to call his frugality, that made him decide not to pay for a sleeping berth on the westbound train. He spent nine days curled up on a seat. Looking out at the prairies, he ate sandwiches out of a basket prepared by his mother. Burbank made his way to Santa Rosa, where one of his brothers had settled.

The plants of California overwhelmed him. The pears were so big that he couldn't finish eating a single one. Yet Burbank struggled to survive even amidst all that plenty. He threshed wheat in the summer and looked for construction work in the winter. Sometimes he found jobs at nurseries. In 1876, Burbank came down with a fever and was bedridden for days in a tiny cabin, where he survived on milk a neighbor provided him from her cow. "These were indeed dark days," Burbank later said.

The following year things improved. Burbank had brought ten of his potato seedlings to California, and his brother let him plant a patch on his land. Burbank put an ad in local newspapers for "this already famous Potato" and found some buyers. His mother and sister moved to Santa Rosa and bought four acres of land, which Burbank began to farm. In his free time, he would hike into the hills, discovering wild plants that botanists had yet to name. Seed companies would pay him for intriguing new species.

After six years in California, Burbank finally got his big break in 1881. A Petaluma banker named Warren Dutton wanted to get into the prune business and was ready to pay a small fortune for twenty thousand plum trees that would be ready to be planted in the fall. It was an absurd demand, but Burbank figured out how to meet it. He bought almonds and planted them on rented land in the spring. The almonds quickly sprouted into seedlings, whereupon Burbank and a hired crew of laborers grafted twenty thousand plum buds onto them. The buds took hold and grew. When their branches became big enough, Burbank cut the almond branches back. Burbank delivered the trees on time, and Dutton proclaimed him a wizard to anyone who would listen. It was the first time someone described Burbank that way, but it wouldn't be the last.

Dutton's praise helped Burbank's business explode. But unlike other nurserymen who prospered in California, Burbank rolled much of his profit into experiments. Following Darwin's guidance, he crossed different varieties to produce new combinations of traits. For his crosses, Burbank used the native California plants that he was becoming familiar with. He also developed a network of contacts in other countries, who supplied him with exotic plants— plums from Japan, blackberries from Armenia—that he could also combine. When he bred them, he would discover variations among their offspring.

"Something must happen to 'stir up their heredities,' as I am fond of saying—to excite in them the variability that normally lies dormant," Burbank later explained. As he ran his experiments, he sometimes felt barely in control of the powers he was summoning. "When you stir up the heredity of any living thing too much it is like stirring up an ant-hill—you find the results much more startling and unsettling than useful or helpful."

Burbank might produce thousands of hybrid offspring from which he might pick just a few to propagate into a new generation. He might breed them for years before reaching the proper form. After a few years of breeding a type of lily, Burbank found a single specimen that met his standards. A rabbit ate it.

Despite these setbacks, Burbank had produced enough varieties by the mid-1880s to start selling them to nurseries. His mysterious power to create new fruits and trees attracted visitors to his farm, to puzzle over his "mother trees"—native plants to which he grafted many different species at once to grow them as quickly as possible.

By 1884, Burbank could advertise a stock of half a million fruit and nut trees. Word of his creations spread—of oranges that could grow in the north, of flowers that would not fade—and before long, newspapers and magazines began publishing profiles of him. They crafted a public persona for Burbank as a botanical alchemist. "In his laboratory garden he has done for Nature in part of one man's lifetime what Nature couldn't do for herself in thousands and thousands of years," one newspaper declared. Others promised his work could feed the hungry and enrich the nation. One reporter wrote that, thanks to a giant prune Burbank developed, "one California town—Vacaville—was literally built by prunes."

Burbank's humble origins helped him become famous. He became an American icon along the lines of Thomas Edison, able to make great discoveries without a college degree. Yet the American scientific community came to admire Burbank as well. They could see (and taste) for themselves that his magic was real.

"In his field of the application of our knowledge of heredity, selection, and crossing to the development of plants," declared David Starr Jordan, the president of Stanford University, "he stands unique in the world."

———

Luther Burbank's self-education in heredity seems to have stopped with reading Darwin. After plowing through *Variation,* Burbank relied on his own instincts to carry out Darwin's vision. As he built his empire in Santa Rosa, he seemed unaware that in the late 1800s, Darwin's theory of pangenesis collapsed.

The early reviews of *Variation* didn't bode well. The psychologist William James dismissed pangenesis as empty speculation. "In the present state of science, it seems impossible to bring it to an experimental test," he said. To James, the book's only value was demonstrating just how baffling heredity remained.

"At the first glance," James wrote, "the only 'law' under which the greater mass of the facts the author has brought together can be grouped seems to be that of Caprice,—caprice in inheriting, caprice in transmitting, caprice everywhere, in turn."

But some scientists stood by Darwin, and none so passionately as his cousin Francis Galton.

Galton, thirteen years his cousin's junior, fashioned his life after Darwin's. After a disappointing stint at Cambridge, Galton led an expedition through southern Africa, and came back a famous geographer. He wrote bestselling travel books and dabbled in many different branches of science, making clever contributions along the way. He attempted to make the first national weather forecasts and designed the first weather maps. In 1859, he began turning his attention to biology, thanks once more to his cousin. Reading *The Origin of Species*, Galton later wrote, "made a marked epoch in my own mental development."

Like Darwin, Galton realized that understanding evolution would depend on making sense of heredity. Half a century later, when Galton wrote his autobiography, he struggled to convey to his readers just how mysterious heredity remained in the 1850s. "It seems hardly credible now that even the word heredity was then considered fanciful and unusual," he wrote. "I was chaffed by a cultured friend for adopting it from the French."

In the early 1860s, Darwin and Galton both investigated heredity, but in profoundly different ways. While Darwin pictured the invisible gemmules, Galton looked for evidence of heredity in the traits that the English upper class valued most. He looked over the biographies of notable men—mathematicians, philosophers, patriots—and was struck by how many of them had notable sons. "I find that talent is transmitted by inheritance to a remarkable degree," he wrote in *Macmillan's* in 1865.

If talent was indeed hereditary, Galton wrote, then it could be bred like the plumage of a pigeon or the fragrance of a rose. In fact, Galton believed England's future well-being depended on a national breeding program to produce more talented humans. He imagined this program as a joyous ritual, bringing gifted young people together to have better and better children. The result would be a species capable of handling all the power that Victorian science and technology was providing it.

"Men and women of the present day are, to those we might hope to bring into existence, what the pariah dogs of the streets of an Eastern town are to our own highly-bred varieties," Galton predicted.

In 1869, Galton published a book-length version of his study, which he entitled *Hereditary Genius*. He declared with remarkable certainty that eight out of a hundred sons of distinguished men were distinguished themselves, a rate far higher than one in three thousand people chosen at random. Here, Galton declared, was proof of the heredity of talent. Yet for all Galton's questionable data, there was a giant void in his book: He had no idea how heredity actually occurred.

With *Variation*, Darwin electrified his cousin a second time. Galton became convinced that pangenesis "is the only theory which explains, by a single law, the numerous phenomena allied to simple reproduction."

Galton set out to prove pangenesis by showing that gemmules existed. Darwin had written that gemmules "circulated freely throughout the system," and so Galton reasoned that if he transfused blood from one animal to another, he should also transfer some gemmules.

Galton wrote his cousin a note: "I wonder if you can help me. I want to make some peculiar experiments that have occurred to me."

He asked Darwin to put him in touch with breeders from whom he could buy rabbits. Over the next few months, Galton had silver-gray rabbits injected with blood from other rabbits of many different colors. He hoped the injected gemmules would change the color of their kits.

"Good rabbit news!" Galton wrote to Darwin on May 12, 1870. "One of the litters has a white forefoot."

But with the birth of more litters, Galton's excitement faded. Injecting blood into rabbits showed no further hint of being able to change their color. The experiments proved "a dreadful disappointment," Emma Darwin wrote to her daughter, and, in March 1871, Galton came before the Royal Society to recount his failure.

"The conclusion from this large series of experiments is not to be avoided," Galton said, "that the doctrine of Pangenesis, pure and simple, as I have interpreted it, is incorrect."

Galton thought he and Darwin belonged to the same team, together searching for heredity. But as soon as Galton gave up on pangenesis, Darwin publicly chided his younger cousin. He wrote a letter to *Nature*, disassociating himself from the rabbit experiments. "I have not said one word about the blood," Darwin declared.

Darwin pointed out that in his own writing, he had talked about pangenesis in plants and single-celled protozoans, which had no blood at all. "It does not appear to me that Pangenesis has, as yet, received its death blow," Darwin protested.

Writing in 1871, Darwin was technically correct. But in the years that followed, another scientist would kill pangenesis for good.

———

That scientist was a German zoologist named August Weismann. Unlike Darwin or Galton, Weismann didn't start his scientific life as they did with an exotic adventure. Rather than sailing around the Galápagos Islands or crossing Namib deserts, Weismann spent his best years squinting through a microscope, observing the fine details of butterflies and water fleas.

Weismann, like many other biologists of his generation, was taking advantage of powerful new microscopes and ingenious chemical stains to document life at the cellular scale. He observed how eggs developed into embryos, how some of their cells turned into eggs or sperm, which came together to make new embryos.

In addition to mapping cells, Weismann and his colleagues could also peer within them. In animals and plants, they could see a pouch inside each cell, which came to be known as the nucleus. Whenever a cell divided, its nucleus turned into a pair as well. But when a sperm fertilized an egg, the two nuclei seemed to fuse into a single one.

What lurked within the nucleus, Weismann and other scientists could not say for sure. It seemed to contain threadlike structures that were duplicated each time a cell divided. But some studies suggested that when eggs developed, they lost half of the normal supply of threads.

Weismann wove together his own observations and those of other scientists into one powerful model of life. He divided the body into two types of cells: germ cells (sperm and eggs) and somatic cells (everything else). Once germ cells developed in an embryo, they carried inside of them a mysterious substance he called germ-plasm that could give rise to new life.

"This substance transfers its hereditary tendencies from generation to generation," Weismann said. Germ cells had a kind of immortality, because their germ-plasm could survive for millions of years. Somatic cells, on the other hand, were doomed to die along with the body in which they were trapped.

If Weismann's so-called germ line theory was right, then Darwin's pangenesis had to be wrong. Darwin envisioned germ cells as wide-mouthed pots into which gemmules from throughout the body could pour. Weismann envisioned a barrier sealing off the germ cells, isolating them from any influence from the somatic cells.

It also meant that the inheritance of acquired traits—taken as a fact by Hippocrates, Lamarck, and Darwin alike—was impossible. An animal's somatic cells might be altered by experiences, but there was no way for those changes to get communicated to its germ cells. "Ever since I began to doubt the transmission of acquired characters," Weismann said, "I have been unable to meet with a single instance which could shake my conviction."

When Weismann turned against the inheritance of acquired characters in the late 1800s, it was still popular. In 1887, a certain "Dr. Zacharias" brought tailless cats to the annual meeting of German Naturalists. Dr. Zacharias claimed the mother of the cats had lost her tail when she was run over by a wagon. Other researchers did surgery on the spinal cord of guinea pigs, causing them to have seizures. Their pups had seizures as well. Mendel's mentor, Carl Nägeli, claimed that the thick coat of mammals in arctic regions had developed in a reaction to the cold air, and then became inherited. Swans and other waterfowl were born with webbed feet thanks to the habit of their ancestors to strike the water with outstretched toes.

To Weismann, none of these stories about acquired characters was proof of inheritance. They could simply be coincidences. The guinea pigs might not have inherited their seizures; instead, they might have developed infections. If a cat lost her tail and then gave birth to tailless cats, the scientific thing to do would be to track down the father and see if he had a tail or not. There was no need to invoke acquired characters to explain why musk ox have thick fur. Natural selection favored individuals that, for whatever reason, had warmer coats that made them less likely to freeze to death.

In 1887, Weismann decided to do what the advocates of acquired characters never did, and run an experiment. He set out to test the idea that mutilations could be passed down. He ran the study on white mice, cutting their tails before letting them mate. The female mice got pregnant and delivered litters. And none of their pups had a shortened tail. Weismann repeated the procedure on their pups, and their grand-pups, and so on over the course of five generations. He produced 901 new mice. They all grew normal tails.

On its own, Weismann admitted, the experiment might not destroy the theory of acquired characters, but it added more weight to all the other reasons to question it. Lamarck's followers claimed proof based on far less evidence.

"All such 'proofs' collapse," Weismann said.

———

Weismann reconfigured how scientists thought about heredity, an accomplishment all the more impressive for all the details of heredity he did not yet know about. After he introduced his germ-line theory, other researchers looked more closely at the multiplying threads in the nucleus of cells. They were dubbed chromosomes.

Researchers determined that a somatic cell carried pairs of chromosomes. (We humans have twenty-three pairs, for example.) A duplicating cell—known as a mother cell—made new copies of all its chromosomes—which it bequeathed evenly between two daughter cells. But when germ cells arose in an embryo, they ended up with only one set of chromosomes. Fertilization brought an egg and sperm together, creating a new set of pairs.

A new generation of scientists then asked how inheriting chromosomes determined the different forms that life could take. Hugo de Vries was among them.

De Vries had trained as a botanist, and at first heredity had meant little to him. He studied how plants grew, stretching their stems and sending out tendrils. His work caught the attention of Darwin, who recounted young de Vries's work in a book about plants. Darwin sent him a complimentary copy and then invited him to visit his estate when de Vries visited England in 1878.

"We talked for a short time about all kinds of things, the country house (which is very large and beautiful), the surroundings (also very beautiful), politics, my journey etc.," de Vries eagerly wrote his grandmother that night. "Thereafter Darwin took me to his room and we talked about scientific subjects. At first about tendrils, in connection with our former correspondence."

Darwin took de Vries on a tour of his garden, handing him a peach along the way. Later, de Vries gushed to his grandmother that he "was received so kindly and cordially as I never had dared hope for."

When de Vries returned home to the Netherlands, he and Darwin kept up the correspondence about plants. But in a letter he wrote Darwin in 1881, de Vries abruptly changed the subject. Now he was consumed with heredity.

"I have always been especially interested in your hypothesis of Pangenesis," de Vries told Darwin, "and have collected a series of facts in favour of it."

De Vries roamed the countryside for "sports of nature"—rare plants that sprouted weird growths or displayed odd colors. He wanted to create an herbarium of monstrosities, he later told a friend. By breeding them, he hoped to prove Darwin's theory of pangenesis right.

When Weismann unveiled the concept of the germ line, de Vries recognized its importance. As a botanist, though, he found it parochial. Plants, like animals, were made of cells that contained nuclei, and inside those nuclei were chromosomes. When plant cells divided, they also made a new set of chromosomes. But plants did not wall off their germ cells early in development. An apple tree would grow for years before producing germ cells that could give rise to pollen grains or seeds. A cutting from a willow could grow into an entire tree, complete with roots, branches, and leaves. A hidden potential to produce new plants must be spread throughout their cells, de Vries thought. While pangenesis might have its problems, he thought it had to be the foundation of any true understanding of heredity.

Darwin died in 1882, leaving de Vries to search for that understanding without the guidance of his guru. He began running experiments with his monsters. He crossed them with ordinary plants, and sometimes their bizarre traits turned up in later generations. De Vries came up with a theory of his own: Every cell contained invisible particles that were responsible for passing traits from one generation to the next. Under some circumstances, the particles in somatic cells could guide the development of a new organism. In honor of Darwin, de Vries called the particles pangenes.

In 1889, de Vries published *Intracellular Pangenesis,* in which he distilled over a decade's worth of work. Hardly any scientists took notice of it. One of the few who did advised de Vries not to mention pangenesis again.

De Vries did not give up. In the 1890s, he noticed that monstrosities crossed with regular flowers produced regular ratios of offspring. De Vries thought that flowers could have different numbers of pangenes in them, and those numbers were what determined traits in their offspring.

Despite his struggles with these ratios, de Vries became convinced that pangenes were real, and that their changes were what made evolution possible. Pangenes could abruptly change in a process he called mutation, and flowers that inherited a mutation abruptly became a new species. De Vries's mutation theory was pushing him far from Darwin, who had argued for the gradual evolution of species through tiny steps.

One day early in 1900, de Vries got a letter from a friend who was familiar with his obsession with hybrid plants. His friend thought de Vries might be interested in a thirty-five-year-old paper by "a certain Mendel." When de Vries scanned the paper, he was stunned that a Moravian monk he had never heard of had found the same patterns he had. He had even come up with a theory of invisible hereditary factors to account for it.

By an unparalleled coincidence, two other scientists studying inheritance, William Bateson and Carl Correns, also stumbled across Mendel's work at about the same time. They all realized that they had been scooped. And they also recognized just how important Mendel's experiments had been. Before 1900, scientists didn't have the right frame of mind to appreciate them. It took Darwin and Galton establishing heredity as a scientific question. It took Weismann and others to look closely at cells to ask how heredity was transmitted.

De Vries, Bateson, and Correns all began sharing the belated news about Mendel. Bateson emerged as the leader of the campaign: He and his colleagues demonstrated that animals could display the same ratios as plants. Even certain hereditary diseases in people fit the pattern. A British doctor named Archibald Garrod noticed that a condition he called alkaptonuria—which turned urine black—tended to run in families. Sometimes when two seemingly healthy parents started a family, about a quarter of their children fell ill. That ratio fit Mendel's predictions: The parents must be carriers, each carrying a recessive factor.

The "whole problem of heredity has undergone a complete revolution," Bateson declared. Mendel's discoveries could at last mature into a true science. Bateson christened it genetics.

———

No sooner was genetics born, however, than it was hurled into battle. Some scientists felt that Mendel must have made a mistake. Some tried to get his neat ratios of hybrids and failed. Other critics found it inconceivable that physical particles could be inherited and give rise to every trait in an organism.

De Vries went his own way. He accepted that Mendel's results were genuine, but he came to doubt they mattered much to big evolutionary changes. Those could only come about through the appearance of major new mutations. Evolution didn't creep forward, de Vries believed. It leaped.

De Vries unpacked this idea in his sprawling two-volume work, *The Mutation Theory*, in 1903. His theory that new mutations could produce new species in a single leap proved sensational. It finally earned de Vries the fame that had escaped him in earlier years. When he came to the United States to give lectures about his mutation theory, newspapers put his face on their front pages. It was on one of those tours that de Vries paid his first visit to Luther Burbank, in 1904.

By then, Burbank no longer considered himself simply a plant breeder. The honors that scientists had heaped on him persuaded him he was a genius of heredity. When scientists visited Burbank, he would regale them with a grand theory—"perhaps as original as Darwin's," he modestly declared—that the universe consisted of what he called "organized lightning." The scientists who listened to Burbank's ramblings politely nodded, said that they were unqualified to judge, and hoped they could gain access to his legendary garden.

De Vries traveled to Burbank's garden to find support for his mutation theory. His own evening primroses produced mutants from time to time, but he had yet to find another species that displayed mutations so clearly. De Vries's gigantic theory had come to rest on precious little evidence, like an elephant trying to ride a bicycle. Maybe Burbank's new varieties were, in fact, a wealth of new mutants.

Between bites of Burbank's stoneless plums, de Vries interrogated his host. Burbank had become wary of sharing his secrets by then. He would sometimes force his workers to empty their pockets to make sure they weren't smuggling out his prize seeds. If they chatted across the picket fence with a passerby, he would fire them. With de Vries, Burbank was more forthcoming. He explained how he had crossed plums, selecting the ones with smaller and smaller stones. He described how he set about breeding cacti without spines as a new source of food for cattle. He searched for varieties to cross, each missing different parts of their spines. Over generations, they became soft enough for Burbank to stroke over his cheek.

De Vries left Santa Rosa impressed by Burbank's passion. "The sole aim of all his labors is to make plants that will add to the general welfare of his fellow beings," de Vries wrote later. As a scientific mission, however, the journey ended up a disappointment. De Vries hoped his visit would shed light on how plants acquired new traits. "Burbank's experience did not throw any light on this question," he concluded.

De Vries's time with Burbank marked the high-water mark in the careers of both men. When he traveled to Santa Rosa, de Vries had become famous as one of the founders of modern genetics and as the author of a controversial new theory about mutations that seemed to overthrow Darwin. Burbank, meanwhile, had become a celebrity as both a mystic of nature and a keen businessman. Things would never be so good for either of them again.

In the years that followed, de Vries would keep fighting for his mutation theory. But the only organisms that had experienced one of de Vries's dramatic mutations were his evening primroses. It turned out de Vries was fooled by an illusion of breeding. What he took to be an entirely new mutation was actually a combination of old genetic variants.

De Vries refused to accept these facts, retiring to the village of Lunteren in the Dutch countryside. For the next sixteen years, the villagers would sometimes spot a tall bearded man walking amidst a garden of primroses.

In December 1904, a few months after de Vries's first visit, Burbank got a letter from the Carnegie Institution. Andrew Carnegie had set up the institution two years earlier to fund important scientific research. Carnegie himself believed that some of the money should go to Burbank, whom he called a genius. The letter informed Burbank he would shortly receive $10,000 "for the purpose of furthering your experimental investigations in the evolution of plants." The institution would send him another $10,000 the next year, and the year after that, with no clear end in mind.

The popular press released a fresh flurry of profiles of Burbank, pointing to the Carnegie cash as science's seal of approval. In 1906, a botanist named George Shull arrived to help Burbank write up scientific reports about his research.

Shull found Burbank to be an artist of nature. As a scientist, however, he was a phantom. When Shull asked Burbank for experimental records, the old horticulturalist might hand him a few sheets of paper on which he had scribbled notes in pencil. "This was a rich, sweet, delicious, superb pear, as good as Bartlett, perhaps much better," he wrote on one sheet. He sliced one of the pears in half and stamped it on the page, letting the juice stain the paper.

Shull tried instead to talk to Burbank to extract useful information. Burbank informed him that he was the greatest authority of plant life that ever lived. He claimed to have already discovered Mendel's results on his own, and yet he also declared that acquired characters could be transmitted from one generation to the next. "Environment is the architect of heredity," Burbank said.

When Shull pressed him for the concrete details of his work, Burbank grew so irritated he started avoiding Shull around the gardens. It wasn't Shull's line of questioning that annoyed him so much as the fact that the young botanist seemed to be preparing to explode his legend. Indeed, Shull reported back to the Carnegie Institution that it would be impossible to use any of the plants to test Mendel's theory of inheritance. In 1910, the Carnegie Institution sent Burbank their last check. Their $60,000 bought them a single report from Shull, about rhubarb.

As the Carnegie money dried up, a swarm of businessmen descended on Burbank, proposing deals to make him staggeringly rich. Some of the hucksters set about publishing a lavish, costly encyclopedia of his life's work. That venture collapsed into bankruptcy in 1916. Other businessmen set up the Luther Burbank Company, to sell his plants directly to customers rather than to nurseries. They mismanaged the venture, unable to align their supply to demand. Things got so desperate that the company started shipping ordinary cacti in place of Burbank's spineless variety. Before putting the plants in the mail, company workers simply scrubbed off the spines with a wire brush. The Luther Burbank Company went bankrupt as well.

Burbank managed to hold on to much of his wealth despite these disasters. But they permanently tarnished his reputation. By the 1920s, Burbank had become an untrustworthy businessman whom scientists no longer revered. He spent his final years puttering around his Santa Rosa farm, cared for by his young second wife, Elizabeth, along with a few assistants. In 1926, Burbank died at age seventy-seven. Thousands of people came to his funeral at a nearby park, and then his body was brought back to his house, where it was buried. Nothing stood over his grave except a cedar of Lebanon. "I would like to think of my strength going into the strength of a tree," he once said. Elizabeth sold off his remaining plants to Stark Bro's, just as Hiatt had sold his Delicious apples three decades before. Burbank's garden tools went to Henry Ford.

After his death, Burbank enjoyed a longer stretch of fame than de Vries had. His face reappeared in popular culture for decades. As late as 1948, the beer company Anheuser-Busch was using his likeness in their ads. In a full-page ad for Budweiser, Burbank stands in his garden, holding out a rose for a mailman to smell. Both Budweiser and Burbank's varieties, the ad declared, were "great contributions to good taste."

In the picture, Burbank has a grandfatherly smile, a shock of gray hair, a starched collar, and a black tie. The image belonged to an earlier chapter in the history of heredity, when breeders could use their intuitions to produce new fruits and flowers, becoming masters of forces they didn't understand. By the 1940s, when the beer ad appeared, heredity meant something very different. It was now a precise molecular science in the hands of some, and a monstrous rationale for oppression and genocide in the hands of others. Even the plants and yeast that went into Budweiser beer in the 1940s had become products of scientific breeding, rather than of Burbank's old wizardry.

There is another picture created after Burbank's death that still feels fresh. The painter Frida Kahlo paid a visit to Burbank's garden in 1930. She had moved from Mexico to San Francisco a few months earlier. Her husband, the artist Diego Rivera, had accepted a commission to paint murals for American patrons, the first of which would capture the spirit of California. Kahlo and Rivera took the short drive from San Francisco to Santa Rosa to visit the home of a hero of the state. Burbank's widow, Elizabeth, gave the couple a tour around the grounds, showed them the cedar under which Burbank was buried, told them stories about her late husband, and gave them some photographs of him to take with them.

Kahlo painted Burbank on a stark, tan California landscape. High clouds moved across the sky, and behind him grew a pair of trees. One tree was small, with oversize fruits. The other grew clusters of balls in different colors, perhaps patterned after one of Burbank's mother trees. From the knees up, Burbank looked like he does in many photographs, with a tranquil expression on his face, wearing a dark suit and holding a plant. In this case, he's holding a philodendron, a vinelike plant with lobed leaves that Kahlo painted to be as big as his chest. Below the knees, Burbank was transformed by Kahlo's powerful imagination. His legs disappeared into the stump of a tree. Kahlo cut away the earth to reveal the tree's roots, which pierced the head, the heart, the stomach, and the legs of a corpse.

Burbank had no children of his own who could carry his hereditary particles after his death. His fame eventually faded. But many of the varieties that he developed continued to grow, to make seeds of their own, and to be replaced by their offspring. Some, like the Burbank potato, bear his name. Others grow namelessly, Burbank's handiwork having been long forgotten. He had found an immortality here on Earth, his work and his plants extending their existence in intimate replication.

A few months before he died, a reporter paid Burbank a visit to ask him about religion. Burbank was such a familiar figure in the United States that reporters would ask his opinion about everything from jazz to crime. At one point in the interview, Burbank said that Jesus had been "a wonderful psychologist," and an infidel to boot. "Just as he was an infidel then, I am an infidel today."

Now the river of letters that poured into Burbank's house turned furious. Prayer groups formed to beseech God to help Burbank see the light. To respond to the attacks, Burbank arranged to give a speech—a sermon, really—at the First Congregational Church of San Francisco on the last Sunday of January 1926. More than 2,500 people crammed the pews.

The seventy-six-year-old Burbank told them that he was no atheist. He subscribed to what he hoped would someday become a religion of humanity, worshiping a God "as revealed to us gradually, step by step, by the demonstrable truths of our savior, science," he said to his audience. Burbank didn't see the point of wasting time pondering hypothetical eternities in heaven or hell. Heredity—the continuity of life through the generations—was vast enough for him. "All things—plants, animals, and men—are already in eternity, traveling across the face of time," he said.

CHAPTER 3

This Race Should End with Them

V INELAND BEGAN as an idea for a perfect city.

In 1861, a businessman named Charles Landis traveled from Philadelphia into the empty Pine Barrens of New Jersey. He bought twenty thousand acres and laid down a map of lots. He called it Vineland. Farmers bought land to grow crops on the fertile soil, and retired Civil War soldiers later came to work in new glass-manufacturing plants. The idea of Vineland endured into the twenty-first century, in the generous width of its main streets, in the triumphant design of its municipal buildings. But a new city has grown on top of Landis's idea: a city that lost its factories, that turned its outlying farms into suburbs, that brought in immigrants not from New England but from Mexico and India.

I came to Vineland on a bright cold day in February, driving along South Main Road, one of the original roads that ran along the city's eastern edge. I passed a bleak, treeless row of gas stations, supermarkets, cell phone shops, and liquor stores. At the intersection with Landis Avenue, I pulled into a Wawa store parking lot and walked inside to buy a bag of peanuts. Car mechanics and home health aides were ordering sandwiches and coffee and lottery tickets. When I came back outside, I looked up at the grumpy, overheated winter sky. The clouds were tormenting the South Main traffic with tantrums of rain. My phone buzzed with a tornado warning for all of South Jersey. I pulled a wool cap onto my head and took a walk, eating peanuts for lunch.

The convenience store driveway curved around a wedge of grass by the intersection. A massive rounded stone stood in the center of the wedge, surrounded by bushes and spotlights anchored into the wood chips. I walked over to inspect it. The stone was inscribed with a name: S. Olin Garrison. No explanation, no date. The drivers of the passing cars and trucks paid the tombstone no notice. I doubt any of them knew who S. Olin Garrison was, let alone why he was buried in front of a Wawa store.

Turning my back to the noisy commercial strip, I looked eastward across a huge, empty field, crossed by a worn concrete path. I walked down the path, under a row of leafless trees that leaned over the left side. The trees had lost some of their boughs, and some were dead. But you could still sense that someone had planted them in grand, rational intervals long ago. The line of trees led my eye across the field to a pair of small, square gazebos in the distance, tilted on the frost-heaved earth. Beyond them was a scattering of old buildings. A late nineteenth-century edifice had a dome sprouting from one corner. Around it huddled a few old houses and outbuildings, falling into disrepair.

I had spent the morning at a nearby historical society looking over photographs of this spot from over a century earlier. Now that I was at the spot itself, I could see it as it looked on an October morning in 1897. There was no Wawa store—no stores at all, for that matter. People passed by on foot, bicycle, or horseback. South Main Road and Landis Avenue bordered a 125-acre farm, with pumpkin patches, apple orchards, and asparagus beds. A high gate stood at the corner, with a name arching overhead: VINELAND TRAINING SCHOOL.

I had come here, and cast my mind back, because the Vineland Training School holds an important place in the history of heredity. Within the walls of the school, Mendel's research was applied to humans, with disastrous consequences. What happened here would influence thoughts about heredity for generations.

In 1897, a path led from the gate into the school grounds, flanked by newly planted trees. The gazebos were plumb and freshly painted. The buildings bustled with two hundred children. S. Olin Garrison, the founder and principal of the Vineland Training School, was very much alive in 1897, and I pictured him in the main school building, working at his desk. I listened to the sweet-toned bell ring from the school's clock tower in the distance.

One morning in October 1897, an eight-year-old girl named Emma Wolverton arrived at the school gate. She was of average height, with a pretty, round face; a wide nose; and thick, dark hair. It's impossible to know what Emma Wolverton was feeling that morning. In later years, she never got the chance to speak publicly for herself about her life. Of the many people who spoke for her, few particularly cared what she had to say. To most of them, she was a cautionary tale about all the ills that heredity could pass down through the generations.

We do know a little about how Emma Wolverton ended up at that corner in Vineland. Her mother, Malinda, grew up in the northern part of the state. At age seventeen, she started work as a servant. Soon Malinda became pregnant with Emma and was thrown out of her master's house. Emma's father, reputed to be a bankrupt drunk, abandoned Malinda, and she ended up in an almshouse, where she gave birth to Emma in 1889.

A charitable family took Malinda and her infant daughter out of the almshouse, and Malinda worked for them for a time. Soon she became pregnant again, and her benefactors insisted she get married to the father. Malinda and her husband had a second child together, after which the entire family moved into a rented house on a nearby farm. When Malinda got pregnant with a third child, her husband denied that the baby was his and abandoned her and the children.

The farm where she rented a house was owned by a bachelor. Not long after her husband left, she moved in with the farmer, and he admitted he had fathered the new baby. Emma's benefactors tried to make things right yet again. They arranged a divorce between Emma's mother and her stepfather, and then a remarriage to the farmer. He consented, but only if Malinda got rid of the children other than his own. It was not long afterward that Emma was delivered to the front gate of the Vineland Training School.

When S. Olin Garrison opened the school in 1888, he originally named it the New Jersey Home for the Education and Care of Feeble-Minded Children. He gave it a motto that would be stamped on their publications for decades to come: "the true education and training for boys and girls of backward or feeble minds is to teach them what they ought to know and can make use of when they become men and women in years." Garrison was determined to provide a more humane place than the typical warehouses where those deemed feebleminded had been abandoned in previous generations. "Our aim is to awaken dormant faculties, to arouse ambition, to inject hope, and develop self-reliance," the school declared in a brochure.

To get Emma admitted into the school, she was provided with a cover story: She didn't get along with other children in her regular school. That somehow raised the worry that she was feebleminded. The definition of *feebleminded* was sprawlingly vague in the late 1800s. People brought children to the Vineland Training School because they suffered epileptic convulsions. Others suffered from cretinism, a combination of dwarfism and intellectual disability. Others had a condition that would later be called Down syndrome. Emma belonged to a class of students who had no obvious symptoms but were still judged unfit for society.

When Emma arrived at the school, the staff gave her a thorough examination to judge whether she should be admitted. They observed "no peculiarity in form or size of head." Emma understood their commands, and she could use a needle, carry wood, and fill a kettle. She knew a few letters, but couldn't read or count. But the staff found her "obstinate and destructive," according to their notes. "Does not mind slapping and scolding."

That was enough. The fact that she had been brought to Vineland because she had become a nuisance at home went unrecorded in her file. Her examiners declared her to be feebleminded. They took her in.

Emma moved into one of the cottages, which she shared with a small group of other children. Every day, the school filled Emma's schedule with classes, duties, and games. Along with reading and math, she was taught about nature on walks through the fields and woods. "We show them the connection between nature and their being," the deputy principal, E. R.

Johnstone, said, "how dependent they are upon the plants and animals for their food and raiment." Emma and the other students spent much of their time singing in music classes. "Proper training will cause these songs of savagery to become the songs of civilization," Johnstone predicted.

"Happiness first and all else follows" was a slogan that hung on the school walls. A team of wealthy Philadelphia women, known as the Board of Lady Visitors, paid for a donkey-pulled streetcar that the children could ride around the perimeter of the farm. The ladies built a merry-go-round at the school, and a zoo stocked with bears, wolves, pheasants, and other creatures. Each year the school put on Christmas plays that residents of Vineland could attend, and each summer the school filled two train cars with students, who traveled to Wildwood Beach for an outing by the sea. One of the earliest photographs of Emma shows her in the back of an open wagon filled with girls and teachers. She sits on a pile of hay, looking back toward the photographer, smiling. The picture is labeled "off for camp."

As an able-bodied child, Emma spent part of each day learning manual trades. She got a garden patch to raise fruits and vegetables. Girls like Emma were instructed in sewing, dressmaking, and woodworking, while the boys learned how to make shoes and rugs. The administrators claimed that this labor prepared the students to someday earn a living. But the school, like many asylums and prisons of the time, also depended on their work for their own income. Between May 1897 and May 1898, the school's records indicated, the students made 30 new three-piece suits, 92 pairs of overalls, 234 aprons, 107 new pairs of shoes, and 40 dressed dolls. They washed 275,130 pieces of laundry. They sold $8,160.81 of produce from the school farm, including 1,030 bushels of turnips, 158 baskets of cantaloupes, and 83,161 quarts of milk. The fact that feebleminded children could do so much skilled labor was a paradox that never seems to have troubled the school's administration. Nor did they feel guilty for the money they made on the children's labor. "We are doing God's work," Johnstone explained.

For evidence of their divine mission, the school pointed to the lives they had saved. They were also sparing society the burden of feebleminded criminals. "The modern scientific study of the deficient and delinquent classes shows that a large proportion of our criminals and inebriates are really born more or less imbecile," declared Isabel Craven, the president of the Board of Lady Visitors.

Feeblemindedness was not just present at birth, Craven believed, but was passed down from parents to children. She shared the standard late nineteenth-century American belief in the heredity of bad behaviors. Somehow, feeblemindedness could be both a medical disorder and the wages of sin, passed down from the sinners to their children. Writing in the school's annual report in 1899, Craven recounted one such story, about an alcoholic woman in Germany in the late 1700s. She had 834 descendants, out of which 7 became murderers, 76 criminals of other sorts, 142 professional beggars, 64 charity cases, and 181 women who led "disreputable lives," as Craven put it.

The Vineland Training School was protecting future generations from this danger by removing feebleminded children from circulation, ensuring that they never got a chance to have children of their own. "What a legacy of crime and expense we may leave to the coming generation in our neglect to care for these incapable ones," Craven warned.

Emma settled into her new home. Her teachers kept track of her progress in their notes. They logged her letters to Santa Claus, in which she asked for ribbons, gloves, dolls, and stockings. She learned to spell and count, although she struggled with arithmetic. She learned how to make a bed. Sometimes Emma's teachers made a note of bad conduct. At other times, they said she marched well. She acted in the Christmas plays. She mastered the cornet and played songs such as "The Star-Spangled Banner" in the school band. She learned to use a sewing machine to make shirt-waists, and then she learned how to build boxes to put them in, complete with paneled tops and mortise-and-tenon joints.

When Emma became a teenager, she was ushered into the school's unpaid labor force. "She is an almost perfect worker," a school administrator noted in her records. Emma waited tables in the school dining room and served as a helper in the woodworking class. She proved herself so capable that Johnstone made her his housekeeper and later put his infant son in her care. For a time, Emma worked as a kindergarten aide at the school, and a visitor to the school mistook her for one of the teachers. It was not the only time that visitors would comment on how normal she seemed.

At age seventeen, Emma met a new member of the Vineland staff: a small, balding man named Henry Goddard. Goddard moved into a new office above one of the workshops, which he filled with strange instruments and machines. He would give children tasks to perform for him, such as having them poke a wand into holes drilled into a sheet of wood as fast as they could.

One day it was Emma's turn to go to Goddard's office.

"I have five cents in one pocket and five in another," he said to her. "How many cents have I?"

"Ten," Emma replied.

Dr. Goddard asked her another sixteen questions about numbers. All told, she got twelve right and five wrong.

Two years later, Goddard summoned her again for another round of questions. Use *Philadelphia*, *money*, and *river* in a sentence. Count backward from twenty.

Goddard's assistants praised her for every answer, although she got a fair number wrong. Later, Goddard reviewed her test and summed up her performance—her entire existence, really—with a single word he had recently invented: *moron*.

Unbeknownst to Emma, Goddard had also started making discreet inquiries about her family. His assistants sought out friends of the Wolvertons for gossip. Goddard was sure of what they would discover: that Emma's family were morons, too.

———

Henry Goddard first came to the Vineland Training School to build a science of childhood, having escaped a disastrous childhood of his own. Around the time he was born in 1869 in Maine, his father was gored by a bull. The injury eventually cost the family their farm, and for a few years Goddard's father eked out a living as a day laborer before dying in

1878. Goddard spent the next three years with his older sister and her family, as his mother, a self-appointed Quaker missionary, vanished for months at a time to preach at Friends meetings across Canada and the Midwest. At age twelve, Goddard was sent to Providence on a scholarship to a Quaker boarding school. "Nobody knew me or cared a whit whether I lived or died," Goddard recalled in his old age.

After finishing his time in "Quaker jail," as he called it, Goddard then went to Haverford College. He wasn't any fonder of that school either, considering it nothing but "a convenient way to keep the sons of rich Philadelphia Quakers out of mischief." He came to hate the very institution of school. It was all a pointless exercise in the rote memorization of Latin and Greek, along with the endless worry that students would fall into sin. It made no difference to how people turned out, Goddard believed; the students from wealthy families went on to prosperous lives, while poorer students like Goddard were left to struggle. "In all my adult life," he later said, "I have felt keenly the defects of my early training."

For all Goddard's scorn of schools, he ended up spending his life around them. He coached football at the University of Southern California for a while before teaching at high schools in Ohio and Maine. But at age thirty, Goddard heard a lecture by the psychologist G. Stanley Hall that changed how he thought about education. Hall told his audience that schools could scientifically liberate the minds of children. Hall's own research had persuaded him that the mental development of children follows a predictable course, just like the metamorphosis of a wingless nymph into a dragonfly. If teachers and psychologists joined forces, Hall said, they could create a new kind of education based on science rather than superstitious traditions.

Goddard immediately quit his teaching job and traveled to Massachusetts to study under Hall at Clark University. After getting his PhD, Goddard moved to West Chester, Pennsylvania, in 1899 to become a psychologist at the state normal school. There, he began gathering the data that psychologists would need to transform teaching. Teachers from across Pennsylvania used Goddard's eye charts to test the vision of their students so that he could figure out how many children were doing badly in school simply because they had trouble reading books and chalkboards. Goddard sent out questionnaires to gauge the moral development of students from one grade to the next. Much as his mother had traveled to Quaker meetings, Goddard went from conference to conference to preach to teachers about the glory of Child Study. He asked his audiences to join him on a quest for "a law of child nature which we can bank upon when once we have comprehended it."

At a 1900 conference, Goddard met E. R. Johnstone, who invited him to visit the Vineland Training School. Goddard was impressed. The Vineland teachers didn't mindlessly deliver the same lessons over and over again. They experimented, revising their teaching based on what helped the students improve. Johnstone insisted that Goddard spend some time during his visit talking with the students themselves. "I never dreaded anything more," Goddard later admitted. But it went better than he had expected, perhaps because Goddard knew what it felt like to be an abandoned child. Afterward, Johnstone congratulated him. "You talked as though you were accustomed to talking to the feeble-minded," he said.

Goddard came away from his visit convinced that Vineland was an exceptional place—"a great family of happy, contented, but mentally defective children," he said. Over the next few years, he stayed in close contact with Johnstone, sharing ideas about using science to bring about a new way of teaching. In 1906, Johnstone invited Goddard to become Vineland's first director of research.

To Goddard, it was a rare scientific opportunity. Vineland could reveal clues about the human mind that studies on ordinary children could not. Anatomists often studied simpler animals—flatworms or sea urchins, for example—to find important lessons that applied to humans as well. Psychologists might gain the same advantage by studying less complex minds. "The training school at Vineland, N.J., is a great human laboratory," Goddard declared.

But when Johnstone announced Goddard's appointment, he also let slip a gloomier motivation for bringing aboard a psychologist. The feebleminded were continuing to have more children, who were inheriting their defects, and society thus faced an impending disaster.

"Degeneracy is increasing, neurotic disease is increasing, defectiveness is increasing," Johnstone warned. Building more Vinelands wouldn't hold back the tide. "By the time more room is made it is filled and the waiting-list is larger than before.

"We must stop the increase," Johnstone warned. "And that means to find where they come from, why they come and what to do to check the stream."

———————

G oddard didn't share Johnstone's bleak view, at least not at first. He hoped that someday his research at Vineland could lead to treatments that could lift up the mental state of the feebleminded. "Suppose we could find some way of exercising these brains so that other cells took up the work of the missing cells!" he mused in 1907. "Would we not find a far greater degree of intelligence than we have ever dreamed of?"

Before unlocking hidden intelligence, Goddard would first need a way to scientifically measure it. He wanted to assign intelligence a number, the way doctors measured blood pressure or body temperature or weight. In Goddard's day, doctors regularly diagnosed children as imbeciles and idiots, but they did so mostly by intuition. Goddard tried to craft a test that could drill down to the biological basis of intelligence. He guessed that the speed of the nervous system was crucial, and so he would put Vineland students in front of an electric key and tell them to tap it with their finger as fast as they could. It worked badly. Some students couldn't even understand what he wanted them to do. Goddard tried other tests. He had the students squeeze a dynamometer as hard as they could, to thread needles, to draw straight lines. But whenever Goddard sat down to analyze the test scores, he found they didn't hang together. A student might do well at one and miserably at another.

"After two years my work was so poor, I had accomplished so little, that I went abroad to see if I could not get some ideas," Goddard later said.

In Europe, Goddard visited universities, schools, and laboratories to observe their research. While he was staying in Belgium, a physician off-handedly gave him a sheet of paper with a series of questions on it. It was a new exam called the Simon-Binet test, named after its creators, the French psychologist Alfred Binet and his assistant, Théodore Simon. At the request of the French government, Binet and Simon set out to design an exam schools could use to identify children who would need special help in class.

Binet recognized he would need a way to measure intelligence, "otherwise called good sense, practical sense, initiative, the faculty of adapting one's self to circumstances," he said. But how could he measure this quality, like a thermometer measures temperature? Instead of trying to measure it directly, Binet decided to measure how each child compared to other children.

Ordinary children got better over time at mental tasks. Highly intelligent ones seemed to Binet to develop faster, while feebleminded children lagged behind. Binet and Simon determined the average score that children of a given age got on a given test. They could then test other children and assign them a mental age based on how well they scored. A feebleminded ten-year-old might have a mental age of five.

It shocked Goddard that a psychologist would try to measure the human mind without some finely machined instrument—a chronoscope, perhaps, or an automatograph. All Binet claimed he needed was for children to answer some questions. Other European researchers warned Goddard that the Simon-Binet test was bogus, but he ended up tucking it into his papers anyway. When he arrived back in Vineland, he discovered it once more and decided to give it a try. After all, he had nothing left to lose.

Goddard administered the test to some of the Vineland students and then looked over the scores. The Simon-Binet test did a remarkably good job of matching the judgments of Vineland teachers. Students who had been determined to be idiots consistently got the lowest scores. The imbeciles did somewhat better, and the children who were simply slow and difficult—like Emma Wolverton—did better still, their mental ages lagging just a few years behind their chronological ones.

Here, Goddard decided, was the measuring tool he had been searching for. Idiots had a mental age less than three; imbeciles, between three and seven. People like Emma Wolverton functioned at a higher level, but lacked a proper label. For those with a mental age between eight and twelve, Goddard reached back to his dreary classics classes and coined the word *moron*, based on the Greek word for "fool." Goddard even sliced each of these new categories into three subdivisions apiece: low-grade, medium-grade, and high-grade.

Once Goddard was done testing Vineland students, he cast his eye over other schools. He managed to get permission to send out five assistants to a nearby school district and test two thousand ordinary students. They found that 78 percent of children had a mental age within a year of their chronological age. Four percent were more than a year ahead, while 15 percent were two to three years behind. Trailing at the rear, 3 percent of the children were three years behind.

"These figures practically amount to a mathematical proof of the accuracy of the Binet tests," Goddard declared. The regularity of the scores, no matter who took the tests, convinced him that they were accurately measuring a biological trait—the mysterious wellspring of intelligence in the brain. They may have also played a part in changing how he thought about intelligence itself. Instead of something malleable, which could be increased by strengthening brain cells, he came to see it as largely determined by heredity.

"It cannot be cured," Goddard concluded. "It is caused, in at least eighty per cent of cases, by disturbances of function in parents or grandparents that might have been prevented."

When Goddard spoke this way, he betrayed his nineteenth-century concept of heredity. He shared the common belief that people who took up a life of crime or alcoholism might somehow taint future generations with their sins. Growing curious about the degeneration of his students, Goddard thumbed through Vineland's admission forms, looking for details about their families. He could find only a little information. To get more, he drafted an "after-admission blank" for parents and physicians to fill out.

Goddard asked whether Vineland students had any relatives who were insane, alcoholic, feebleminded.

When the blanks came back filled out, Goddard was surprised at how many relatives suffered from these weaknesses. To gauge the full scope, Goddard wanted to hire a team of skilled assistants to "collect data on heredity."

How he would pay for the project, Goddard couldn't say. In the midst of this uncertainty, a letter arrived at the school in March 1909 like an answered prayer. One of the country's leading scientists, a geneticist by the name of Charles Davenport, wanted to know if anyone at Vineland had data about the heredity of feeblemindedness.

———

D avenport had leaped to fame only a few years before writing to Vineland. He had earned a PhD in zoology at Harvard in 1892, going on to a solid but obscure career studying scallops and other marine animals. He moved to Cold Spring Harbor, a Long Island village, where he ran a summer school for biology teachers.

But Davenport had great ambitions far beyond beachcombing. He pioneered new statistical methods to make precise comparisons between animals, based on their size and shape. Once these methods had matured, Davenport predicted, "biology will pass from the field of speculative sciences to exact sciences." He struggled to use statistics to understand heredity, comparing parents to their offspring. When Mendel's work came to light in 1900, with its concepts of dominant and recessive characters, it hit Davenport like a lightning bolt to the skull.

Davenport persuaded the Carnegie Institution to turn Cold Spring Harbor from a sleepy summer school into a full-time research station for genetics. In 1904 the Station for Experimental Evolution opened its doors. Hugo de Vries traveled by train to Cold Spring Harbor and gave a speech to celebrate the event. He celebrated Davenport in particular as its director. "With him it will open up wide fields of unexpected facts, bringing to light new methods of improvement of our domestic animals and plants," de Vries said.

For his first few years as director, Davenport fulfilled that prediction. He brought together a team of scientists who embarked on studies on heredity, investigating flies, mice, rabbits, and ducks. George Shull, the botanist who would later inspect Luther Burbank's gardens, grew corn and primroses in the Cold Spring Harbor fields. Davenport himself studied chickens and canaries. Inspecting his canaries, he concluded that the crest of feathers on their head was a dominant Mendelian trait.

But Davenport wasn't content with canaries. He wanted to decipher human heredity, too. Davenport couldn't study human heredity by raising experimental families. Instead, he set out to create a science of pedigrees. For centuries, people had been recording their genealogy, and sometimes those trees offered hints of heredity. The Habsburg jaw reappeared in generation after generation of royal portraits. In the nineteenth century, asylums kept records hinting that insanity tended to run in families. Davenport realized that if pedigrees were detailed enough, they might reveal Mendel's signature over many human generations.

Working with his wife, Gertrude, a zoologist, Davenport started with simple studies on the color of people's eyes and their hair. He then expanded his research, training a team of fieldworkers to search across New England for families with hereditary disorders such as Huntington's disease. Davenport also wondered if American asylums and other institutions—homes for the deaf and blind, insane asylums, prisons—might already have the information he was looking for. When he wrote to the Vineland Training School, he was stunned to get a letter back from Goddard, explaining all the work he had already done.

"I can hardly express my enthusiasm over these blanks," Davenport told Goddard, "and my enthusiasm that you are planning, I trust, extensive work in the pedigree of feeble minded children."

Davenport traveled to Vineland to meet Goddard and help launch the project. He showed Goddard how to manage field researchers and analyze the data they brought back. Most important of all, Davenport gave Goddard a crash course in genetics.

By 1909, a growing number of biologists had come to accept Mendel's findings. But none of them could yet say for sure what was responsible for his patterns. The Danish plant physiologist Wilhelm Johannsen gave Mendel's factors a new name: genes. "As to the nature of the 'genes,'" Johannsen warned, "it is as yet of no value to propose any hypothesis."

Under Davenport's guidance, Goddard swiftly embraced Mendelism. It remained to be seen whether feeblemindedness was a recessive trait, arising in children when they inherited the same gene from both parents. To search for evidence, Goddard talked a Philadelphia philanthropist into paying for a study on heredity. He built up his field team, choosing only women, who he required to have "a pleasing manner and address such as inspire confidence," he said, along with "a high degree of intelligence which would enable her to comprehend the problem of the feeble-minded." Goddard would come to depend most of all on his top fieldworker, a former school principal named Elizabeth Kite, who had studied at the Sorbonne and the University of London.

Kite and the other fieldworkers began traveling to meet the families of the Vineland students. Within a matter of months, Goddard claimed he saw patterns that "seem to conform perfectly to the Mendelian law."

Writing about the results in the school's annual report, he predicted great things for Vineland. "Once we prove that the law holds true for man we shall be in the possession of a powerful solvent for some of the most troublesome problems," he said. "We are within reach of a great contribution to science that would make the New Jersey Training School famous the world over and for all time."

———

Fanning out from Vineland across New Jersey and neighboring states, Goddard's fieldworkers gathered data on 327 families of students. In a few cases, the families had normal intelligence. The feeblemindedness of the students seemed to arise from some unknown source. It was far more common, however, for the fieldworkers to find families with many feebleminded members, not to mention alcoholics and criminals.

Back in Vineland, Goddard gathered what he believed to be more evidence that feeblemindedness was inherited like Mendel's wrinkled peas. If two feebleminded parents had children, the school records seemed to show, much of their family might end up feebleminded, too. Based on his pedigrees, Goddard estimated that about two-thirds of feebleminded people owed their condition to heredity. "They have inherited the condition just as you have inherited the color of your eyes, the color of your hair, and the shape of your head," he said.

This dawning realization revolted Goddard. He felt as if he was pulling a scrim away from American society, revealing a hidden rot. And none of the stories gathered by his fieldworkers appalled him more than that of Emma Wolverton.

When Goddard first examined Emma, she seemed just one of many morons in the school's care. She was a pleasant enough student within the confines of Vineland. But she would be doomed if she stepped off the property. "She would lead a life that would be vicious, immoral, and criminal, though because of her mentality she herself would not be responsible," Goddard predicted.

Goddard's curiosity about Emma sharpened when Kite dug into the history of the Wolverton family. Kite first managed to track down Emma's mother, Malinda. By then, Malinda had eight children and was earning money by working as a farmhand and selling soap. Kite told Goddard that Malinda seemed indifferent to her family—even to herself. "Her philosophy of life is the philosophy of the animal," Goddard later declared.

Kite pushed further back into Emma's pedigree. She investigated Emma's aunts and uncles and cousins. She traveled to the reaches of New Jersey—to slums, to farms, to mountain cabins—and came back with more disturbing tales of filthy half-naked children, of unheated tenements, of mothers covered in vermin, of incest.

Kite sometimes proffered a letter from the school to get into people's houses, but other times she hid her mission, sweetly asking if she could get shelter from impending storms or pretending to be a historian researching the Revolutionary War. She asked old people about their dim memories of long-dead relatives. They told of horse thieves, of young women seduced by lawyers, of an old drunk nicknamed "Old Horror" who would show up at the polls on Election Day to vote for whoever would pay him.

Kite eventually traced 480 Wolvertons to a single founding father, named John Wolverton. She claimed conclusive proof of feeblemindedness in 143 of his descendants. But Kite also encountered descendants of John Wolverton who were doctors, lawyers, businessmen, and other respectable citizens. Their intelligence seemed utterly different from Emma's relatives. The two branches of the family, the high and the low, didn't seem to know of each other.

Kite was confused until an elderly informant dispelled the fog. John Wolverton, Kite learned, had been born into an upstanding colonial family. At the outbreak of the Revolutionary War, he joined a militia, and when the militia stopped one night at a tavern, he got drunk and slept with a feeble-minded girl who worked there. John promptly got back on the respectable path, married a woman of good Quaker stock, and went on to have a happy family, with many descendants who rose to prominence.

John had no idea that he had impregnated the tavern girl, who gave birth to a feebleminded son. She named him John Wolverton after his missing father. When John the younger grew up, he turned out an utterly different man—depraved enough to earn the nickname "Old Horror." He started a family of his own, and the two lines of Wolvertons veered in different directions over the next 130 years—one to greater respectability, the other into feeblemindedness and crime.

"The biologist could hardly plan and carry out a more rigid experiment," Goddard said. The data flowing into Vineland "was among the most valuable that have ever been contributed to the subject of human heredity," he said.

———

G oddard convinced himself the United States was sliding into a crisis of heredity. "If civilization is to advance, our best people must replenish the Earth," he said. To Goddard, the best people in the United States were his fellow New Englanders, "the stock than which there is no better."

But one by one, the great New England families were disappearing for lack of children. Meanwhile, the feebleminded were multiplying at over twice the average rate, according to Goddard's estimates.

Goddard was hardly the first person to contemplate controlling human heredity. Four centuries beforehand, Luis Mercado had advised people with hereditary disorders to avoid having children together. In the early 1800s, alienists urged that the insane be prevented from starting families. Francis Galton turned these concerns into something far more extreme: a call to governments to breed their citizens like cows or corn. Galton recognized that in order to win people to his cause, he would need, as he put it, "a brief word to express the science of improving stock." In 1883, he came up with an enduring term: *eugenics.* To Galton, eugenics was full of happy visions of arranged marriages that would lead to ever-better generations of humans. "What a galaxy of genius might we not create!" Galton promised.

Galton's enthusiasm attracted some noteworthy English biologists, who formed the Eugenics Education Society. But they never gained much power or influence over British affairs. By the dawn of the twentieth century, eugenics had begun taking root in the United States, and there it flowered into darker blooms. American eugenicists wanted to prevent people with bad traits from having children. Some argued for institutionalizing the feebleminded to stop them from having sex. Some called for sterilization. In 1900, an American physician named W. D. McKim went so far as to call for "a gentle painless death." He envisioned the construction of gas chambers to kill "the very weak and the very vicious." It would be pointless to try to improve these people through experience, because, McKim declared, "heredity is the fundamental cause of human wretchedness."

Davenport embraced eugenics without any hesitation, and he argued that Mendel's rediscovery only strengthened the case for it. If genes were carried in the germ line, there was nothing to be done about the bad ones except to keep them from poisoning the next generation. Davenport believed that eugenics would have to be carried out based on a thorough knowledge of hereditary traits, and so he established a repository of data— the Eugenics Record Office—next to the research station at Cold Spring

Harbor in 1910. Ultimately, Davenport predicted, eugenics would provide "the salvation of the race through heredity."

Under Davenport's sway, Goddard quickly became a eugenicist, too. In 1909, he joined Davenport on a prominent committee of eugenicists, and two years later he published a manifesto entitled "The Elimination of Feeble-Mindedness." Goddard wrote that it was possible for environmental causes, such as an illness during pregnancy, to cause feeblemindedness, "but all these causes combined are small compared to the one cause— heredity."

To eliminate feeblemindedness, Goddard rejected the calls of people like McKim to kill the feebleminded. But he did want to make sure they didn't get to have children. And by "they," Goddard mostly meant women.

Goddard conjured up a specter of attractive, feebleminded women wantonly seducing decent men. He warned that the country's reformatories were full of feebleminded girls who "do not conform to the conventions of society," who were "boy crazy" or, worst of all, "preferred the company of colored men to white." These feebleminded girls "in many instances are quite attractive," Goddard warned, requiring them to be put "under the care, guidance, and direction of intelligent and humane people, who will make their lives happy and partially useful, but who will insist upon the one important thing, and that is that this race should end with them; they shall never become the mothers of children who are like themselves."

Institutionalization wasn't the only way to keep women from becoming mothers. Goddard joined the movement to sterilize women deemed unfit. In the early 1900s, an Indiana prison surgeon named Harry Sharp performed vasectomies to stop men from transmitting defective "germ plasm," and in 1907 the Indiana legislature made sterilization a state policy. In New Jersey, Goddard lobbied for a similar bill, which Governor Woodrow Wilson signed in 1911. The first woman slated to be sterilized took her case to New Jersey's Supreme Court, which ruled it unconstitutional in 1913 as cruel and unusual punishment. Goddard responded to the defeat by redoubling his efforts. He joined new committees with ominous names, like the Committee for the Heredity of Feeble-mindedness, and the Committee to Study and to Report on the Best Practical Means of Cutting Off the

Defective Germ-Plasm in the American Population. "There is no question that there should be a carefully worded sterilization law upon the statute book of every State," Goddard said.

———

L obbying governments and publishing reports would not be enough for Goddard. He wanted to win over public opinion. The heap of data he was collecting from hundreds of families would not make the country as a whole appreciate the threat of feeblemindedness. He needed to find a parable to illustrate the destructiveness of feeblemindedness through a single family. The choice was obvious: Emma Wolverton and her ancestors.

Goddard began work on a book, his first. He used the school's notes about Emma to put together a short biography up to age twenty-two. To protect her identity, he referred to her as Deborah Kallikak. Her last name was another one of Goddard's Greek creations—a combination of the words *kalos* ("good") and *kakos* ("bad"). Yet he felt no compunction about adding photographs of Emma to the book. In one picture she posed at her sewing machine. In another, she held a book open in her lap, her thick black hair kept neat in a bow. Casual readers might not see anything amiss with this young woman, but Goddard was quick to set them aright: Intelligence tests showed that she had a mental age of a nine-year-old.

"The question is, 'How do we account for this kind of individual?'" Goddard asked. "The answer is in a word 'Heredity,'—bad stock."

To prove his point, Goddard used Kite's research to tell the story of the Wolvertons. He started with John Wolverton, renaming him Martin Kallikak. Interspersed with the tales of drunks and horse thieves, Goddard's book included photographs Kite took of Emma's relatives—old women and dirty children scowled at the camera, standing in front of sheds or sitting on sagging porches. Goddard also added family trees to the book, drooping with squares and circles, some of which were colored black to indicate feeblemindedness. The defect flowed down through six generations of the trees, demonstrating the power of heredity.

The story of the Kallikaks, Goddard concluded, was a powerful argument for rounding up the feebleminded and putting them in colonies, at least until a better solution could be found. Sterilization might turn out to be that solution, but Goddard warned against simply operating on every member of feebleminded families. To Goddard, his pedigrees seemed to show that feeblemindedness was a Mendelian trait, carried on a gene. If that was true, then it was entirely possible for a moron to have some children who were feebleminded and others who were of normal intelligence. To sterilize them all would be like using a hatchet when a scalpel would do.

The one thing that would *not* save the country from feeblemindedness was naive hope. "No amount of education or good environment can change a feeble-minded individual into a normal one," Goddard warned, "any more than it can change a red-haired stock into a black-haired stock."

———

I n 1912, Goddard published *The Kallikak Family*. It gave a modern, Mendelian polish to old beliefs about feeblemindedness as a punishment for sin. The *Evening Star*, a Washington, DC, newspaper, reprinted large excerpts from *The Kallikak Family*, accompanied by a shuddering commentary: "I doubt if there is in all literature a more damning presentation of how one single sin can perpetuate itself in generations of untold misery and suffering, to the end of time."

The book became a bestseller, turning Goddard—a psychologist at a little-known backwoods institution—into one of the most famous scientists in the United States. His fame helped attract more attention to his imported intelligence tests. The New York City school system adopted them, administering them to all their students, and soon other school districts across the country followed suit. The United States Public Health Service reached out as well. They didn't need his help to teach students. Rather, they wanted to test the flood of immigrants arriving in the United States.

Between 1890 and 1910, more than twelve million immigrants traveled from Europe to Ellis Island. Doctors inspected thousands of people arriving there each day to make sure they were in good physical health. In 1907, Congress passed a law to also exclude "imbeciles, feeble-minded and persons with physical or mental defects which might affect their ability to earn a living." The new law meant that the doctors on Ellis Island had to inspect the minds of immigrants as well as their bodies. Congress gave them no guidance, and so the Health Service asked Goddard if he could adapt his test to find the feebleminded among the immigrants.

"We were in fact most inadequately prepared for the task," Goddard later admitted. He knew that a test designed for American children might not work well on adults who didn't speak English or understand anything of American culture. But Goddard accepted the request, unwilling to pass up the opportunity, and created a new test for immigrants.

Goddard brought his team of fieldworkers to Ellis Island on a series of trips, starting in 1912. When ships docked and immigrants shuffled into the main building on the island, Goddard's fieldworkers scanned them. They pointed out those who looked like they might be feebleminded. The selected immigrants were pulled out of the crowd and taken to a side room. There, another fieldworker and an interpreter would give each immigrant a series of tasks, such as fitting blocks into holes or telling them what year it was.

Goddard's staff kept careful records of the tests, which he analyzed back in Vineland. The results stunned him: A huge proportion of the immigrants tested as feebleminded. Goddard broke down the results by ethnic group: 79 percent of Italians were feebleminded, 83 percent of Jews, 87 percent of Russians.

When Goddard published the figures, they were seized upon by opponents of immigration. For years they had been claiming that the new wave of immigrants from eastern and southern Europe was a burden to the country. More recently, they translated their bigotry into the language of eugenics. In 1910, Prescott Hall, a leader of the Immigration Restriction League, made the connection clear. "The same arguments which induce us to segregate criminals and feebleminded and thus prevent their breeding," he said, "apply to excluding from our borders individuals whose multiplying here is likely to lower the average of our people." Goddard now handed them seemingly hard numbers, which they would use to justify slashing immigration quotas.

Goddard himself was more suspicious of his own results. "They can hardly stand by themselves as valid," he said. Immigrants might score badly on tests for all sorts of reasons. A Russian peasant might never get the chance to learn how to count; calendars might be useless to him, as he worked on a farm. Goddard worked through the numbers again, using a more lenient cutoff for feebleminded, and found that the fraction dropped by half.

On reflection, Goddard seemed comfortable with the notion that 40 percent of immigrants were morons. "It is admitted on all sides that we are getting now the poorest of each race," he said. But Goddard didn't argue that any race was inherently less intelligent. He did suspect that some immigrants inherited their feeblemindedness—"Morons beget morons," Goddard said—but poverty might be to blame for the low test scores of many other immigrants. "If the latter, as seems likely, little fear may be felt for the children," Goddard said.

Goddard's team was now overwhelmed with work. In addition to studying immigrants, he was continuing to analyze the data from hundreds of families that Kite and others had interviewed. Goddard was also training psychologists at Vineland in mental testing. But the work at the lab came almost entirely to a halt when the United States entered World War I and much of his staff enlisted. Goddard decided he could help the cause in his own way as well. He warned the army that it might risk losing the war by unwittingly drafting hundreds of thousands of morons.

The army had Goddard and a group of his fellow intelligence experts draw up a test they could give to draftees. In 1917, he hosted a meeting at Vineland, where they adapted their tests to examine young men. The army then hired four hundred psychologists, who administered the new test to 1.7 million soldiers. It was an intelligence study thousands of times bigger than anything ever attempted before.

"The knowledge derived from testing of the 1,700,000 men in the Army is probably the most valuable piece of information which mankind has ever acquired about itself," Goddard later declared. The soldiers followed the same swelling curve that Goddard had seen when he had tested New Jersey schoolchildren six years earlier. Most of the scores were close to the overall average, while a few soldiers scored exceptionally far above or below the rest. Goddard saw the army results as a vindication of everything he had been saying about the biological nature of intelligence.

Yet the average score of the soldiers was startlingly low. According to Goddard's standards, 47 percent of the white soldiers and 89 percent of the blacks should be categorized as morons. The average white soldier, the psychologists found, had a mental age of thirteen years, just barely above the cutoff for feeblemindedness. The majority of Americans, in other words, was feebleminded or close to it.

When news of the results got out, it caused many Americans to look at their country with a new sense of self-loathing. "We have a working majority of voters who have children's minds," a prominent newspaper editor named William Allen White declared.

White was convinced that the "moron majority," as he dubbed it, must be a recent development. "A new biological condition faces us," he warned. The new immigrants from southern and eastern Europe lacked the mental grasp of the colonists who fought in the revolution. "Our darker-skinned neighbors breed faster than we," he explained, and their descendants inherited their feeblemindedness. "The plasm of the lame brain keeps right on producing lame brains," White concluded.

To Goddard, the army test results demanded a new form of government. Only about 4 percent of soldiers got an A on the test, meaning that they possessed "very high intelligence." The top 4 percent of the country must be allowed to rule over the remaining 96. The fact that the United States was a democracy might make this arrangement hard to achieve, but Goddard believed that if the most intelligent came to understand how to make other Americans comfortable and happy, they would be elected to rule. "And then will come perfect government," Goddard declared in a 1919 lecture at Princeton.

To put it another way, Goddard had decided that the entire country had to be turned into a giant Vineland Training School. The children at the school had not voted to put Goddard and the rest of the administration in charge of their care, of course. "But they would do so if given a chance because they know that the one purpose of that group of officials is to make the children happy," Goddard said.

—————

A lmost no one outside of Vineland knew that Emma Wolverton was Deborah Kallikak. But in the tiny world of the training school, everyone was aware, Emma included. Yet her local fame did not protect her from the brutal indifference of institutional life. Two years after the publication of *The Kallikak Family*, Johnstone summoned her to his office to tell her that she was going to have to leave.

Rich children could stay for life at the Vineland Training School if their parents paid a onetime fee of $7,500. Poor children, whose care was paid for by the state of New Jersey, had to move out when they grew up. By the time they became adults, only a few of Vineland's students could be trusted to live on their own. The rest needed to be moved elsewhere. Now twenty-five, Emma Wolverton walked back out the gate she had entered seventeen years before. Garrison had died in 1900, and now his tomb stood at the corner just outside the gate. She stopped to thank him for her time there. "The Training School," she whispered. "My home."

Her trip was short. Emma was moved across Landis Avenue to the New Jersey State Institution for Feeble-Minded Women. Its mission was to keep its inmates from "propagating their kind."

Across the street, the institution staff also knew that Emma was the real Deborah Kallikak. While she might have been famous for her monstrous family, they found Emma capable and well trained. She got to work with a "dignified courtesy," according to a social worker there named Helen Reeves. She cared for the children of the institution's staff, including that of the assistant supervisor. The children adored her and would send letters to her for the rest of her life. Emma also worked in the institution's hospital, even serving as a special nurse during an outbreak in the early 1920s. One day a patient bit one of her fingers so badly it had to be amputated. She sported the injury with pride.

Emma discovered plays to perform in her new home as well. Once, when she played Pocahontas in a play at the institution, she had to throw herself on a dummy that represented Captain John Smith.

"You could put more pep in it," the superintendent shouted during rehearsal.

"If it were a real man, I would," she replied.

Emma even managed to find a few real men. While she worked as a nurse during the epidemic, she moved into a room near the patients where she was under less monitoring. Using her skills at woodworking, she tinkered with the window screen so that she could slip in and out at night unnoticed. In the moonlight, she would meet a maintenance worker. They were eventually caught, and her suitor was "kindly dismissed by a lenient justice-of-the-peace," as Reeves later put it.

Emma became involved with at least two other men, but each time the authorities broke it off. Only a few clues about those relationships survive. In 1925, the institution hired Emma out as a maid, but her service was cut short after less than a year. Over thirty years later, Emma met a psychology intern named Elizabeth Allen. Allen later recalled the stories told about Emma at the institution. "Apparently every time she was released to work on the 'outside' she would return pregnant," Allen wrote. If Emma did indeed become pregnant, there's no record of a child, an abortion, or sterilization.

"It isn't as if I'd done anything really wrong," Emma later complained. "It was only nature."

―――――

Only four years after Emma Wolverton was forced out of the Vineland Training School, Henry Goddard was forced out as well. Johnstone shut down Goddard's laboratory in 1918, but the documents that have survived don't offer many clues as to why things ended so badly. Writing to one of his funders, Goddard condemned the decision as a "fatal error."

Perhaps the parents of Goddard's subjects grew weary of him using them as psychological guinea pigs. Whatever the reason, Goddard abruptly left the Vineland School for Ohio. His celebrated work on eugenics and intelligence came to an end. In Ohio, he worked in relative obscurity, studying how to prevent juvenile delinquency and to help gifted children thrive.

The Kallikak family had gained so much strength in the popular imagination that they no longer depended on Goddard. They endured without him. Paul Popenoe, the editor of the *Journal of Heredity*, recounted their story as he lobbied for more states to sterilize the feebleminded. "Such children should never be born," Popenoe declared. "They are a burden to themselves, a burden to their family, a burden to the state, and a menace to civilization." In 1927, the Supreme Court heard a case about a young Virginia woman named Carrie Buck who had been scheduled for sterilization. The eugenicists submitted *The Kallikak Family* as evidence that Buck's children would be doomed. The Supreme Court approved the state's petition, and Buck was sterilized. The court's decision led to a boom in sterilizations in the years that followed.

In the 1920s, Goddard's work with the US Army also continued to fuel scientific racism. Eugenicists pointed to the difference between black and white soldiers on the army tests as proof of hereditary differences in intelligence between the races, and that the races should not be allowed to intermarry. The eugenicist Madison Grant declared that miscegenation was "a social and racial crime of the first magnitude."

American racism of the 1920s divided humanity into far thinner slices than just black and white, though. Eugenicists declared that northern Europeans were superior to people from the rest of the continent. They pointed once more to Goddard's work on Ellis Island, as well as to the army intelligence tests, on which immigrant Italians, Russians, and Jews did poorly. They also ignored the fact that these soldiers came from families that had only recently arrived in the United States.

Harry Laughlin, who worked for Davenport at the Eugenics Record Office, testified to Congress that immigration threatened to pollute the American gene pool. "The lesson is that immigrants should be examined, and the family stock should be investigated, lest we admit more degenerate 'blood,'" he said. In 1924, Congress tightened immigration with the passage of the National Origins Act, keeping out undesired races.

The Kallikaks became celebrities far beyond America's shores as well. In 1914, Goddard's book was published in Germany to great acclaim. For years, many German doctors and biologists had been calling for a government-run program to breed the best parents, along with sterilization of the unfit. When Adolf Hitler was imprisoned in 1924, he learned of the Kallikaks in a book he read about heredity. Soon after, Hitler wrote *Mein Kampf*, in which he mimicked the language of American eugenicists, declaring that steriliza-tion of defective people "is the most humane act of mankind."

When Hitler came to power, an appalling number of German scientists and doctors heartily joined him in his campaign to alter humanity. "The head of the German ethno-empire is the first statesman who has made the tenets of hereditary biology and eugenics a directing principle of state pol-icy," declared the geneticist Otmar von Verschuer. In 1933, the year Hitler seized power, a new German edition of *The Kallikak Family* was published. In his introduction, the translator, Karl Wilker, made clear just how impor-tant Goddard's work had been to the Nazis.

"Questions which were only cautiously touched upon by Henry Herbert Goddard at that time . . . have resulted in the law for the prevention of sick or ill offspring," Wilker wrote. "Just how significant the problem of genetic inheritance is, perhaps no example shows so clearly as the Kallikak family."

The Nazis used the Kallikaks as a teaching tool. In 1935, the govern-ment released an educational film called *Das Erbe* ("Inheritance"). It begins with two older male scientists explaining to their eager young female as-sistant about the laws of heredity. Over a montage of flowers and birds, of racehorses and hunting dogs, they talk about how to produce new breeds of animals and plants. A breeder's success depends on picking the right indi-viduals to produce the next generation. The same is true for people. No better example of the harm of poorly planned families is the Kallikak fam-ily, "the work of American eugenicist Henry Goddard," one of the German scientists says.

The screen turns black, and a title appears across the top: "The Descen-dants of Lieutenant Kallikak." The lieutenant is marked by a circle, from which springs downward branches—493 "superior offspring" from a woman of healthy stock, along with 434 "inferior offspring" from a woman with a hereditary disease.

"A single ancestor with hereditary disease was enough to leave a large number of unfortunate descendants," one of the scientists explains. "This is just one example among thousands." Sympathy for the suffering of such people required preventing them from reproducing—"by all means."

After the pedigree appears in full, it is replaced on the screen by a quotation from Hitler: "He who is not healthy and dignified in spirit can not perpetuate his suffering in the body of his child."

In the same year that *Das Erbe* was released, the Nazis put on the Exhibition for Hereditary Care, where visitors could look at exhibits on the many disabilities that needed to be eradicated. A doctor got into a conversation with a skeptical visitor. To persuade him of the importance of eugenics, the doctor recounted the story of the Kallikaks. "This examination was initiated and directed by the American Professor Goddard," the doctor assured the visitor. "There is even a book about it."

The visitor was persuaded, asking the doctor if all the "cripples and idiots" shown at the exhibition were due to the same cause.

"Yes," the doctor replied. "There is only one answer: heredity."

Hitler followed up on this propaganda by establishing a new set of "racial hygiene" laws. Hereditary health courts accepted applications from doctors for people who were so unfit they should not be allowed to have children. The feebleminded made up the majority of the approvals. Psychiatrists devised intelligence tests for the courts. In one exam, they gave subjects a suitcase, books, bottles, and other objects. They had to pack the suitcase so that the lid could be easily closed. Their lives might depend on that suitcase.

Within a year of the passage of the first racial hygiene law, the hereditary health courts approved more than 64,000 sterilizations, and by 1944, Germany sterilized at least 400,000 people, including the mentally ill, the deaf, Gypsies, and Jews.

In 1939, Hitler expanded his campaign against the feebleminded, launching a program to kill children judged to be idiots, along with those suffering deformities. Their parents were told that they had died during surgery or due to an accidental overdose of sedatives. Soon children were being killed for being teenage delinquents, or just for being Jewish. Hitler then added yet another program to kill adults who were institutionalized for feeblemindedness or other defects. Before extermination, children would be asked questions that wouldn't have been out of place at Vineland, such as "Can you name the four seasons?"

The program, known as T4, would ultimately claim 200,000 lives. It operated on a scale so far beyond what the Nazis had attempted before that they had to invent new technology for the slaughter—including gas chambers. McKim's eugenic dream had become real.

———

A few people saw straight through the Kallikak story right away. In 1922, the journalist and political commentator Walter Lippmann delivered an attack in the *New Republic*. He granted that Binet's original tests had some value as a way to identify children in need of special education. But since then, in the hands of people like Goddard, they had been used to promote monstrous distortions. "The statement that the average mental age of Americans is only about fourteen is not inaccurate. It is not incorrect. It is nonsense," he wrote.

It was nonsense, Lippmann declared, to treat intelligence as something as straightforward as height or weight, when psychologists had yet to actually define it. Until that day, intelligence would remain simply the thing that intelligence tests measure. But those tests were constantly in flux, as their designers adjusted their thresholds to produce results that satisfied their expectations. To conclude from these tests, then, that intelligence was a hereditary trait was downright pernicious. "Obviously this is not a conclusion obtained by research," Lippmann declared. "It is a conclusion planted by the will to believe."

To reach that conclusion, testing advocates had to ignore all sorts of experiences that could influence the scores—especially those in early childhood, when the brain is still developing. And they had to embrace stories like that of the Kallikaks without any healthy skepticism.

In fact, Lippmann warned, there was "some doubt as to the Kallikaks."

Even if the story was true, it wouldn't be as compelling an experiment as Goddard claimed. To see how powerful heredity really was, it would have been necessary for Martin Kallikak to have fathered an illegitimate child with a healthy (but poor) woman. Likewise, his respectable marriage would need to be with a feebleminded woman from a prosperous family. "Then only would it have been possible to say with complete confidence that this was a pure case of biological rather than of social heredity," Lippmann said.

Some scientists questioned the Kallikak story as well. In 1925, a Boston neurologist named Abraham Myerson mocked the lurid tale of Martin Kallikak's disastrous dalliance with a feebleminded girl, after which he "used his germplasm in orthodox fashion by marrying a nice girl who bore him nice children and started a row of nice people—all nice, no immoral, no syphilitics, no alcoholics, no insane, no criminals."

Myerson found it ridiculous that Goddard thought he could diagnose generations of Kallikaks based on the stories collected by Elizabeth Kite. "I cannot get any definite information about my great-great-grandfather, much as I have tried," Myerson joked, "but a girl who left so little impression on her times as to be 'nameless' is positively declared to be feeble-minded."

Perhaps the most important opponent of the Kallikaks was a biologist who spent much of his time in a lab full of milk bottles packed with rotting bananas. Thomas Hunt Morgan didn't know much about psychology, and yet his attack on *The Kallikak Family* was the most profound of all. More than anyone, he could see how weak the foundations were on which Goddard built his story.

Morgan kept rotting bananas in his lab at Columbia University in New York City in order to feed a species of fly called *Drosophila melanogaster*. He had begun studying them in 1907, hoping to catch one of de Vries's species-creating mutations. But Morgan came to realize that no single mutation could create a new species. It could give rise to a new trait, however. One day, Morgan and his colleagues spotted a male fly that grew white eyes instead of the normal red. The scientists put the white-eyed male together with a red-eyed female and the insects mated. The female then produced healthy eggs, which developed into red-eyed offspring. Morgan's team then

bred those flies with each other, and found that in the following generation, some of the male insects had white eyes. It was puzzling that only
males could inherit white eyes, but could not pass them down to their own
sons. In search of an explanation, Morgan and his colleagues made a major
discovery about the nature of genes.

Morgan's flies, like all animals, had chromosomes in their cells. Chromosomes usually came in identical pairs, with one exception—a mismatched set of chromosomes that came to be known as X and Y. Studying
insect cells, scientists discovered that males carried one X and one Y, while
females carried two Xs. This discovery raised the possibility that the X and
Y chromosomes carried hereditary factors—what came to be known as
genes—that determined which sex an insect would be. The fact that Morgan's male flies could develop white eyes might mean that a gene located on
the X or Y chromosomes determined eye color.

After many experiments with the flies, Morgan's team figured out that
this was indeed the case. White eyes are produced by a recessive mutation
on a gene located on the X chromosome. Females with one copy of the white-
eye mutation can have red eyes anyway, because their other X chromosome
is normal. But since males have only one X chromosome, they can't compensate for the mutation and develop white eyes. Further experiments in
Morgan's lab revealed that the sex chromosomes could also carry mutations
to other traits, such as one that turns the bodies of flies yellow or shrinks
their wings. It became clear from experiments like these that chromosomes
carried genes, and that a single chromosome could carry many of them.

As Morgan's team pinned down the location of more genes, they came to
realize that heredity was a lot more complicated than scientists had previously thought. When Mendel's work was initially rediscovered, many geneticists assumed that each trait was controlled by a single gene. Morgan's team
found that many genes could influence a single trait. For example, they identified twenty-five different genes that could change the color of a fly's eyes.

"It is of the utmost importance that this hypothesis be understood," the
Journal of Heredity declared when Morgan published some of his findings
in 1915. If genes worked in such an intricate way in flies, the story in

humans had to be far more complex. "Those who accept it must give up talking about, e.g., Roman nose being due to a determiner for Roman nose in the germplasm. The modern view would say that the 'Romanness' of the nose is due to the interaction of a very large number of factors."

Early in his career, Morgan had started out on good terms with Charles Davenport and other American eugenicists. But he was appalled to see how desperately they clung to a Roman-nose view of heredity, even as the evidence piled up against it. In a 1925 book, Morgan spelled out all that was wrong with their approach to human nature.

It was true that individual genes might play a small part in explaining behavior, Morgan granted. Davenport and other scientists had gathered compelling evidence that a single dominant mutation caused Huntington's disease, for example. But Morgan doubted Goddard's claim that something as amorphous as "feeblemindedness" could have such a simple hereditary explanation.

"It is extravagant to pretend to claim that there is a single Mendelian factor for this condition," Morgan wrote.

Morgan didn't think it would be possible to really begin to study the heredity of feeblemindedness until scientists decided what they actually mean by intelligence itself. "In reality our ideas are very vague on the subject," he wrote. Scientists would also have to give more credit to the ways in which the environment influenced the human mind. In Morgan's own research on flies, he had learned to respect the power of the environment. His students discovered one strain of flies that developed normally if they were born in the summer but tended to sprout extra legs if they were born in the winter. It turned out that the researchers could get the same outcomes in their lab simply by changing the temperature in which they reared the fly eggs. It was thus meaningless to talk about the effect of their mutation without taking into account their environment.

When Morgan looked at the pedigrees of families like the Kallikaks, he did not see undeniable proof of the heredity of feeblemindedness. He saw instead many generations of poor people suffering enduring hardships. "It is obvious that these groups of individuals have lived under demoralizing social conditions that might swamp a family of average persons," Morgan wrote. "The effects may to a large extent be communicated rather than inherited."

If that was true, Morgan argued, it was patently ridiculous to turn to eugenics to try to improve humanity's lot. "The student of human heredity will do well to recommend more enlightenment on the social causes of deficiencies," he concluded.

———

B y the 1930s, many other geneticists had followed Morgan's example and repudiated eugenics, as both bad science and bad policy. The Eugenics Record Office, the hub of research and social policy based on human heredity, sank into disrepute. In his testimony to Congress, Harry Laughlin offered statistics that supposedly showed the intellectual superiority of northern Europeans. They turned out to be full of glaring errors. The Carnegie Institution, which gave much of the money to run the Eugenics Record Office, realized that its fieldworkers had been gathering sloppy, subjective data that would be useless for scientific research. Even the organization of the files turned out to be a "futile system." The office was shut down in 1939, having been judged "a worthless endeavor from top to bottom."

American eugenicists lost more followers as they cozied up to the Nazi government, pleased to see their policies put so aggressively into action. Laughlin even traveled to Germany to accept an honorary degree. Once the full scope of the Holocaust emerged, the eugenics of people like Laughlin and Davenport would never be able to separate itself from genocide.

The Kallikak Family finally went out of print in 1939. By then it had worked its way into psychology textbooks, where it could terrify college students. A psychologist named Knight Dunlap complained of having to talk one of his students out of committing suicide for fear of having inherited a mental defect from her family. Fortunately, as he later recalled, he was able to ease her anxiety by promising that "her chances of going insane were no better than my own." In 1940, Dunlap published a blistering attack on The Kallikak Family in the journal Scientific Monthly. "Even in books written by psychologists who ought to know better, the Kallikaks skulk in the corners of the pages, and leap out upon unwary students."

In 1944, a doctor named Amram Scheinfeld published a harsh memorial to mark the thirtieth anniversary of *The Kallikak Family*. Writing in the *Journal of Heredity*, Scheinfeld scoffed at the idea that a single mutant gene could have worked its way through one branch of the Kallikak family, causing feeblemindedness and other attendant ills along the way. He skewered Goddard for ignoring the possibility that what he thought was inherited behavior was the result of growing up in grinding poverty. The only reason that the Kallikak study had become so well-known, Scheinfeld said, was because it "would permit those on top to smugly keep their place, while relieving them of the necessity of doing very much for those at the bottom." And its legacy had been dreadful, not just for genetics but for human society in general. The idea at the core of *The Kallikak Family*, that some people were genetically superior to others, Scheinfeld said, "helped to bring on the present war."

These attacks—Dunlap, for example, declaring that "the Kallikak phantasy has been laughed out of psychology"—galled Goddard. The rising generation of psychologists were creating a caricature of him and his ideas. In the years after he was forced out of Vineland, Goddard drifted away from the eugenics movement. Rather than figuring out how to keep the feebleminded from having children, Goddard spent his time trying to find ways to help children, no matter their condition. "As for myself," Goddard once said, "I think I have gone over to the enemy."

In truth, Goddard moved only a bit closer to the enemy. In 1931, he traveled from Ohio back to Vineland to speak at a meeting celebrating the twenty-fifth anniversary of the research laboratory. As he spoke, it became clear that Morgan's genetics lessons had not sunk in. Goddard granted that perhaps feeblemindedness depended on more than one gene. But he still believed it was overwhelmingly hereditary. Sterilizing a feebleminded woman would very likely prevent the birth of more feebleminded babies. The Great Depression was reaching its depths when Goddard came back to Vineland, and he blamed it largely on America's lack of intelligence: Most of the newly destitute didn't have the foresight to save enough money. "Half of the world must take care of the other half," Goddard said.

Goddard also defended the data he had collected at Vineland against the growing number of critics. "No one has shown where the Vineland figures are in error," he declared in his 1931 speech. But privately, Goddard had an inkling that something was wrong.

The attacks on *The Kallikak Family* led him to write to Elizabeth Kite about her fieldwork. Kite confessed that she had never bothered to find out the name of the girl in the tavern. Her excuse for this lapse was that discovering the tawdry origin of the feebleminded Kallikak line had left her stunned. "That was all I could stand for one day!" Kite told Goddard.

In 1942, when Goddard published a defense of the Kallikak research, he lied about Kite's lapse. He said that he knew the woman's name but had withheld it for the sake of privacy. The only flaw Goddard could see in his work was that it was ahead of its time. "Much in the way of polish is lacking in this pioneer study," he said.

That marked the end of Goddard's attempts to salvage his reputation. Soon afterward he retired from Ohio State University and published a guide to parenting called *Our Children in the Atomic Age*. He thought about writing an autobiography, but he only got as far as a decidedly un-eugenic title: *As Luck Would Have It*. In 1957, Goddard died at age ninety. In their obituary, the Associated Press remembered him for two accomplishments: coining the word *moron* and discovering the Kallikaks. "The author's conclusion was that 'the Kallikak family presents a natural experiment in heredity,'" the obituary writer reported. "Later some other psychologists cast some doubt on his deductions."

Even after Goddard's death, the Kallikak family lived on. Henry Garrett, a psychologist at Columbia University who served for a time as president of the American Psychological Association, would retell the story for decades. In 1955, he published a textbook called *General Psychology* that included a full-page illustration of the Kallikak genealogy. Martin Kallikak stands like a towering colonial colossus. His arms are akimbo, and the left half of his body shaded. Down his left side spills a cascade of demonic faces.

"He dallied with a feeble-minded tavern girl," Garrett wrote alongside the illustration. "She bore a son known as 'Old Horror' who had ten children. From 'Old Horror's' ten children came hundreds of the lowest types of human beings." Their hair was swept back like demon horns.

On his right side, Kallikak was white, flanked by tranquil faces of men and women in proper hats. "He married a worthy Quakeress," Garrett wrote. "She bore seven upright worthy children. From these seven worthy children came hundreds of the highest types of human beings."

The textbook would go through many editions, and students would still be looking at the Kallikak family in the 1960s. In 1973, the year of his death, Garrett railed against the constitutional right to vote, complaining how "the vote of the feeble-minded person counts as much as that of an intelligent man."

———

In the 1980s, curious investigators uncovered Deborah Kallikak's real name. A pair of genealogists, David Macdonald and Nancy McAdams, worked back through Goddard's account, determining the true identity of Emma Wolverton's relatives. In the process, every piece of Goddard's book—the founding testimony of modern eugenics and an inspiration for one of the greatest crimes in history—simply vanished.

It turned out that Elizabeth Kite had misunderstood an old woman she interviewed in 1910. Kite got the impression that a soldier named John Wolverton had a bastard son named John Wolverton. In fact, the two John Wolvertons were second cousins. In other words, Goddard's natural experiment in heredity never happened.

The bad branch of the Wolverton clan turned out not to be a horde of feebleminded monsters. John Wolverton—whom Goddard called Martin "Old Horror" Kallikak—was not an unwashed drunk who rolled off porches after too much cider. Public records show he was a landowner, and that he eventually transferred his property to his children and grandchildren. The 1850 census indicates that he lived with his daughter and her children, all of whom could read. Just before his death in 1861, his property was valued at the respectable sum of $100. Old Horror's descendants didn't match Goddard's grotesque portraits either. Their ranks included bank treasurers, policemen, coopers, Civil War soldiers, schoolteachers, and a pilot in the Army Air Corps.

Emma happened to have the bad luck to be born into a Wolverton family that was ripped apart in the great migration of American farmers into cities in the late 1800s. Her maternal grandparents moved to the outskirts of Trenton, where her grandfather worked as a laborer. There were eleven children in the family, six of whom died young. Life for the remaining five was hard, and at times unbearable. Emma's grandfather appears to have been a menace to his children, who were all removed from the household. Emma's aunt Mary visited her parents in 1882 at age twelve. Her father attacked her, and she gave birth to a child, who soon died. Emma's grandfather was prosecuted for incest a few months later, but there's no record that he served time in prison.

Despite growing up in a poor, uneducated, violent family, Emma's relatives endured. Emma's aunt Mary returned to her foster family for the rest of her childhood, and later in life she got married. Emma's uncle George, whom Goddard described as a feebleminded horse thief, actually made a living as a farmhand and was a member of the Salvation Army. Emma's uncle John held jobs as a millworker and rubber worker in Trenton.

Even Emma's mother, Malinda, eventually found a stable life. After she married her second husband, Lewis Danbury, in 1897, they stayed together for thirty-five years, until her death in 1932. Lewis was later buried next to her. Emma's half brothers and sisters, dismissed by Goddard as feebleminded, were nothing of the sort. Fred Wolverton fought in World War I and worked as a car mechanic. One of Emma's nephews became a career army man, while another worked as a golf pro.

By the time Emma Wolverton's true history came to light, she had been dead for years, buried on the institution's grounds. She had lived there for fifty-three years. In her later years, she worked in the institution's gymnasium, producing plays performed by the inmates. Emma would sew the costumes and build the sets. She filled her spare time reading books and magazines or wrote letters to friends. She even left the institution from time to time, accompanying the staff on outings. She wandered among the dinosaurs at the American Museum of Natural History and fed bits of bread to the squirrels in Central Park.

In 1957, the year that Goddard died, Emma met the intern Elizabeth Allen. "Emma was tall and reticent," Allen later recalled. "She reminded me of anyone's elderly aunt."

Emma was sixty-eight. She had stopped producing plays, but she still worked, ironing institution uniforms. A space at the institution was converted into a tiny apartment where she could live by herself. Allen was shocked when Emma told her that she was Deborah Kallikak. The story of the Kallikaks was well-known to all psychologists in the 1950s, and Allen found it hard to believe Emma was the dangerous moron of Goddard's description.

"I found her to be informative and interesting to talk with," Allen said. "She was considerate and personable and certainly not what I would think of as a retarded person. It was said that her judgment was not fully developed— understandable for someone practically raised in an institution."

In later years, Emma developed arthritis. She stopped sewing and woodworking. Instead of writing letters, she dictated them. But even in her eighties, confined to a wheelchair, she still sang songs from the plays she had performed in.

I'm a gypsy, I'm a gypsy
Oh I am a little gypsy girl
The forest is my home
And there I love to roam
For I am a little gypsy girl.

She never did roam. Capable as Emma proved herself over the decades of hard work, she came to believe that she deserved to remain, in effect, a prisoner. "I guess after all I'm where I belong," she told Helen Reeves. "I don't like this feeble-minded part but anyhow I'm not idiotic like some of the poor things you see around here." In her old age, she was offered the chance to leave the institution, but declined. She lived out her days there, dying at age eighty-nine in 1978. She was buried on the institution grounds.

After Emma left the Vineland Training School, she never saw Goddard again. But she once told Reeves that she had named one of the cats Henry, "for a dear, wonderful friend who wrote a book. It's the book what made me famous."

"She was devoted to the people who conducted the study, as though they were her family," Allen recalled. When Goddard sent Emma a Christmas card in 1946, Reeves wrote back to him to let him know how happy Emma was to receive it.

"The nicest thing about it," Emma told Reeves, "is that he thought I have the brains to understand it which of course I do."

CHAPTER 4

Attagirl

NINE YEARS before she enrolled her daughter at the Vineland Training School, Pearl Buck woke up out of an ether sleep and saw a bloom of plum blossoms on the table by her bed. She turned her head to see a nurse holding her newborn baby in a pink blanket. Pearl looked into the girl's eyes.

"Doesn't she look very wise for her age?" she asked the nurse.

It was a warm day in March 1920. Pearl Buck was twenty-eight, an American-born teacher living in northern China. She had grown up in China, her missionary parents having brought her there as a baby. After four years of college in the United States, she returned to care for her ailing mother. Soon afterward, she met an expat agricultural expert named John Lossing Buck, whom she married in 1917. For the first three years of their marriage, they lived in a remote town called Nanhsuchou. From the windows of the house, she could see miles of flat farmland. Over the green wheat, mirages of lakes and mountains flirted with her eyes. She and Lossing named their girl Caroline.

Carol, as she quickly came to be known, was a fair-haired, blue-eyed baby. A few things caught Pearl's attention, but she didn't give them much mind. Carol had eczema that made her scratch. Her skin gave off a peculiar musty smell. Pearl had more important things to worry about. A few weeks after Carol's birth, Pearl's doctor told her that she had a tumor in her uterus.

She took the long journey back to the United States to have it surgically removed. The tumor proved to be benign, but her American doctors informed Pearl she would be unable to have any more children.

The Bucks moved from Nanhsuchou to the city of Nanking, where Lossing got a job teaching agriculture at the university. Pearl taught English, while Carol played in the gardens and bamboo groves surrounding their house. As Carol grew, Pearl began to worry. The babies of her friends were beginning to walk. Carol still crawled. They began to speak. Carol babbled. Her eczema grew so bad that Pearl would sometimes put bandages on her hands so that she wouldn't rake her skin.

Pearl kept her worries to herself, partly out of shame, and partly out of her knowledge that her family would have little sympathy. Pearl's father was a rigid fundamentalist who cared only about tallying up the souls he saved. Her mother, suffering from a lethal digestive disorder called sprue, had abandoned Christianity as she waited to die. And Lossing, Pearl discovered after they got married, was a hollow man. "He has never seen or understood anything," she would later say.

Carol eventually learned to walk, but she still wasn't learning to talk. She was big for her age, restless, and demanding, making her desires known with jabbering and grunts. She sniffed at visitors and jumped up on them as a friendly dog would. The things that made other children laugh or cry drew only a blank stare from Carol. Pearl's friends assured her that everything was fine, that children begin speaking at different ages. Years later, they would confess to Pearl that they shrank from speaking the truth. They knew something was wrong.

That summer, Pearl took Carol to the seashore to play on the beach and ride donkeys through the nearby valleys. She even managed to teach Carol to speak a few words. One day that summer, Pearl went to a lecture by a local pediatrician about the health of young children. The pediatrician described some warning signs of psychological disorders, such as incessantly running around, and it sounded to Pearl as if she was talking about Carol. The next day, the pediatrician paid Pearl a visit with some other doctors. Examining Carol, they could tell that something was indeed wrong, but they couldn't say what. For a firm diagnosis, Pearl would need to take Carol to the United States.

The Bucks already had a trip back home in the works so that Lossing could pursue a master's degree at Cornell. He and Pearl settled into a cramped two-room apartment in Ithaca, New York, and from time to time Pearl would take Carol around the country to see doctors—psychologists, pediatricians, gland specialists. They all told her something was wrong, but none could give her a diagnosis. Yet she always left the exams with an unfocused hope that Carol would get better.

Pearl's last trip took her to the Mayo Clinic in Minnesota. There, a young doctor gently broke the news to her: Carol had stopped developing mentally.

"Is it hopeless?" Pearl asked him.

"I think I would not give up trying," the doctor said.

Pearl and Carol walked out of the doctor's office and made their way down an empty hall. A small, bespectacled doctor with a clipped black mustache emerged from a room, and he asked in a crisp German accent if the other doctor had said Carol could be cured.

Pearl said he didn't rule it out.

"She will never be well—do you hear me?" the second doctor said. "Find a place where she can be happy and leave her there and live your own life."

Pearl staggered out of the clinic. Carol, happy to be done with the strangers, danced ahead. When she noticed that her mother had started to cry, Carol laughed.

For the rest of her time in the United States, Pearl struggled to make the best of her life. She earned a master's degree of her own in English and wrote a few articles about China. To most Americans in the 1920s, the country was an alien giant, and so editors were happy to publish stories from someone with such deep knowledge of the place. Pearl discovered that she enjoyed writing and that she was good at it. Before returning to China, she and Lossing visited a New York orphanage and adopted a three-month-old girl they named Janice.

Back in China, Pearl became overwhelmed by sadness over Carol. She couldn't even bear to listen to music. When guests came to the house, she would put on a brave face, but as soon as they left, she would let her sorrow have its way. Pearl began writing stories along with her essays, imagining the lives of Chinese people around her. Carol would become intensely jealous as Pearl became absorbed in her work. She threw porridge at her mother and used handfuls of potting soil to clog the keys of Pearl's typewriter.

While the Bucks had been away, China had grown far more dangerous. The Kuomintang and its enemies had begun battling for control of different pieces of the country. For two years, the fighting remained far from Nanking, but in 1927 it reached the city. As foreigners were shot and raped, the Bucks hid in the hut of a Chinese woman Pearl knew. Pearl kept Carol and Janice quiet so that they wouldn't draw the attention of nearby soldiers. She vowed to herself to kill her girls before letting the soldiers take them away.

The attacks subsided after American and British gunships arrived in Nanking and fired on the city. The Bucks took the opportunity to flee, making their way to Shanghai. Shanghai proved only a brief stopover for them; the fighting drove the Bucks out of China altogether. They ended up in Japan, surviving in a remote forest cabin for months on fish, fruit, and rice.

Once China settled down again into a relative calm, the Bucks returned. Pearl now became painfully aware of how the children of her friends were developing and thriving while Carol, now eight, still acted like a toddler. When she tried to teach Carol to write, her daughter managed to learn only a few words. During one of their lessons, Pearl took the pencil from Carol and was startled to discover that her daughter's hand was drenched in sweat from all her effort. Pearl was ashamed that she had made Carol so miserable and decided to stop forcing her to try to become like other girls. As her mother, Pearl would only try to make Carol happy.

"I realized I must leave her in some place," Pearl later recalled, "and my heart is wrenched in two at the thought."

Aside from the dread of separation, Pearl also recognized that she faced some grim economics. Lossing thought Carol should go to a state institution. The idea terrified Pearl, but she knew they didn't have the money to pay for a private school. Pearl realized she would have to find the funds on her own. "I had found out enough to know that the sort of place I wanted my child to live in would cost money that I did not have," she later wrote.

Her income from teaching was meager, and she made even less from writing articles for American magazines. She wondered if fiction might pay better. By then she had finished her first novel, which she called *East Wind: West Wind*. She got an idea for a second novel that might sell well. Whenever she found a free ten minutes between chores or caring for Carol, she would sit down to her typewriter and write about the adventures of a Chinese farmer she named Wang Lung.

In 1929, the Bucks traveled back to the United States. As Lossing negotiated a new grant for his work on Chinese agriculture, Pearl searched for a place where Carol could live. Many of the visits left her chilled. At one institution, the children were clothed in burlap and herded like dogs. Eventually, Pearl ended up in southern New Jersey, at a farm where the children seemed happy.

"I saw children playing around the yards behind the cottages, making mud pies and behaving as though they were at home," she later recalled. "I saw a certain motto repeated again and again on the walls, on the stationery, hanging above the head's own desk. It was this: 'Happiness first and all else follows.'"

In September 1929, Pearl Buck enrolled her daughter at the Vineland Training School. Emma Wolverton had been taken out of the school fifteen years earlier, and a decade had passed since Henry Goddard had left. The enthusiasm for eugenics had left the place as well. In the 1920s, Vineland psychologists did important research on classifying different forms of feeblemindedness—what are now known as intellectual developmental disorders. They created a test to track the social development of children that's known today as the Vineland Social Maturity Scale.

Pearl stayed with friends for a month while Carol settled in at the school. It was the first time they had been separated in her life, and for Pearl it was torture. She listened for her daughter's calls for help in the night, her steps on the stairs. "Only the thought of a future with the child grown old and me gone kept me from hurrying to the railway station," she said.

Pearl went to New York to show the manuscript for *East Wind: West Wind* to a publisher named Richard Walsh. He bought it, along with the new novel she was in midst of writing. When she and Lossing returned to China in 1930, she worked on nothing else, losing herself in the story of Wang

Lung to keep her pain at bay. When she sent the book to Walsh, he gave her a name for it: *The Good Earth*.

Pearl's gritty story about a poor Chinese hero was an unfamiliar one to American readers. If they had read any fiction from China, it was classical tales about the country's elite. *The Good Earth*, published in the midst of America's Great Depression, felt like an Asian parallel to *The Grapes of Wrath*. In 1932, it earned Pearl the Pulitzer Prize, and it also proved a smashing commercial success. In just the first eighteen months after publication, Buck earned $100,000, and the book would earn hundreds of thousands more during her lifetime.

Pearl had only wanted to pay for a home for Carol. Instead, she became a celebrity. In quick succession, Pearl moved back to the United States, got a Nevada divorce from Lossing, married Walsh, bought a farm in Pennsylvania, and adopted more children. Hollywood turned *The Good Earth* into a box-office hit, while Pearl found herself in fierce demand for lectures around the country.

Pearl made savvy use of her new fame to champion political causes, especially civil rights. Growing up in China, she became keenly aware of the contempt some Chinese had for her simply because she was white. When she returned to the United States, she scoffed at the idea that the country's blacks and whites were biologically distinct in any meaningful way, calling humanity "a creature hopelessly mongrel." In 1938, just seven years after publishing *The Good Earth* in the hope of taking care of Carol, Pearl S. Buck won the Nobel Prize in Literature. When she got the news, she responded in Chinese: "*Wo pu hsiang hsin.*" (I don't believe it.)

The more stories that Pearl told, the more the world clamored to hear her own. But she refused to reveal Carol's secret. "It is not a shame at all but something private and sacred, as sorrow must be," Pearl wrote to a friend. When reporters asked about her family, she would say she had two daughters, one of whom was away at school. An old friend from Nanking was interviewed by an Ohio newspaper and recalled Pearl's suffering over Carol. Pearl got wind of the story and arranged to have it quashed. She wanted to protect Carol, but herself as well. "I would gladly have written nothing if I could have just an average child in Carol," she once said.

From the profits on *The Good Earth*, Pearl gave the Vineland School $40,000, guaranteeing Carol a lifetime of care. Pearl later paid for the construction of a new two-story cottage where Carol could live with fifteen other girls, complete with a French provincial bedroom set, a phonograph, and a collection of records. (Carol liked hymns, hated jazz.) Once Pearl returned to the United States, she would visit Carol as often as she could—sometimes as often as once a week—and sometimes brought her back to her farm in Pennsylvania for a few days. Pearl thus got to watch Carol grow up. She began to bathe and dress herself, even to tie her shoelaces. She learned to eat with a fork and spoon, to sew, and to use words to tell others what she needed. She roller-skated. She loved to ride a tricycle around the school grounds. Decades later, people would sometimes see a gray-haired woman pedaling still.

By 1940, Pearl had reached a kind of melancholy peace with Carol's fate. "All sense of flesh, of my flesh, is gone," she wrote in her journal. "I feel toward her as tenderly as ever, but I am no longer torn. I am, I suppose, what may be called 'resigned' at last. Agony has become static—it is true but I will not disturb it or allow it to move in me."

Pearl continued publishing at an industrial pace. But her literary reputation had grown dim. The men who came to dominate American mid-century literature treated her writing merely as women's novels. Pearl tried to write about life in the United States, but readers thought of her only as a chronicler of China. She also made a growing number of enemies with her political activity. Even at the height of World War II, she criticized the American government, asking how the United States could fashion itself as the enemy of fascism when it accepted white superiority at home and promoted imperialism abroad. After the war, the FBI decided she was a Communist in spirit if not in party membership.

Pearl could sense that hostility was growing around her, but she didn't stop working for her causes. She even took up new ones. Having adopted five children, she spoke out against orphanages and foster homes. Before she knew it, a desperate mother had dropped off a child at her farm. Pearl responded by creating a private adoption agency specializing in finding homes for Amerasian children who were rejected by both sides of their family. Pearl raised money for the research at Vineland, and by the 1940s she was in charge of fund-raising for the entire school.

One of the fund-raisers she worked with urged her to publish something about Carol, to help draw attention to the school. At first, Pearl found the requests intensely annoying. But eventually the fund-raiser won her over. She sat down and began to write about Carol. "I have been a long time making up my mind to write this story," she began.

Pearl presented a clear-eyed account of Carol's childhood and her own pain, shame, and reconciliation. She confessed to thinking how her daughter might be better off dead. She recounted how she learned to stop blaming Carol for what was not her fault, and to recognize her right to develop her mind as far as nature would allow.

"It was my child who taught me to understand so clearly all people are equal in their humanity and that all have the same human rights," Pearl wrote. "Though the mind has gone away, though he cannot speak or communicate with anyone, the human stuff is there, and he belongs to the human family."

Pearl published her essay in *Ladies' Home Journal* in May 1950, and it was later released as a short book called *The Child Who Never Grew*. All the royalties went to the Vineland Training School. In 1950, when intellectual developmental disability was still a source of shame and confusion, her frankness was nothing short of astonishing, especially coming from a bestselling, Nobel Prize–winning writer. *The Child Who Never Grew* was translated into thirteen languages, and Pearl got mailbags full of letters from parents of children like Carol. She answered every one.

At the end of the book, Pearl called for better care for people like Carol and urged that more research go into understanding intellectual developmental disorders, pointing to the work carried out at Vineland as an example of what needed to be done. She highlighted Goddard's intelligence testing and the Vineland Social Maturity Scale.

It's telling that Pearl didn't mention the research that first brought the Vineland Training School to international attention: Goddard's study of heredity. In fact, Pearl took great pains to scrub Carol's story clean of any possible hereditary taint. She declared that there was no trace of mental retardation in her own family or in Lossing's. Carol's story, in other words, had nothing to do with the other famous tale of a Vineland student, *The Kallikak Family.* "The old stigma of 'something in the family' is all too often unjust," Pearl wrote.

Unbeknownst to Pearl, there was something in the family after all. It was not an inheritance of sin or degeneration, however. It was a hereditary disease. In fact, a doctor had come to Vineland a decade before Pearl published *The Child Who Never Grew* and correctly diagnosed Carol with the disorder. No one had told Pearl, and she would have to wait another decade to find out for herself.

———

E ight years after Pearl Buck gave birth to Carol, a woman in Oslo named Borgny Egeland had a girl of her own. Liv Egeland seemed a healthy baby at first, although Borgny was puzzled by the odor of her hair, skin, and urine. It reminded her of a horse stable. Her puzzlement turned to worry as Liv reached age three unable to utter a single word. Yet her doctor, finding nothing wrong with Liv, told Borgny to give her more time.

Unlike Pearl Buck, Borgny Egeland was able to bear another child. In 1930, she gave birth to a son, Dag, who gave off the same musty odor as Liv. And later he also failed to learn how to speak. Borgny searched for a doctor who could explain this bizarre coincidence. By the time Liv was six, she could say only a few words and had trouble walking. Dag, now four, couldn't talk at all. He was unable to eat, drink, or walk on his own.

The doctors Borgny consulted had no explanation for why both of her children had developed the same symptoms. Nor could they offer any treatment. Borgny refused to share their resignation. She kept visiting doctors until she ran out of names, and then she paid a woman to give her children baths in herb-soaked water. She sought help from a psychic. Finally, Borgny learned that her sister was acquainted with a doctor at Oslo University Hospital who was an expert on metabolic disorders. She asked her sister to contact the doctor, named Asbjørn Følling, to see if he thought their odor and their intellectual development were linked.

Følling had never heard of such a thing. He doubted he could help, but he didn't want to disappoint Borgny after she had suffered so much. He invited her to bring the children to see him. The exam revealed nothing new. But Følling also asked Borgny to bring him some of Liv's urine so that he could carry out some chemical tests to track down the source of the odor.

Følling carried out his experiments in a makeshift lab in the attic of the medical ward. He added drops of ferric chloride to Liv's urine to test for diabetes. If Liv had the disease, it would turn purple. Instead, her urine turned green. Følling had never seen such a thing. He hadn't even heard of such a thing happening before. Baffled, Følling asked Borgny to bring him some of Dag's urine. When Følling ran the test again, the urine shimmered green once more.

Følling searched through the medical literature for an explanation, but no one had ever observed the reaction. He wondered if Borgny was giving her children aspirin or some other medicine that was tinting the urine. As a test, he asked Borgny to keep her children off any medication for a week. When he experimented on their urine again, it still turned green.

It took two months of experiments—and twenty-two liters of Egeland urine—for Følling to finally find the cause. The children's urine was loaded with a compound not found in healthy people—a cluster of carbon, oxygen, and hydrogen atoms known as phenylpyruvic acid.

Based on his deep knowledge of human metabolism, Følling came up with a hypothesis to explain the strange chemistry. Proteins are made of building blocks called amino acids. One amino acid is called phenylalanine, which people must get from their food. Any extra phenylalanine people don't use to make proteins gets broken down by enzymes in the liver. Følling reasoned that the Egeland children were not breaking down their phenylalanine. Somehow, the rising level of phenylalanine harmed the children. Some of it was converted into a similar molecule, phenylpyruvic acid, and washed out of their bodies in their urine.

To test his idea, Følling examined other children with similar symptoms. He ended up finding the green signature in the urine of ten patients in total. They included three pairs of siblings—a coincidence that led Følling to suspect the condition was a hereditary disorder.

Yet Følling could plainly see that Borgny Egeland and all the other parents of these children were healthy. Some of them had other children who were healthy as well. The disorder must be caused by a recessive factor, Følling reasoned. Each parent was a carrier, with one defective copy of some unknown gene, and some of their children had the misfortune of inheriting a bad copy from both of them.

Følling found support for this hypothesis when he followed up with two parents, each of whom had gotten remarried. Between them, they had twelve more children. All of their offspring from their new marriages were healthy, and none of them had urine that turned green. The recessive factor was probably very rare in Norway, Følling reasoned, meaning that the odds of marrying two people who were carriers was next to zero. The children of the second marriages might inherit one recessive factor at most, meaning that they could not develop the disease.

Følling quickly wrote up his discovery and gave the disease a name: imbecillitas phenylpyruvica. Not since Archibald Garrod had discovered that the black urine of alkaptonuria was a hereditary disorder had someone found such a clear-cut case. Yet few scientists paid Følling's 1934 paper much notice. He could not say precisely what was wrong in people with the disease. Nor could he account for how a problem with phenylalanine could affect the brain.

Only a small circle of scientists who studied intellectual developmental disorders recognized how important his findings were. Even if imbecillitas phenylpyruvica was rare, it still represented what Henry Goddard had been chasing after: a hereditary cause of feeblemindedness. Følling's study was even more significant because he had invented a straightforward way to give a precise diagnosis.

One of the first doctors to take up Følling's test was a British doctor named Lionel Penrose. Although only in his mid-thirties at the time, Penrose had already become a leading expert on intellectual developmental disability in Britain. He had climbed the ranks swifly, having come late to medicine. Penrose had started out studying mathematical logic at Cambridge, and then he traveled to Vienna to investigate the psychology of mathematical thinking. When that work hit a dead end, Penrose got curious about mental disorders and what they might reveal about the mind. At age twenty-seven, he returned to Cambridge to study medicine. Four years later, now a freshly minted MD, Penrose became a medical research officer at the Royal Eastern Counties Institution at Colchester, a home for "mental defectives."

Penrose entered the profession as a passionate critic of eugenics, dismissing it as "pretentious and absurd." In the early 1930s, eugenics still had a powerful hold on both doctors and the public at large—a situation Penrose blamed on lurid tales like *The Kallikak Family*. While those stories might be seductive, eugenicists made a mess of traits like intelligence. They were obsessed with splitting people into two categories—healthy and feebleminded—and then they would cast the feebleminded as a "class of vast and dangerous dimensions."

Penrose saw intelligence as a far more complex trait. He likened intelligence to height: In every population, most people were close to average height, but some people were taller and shorter than average. Just being short wasn't equivalent to having some kind of a height disease. Likewise, people developed a range of different mental aptitudes.

Height, Penrose observed, was the product of both inherited genes and upbringing. He believed the same was true for intelligence. Just as Mendelian variants could cause dwarfism, others might cause severe intellectual developmental disorders. But that was no reason to leap immediately to heredity as an explanation.

"That mental deficiency may be to some extent due to criminal parents' dwelling 'habitually' in slums seems to have been overlooked," Penrose said. He condemned the fatalism of eugenicists, as they declared "there was nothing to be done but to blame heredity and advocate methods of extinction."

The wrongheaded ideas of eugenicists led them to wrongheaded solutions, such as sterilization. Even if a country did sterilize every feebleminded citizen, Penrose warned, the next generation would have plenty of new cases from environmental causes. "The first consideration in the prevention of mental deficiency is to consider how environmental influences which are held responsible can be modified," Penrose declared. He suspected that many cases of mental deficiency were caused by a mother's syphilis or X-ray tests during pregnancy.

At Colchester, Penrose launched a study he hoped would lead to more humane, more effective treatments for intellectual developmental disorders. He set out to classify the disorders and determine some of their causes. Over the course of seven years, he examined 1,280 subjects and carefully studied their families as well. Drawing on his expertise in mathematics, Penrose developed sophisticated statistical methods to search his data for links among mental deficiency, heredity, and the environment.

As soon as Penrose heard of Følling's discovery, he wanted to try it out for himself. It was so simple, he later wrote, that it was puzzling no one had discovered it before. Penrose ordered that urine from 500 patients at Colchester be put to Følling's test. Out of those samples, 499 did not change color. But a single sample turned green.

The emerald urine belonged to a nineteen-year-old man who had never walked or talked. He spent his days rocking back and forth, his wasted arms and legs bent close to his body. After the test, Penrose paid a visit to the man's family. His parents were hardworking and healthy, although his father was convinced that people were poisoning him. Their other children were all relatively normal, except their five-year-old son. Like his older brother, the boy could not walk or speak. Penrose tested the urine of the children and found that they were all normal—except the five-year-old boy.

Studying these and other cases, Penrose proposed that a single hereditary factor was responsible for the disorder. While people with two copies of the recessive factor might be rare, he suggested that many more people might have a single copy. When Penrose published his research, he decided not to use the original name for the disease, imbecillitas phenylpyruvica. He preferred a new name coined by his collaborator, Juda Quastel: phenylketonuria. It was, Penrose boasted, "preferable to the original more cumbersome designation." That name has stuck ever since, although it's often shortened to PKU—which Penrose called "an abominable abbreviation."

Over the next few years, an American researcher named George Jervis confirmed Penrose's hypothesis and worked out the chemistry of the disease.

Normally, an enzyme known as phenylalanine hydroxylase breaks down the body's extra phenylalanine. In people with PKU, the enzyme doesn't work. The body's phenylalanine reaches toxic levels and spreads throughout the body, wreaking havoc.

As the biology of PKU became clearer, Penrose realized that it might not be inevitable, even if it was hereditary. Penrose reasoned that a diet low in phenylalanine might prevent people with PKU from becoming poisoned.

But because phenylalanine is so abundant in food, Penrose found it difficult to draw up a diet for his patients. He restricted the diet of one patient to only fruit, sugar, and olive oil, supplemented with vitamin pills. It lowered his patient's phenylalanine levels for a couple of weeks, but they bounced back up. Seeking help, Penrose contacted Frederick Gowland Hopkins, a Cambridge biochemist who had won the Nobel Prize in 1929 for the discovery of vitamins. When Penrose told Hopkins about PKU, Hopkins declared that a diet for the disorder would cost a thousand pounds a week.

Penrose abandoned a search for a diet, but he continued to study people with PKU. Whenever he visited a new institution, he would sniff the air for a musty odor. If he discovered patients who he suspected of having PKU, he would examine them for other telltale features of the condition, such as fair hair and blue eyes. Then he would order a simple urine test.

In 1939, while on a trip through the United States, Penrose paid a visit to the Vineland Training School. There he met the nineteen-year-old Carol Buck. "I was informed that this patient was the daughter of a distinguished writer but that, in spite of obtaining all the best opinions in the United States, no cause for the defect had been found," Penrose later wrote.

Penrose met Carol at the cottage her mother, Pearl, had built for her. "Everything was beautifully appointed," he recalled. But when Penrose sniffed the air, he detected the familiar mustiness. He noticed Carol's blue eyes and fair hair. He checked her reflexes. "I felt quite certain of the diagnosis and told my hosts what I thought," Penrose said.

Penrose was dismayed that his hosts didn't know what he was talking about. It had been five years since Følling had published the first account of PKU. Even at an advanced institution like Vineland, however, no one recognized it as a possible cause of retardation. "'Impossible,' they said. 'How can you come here and in a few minutes find something which all our best clinicians have missed?'" Penrose wrote.

The next morning, Penrose tested Carol's urine. He saw "the wonderful green color." But no one at the school ever told Carol's mother about Penrose's diagnosis.

———

Penrose, a lifelong pacifist, sat out World War II in Canada. In 1945, he got an invitation back home, to become the next Galton Professor of Eugenics at University College London and director of the Galton Laboratory. The irony of the titles was not lost on him.

Francis Galton, the scientist who had coined the term *eugenics*, had left some of his family fortune to pay for a professor to run a eugenics research lab, gathering data about heredity in the hopes of improving the human race. After Galton's death in 1911, the lab buzzed with research for three decades, until it fell to German bombs. Penrose agreed to rebuild it, but it would not be the same when he was done. He sought to wipe eugenics away. He even changed the name of his position to Galton Professor of Human Genetics—but only after a legal battle that lasted until 1963.

As the new Galton Professor, Penrose was required to give an inaugural lecture. He used the opportunity to let the world know that things had changed, and he used PKU as a case study. The title of his talk was "Phenylketonuria: A Problem in Eugenics."

As Penrose drafted his lecture in 1945, the memories of the Holocaust were still horrifically fresh. It had been less than a year since Auschwitz, Dachau, and Bergen-Belsen had been liberated. The Nazis had justified the horrors of their "race hygiene" by pointing to the work of eugenicists. In the postwar years, Penrose now worried that eugenics might survive their defeat. Leading eugenicists in England and other countries were still pushing their agenda. In the United States, sterilization laws justified on the basis of eugenics remained on the books, and people were being regularly robbed of the chance to have children.

In his lecture, Penrose directed his wrath at lingering eugenicists, showing how their calls to manage human reproduction for the betterment of the species were absurd—"pernicious ideas based upon emotional bias," as he put it. And Penrose used PKU as a case study for why the eugenics agenda should be thrown out.

By 1946, scientists had studied some five hundred people with PKU, and their family histories clearly demonstrated that the disease was hereditary. In other words, children had to inherit the same version of a gene from both parents. Scientists still didn't know what genes were, but to a eugenicist, Penrose speculated, that wouldn't matter. To get rid of PKU, all that would be required would be to stop people from passing the gene down to future generations.

"This view, however, is incorrect," Penrose said. "We cannot take the same attitude here that we might with regard to some noxious pest and simply ask to have the offending genes exterminated."

PKU was a recessive condition, meaning that a child had to inherit two faulty copies of the same gene to develop the diseases. As far as Penrose and other scientists could tell, people with a single copy of the defective gene were healthy—so healthy, in fact, that it was impossible to identify carriers until they had children with PKU. Based on the number of cases he had found, Penrose estimated that 1 percent of people in Great Britain were carriers. (Later research would indicate that the true figure is probably twice that.)

"To eliminate the gene from the racial stock would involve sterilizing 1% of the normal population, if carriers could be identified," Penrose declared. "Only a lunatic would advocate such a procedure to prevent the occurrence of a handful of harmless imbeciles."

When Penrose treated people with PKU, their relatives would anxiously ask him how likely it was that they might be carriers. Should they not have children? Penrose worked through the odds. The chances of a sibling of someone with PKU being a carrier is two in three. Penrose estimated that the chances of a prospective mate also being a carrier was one in a hundred. And the chance of a child of two carriers inheriting PKU was one in four. Multiplying all those probabilities together led Penrose to conclude that the chance of a relative of someone with PKU having a child with PKU was only one in six hundred.

"In my opinion," Penrose said, "this risk is no adequate ground for discouraging the union."

In a sly aside, Penrose also noted that PKU undermined the Nazi myth of an Aryan race that was superior to races of Jews or blacks. In the United States, Jervis had not found any Jews or blacks with PKU. Instead, many of the people with the disease were Germans and Dutch. "A sterilisation programme to control phenylketonuria confined to the so-called Aryans would hardly have appealed to the recently overthrown government of Germany," Penrose said.

To finish up his lecture, Penrose predicted that the story of PKU would turn out to be similar for many other diseases. "Many rare recessive disabilities have been identified in man, and doubtless many more lie awaiting detection," he said. "Not improbably, about two people out of every three are carriers of at least one serious recessive defect."

Humanity, in other words, was not some genetically uniform stock that could be purged of a few defectives. Penrose saw our species as rich with genetic diversity, and forever falling short of genetic perfection. To eliminate imperfection would demand eliminating humanity itself.

———

After his attack on eugenics, Penrose went on to build the first large medical genetics program, designed to identify new hereditary disorders. The geneticists under Penrose's leadership in the early 1950s examined patients, ran blood tests, and drew pedigrees. They traced the inheritance of genes, despite still not knowing what genes are. But if they had taken a stroll down Bloomsbury Street to King's College London, they could have watched a woman take X-ray pictures that would soon start to unravel that mystery.

By the 1920s, Thomas Hunt Morgan and his colleagues had persuaded their fellow scientists that genes were physical things, located in chromosomes. Chromosomes were chemical mixtures, including proteins as well as a mysterious molecule called deoxyribonucleic acid, or DNA for short. By the early 1950s, researchers had performed some elegant experiments with bacteria and viruses that made it clear that DNA, not proteins, was the stuff of genes. When viruses infected bacteria, for example, they only injected DNA; none of their proteins made it into the cells.

In 1950, a thirty-year-old scientist named Rosalind Franklin arrived at King's College London to study the shape of DNA. She and a graduate student named Raymond Gosling created crystals of DNA, which they bombarded with X-rays. The beams bounced off the crystals and struck photographic film, creating telltale lines, spots, and curves. Other scientists had tried to take pictures of DNA, but no one had created pictures as good as Franklin had. Looking at the pictures, she suspected that DNA was a spiral-shaped molecule—a helix. But Franklin was relentlessly methodical, refusing to indulge in flights of fancy before the hard work of collecting data was done. She kept taking pictures.

Two other scientists, Francis Crick and James Watson, did not want to wait. Up in Cambridge, they were toying with metal rods and clamps, searching for plausible arrangements of DNA. Based on hasty notes Watson had written during a talk by Franklin, he and Crick put together a new model. Franklin and her colleagues from King's paid a visit to Cambridge to inspect it, and she bluntly told Crick and Watson they had gotten the chemistry all wrong.

Franklin went on working on her X-ray photographs and growing increasingly unhappy with King's. The assistant lab chief, Maurice Wilkins, was under the impression that Franklin was hired to work directly for him. She would have none of it, bruising Wilkins's ego and leaving him to grumble to Crick about "our dark lady." Eventually a truce was struck, with Wilkins and Franklin working separately on DNA. But Wilkins was still Franklin's boss, which meant that he got copies of her photographs. In January 1953, he showed one particularly telling image to Watson. Now Watson could immediately see in those images how DNA was shaped. He and Crick also got hold of a summary of Franklin's unpublished research she wrote up for the Medical Research Council, which guided them further to their solution. Neither bothered to consult Franklin about using her hard-earned pictures. The Cambridge and King's teams then negotiated a plan to publish a set of papers in *Nature* on April 25, 1953. Crick and Watson unveiled their model in a paper that grabbed most of the attention. Franklin and Gosling published their X-ray data in another paper, which seemed to readers to be a "me-too" effort.

Franklin died of cancer five years later, while Crick, Watson, and Wilkins went on to share the Nobel prize in 1962. In his 1968 book, *The Double Helix*, Watson would cruelly caricature Franklin as a belligerent, badly dressed woman who couldn't appreciate what was in her pictures. That bitter fallout is a shame, because these scientists had together discovered something of exceptional beauty. They had found a molecular structure that could make heredity possible.

DNA, they discovered, is a pair of strands twisted into a double helix. Between the strands, a series of compounds called bases bonded to each other. Over the next thirty years, scientists worked out how this structure allowed DNA to carry genes. Each gene is a stretch of DNA, made up of thousands of bases. Each base can take one of four different forms: adenine, cytosine, guanine, and thymine—A, C, G, T for short. A cell carries out a series of chemical reactions to translate a gene's sequence of bases into a protein. A cell first makes a copy of the gene, creating a single-stranded series of bases called ribonucleic acid, or RNA. That RNA molecule is taken up by a molecular factory called a ribosome, which reads the sequence of RNA and builds a corresponding protein.

The discovery of DNA seemed to reduce heredity to a reliably simple recipe. It came down to turning one DNA molecule into a pair. A cell's molecular machinery pulled apart the two strands of a DNA molecule and then assembled a new strand to accompany each of them. Each base could bond only to one other: A to T, C to G. The cell could thus build two perfect copies of the original DNA—like engendering like, but on an atomic scale.

Sometimes cells make mistakes, however. These errors leave one of the new DNA molecules altered. A single base may change from A to C. A stretch of a hundred bases may be accidentally copied out twice. A thousand bases may be cut out altogether. These are the mutations that scientists like Hugo de Vries and Thomas Hunt Morgan spent years trying to figure out. Mutations can produce new versions of genes—alleles, as they came to be known. Sometimes alleles work the same as before. But, in cases such as PKU, they fail to work at all.

Later generations of scientists would use this discovery to determine the molecular details of PKU. The enzyme Jervis had discovered, phenylalanine hydroxylase, is encoded by a gene called PAH. In our livers, cells translate the PAH gene into the enzyme, which can then break down phenylalanine. In carriers, such as Pearl and Lossing Buck, one copy of the PAH gene carries a mutation that prevents cells from making the enzyme.

Pearl and Lossing had no idea that anything was wrong in their DNA, because their other copy of the PAH gene lacked the mutation. They could make enough phenylalanine hydroxylase for their metabolism to run properly. But when a child like Carol inherited a faulty copy of the PAH gene from both her parents, she could not make any working enzymes and suffered the consequences.

Fifty years would pass after Følling and Penrose proposed that PKU was caused by recessive factors before scientists finally saw the factors with their own eyes. By then, however, the lives of people with PKU had already dramatically improved. A child born with PKU, if properly cared for, would never have to face a future like that of Carol Buck.

The journey to a treatment started in 1949, when a British woman named Mary Jones brought her seventeen-month-old daughter, Sheila, to a Birmingham hospital. Sheila couldn't stand or even sit up. Nor did she take an interest in her surroundings. A doctor at the hospital named Horst Bickel examined Sheila and informed Jones that she had PKU. "Her mother was not at all impressed when I showed her proudly my beautiful paper chromatogram with the very strong phenylalanine (Phe) spot in the urine of her daughter proving the diagnosis," Bickel later recalled.

Jones wanted to know what Bickel was going to do now that he had discovered Sheila's disease. There was nothing to do, Bickel explained.

Jones rejected his answer. She came back the next morning to demand help. When he turned her down, she came back every morning with the same demand.

"She was very upset and did not accept the fact that at the time no treatment was known for PKU," Bickel said. "Couldn't I find one?"

At the time, Bickel had little reason to think he could. Lionel Penrose had already tried to design a diet for PKU, without any results to show for it. Penrose became convinced that mental retardation wasn't caused by the inability to convert phenylalanine. Instead, he thought, the two symptoms both arose from an unknown source. A diet was no more likely to cure PKU retardation than eyeglasses would make an old man's wrinkles disappear.

Jones was so insistent, though, that Bickel decided to talk to some of his colleagues about a diet for PKU. He learned that a biochemist in London named Louis Wolff had tried concocting a broth that could provide protein to people with PKU without poisoning them with phenylalanine. When he proposed feeding his broth to patients, his superiors at Great Ormond Street Hospital told him his job did not involve crazy treatments for the incurable. Wolff gave his recipe to Bickel, who followed the directions, working in a frigid lab kept cold to prevent the concoction from spoiling.

Eventually, Bickel prepared enough of the stuff for Sheila. He instructed Jones that the girl was to eat nothing else. To his delight, the phenylalanine in Sheila's bloodstream dropped, and did not bounce back the way it had in Penrose's experiments fifteen years earlier. The diet even showed signs of improving her brain. Within a few months she began to sit up, then to stand, then to walk with assistance. Her musty odor even disappeared. But when Bickel told his colleagues at the hospital, they scoffed. They were sure Sheila had improved merely thanks to the extra attention she was getting. Bickel decided there was only one way to persuade them: take Sheila off of the diet.

Without telling Jones, Bickel secretly added phenylalanine to the formula. Within a day on the altered diet, Sheila started deteriorating. Soon she stopped smiling, making eye contact, or even walking. Bickel and his coworkers told Jones of their secret maneuver, and put her back on the low-phenylalanine formula. While the transformation was enough proof for Bickel, he didn't think it would be enough to persuade skeptical colleagues. He got Jones's permission to bring Sheila into the hospital and feed her phenylalanine again. This time, Bickel captured her decline by filming a silent movie.

In the first scene in the movie, Sheila is on Bickel's phenylalanine-free diet. She looks healthy and alert. She sits in a high chair, and behind her is draped a curtain covered in fleurs-de-lis. An arm, swathed in a lab-coat sleeve, moves into the frame, dangling a ring of keys. Sheila looks up at the keys. She studies them, and then reaches upward. As she taps the keys, she watches them swing back and forth. Sheila then grasps one of the keys in her own fingers. Now another lab-coat arm moves into the frame, bringing forward a rattle. She makes the difficult choice between keys and rattle calmly. She takes the keys and flings them to the floor.

The next scene was shot after Sheila went back on an ordinary diet for three days. She is a profoundly different child. She sits on the floor, gazing into middle space, her hair a chaotic tuft. When someone shows her keys, she takes several seconds to notice them. She reaches slowly, drooling, but can't grasp them.

The movie jumps ahead two days. Now Sheila doesn't even bother to reach for the keys. She just looks at them and cries. The screen turns black again: "Four weeks after resuming her low-phenylalanine diet," a card reads. In the next scene, Sheila is walking, pushing a chair across the room with stubborn determination. She looks up with an intense gaze—not sad, not happy, perhaps just wondering what she's been put through.

Bickel's movie was impressive enough to change the minds of doctors at Great Ormond Street. Wolff, Bickel, and their colleagues got the green light to put more children on the low-phenylalanine diet. In every case, they saw significant improvements. The diet wasn't a panacea by any means. While the children scored better on intelligence tests, they remained far below average because they had already suffered so much irreversible brain damage. The researchers also saw that the benefits could vanish if parents didn't sustain the diet every day. Sheila Jones kept getting better, learning to scribble with a crayon and build a tower of bricks. But her mother, a single parent struggling with mental illness, couldn't keep up Sheila's demanding diet. Eventually, Mary Jones ended up in an institution, and Sheila had to be put in one as well. Without the diet Bickel and Wolff had invented, Sheila Jones was doomed to live there the rest of her life. She learned to feed and dress herself, but she never learned to speak.

Bickel and Wolff's breakthrough inspired other scientists and pharmaceutical companies to concoct better formulas. As scientists studied how children on these diets turned out, they found that the earlier they got away from phenylalanine the better off they were in the long run. In the 1950s, however, doctors were still using Følling's test to detect PKU, which works only after children have built up relatively high levels of phenylpyruvic acid in their urine. To make the diets more effective, they'd need an earlier test.

At the time, scientists knew that PKU was caused by a recessive gene, and they also knew that the gene must be a specific sequence of DNA on a chromosome. But no one knew where it was. Even if they had known, they wouldn't have been able to sequence it, because the technology required would not become available for many decades. Instead, researchers tried to invent new tests for PKU that could detect lower levels of phenylalanine.

In 1957, a California pediatrician named Willard Centerwall figured out how to diagnose PKU by dabbing a child's diaper with ferric chloride. His test made it possible for doctors to identify children with the disease when they were still just a few weeks old. Soon afterward, an American medical researcher named Robert Guthrie developed a test that used blood rather than urine. Guthrie's test was quick, reliable, and cheap. Even better, it could detect PKU in a newborn with just a pinprick's worth of blood.

These advances were celebrated in the *Saturday Evening Post*, *Time*, and the *New York Times*. Before 1960, only 25 percent of people with PKU lived to the age of thirty, the majority dying young of infections in institutions. But now doctors could detect it and then treat it. Although PKU affected only a few hundred Americans, the press hailed the work of Guthrie and others as an unprecedented victory over heredity.

At the same time, thanks in part to Pearl Buck's *Child Who Never Grew*, many parents of retarded children had cast off their shame and were organizing. While there were many causes of retardation, the parents put the spotlight on PKU to inspire more support for care and research. As part of the 1961 National Retarded Children's Week, President John F. Kennedy welcomed two sisters with PKU, Kammy and Sheila McGrath, to the White House.

Both girls had PKU, but it had affected their lives in fundamentally different ways. Sheila, the older sister, had been diagnosed with PKU when she was a year old. By then she had suffered so much brain damage that at age seven she was now living in an institution. When the McGraths had Kammy two years later, their doctor used Centerwall's diaper test to diagnose her with PKU at three weeks. The McGraths immediately put her on a diet of special protein powder and low-protein foods. She had avoided Sheila's toxic exposure, and now, at age five, she was healthy and living at home.

When the McGrath family came to the Cabinet Room of the White House, Kennedy greeted them personally. He led Kammy to a rocking horse and watched her rock on it.

"Attagirl," the president said. "They are the best behaved children we've had in the White House—and that includes those who live here."

The McGrath family visit is memorialized by an official White House photograph. Kammy and her parents stand by the president, looking at Sheila. She sits on a rocking horse, gazing away. In May the following year, Sheila and Kammy appeared in *Life*, posing for a photo essay about Guthrie's test. Kammy, her hair in pigtails, grins over a mountain of protein powder poured onto a table. Sheila, her hair chopped short, is wearing a dark dress and sitting in a rocking chair set back from the table.

The wordless image delivered a clear message: Modern medicine had allowed Kammy to avoid Sheila's fate. "The sentencing is not mandatory," the *New York Times* declared. "Phenylketonuria can be kept in check, if diagnosed early enough, and a child can live a normal life."

In December 1961, the Kennedy administration panel moved to seek mandatory testing for PKU of all newborns. In 1963, Massachusetts passed a law that required screening for the disease, and other states soon followed. Within ten years, 90 percent of American children were being screened, and Guthrie and other researchers set up PKU testing programs in other countries as well. In later years, other hereditary disorders would be added to newborn screening, giving children as much of a head start as possible. By the 1970s, the first generation of people treated for PKU since birth reached adulthood. They could finish school, hold jobs, have ordinary lives. In 2001, a graduate student named Tracy Beck became the first person with PKU to gain a PhD. She became an astronomer, helping to build the James Webb Space Telescope. For thousands of years, people who inherited the mutations in Beck's PAH genes would have looked to the sky and not known the word for the lights they saw. Now Beck was helping to extend humanity's gaze to the farthest edges of the universe.

I n 1957, the Vineland Training School decided to test all their students for PKU. One of the few to test positive was Carol Buck.

In one sense, the result was nothing new: Penrose had made the same diagnosis twenty years before, using Følling's crude test. But this time the school told Pearl Buck. She could finally give Carol's condition a name, nearly four decades after it had altered her own life.

The name was new to Buck. She studied up on it, and when she traveled to Norway in 1958, she sought out Følling himself. Buck learned as much as she could from the seventy-year-old doctor. Soon afterward, she wrote a letter to her ex-husband, Lossing. She explained to him that they shared an invisible bond, one that neither had known about. After Pearl and Lossing had divorced, he had remarried. He and his second wife had two healthy children. In her letter, Pearl warned Lossing that they may have inherited his dangerous legacy.

"In Carol's case nothing matters, it is too late," she wrote. "But I think of your children, who carry the genes in their bodies. It is essential before they marry, that this blood is tested, and the blood of the person they marry."

In 1960, Willard Centerwall paid Pearl Buck a visit at her home in Pennsylvania. She confided in him that Carol had recently been diagnosed with PKU. From a pocket, Centerwall produced a vial of phenylacetate crystals. He invited Buck to sniff it.

"Immediately she recalled that Carol, as a child, had the same unusual odor," Centerwall later remembered.

Buck did not write anything about Centerwall's visit, about smelling an odor that took her back forty years to Nanking, to the bamboo garden where she watched her daughter play. We can't know what it was like for Buck to suddenly learn that this odor had actually been a signal. It might have told her about her own genetic makeup, about a rare genetic variant that she had inherited from her mother or her father, a variant that Lossing had also acquired from his own ancestry, which they had combined in their child. We don't know what it was like to discover all this at the very point when children with PKU could at least be treated, to learn that every meal she made for Carol was unwitting poison.

What little we do know comes from her other daughter, Janice. In 1992 Janice recalled that Pearl "had trouble accepting that her family's genes may have contributed to this disorder."

In the 1960s, as the first generation of PKU children got to grow up with healthy brains, life went on for Carol and Pearl much as it had for decades. Every December, Pearl wrote a letter to the Vineland Training School, with a list of gifts to be purchased for Carol, who was now in her forties. Crayons and coloring books, beads, glazed fruit, candy, doll blankets, a musical top. The list didn't change from one Christmas to the next.

In 1972, Pearl paid her last visit to Carol. She had been diagnosed with lung cancer, and her treatments would keep her alive only a few months after the diagnosis. Carol Buck outlived her mother by another twenty years. She, too, was diagnosed with lung cancer. She died at age seventy-two in 1992, and was buried on the grounds of the Vineland Training School, across the street from Emma Wolverton's grave. Neither Carol nor her mother smoked, raising the possibility that they shared a different mutation that raised their risk of the disease.

PKU is rare, but its story has been told many more times than far more common disorders. It has a powerful moral, but the moral depends on who tells the story.

For some, the story of PKU embodies the triumph of genetics. Mendel's early followers had been mocked by those who couldn't believe that experiments on peas could account for why like engenders like. Mendel's work opened the way to the discovery of genes, and now scientists were finding precise effects of genes on health. Genetics not only explained how PKU arose but allowed doctors to tame it.

In the mid-1980s, a gigantic project took shape that would allow future generations of researchers to quickly find the mutations behind any hereditary disease. Rather than examine the DNA of a single gene, they wanted to sequence every bit of DNA in all forty-six human chromosomes—the entire human genome. "The possession of a genetic map and the DNA sequence of a human being will transform medicine," promised the Nobel Prize–winning biologist Walter Gilbert.

To show how this transformation would happen, Francis Collins, the director of the National Center for Human Genome Research at the time, offered the PKU story. Scientists found the inherited flaw and then devised a rational treatment for it. "If you simply remove foods with phenylalanine from the child's diet, he or she will live a normal and healthy life," Collins declared. Sequencing the entire human genome would make it possible for scientists to pinpoint mutations that caused thousands of other diseases, and potentially open the way to treatments for them as well. "PKU is the example where the paradigm was proven," Collins said.

To other scientists, however, PKU demonstrated the deep flaws in such gene-centered research. From the earliest days of genetics, researchers recognized that it was a fallacy to talk about a gene being "for" a trait or a disease. Genes don't have so much power. They exist in an environment, and their effects may be very different in different surroundings. Thomas Hunt Morgan, for example, had observed how a mutation in his flies made them sprout extra legs—but only in cold temperatures.

Once researchers discovered a diet for PKU, it became an even better illustration of the malleability of genes. In 1972, the British biologist Steven Rose declared that PKU demonstrated how pointless it was to talk about something like a "high I.Q." gene. A variant of the PAH gene could lead to low intelligence test scores if a child was left untreated. Or the same child could score in the normal range if given the right diet.

"Hence the environment has 'triumphed' over the genetic deficiency of the individual," Rose said. "To talk of 'high I.Q. genes,' or to try to disentangle the genetic programme from the environment in which it is expressed is both disingenuous and misleading."

No matter which moral people drew from PKU, their stories had one thing in common: Science had triumphed utterly over the disease. In 1995, the journalist Robert Wright told his own PKU story as a way to attack the idea that our intelligence is fixed by the genes we inherit. In the absence of any treatment, Wright observed, PKU mutations will reliably cause children to have devastating intellectual disabilities. "It turns out," he cheerfully wrote, "that if you put all infants on a diet low in the amino acid phenylalanine, the disease disappears."

It should come as no surprise that neither Wright, Rose, nor Collins themselves had PKU, or ever had to care for a child with it. Even with the most sophisticated diets and supplements medicine can offer, PKU never disappears. Starting in the 1950s, children with PKU began to escape the devastation of brain damage, but only if they stuck relentlessly to the dreary regimen of foul-tasting concoctions. Over the years, the PKU foods became tastier, but children growing up on a low-phenylalanine diet still had to watch their friends gorge themselves on pizza and ice cream, sometimes ending up feeling isolated from society.

When the first generation of children with PKU grew up, doctors allowed them as adults to switch to a regular diet. They soon suffered a new round of symptoms as the phenylalanine surged back into their bodies. Now people with PKU are urged to stay on the diet for their entire lives. It's often a struggle to get the right balance of nutrients each day while avoiding even the slightest trace of phenylalanine. For now, the experience of the disease is a tense negotiation between heredity and the world in which it unfolds.

PART II

Wayward DNA

CHAPTER 5

An Evening's Revelry

I N 1901, WILLIAM BATESON sent an urgent report to the Royal Society on "the facts of heredity." Those facts, Bateson explained, had just been thrown into sharp relief with the rediscovered, newly appreciated work of Gregor Mendel. Bateson and other scientists were confirming the patterns that Mendel had observed. Those patterns were so trustworthy and so profound, Bateson said, that they deserved one of the loftiest titles in science: "Mendel's Law."

A scientific law predicts some aspect of the universe, usually with a short, sweet equation. Isaac Newton discovered the laws of motion that came to bear his name. Robert Boyle is memorialized with Boyle's law, which predicts the pressure of a gas from its volume. Mendel's work likewise gave heredity a numerical clarity. Parents have a fifty-fifty chance of passing down either of their two copies of a given gene. Mendel's Law ensures a three-to-one ratio between dominant and recessive traits. It doesn't matter if the trait is a wrinkled coat on a pea or PKU in humans. The numbers stay the same.

Mendel's discovery was indeed one of the most important in the history of science. But the patterns he saw aren't really a law. Newton's laws of motion are as true in a distant galaxy as they are here on Earth. They were as true thirteen billion years ago in the universe's infancy as they are true today. Mendel's Law has far narrower boundaries. It is only relevant to places where life exists—in other words, as far as we know, only on Earth. Even when life first emerged some four billion years ago as single-celled microbes, Mendel's Law did not yet exist. Microbes are not like pea plants or people, and, as a result, they don't have dominant or recessive characters.

Mendel's Law would have to wait for a couple billion years or so for a new lineage of life to emerge—one that would give rise to plants, fungi, and animals like us. Mendel's Law, in other words, is less like Boyle's law than it is like our spleens or our retinas: It emerged as life evolved. Earth is actually home to many different kinds of heredity, each arising through a combination of natural selection and lucky flukes.

———

L ife likely emerged as its early, simple chemistry got complicated. Amino acids, bases, and other molecular building blocks were present on the early Earth. Short chains of these compounds may have concentrated together, perhaps trapped in oily films on the seafloor or encased in cell-like bubbles. Crowded into these cramped spaces, their chemistry may have accelerated, pushing them over the border dividing nonlife from life.

It's likely that the first life-forms were profoundly unlike life today. Today, animals, plants, and bacteria—all cellular life, in fact—encode their genetic information in DNA. But DNA would be an unlikely candidate for the first hereditary molecule, because it's both helpless and demanding.

In order for a cell to read the information stored in its DNA, it needs to deploy many proteins and RNA molecules at once. When a cell divides, it needs another army of molecules to make a second copy of its DNA. The first life on Earth must have had a simpler beginning.

One possibility is that life started out without DNA or proteins. Instead, it relied solely on RNA molecules. A primordial cell might have contained a few different types of short RNA molecules that helped each other replicate.

Experiments with RNA molecules suggest how this would have unfolded. One RNA molecule might grab bases and weld them together, using a second RNA molecule as a template. That second molecule might do the same for a third. If the last RNA gene in the line turned around and helped the first one, the entire circle could feed back on itself. These primordial RNA molecules would have had a twofold form of heredity: They inherited the genetic information from their ancestor, and also the same twisted shape that allowed them to help build new molecules.

This first heredity would have also been sloppy. Sometimes a new RNA molecule would turn out to be slightly different from its template. This error would often be fatal, making it impossible for the RNA molecules to copy themselves any further. But in a few cases, it would actually improve the chemistry. Faster-replicating cells would have outcompeted their slower rivals.

RNA-based life, living in an ocean or a tide pool, may well have lived amidst loose amino acids. As their RNA molecules evolved into more sophisticated forms, some of them may have begun connecting amino acids into short chains, called peptides. These peptides may have been able to do jobs of their own inside of cells. And with time the peptides may have grown into large, complex proteins.

It's also possible that some RNA-based life evolved to make DNA as well. The double-stranded DNA molecules would have proven more stable than single-stranded RNA, and also less prone to damage. When the early DNA-based organisms copied their genes, they made fewer mistakes. Their newfound accuracy could have opened the way for more complexity in life, since they had a lower risk of ending up with a life-stopping mutation.

Once DNA-based life took hold, it overran the planet. By about 3.5 billion years ago, these single-celled microbes had diverged into two great evolutionary branches, known as bacteria and archaea. They're impossible to tell apart under a microscope, but they have some important differences in their biochemistry. Bacteria and archaea use different molecules to build their cell walls, for example, and use different molecules to read their genes.

But both lineages of microbes proved astonishingly versatile, adapting to just about every bit of Earth where they could get water and energy. Microbes adapted to grow on the sea surface, catching sunlight; on the seafloor, where they ate sulfur and iron; deep in the Earth, where they harnessed the energy of radioactivity. Scientists estimate that Earth is home to about a million billion billion microbes, which may belong to a trillion different species.

But none of them follow Mendel's Law.

A typical microbe—say, the *Escherichia coli* dwelling in your gut—has only a single chromosome: a long circle of DNA. Arrayed along that loop are several thousand genes. If *E. coli* can draw in some glucose or another sugar from your breakfast, it can grow until it's ready to replicate. It elegantly unwinds the two strands of the circle. Onto each strand, the cell builds a second one, creating two nearly identical chromosomes. The cell then cuts itself in two. It drags its two chromosomes to opposite sides and then builds a wall down its middle. Each of the new microbes is a near-perfect copy of its ancestor, inheriting a chromosome as well as about half of the molecules in the ancestor cell.

We humans can have the opportunity to get to know our parents. For microbes, that chance never comes, because their ancestors vanish—or, to put it another way, split into their daughter cells. Mendel's Law describes how hereditary factors from two parents combine to produce an organism. To a microbe, it's meaningless.

The heredity of microbes is different from ours in another important way: They can inherit genes along many different routes. They can gain genes as we do, as copies of the genes of their direct ancestors. This process is known as vertical inheritance. But they can also inherit genes from unrelated microbes, through horizontal inheritance.

Horizontal inheritance helped scientists discover what genes were made of. In the 1920s, researchers discovered that if they killed deadly strains of bacteria and mixed them with harmless ones, the harmless strains turned deadly. What's more, when the transformed bacteria divided, their descendants inherited their deadliness. Later, a microbiologist named Oswald Avery and his colleagues isolated the different kinds of molecules inside bacteria to figure out which was the mysterious "transforming principle." Through many rounds of experiments, they came down in favor of DNA.

It turned out that the bacteria Avery studied were being transformed by taking up loose DNA and incorporating some of it into their own chromosome. They gained genes that they could use to make their hosts sick. But later research has revealed that horizontal inheritance can also take place by other means. Along with their main chromosomes, for example, microbes often carry ringlets of DNA, called plasmids, with genes of their own. Microbes will sometimes build tubes that they stick into other microbes, pumping in their plasmids. The plasmid may then float in its new host, or it may paste itself into the chromosome.

Horizontal inheritance may seem bizarre, but it happens all around us. It even happens inside us. An experiment carried out in 2004 by a team of Danish scientists showed how a species called *Enterococcus faecium* horizontally inherits DNA within our own bodies. Over the past few thousand years, the species has evolved into strains that colonize the human gut and skin, and others that prefer living in animals. Most strains of *E. faecium* are harmless, but some can cause potentially fatal infections in the blood and bladder.

The standard treatment for an *E. faecium* infection is a dose of antibiotics. There was a time when that treatment always got the job done. But by the early 2000s, *E. faecium* had evolved into a medical nightmare. More and more often, doctors found that the bacteria carried genes that allowed them to resist drugs. When a resistant strain takes hold in a patient, the bacteria multiply without check, passing down their resistance genes vertically to their descendants.

In 2004, half a dozen brave souls agreed to drink two cups of milk. In the first cup were a billion *Enterococcus faecium*. They belonged to a strain isolated from humans and could be easily killed by an antibiotic called vancomycin. Three hours later, the six volunteers drank a second cup containing another billion *E. faecium* that came from chickens, carrying a gene that made them resistant to vancomycin.

This milk drinking was part of an experiment at Denmark's National Center for Antimicrobials and Infection Control. Over the following month, the Danish scientists collected stool samples from the six subjects and surveyed them for the two strains of E. faecium. The chicken strain quickly became rare and then disappeared after a few days. The human strain, better adapted to its new home, lasted longer.

But in three of the six subjects, the scientists found that the human strain had changed. Now each generation of bacteria passed down a new gene they didn't have at the beginning of the experiment. They inherited the chicken strain's gene for vancomycin resistance.

Microbes can even inherit genes horizontally from their greatest enemies: viruses. Viruses—protein shells containing genes—have a form of heredity distinct from that of cellular life. A virus does not reproduce by copying its own genes and dividing in two. Instead, it invades a host cell. A virus that attacks bacteria—known as a bacteriophage—typically lands on the cell wall of a host and injects its DNA inside, like shooting a piece of spaghetti out of a syringe. Bacteria have several ways of recognizing this DNA and destroying it. But none of them are foolproof. If the virus's genes survive long enough, they commandeer the cell. The cell makes proteins from some of the virus's genes, which then drive the cell to make new viruses, complete with new copies of the original virus's genes.

When it comes to viruses, heredity is almost an abstraction. They have no material bond to their ancestors, since all the atoms in a new virus come from the host cell where it formed. For viruses, heredity is an invisible thread of information joining one virus to its progeny.

As viral genes get packaged into new viruses, sometimes things go awry. A gene from their microbial host may get swept up inside a viral shell. The new viruses that leave the microbe carry that host gene with their own, and they may later inject it into a new host. In some cases, the microbial gene may end up in its new host's chromosome. Viruses can thus act like accidental ferries, transporting microbial genes from one organism to another— sometimes even moving them between species.

As scientists examine microbes more closely, they have discovered still more strange forms of heredity. One particulary weird kind of microbial inheritance came to light in the early 2000s as scientists were investigating how bacteria fight against viruses.

It turns out that when many species of microbes are exposed to a new virus, they can learn how to stage a swift, precise attack against it. Vertebrate animals like ourselves have the same capacity. When we get attacked by influenza or a cold virus, our immune system can build antibodies that will wipe these strains out as soon as they try to attack us again. Bacteria can't use an immune system made up of billions of cells—each microbe is one cell that has to fend for itself. But they manage this feat all on their own, using a system of molecules called CRISPR-Cas.

When viruses infect bacteria, they typically land on their victim and inject a string of DNA inside. Many microbes can chop off the tip of this incoming DNA and insert it into a stretch of its own DNA, called a CRISPR region. (*CRISPR* is short for *clustered regularly interspaced short palindromic repeats*.)

If microbes manage to survive this initial attack from a virus, they will be equipped to resist the next one. They prepare for a subsequent infection by building a short RNA molecule that matches the bit of viral DNA they grabbed from the first attack. A protein called a Cas enzyme folds itself around the RNA molecule, and the two together float off through the cell.

If the same strain of virus tries to inject its DNA into the microbe, the CRISPR-Cas system latches onto the incoming genes. The Cas enzyme pulls the viral DNA strands apart and chops them into pieces. Shredded into harmless debris, the virus cannot take over the microbe.

As a microbe battles virus after virus, it may store away samples from a dozen of its enemies. And when it divides, it passes down this accumulated knowledge to its descendants. When the microbe copies its chromosome, it copies its CRISPR region along with the rest of its DNA. August Weismann's germ line barrier may prevent the experiences of animals from altering their germ cells. But for bacteria, no such barrier exists. In a sense, soma and germ are bound up in a single cell.

Some scientists have argued that CRISPR is a genuine case of Lamarckian heredity. Of course, virus-fighting bacteria are a far cry from the leaf-plucking giraffes of Lamarck's imagination, and so the question can descend into a squabble over semantics. What's indisputably clear, however, is with CRISPR scientists have found yet another channel of heredity beyond Mendel's Law.

———

A bout 1.8 billion years ago, a new form of life evolved on Earth. Its cells were much larger than those of bacteria and archaea. Its DNA was tucked with exquisite care inside a pouch called the nucleus. It generated abundant amounts of fuel in special pods called mitochondria. Among the many forms this new kind of life would take would be our own.

These microbial monsters were eukaryotes. Their descendants would give rise to protozoans, the predators of the microbial world that hunt through soil and sea for single-celled prey. Eukaryotes evolved into all multicellular life on Earth as well, including fungi, plants, and animals like us. Along with their nucleus and large size, eukaryotes share many other traits that bacteria and archaea lack. But one of those traits matters most to heredity: Eukaryotes pass down their genes to their offspring in a unique way, one that allowed Mendel's Law to emerge.

While bacteria and archaea have a single chromosome, eukaryotes carry pairs of them. Different species have different numbers of pairs. We humans have 23 pairs, but pea plants have only 7. Yeast have 16. Some butterflies have 134.

When our somatic cells divide, they copy all their chromosomes, creating an extra pair for each one. They tear down the nucleus, pull half the chromosomes to each side, and split themselves down the middle. Each new cell now has its own 23 pairs. This kind of division—called mitosis—is fundamentally similar to what bacteria do: turn one cell into two identical cells.

Our bodies use mitosis to grow and rejuvenate themselves. But in order to make germ cells, we have to make sperm or eggs that have only one set of chromosomes rather than a pair. The simplest way to make sperm and eggs would be to simply pull apart the pairs of chromosomes in a somatic cell and allot one set to each germ cell. But our bodies do not do that. Instead, they indulge in a process called meiosis, which is laughably baroque.

In men, meiosis takes place within a labyrinth of tubes coiled within the testicles. The tube walls are lined with sperm precursor cells, each carrying two copies of each chromosome—one from the man's mother, the other from his father. When these cells divide, they copy all their DNA, so that now they have four copies of each chromosome. Rather than drawing apart from each other, however, the chromosomes stay together. A maternal and paternal copy of each chromosome line up alongside each other. Proteins descend on them and slice the chromosomes, making cuts at precisely the same spots.

As the cell repairs these self-inflicted wounds, a remarkable exchange can take place. A piece of DNA from one chromosome may get moved to the same position in the other, its own place taken by its counterpart. This molecular surgery cannot be rushed. All told, a cell may need three weeks to finish meiosis. Once it's done, its chromosomes pull away from each other. The cell then divides twice, to make four new sperm cells. Each of the four cells inherits a single copy of all twenty-three chromosomes. But each sperm cell contains a different assembly of DNA.

One source of this difference comes from how the pairs of chromosomes get separated. A sperm might contain the version of chromosome 1 that a man inherited from his father, chromosome 2 from his mother, and so on. Another sperm might have a different combination. At the same time, some chromosomes in a sperm are hybrids. Thanks to meiosis, a sperm cell's copy of chromosome 1 might be a combination of DNA from both his mother and father.

The basic biology of meiosis is the same inside a woman's body, but the timing is very different. The first steps take place while she's still an embryo in her mother's womb. A group of cells inside a female embryo take on a new identity as egg precursors, moving together to where the ovaries will later develop. When the embryo is seven months old, the precursor cells begin meiosis, doubling their chromosomes, pairing some of them together, and exchanging some pieces of DNA. But the chromosomes then freeze in place midway through meiosis. They stay that way for years, until girls reach adolescence and start to ovulate.

During each ovulatory cycle, a single egg precursor turns on its meiosis and completes the cycle. As with sperm, a woman's meiosis produces four new cells, each with only twenty-three chromosomes. But only one of those cells matures into an egg. The other three cells wither down to vestiges, known as polar bodies.

Scientists can now see how meiosis drove the patterns that Mendel observed in his garden. When Mendel crossed tall and short pea plants, for example, he grew hybrids that were all tall. But when he crossed them, a quarter of the next generation turned out short again. Now scientists know the genes responsible for those differences. Known as LE, it makes a protein that triggers peas to grow. His short pea plants carried two copies of a mutant form of the LE gene. The LE proteins in these plants didn't work properly, stopping their growth. The hybrids had one working copy of LE, which was enough to grow normally.

When a hybrid pea plant matured, some of its cells went through meiosis before producing pollen and ovules. The cells duplicated their chromosomes, shuffled some genes from one chromosome to its partner, and then pulled them apart into four sets. It was a matter of chance whether a pollen grain ended up with the chromosome carrying the normal version of the LE gene or the mutant form. As a result, half the germ cells produced by each pea plant had each copy of the gene.

The biologist Laurence Hurst once wrote that meiosis takes place "in a manner reminiscent of drunkards returning from an evening's revelry: one step backwards, two steps forward." Yet this strange stumbling is also responsible for heredity's most elegant patterns.

———

Scientists first spotted chromosomes in the mid-1800s, but meiosis didn't come to light until decades later. In the early 1900s, a Belgian priest named Frans Alfons Janssens stained fertilized salamander eggs so that he could observe their chromosomes through a microscope. The stains captured them at different stages of meiosis, like frames from a movie. It looked to Janssens as if the chromosomes were intimately interacting with each other and then pulling apart.

In the brief report he published on his discovery in 1909, Janssens didn't try to draw any profound lessons about heredity. But he had a hunch it would turn out to be important. "Are we being presumptuous?" Janssens asked. "Time will tell."

It didn't take much time at all. While Janssens was peering at salamander cells in Belgium, Thomas Hunt Morgan was breeding white-eyed flies in New York. Morgan and his colleagues first discovered that the hereditary factor for red or white eyes was located on a chromosome. (Today, we'd say that the gene for eye color is a stretch of DNA on the chromosome.) Morgan's team also found another factor, which produced short wings on flies, on the same chromosome.

Because that chromosome happened to be the X, Morgan and his colleagues could study these factors by breeding flies. They took advantage of the fact that males have one X chromosome and one Y, while females have two X's. Morgan and his students used breeding to produce female flies with both white eyes and short wings. One of their X chromosomes carried the factor for white eyes in flies, while the other carried the one for short wings. Then the scientists bred these females with red-eyed males.

The sons of these female flies inherited only one X chromosome, all getting it from their mother. It was thus no surprise to the scientists that some of the sons had red eyes, while others had short wings. But Morgan and his students also found something extraordinary: A few sons ended up with white eyes *and* short wings. A few other sons developed red eyes and long wings. The X chromosomes of their mothers were trading hereditary factors, creating new combinations of traits.

In later studies, Morgan's team showed they could also take two factors sitting on the same chromosome and split them apart. They reared flies in which the same X chromosome carried the factors for short wings and a yellow body. Sons that inherited that particular chromosome from their mother developed both traits. When Morgan's team bred those flies, however, a fraction of sons ended up with yellow bodies but normal-size wings. Others had normal bodies with short wings.

Morgan didn't quite know how to make sense of these results at first. By good fortune, he happened to stumble across Janssen's report. He realized that Janssen had unwittingly found the physical solution to his own experiments. Morgan and his colleagues quickly wrote up a new hypothesis that combined both sets of results. Each chromosome, they argued, carried a set of factors arrayed in a line like beads on a string. When female flies developed their eggs, their X chromosomes crossed over each other and traded segments.

The joining and splitting of traits happened only rarely, but Morgan and his students noticed that they occurred with striking regularity. A particular trait might get split from a second one in 1 percent of offspring. But it might get split from a third trait 2 percent of the time. Morgan's student Alfred Sturtevant realized that the reason for this puzzling pattern had to do with where genes sat on their chromosomes.

When chromosomes get broken into segments during meiosis, the genes that are close to each other tend to stay on the same segments. Distant genes are more likely to get separated. If someone starts ripping dictionaries apart at random places and handing you the pieces, you can bet that the chunk that contains *meiosis* will be more likely to contain *mitosis* than *chromosome*. Sturtevant's insight led the way to genetic maps, which marked how far apart genes were from each other. Heredity now gained a geography.

––––––––

Time and again, the principles of heredity that Morgan's group discovered in flies proved true in other species. Meiosis was no exception. We humans, along with other animals, also turned out to be the products of meiosis. The slimy kelp beating in the tides carry out meiosis, too, as do groves of bamboo clattering in the wind, and stinkhorns heaving out of the ground. While scientists have put forward a number of explanations for why meiosis evolved, one has gained a lot of evidence in recent years: Meiosis lets evolution do its job better.

Consider what meiosis does inside of one of Morgan's *Drosophila* flies. Like other flies, it has a collection of traits—let's say those traits include short wings, a strong immune response, and the ability to make lots of eggs. And let's say the genes for those three traits—one bad and two good—all sit on the same chromosome. Without meiosis, that fly would only be able to pass down its three alleles in one bundle, since the three genes all sit on the same chromosome. What's more, if any new harmful mutations arose on that chromosome in later generations, it would also get passed down along with the other alleles. Over the generations, the fly's descendants would sink under a burden of bad mutations.

Give the fly meiosis, and everything changes. Its descendants are no longer doomed to inherit a particular combination of alleles on each chromosome. Meiois shuffles the alleles into new combinations. Some of the fly's descendants may inherit the alleles for frail wings and a weak immune system. But meiosis also allows other descendants to end up with powerful wings and a strong immune system. These stronger flies can reproduce, and their offspring will sustain the population into future generations. The population of flies ends up with combinations of superior genetic variants, while many harmful mutations disappear into oblivion.

Michael Desai, a biologist at Harvard, tested this idea by staging a competition among yeast. He chose these single-celled fungi for their flexibility when it comes to reproducing. Yeast can either clone themselves or have sex. To clone itself, a yeast cell grows a bud that bulges from its cell wall. It copies its chromosomes and stuffs the new copies into the bud, which can then break off to become a cell of its own.

Sometimes, yeast have sex instead. The strain Desai studied exists in two so-called mating types, known as *a* and α. Each type releases a chemical that lures yeast of the other type. The *a* and α cells approach each other and fuse into one. The merged cell, which now contains a double set of chromosomes, can then multiply into new cells. But if it runs out of food, it responds by carrying out meiosis between its *a* and α chromosomes.

The yeast cell partners its chromosomes together and shuffles DNA. It then separates its chromosomes into two sets, each of which get stored inside a spore. Those tough-coated spores can drift away, taking their mixed-up genes to a better place where they may be able to grow again.

In his experiment, Desai allowed some of his yeast to have sex every ninety generations. The rest of the yeast could only clone themselves. Desai let the clones and the sexual yeast compete in test tubes for food. Sometimes new mutations arose that made a yeast cell do better than the rest of the population, allowing it to produce more offspring. For a thousand generations, Desai and his colleagues kept track of how each group of yeast fared in the evolutionary race.

The differences between yeast that could have sex and those that couldn't were clear. Sometimes a beneficial mutation would arise in the cloning yeast, letting them reproduce faster than the clones that lacked it. But along with that good mutation, the clones passed down bad mutations. The yeast that Desai allowed to have sex could separate good mutations from bad ones, thanks to meiosis. And when more good mutations emerged, meiosis was able to bring them together in new combinations, to produce even better yeast. At the end of the experiment, the yeast that could have sex had evolved to grow much faster than the clones.

———

This ancient shuffling is the answer to some of the most common questions about heredity. When Grace gave birth to our second daughter, Veronica, we watched her grow and wondered how much she would turn out like her older sister, Charlotte. After all, they had the same parents, meaning that they had inherited DNA from the same two genomes. They were raised in the same house, eating the same food. But Charlotte and Veronica turned out to be far from clones. Charlotte is luminously pale, with freckles, greenish eyes, and strawberry-blond hair. Veronica has a deeper tone to her skin and mahogany-colored irises. Charlotte grew to five foot six, a fairly average height. Veronica has always been off the charts, making people assume she's a couple of years older than she really is. As a child, Charlotte would hold back when we introduced her to new people, sizing them up. Veronica, standing next to her, would launch herself into the air and shout her name. At age twelve, Charlotte became obsessed with galaxies and dark matter. Veronica didn't care much what the universe is made of. She'd rather sing, or read Jane Austen.

The experiences our daughters have had probably account for some of their differences. But so does meiosis. Grace and I gave each of our children different combinations of the DNA we inherited from our own parents. The unique combination of alleles that each of our children ended up with had a unique influence on how she grew up.

Yet meiosis also works in strange ways that defy our intuitions. Parents pass down one copy of each chromosome to each child; which chromosome is inherited is a fifty-fifty matter of chance. The DNA in any pair of siblings, statistics would suggest, should be 50 percent genetically identical. Identical twins, by contrast, are 100 percent identical, because they are the product of a single fertilized egg. First cousins, who have only one set of grandparents in common, are on average 12.5 percent genetically identical.

All this is true—but only on average. It's just as true to say that if you roll a pair of dice, they'll turn up close to a seven. Yet of any particular roll may still turn up snake eyes. After meiosis shuffles DNA between chromosomes, it's possible for a woman's eggs to end up with more DNA from her father than her mother, or vice versa. Two siblings might arise from eggs that happen to have more DNA from their maternal grandmother than their maternal grandfather. The reverse may be true for other siblings. Meiosis can thus make two siblings more genetically similar to each other than to the rest of their siblings.

The ability to read DNA allowed scientists to measure this genetic similarity in real people. In 2006, Peter Visscher, a geneticist at the Queensland Institute of Medical Research in Australia, and his colleagues studied 4,401 pairs of siblings, examining several hundred genetic markers in each volunteer. The siblings often had a series of identical genetic markers along a chromosome—segments they inherited from one of their parents. On average, they found about half of the DNA in the siblings was made up of these identical stretches. But many of the siblings deviated from a perfect 50 percent. At the high end, the researchers found a pair of siblings who shared 61.7 percent of their DNA. At the low end was a pair of siblings who shared only 37.4 percent. Along the spectrum of inheritance, in other words, some of our siblings are more like our identical twins, others more like cousins.

———————

Once Mendel's so-called laws evolved in the first eukaryotes, they passed them down to their descendants. It has endured in most of the lineages even till today. Nearly two billion years later, tarantulas use meiosis to mix chromosomes and shuffle genes. So do hummingbirds, roses, and death cap mushrooms. But for all the enduring advantages that meiosis may offer, under the right circumstances it can fade and vanish.

In thousands of species of plants, for example, meiosis has crumbled away. Their ovules do not develop from precursor cells shuffling the DNA and then pulling apart pairs of chromosomes. Instead, these plants can produce ovules through a fairly ordinary division of cells. Mother cells with pairs of chromosomes produce daughter cells with precisely the same pairs.

Although these plants evolved a way to give up meiosis, they still cling to some vestiges of their history as sexual species. They can develop their ovules only if pollen grains settle on their flowers and deliver the right molecular signals. But all they need from the pollen are these signals. They make no use of the male DNA.

One of these odd plants happens to be hawkweed, the plant Mendel chose to study as a follow-up to peas. His peas reliably carried out meiosis, producing a three-to-one ratio of dominant and recessive traits. It was his bad luck to then pick hawkweed—a plant that had evolved away from that sort of heredity—to search for those same ratios. When Mendel painted pollen onto hawkweed flowers, he usually triggered them to make seeds containing an identical copy of their own DNA, and taking in none of the DNA from the pollen. Only after geneticists learned to trace the path of genes from one generation of hawkweed to the next did they realize Mendel's great misfortune.

Plants and other eukaryotes lose meiosis when the evolutionary benefits no longer outweigh the costs. In certain situations, organisms can reproduce more successfully if they simply duplicate their own DNA rather than combine them with the opposite sex and break apart the links between their genes.

But there are other ways for them to break Mendel's Law as well. Sometimes individual genes take over heredity for their own evolutionary benefit.

These molecular hackers first came to light in the 1920s with the discovery of flies with too many daughters. A Russian biologist named Sergey Gershenson went into a forest to trap a species of fly called *Drosophila obscura*. When he brought the flies back to the Institute of Experimental Biology in Moscow, he figured out how to keep them alive on a diet of fermented raisins, potatoes, and water. Some of the female flies he trapped were carrying fertilized eggs, which they then laid by the thousands. Gershenson picked some of their offspring to breed new lines he could study for inherited traits.

There was something peculiar about two of the lines, Gershenson noticed. Typically, a batch of eggs produced by *Drosophila obscura* contains an even balance of males and females. But in two of Gershenson's lines, the mothers tended to produce far more daughters than sons. Sometimes they had no sons at all. The ratios were so extreme, Gershenson said, "that it seemed impossible to explain them by accidental causes."

To find the true cause, Gershenson carried out a series of breeding experiments. The penchant for daughters could be passed down like a simple genetically encoded trait. Eventually, Gershenson figured out that it was determined by a gene on the X chromosome. But he couldn't understand how the gene tilted the balance away from sons and toward daughters. Whatever its particular trick might be, Gershenson realized it had slipped through a loophole in Mendel's Law.

Flies normally have a 50 percent chance of becoming male or female, because sperm have a 50 percent chance of acquiring an X or a Y chromosome. As a result, a normal gene on the X chromosome will end up in about half of a male fly's offspring. In Gershenson's flies, on the other hand, the math is different. If a male fly carried the mysterious mutation he discovered, most—or even all—of his offspring inherited his X chromosome. Few if any inherited his Y. Those flies could then pass the daughter-producing gene down to their own offspring. Overall, the odds of flies inheriting the daughter-favoring mutation would be much higher than 50 percent. As a result, it would become more common in a population.

"This," Gershenson concluded, "favors its extension."

———

At first, Gershenson's discovery might have seemed like an oddball exception to the rules of heredity. But it didn't take long for scientists to find other cases where genes were fixing Mendel's dice in their own favor. Collectively, these violations came to be known as gene drive. Gene drive is so powerful that it can spread a gene like an intergenerational epidemic, until it dominates an entire population. Today, the gene drive catalog has many entries, not just in flies, but in plants, fungi, mammals, and— perhaps—even humans.

In some cases, a gene drive spreads itself by encoding a toxin. Sperm cells carrying it make the toxin, which then spreads to other sperm. The other sperm die—unless they have the gene drive element, which also contains an antidote to the toxin. In other cases, the gene drive waits until male embryos start developing before switching on and killing them.

Gene drive can break Mendel's Law in females, too. When a precursor egg cell develops, it divides into four cells. One becomes the egg, while the other three become polar bodies—in other words, three reproductive dead ends. A given copy of a gene normally has a fifty-fifty chance of ending up in the egg rather than in a polar body. Some genes have evolved the ability to manipulate those odds. They're more likely to end up in the egg—and thus to get passed down to future generations of daughters.

With so much evidence for the power of gene drive among other eukaryotes, it stands to reason that we might be subject to it as well. But the evidence for cheating on Mendel is still unclear in humans. It's not surprising that it would be hard to study gene drive in our own species. Scientists can breed flies and fungi, inspecting every step of reproduction to catch gene drive in the act. When it comes to humans, geneticists must make the best of uncontrolled history.

The most obvious sign of a human gene drive would be a human version of what Gershenson observed: families full of daughters. But the relatively small size of human families makes it hard to know if such families are the result of gene drive. Just because I have two daughters doesn't mean Grace and I might not have had sons if we had ten children.

One way to search for it in humans is to step back from individual families and combine thousands of them into one big analysis. Even if each family is relatively small, they can add up to a horde of people big enough to let scientists distinguish between chance and drive. Some of those databases include genetic markers. It should be possible to find some genetic markers that are passed down from parents to children more often than you'd expect based on Mendel alone.

While the concept is sound, scientists are struggling to get a clear picture of gene drive in our species. A few promising genes have turned up in recent studies. But when scientists have tried to replicate those studies in other groups of people, they have not found an effect. It's possible that we need to wait for more accurate and detailed DNA sequences before scientists can find a clear sign of gene drive rampaging through our species.

It's also possible that our ancestors were besieged by gene drives, but overcame them. Gene drives are shortsighted in their victory. They can sweep quickly through a population, but in the process they can put a species at grave risk. If a gene kills off sperm with Y chromosomes, males can become dangerously rare in a population. More and more females never encounter a male in their life, and die without offspring. The population shrinks and then collapses. In some cases, it may take just a few dozen generations for a gene drive to push a population to extinction.

While gene drive extinctions can theoretically happen, no one has yet seen one unfold in the wild. Many gene drives may fall short of total oblivion because organisms evolve defenses against them. Animals and plants will sometimes evolve special RNA molecules that can interfere with gene drives, blocking the production of new proteins. Mutations may then disable gene drives, rendering the defenses no longer necessary. The genes for these defenses may mutate as well. Yet even after millions of years, the vestiges of these defenses can still be recognized.

It turns out our own genome is littered with relics of this conflict. Even if gene drives are not exploiting us today, they have been an important part of our history. And today we inherit the genetic scars of ancient struggle. What Mendel discovered was not a law so much as a battleground.

CHAPTER 6

The Sleeping Branches

I DOUBT MANY CHILDREN give much thought to meiosis. But there comes a point early in the life of all children when they realize that they weren't simply brought into existence by their parents. They get up on their toes and peer beyond their mother and father, back into their genealogical past. They realize that their parents have parents of their own, who have parents, too, and so on back along family branches that stretch over memory's horizon. They realize all those ancestors are part of the reason they are alive. They wonder what would have happened if one great-great-great-grandmother decided to turn down the marriage proposal of a great-great-great-grandfather. Somehow, through an improbable flow of heredity down merging streams, they all converged on one baffled child.

I can remember my own first bafflement. When I interrogated my parents about their ancestry, I was amazed at how quickly they ran out of answers. My father, who was born in 1944 in Newark, told me about his parents. William Zimmer had been a doctor, and Evelyn Rader a librarian. They were both Reform Jews and dedicated socialists who played Paul Robeson records around the house when my father was a boy. It took me years to notice those Robeson 78s tucked away on a shelf in my parents' house. Those licorice slabs are among the few points of contact I have ever had with my paternal grandparents. My grandfather died when my father was three, my grandmother the summer before he went to college. I never got to see them get into a Passover argument about politics with their son, who turned Republican in college and later became a congressman. When I pushed my father to tell me about my older ancestors, his genealogical knowledge sputtered out quickly, leaving me with a blurry origin story that put his ancestors somewhere in the neighborhood of Germany, or Ukraine, or somewhere in between.

My mother came from shiksa stock: a German-Irish Catholic mother, Marilou Pohl, and an English Protestant father, Harrison LeGrande Good-speed, Jr. Her parents met at a tennis match as teenagers growing up in Grand Rapids, Michigan. Things then moved fast, as they often did in the 1940s. My grandfather, whom everyone called Peter, converted to Catholi-cism, married Marilou, headed off to Germany to fight Nazis, and then re-turned a year later to his wife and daughter—my mother. They went on to have three more children, whom they raised in a world of optimistic little businesses, tidy bowling lanes, tipsy games of bridge, and endless rounds of golf. The first time my father stepped aboard a commercial airplane was to fly to Michigan to marry my mother at her parents' house in 1965. It must have felt like an alien planet to him. To the Goodspeeds, the twenty-one-year-old Jew from New Jersey may well have seemed like an extraterrestrial.

I was fortunate to know my mother's parents for several decades, but beyond them, the maternal line dims as well. On the Pohl side, my great-grandparents died in the fifties, leaving behind only vague tales of sadness and early death. Harrison Goodspeed's father, on the other hand, lived long enough to give me a purple toy car for an early birthday, and to puzzle me by disappearing from life. I never met my great-grandmother Dorothy Rankin. What little I know of her comes from two photographs. In one, she poses in a flapper dress and necklaces; on the back of the picture, she scrib-bled a note to someone back in Michigan, explaining how grand Paris is. In the other picture, she stands in a shady front yard next to my great-grandfather, cradling my grandfather. Dorothy Rankin died a few months after that picture was taken.

In her thirties, my mother began investigating our ancestors, leading us on trips to rub gravestones in old New England cemeteries. Seeing her interest in the family history, my great-grandfather decided she should inherit a book of his about the family, published in 1907. One day the old leather volume appeared on a shelf in our living room: *History of the Goodspeed Family, Profusely Illustrated, Being a Genealogical and Narrative Record Extending from 1380 to 1906, and Embracing Material Concerning the Family Collected During Eighteen Years of Research, Together with Maps, Plates, Charts, Etc.*

1380? I was binge-reading *Lord of the Rings* at the time. Seeing my genealogy pushed into the Middle Ages felt like gaining citizenship in Gondor. When my mother explained to me that the name Goodspeed came from the early English exclamation *Godspeed,* I saw knights bidding each other well as they rode off to fight orcs.

When I got around to dipping into *History of the Goodspeed Family,* my medieval ancestors didn't live up to my hopes. The Goodspeeds first appear in the historical record in 1380, when one John Godsped was sued "touching a trespass." In 1385, another Goodspeed failed to pay a debt. In 1396, Robert Godsped killed a man named John Archebaud, but was pardoned "out of regard for Good Friday."

The author of *History of the Goodspeed Family,* a distant cousin named Weston Arthur Goodspeed, downplayed our family's criminal debut. "All of these offences, except the one causing the death of John Archebaud, were trivial and would have no standing in the courts of today, except in civil suits," Weston sniffed. I could imagine him then giving a careless shrug, adding, "Besides, does anybody *really* miss John Archebaud?"

In his eighteen years of research, Weston Goodspeed looked hard for signs of nobility. He found nothing. "A thorough examination of the English books on peerage fails to reveal the name Goodspeed," he admitted. "To those of our great family who will regard this as a serious social blow, the author of this volume extends his profound pity, sympathy and commiseration."

But what did that really matter, Weston asked, since all of the coats of arms displayed by American families were fake? "Some were fictitious or fraudulent," he declared. "Some were even ludicrous in their pretensions."

The Goodspeeds should be proud of their humble origins, of the fact that the original Goodspeed who came to America, my great-great-great-great-great-great-great-great-great-grandfather Roger Goodspeed, was just a yeoman. "The undoubted respectability and sterling qualities of the English yeomanry may be considered in democratic America as far superior to a coat of arms thus bought and unearned," Weston declared.

Roger Goodspeed was born in 1615 in Wingrave, England, and sailed to Massachusetts in his early twenties. There's no evidence that he took the journey as a Puritan fleeing persecution. He "merely wished like thousands of others to improve his surroundings and America seemed to offer the best opportunity," Weston wrote. Roger Goodspeed's name first pops up in historical records in 1639 as one of the first farmers to settle in Barnstable, a town on Cape Cod. A decade later, he built a new farmhouse a few miles away on the bank of the Herring River, which came to be known as Goodspeed's River. There he lived till his death in 1685. Over the course of his entire life, Roger Goodspeed made only a few ripples in written history: accusing a neighbor of stealing a goat, signing his will with a single letter, *R*.

Roger Goodspeed had three daughters and four sons. They inherited his DNA and his name. Later, they also inherited his bridles and saddles, his trenchers and his spinning wheel. They bore him twenty-two grandchildren, and in later generations his descendants spread through the colonies and then across the United States. About 250 years after Roger Goodspeed's arrival in Massachusetts, Weston Goodspeed started gathering information about his descendants, writing letters to relatives and searching archives, and eventually amassed biographical details on 2,429 Goodspeeds.

History of the Goodspeed Family ended up stretching to 561 pages. But Weston didn't treat it as the last word. It was supposed to be the opening salvo of a long campaign. Weston cataloged only American Goodspeeds through the male line; he promised to add the female branches in a future edition. He even dreamed the book would inspire yearly Goodspeed conventions. "It is the intention to call the first general assembly of the Goodspeeds," he declared, "for the purpose of effecting an organization which thereafter, it is hoped, will be permanent, will hold annual meetings, will continue the publication of these records in the future, and will take any other steps that shall be in the interest of the family and agreeable to all."

The Goodspeed meetings never came to pass, nor did Weston ever expand the family tree. The scraps of information that survive about Weston suggest a life marinated in disappointment. He worked at a small publishing company run by his brothers until it shut down near the end of the 1800s. The 1900 census listed Weston Goodspeed as unmarried and unemployed at age forty-eight. Seven years later he published *History of the Goodspeed Family*, and by 1910 the census showed he had moved to a Chicago boardinghouse run by a widow. Weston died in 1926, at age seventy-four, without producing a new volume of his genealogy, not to mention any heir to the Goodspeed name.

I still sometimes take the *History of the Goodspeed Family* down from the shelf on visits back home. Scanning its parade of wills, court records, and inventories of children, I puzzle about the genealogical drive that propelled its creation, the force that made Weston spend a large fraction of his time on Earth building a catalog of 2,429 people—people who were mostly unaware of one another.

Weston left a clue at the beginning of his book. He dedicated it "to the rapid, symmetrical and beautiful growth of the family tree; to the avoidance of all wind-storms likely to damage the orchard; to the eradication of the insects of ignorance and immorality certain to contaminate the fruit; to the transplantation of buds and scions in all agreeable soils; to the awakening of the sleeping branches to bright foliage and sweet blossoms; and to plenteous harvests of golden children grown in the sunshine of love, liberty and law."

Weston saw himself as a naturalist, in other words. He was describing an organism that extended itself seamlessly through the United States—a tree of heredity that sprouted from Roger Goodspeed, the Adam to all American Goodspeeds.

Yet Weston didn't do a very good job of showing what, if anything, binds the branches of the Goodspeed tree together, what made that tree a thing worth documenting in such painstaking detail. The Goodspeeds had no crown to pass down from king to prince, realigning the world along the way. We're not Rockefellers, with a vast fortune carried down through generations. In all honesty, American history would not have been any different if Roger Goodspeed's ship sank halfway across the Atlantic.

As far as I can tell, Weston believed what bound the Goodspeed family together, what he believed was inherited by every new generation, was goodness. A number of Goodspeed men fought in the Civil War—not as generals or colonels, granted, but as valiant Union Army soldiers nonetheless. "The splendid military record of these men will ever be a heritage of pride and glory for all who bear the family name," Weston declared. Of course, it would be hard to find a family in the United States in the 1860s that didn't send some sons to war. Having never served in the military myself, I don't see how I'm entitled to bask in that heritage of Civil War bravery.

Most Goodspeeds didn't fight in wars, but Weston still found some goodness in them as well. Francis Goodspeed, Weston wrote, "even as a boy was broad-minded and loved his books." John F. Goodspeed "was engaged in the furniture business; he devised 'Goodspeed's Superior Polish.'" Seymour Goodspeed "has accumulated a comfortable competence, reared a large family to correct and useful lives, is passing a clean and honorable career, and has the respect of all who know him." Thomas Goodspeed "has never failed to vote at State and National elections." Of one family of Goodspeeds, Weston simply noted, "All became good citizens."

Not long ago I discovered that Google put *History of the Goodspeed Family* online. I decided to play a game, seeing if I could find any keywords of a scandal. I tried *murder, bribery, illegitimate, alcohol.* I have yet to win. At best, I can only find faint shadows cast across the inherent goodness of Goodspeeds. Riland Goodspeed, born in 1841, became the manager of a California ranch—an "immense and beautiful ranch," of course. Eventually he fell in love with the owner's daughter—"a gifted and most fascinating woman," of course. Then Cousin Weston gets cryptic. Riland and his wife got married "under romantic circumstances and after several notable escapades." As for the rest of the marriage, Weston simply noted that "after many years they were divorced, largely upon whimsical grounds."

Compare the unblemished saga of the Goodspeeds with *The Kallikak Family*, which Henry Goddard published only five years later. They're both quintessentially American expressions of our beliefs about heredity. Goddard envisioned a pure line of crime and feeblemindedness. Weston presented a pedigree of middling Protestant prosperity. While Goddard envisioned some Mendelian factor poisoning the Wolvertons, Weston Goodspeed seems to have believed that the Goodspeeds inherited a moral factor, perhaps acquiring it from their parents among the lessons they got about democracy and furniture polish.

The American obsession with genealogy was caused by a case of transoceanic amnesia. Roger Goodspeed, born and raised in seventeenth-century England, was steeped in traditional European customs for remembering ancestors. The Bible's genealogies linked Jesus by blood back to the Old Testament patriarchs. Kings and noblemen justified their power with hereditary chains linking them back to the mythic past. William the Conqueror's genealogy reached all the way back to the warriors of ancient Troy.

By the Renaissance, rich merchants were hiring genealogists, too, in order to track their investments and determine how to wed their children so as to keep the wealth within the family. A yeoman like Roger Goodspeed couldn't afford to hire a professional London genealogist. Judging from the *R* he wrote for his signature, he probably couldn't have read a genealogist's report anyway. Nevertheless, Roger probably carried family stories in his mind from England to America, where he transmitted them to his children; they told his grandchildren in turn.

Roger Goodspeed's stories probably all took place within a few miles of his birthplace in Wingrave, since people in earlier generations had rarely moved far from their home villages. To travel more than three thousand miles from home in the 1630s, as Roger did, was a radical dislocation. It dropped the Atlantic between him and the wellspring of his stories. In later years, as the Goodspeed family tree branched across the colonies, those old stories grew blurry. Cousins were forgotten, myths took over.

By the 1700s, some American families were already trying to anchor their genealogy back to Europe. In 1771, Thomas Jefferson wrote to an acquaintance preparing to sail for London, asking if he could research the Jefferson coat of arms. "I have what I have been told were the family arms, but on what authority I know not," Jefferson complained. Another Founding Father, Benjamin Franklin, traveled in 1758 to the English village of Ecton, where the Franklin family had lived for centuries. Determined to uncover his genealogy, he perused the parish registers, inspected the moss-covered gravestones of his ancestors, and chatted with the rector's wife about the Franklin family. The rector later sent him a hand-drawn family tree stretching back to 1563.

"I am the youngest Son of the youngest Son of the youngest Son of the youngest Son for five Generations," Franklin wrote to a cousin, "whereby I find that had there originally been any Estate in the Family none could have stood a worse Chance of it." Yet Franklin also came away from his research convinced that he inherited the temperament of his ancestors, "for which double Blessing I desire to be ever thankful."

Franklin and Jefferson helped forge a new country that rejected the ancient power of heredity. "One of the strongest natural proofs of the folly of the hereditary rights of kings is that nature disproves it," Thomas Paine declared in *Common Sense*. Kings often turned out unfit to rule, Paine observed, as if nature produced an ass instead of a lion.

Yet the Revolutionary War did not destroy heredity's allure. Old colonial families tried to cling to their high status in the new republic by flaunting their European origins. They put coats of arms on their silverware, their hearses, and their gravestones. Newly rich bourgeois families used genealogy to buy some respectability of their own. Some spent time and money doing research or hiring one of America's new professional genealogists to do their work for them, uncovering connections to aristocracy and supplying coats of arms, even if their newfound heraldry often turned out to be fake.

Families of lesser means kept track of their families as well, sewing needlepoint genealogical trees and writing names in family Bibles. If they couldn't prove they inherited noble blood, at the very least they could feel some pride in virtuous blood. In the early 1800s, a Massachusetts woman named Electa Fidelia Jones investigated her roots, celebrating the Puritan blood that ran through her like a "magnetic wire," vibrating two centuries later with a message for anyone who could appreciate it. She was thrilled to discover some of her fourth cousins through her research; the find was a better inheritance than any ancient fortune, she said.

But other kin did not please Jones. She uncovered a female relative and her husband from the 1750s who were "so near idiocy that it was said at the time of their marriage that laws *ought* to be enacted to prevent the marriage of those so unfit to sustain the relations which they assumed." Among the children this unfit couple had, Jones complained, some were "so low in the scale of being that I do not wish to make their acquaintance so far as to ask after their name & age."

As she drew her family tree, Jones left those branches hidden. Undistracted by her disreputable kin, she could spend her time dreaming of visiting her Puritan ancestors. "I love to go back in imagination to those old firesides," she said.

While Americans eliminated embarrassing relatives from their genealogies, they also tried to link themselves to famous figures. John Randolph, an early US senator from Virginia, boasted that he was a direct descendant of Pocahontas. Shortly before he died in 1833, he regaled a visitor with a detailed account of his genealogy that took him all the way back to William the Conqueror. Tracing his ancestry to a king didn't mean Randolph could inherit the throne of England. But it did let him enjoy a little rubbed-off glory.

Randolph's obsession endures today. Every April, a few dozen people gather in a Washington, DC, club for the annual dinner and meeting of the Order of the Crown of Charlemagne in the United States of America. To be invited to dinner, people must prove that they are direct descendants of the eighth-century ruler of the Holy Roman Empire. To make the task easier for Charlemagne's descendants, the order will be satisfied if you can just link your genealogy to someone on their list of "Gateway Ancestors," such as James Claypoole of Philadelphia and Agatha Wormeley of Virginia. On its website, charlemagne.org, the order declares that its objective is "to maintain and promote the traditions of chivalry and knighthood."

By the mid-1800s, the search for celebrity, nobility, and virtue had turned American genealogy into a full-blown industry. Guilds formed, publishing official journals of their research. Ralph Waldo Emerson found the new enterprise decidedly un-American. It was a turn to the past in a country that should have been looking toward the future.

"When I talk to a genealogist," Emerson wrote in his journal in 1855, "I seem to sit up with a corpse."

———————

S ome of the ships that sailed into Massachusetts Bay in the 1630s were delivering settlers from England, including my own ancestor Roger Goodspeed. But in 1638 a ship called the *Desire* arrived from the West Indies carrying passengers from another land. The governor of Massachusetts, John Winthrop, recorded the ship's contents: "some cotton and tobacco, and negroes, etc."

The *Desire* delivered the first recorded shipment of African slaves to New England. Unlike Roger Goodspeed, the men and women stowed in the *Desire* would not pass down their goods to their children, or even their names. In their new home, American slaves sustained their genealogies as best they could by telling their children about their ancestors, but much was lost. The abolitionist Frederick Douglass, born in 1818, lived his first seven years with his maternal grandparents. He never learned much more about his ancestry. Without any records of families, marriages, births, or deaths, such knowledge was impossible to gain.

"Genealogical trees do not flourish among slaves," Douglass later wrote.

As some slaves gained their freedom, they began to sketch trees. Henry Highland Garnet's family escaped from slavery when he was nine; he went on to become a prominent abolitionist minister and served as US minister to Liberia. Garnet's ancestors had been slaves for generations, but he once said that "his great grandfather was the son of an African Chief, stolen from his native country in his youth and sold into Slavery on the shores of Maryland."

Garnet was part of an elite layer of nineteenth-century African American society, made up of college-educated professionals—ministers, doctors, government workers—who developed an interest in genealogy as keen as that of their white counterparts. They also used it to celebrate their superiority. The poet Langston Hughes first encountered this obsession when he moved to Washington, DC, in 1924 at age twenty-two. He lived there with his cousins, "who belonged to the more intellectual and high-class branch of our family," Hughes wrote. Hughes's cousins introduced him to "the best colored society," and in those circles, Hughes was both exasperated and amused to hear people boast that they descended from the leading Southern white families "on the colored side." Which, Hughes observed, "of course meant the *illegitimate* side."

Eventually, Hughes got so sick of high society that he began spending most of his time on Seventh Street, "where the ordinary Negroes hang out, folks with practically no family tree at all."

But just because the people on Seventh Street didn't have a family tree didn't mean they didn't want one. And as the civil rights movement gained strength over the course of the twentieth century, some African Americans tried to reclaim their ancestry with genealogy. They had to travel a far rougher trail than their white counterparts. Slaves did not leave wills; they were listed in them, alongside oxen and pewter. Some of the branches of African American family trees led to white planters who raped their female slaves, usually without acknowledging their paternity. The erasure of African genealogy reached down all the way to their names. In 1679, a New York mariner named John Leggett bequeathed to his son "a negro boy . . . known by the name of 'You-Boy.'"

When the journalist Alex Haley was growing up in 1920s Tennessee, he would listen to his older female relatives talk about their slave ancestors. As they spat tobacco off their porch, they told him stories that reached all the way back to Haley's great-great-great-great-great-grandfather, whom his grandmother simply called "the African."

The African was captured and shipped to the American colonies, where he was sold to a Virginia planter and renamed Toby. But he demanded to be called Kin-tay. The old women would sometimes recite a few African words to Haley that Kin-tay had taught the family, their meaning now lost.

Playing with his friends—both black and white—Haley would recount the stories. When the parents of his white friends got wind of his tales of whippings and beatings, they disappeared. Those stories stayed with Haley for years, through college and a career in the Coast Guard, and into the early 1960s as he became a reporter. On a trip to London in 1964, Haley stopped in at the British Museum and saw the Rosetta stone. He thought back to the impenetrable words of the African. The next year, he was in Washington, DC, and visited the National Archives. There he found the names of his emancipated slave ancestors in North Carolina, just as his relatives had recounted. Haley decided to use genealogy to find the African and then write a book about the experience.

"In America, I think, there has not been such a book," Haley told his editor. "'Rooting' a Negro family, all the way back."

Haley's research led him to conclude that the African in his past had been brought from the Gambia. He flew there and made inquiries, which ultimately led him to a traditional historian known as a *griot*. The *griot* looked Haley over and said he resembled a people called the Kinte. To Haley, the name sounded suspiciously like Kin-tay. The *griot* told him about one man from that group, named Kunta Kinte. His biography seemed to fit what Haley knew of the African.

Haley declared that he had discovered his kin. The news that an American cousin had returned raced through the Kinte villages. When Haley drove to see them, children shouted to him in greeting, "Meester Kinte!"

In 1976, Haley published an account of his ancestry, called *Roots: The Saga of an American Family*. It opened with Kunta Kinte's life in Africa, and then followed him to the American colonies, where he became a slave and started the family line that would lead to Alex Haley himself. *Roots* was something that African Americans had never encountered before: Haley was excavating hidden cables that connected living African Americans to their slave ancestors and all the way back to particular people on the mother continent. And it was irresistible—not just to black audiences but to white ones as well. *Roots* sold 1.5 million copies in hardback in its first eighteen months and was turned into a television miniseries that drew an estimated 130 million viewers.

The emotional power of *Roots* was impossible to deny, but when the historian Willie Lee Rose read the book, something didn't seem quite right. Or, rather, many little things seemed wrong. Haley wrote that Kunta Kinte picked cotton in northern Virginia in the 1760s. Cotton was never grown so far north. Kunta Kinte supposedly put up wire fencing on his plantation. Wire fencing only came into general use a century later.

"These anachronisms are petty only in that they are details," Rose wrote in the *New York Review of Books* in 1976. She worried that they were symptoms of a more profound flaw running through the entire book. "They are too numerous and chip away at the verisimilitude of central matters in which it is important to have full faith," Rose warned.

At first, Haley shirked off such criticisms, but the questions didn't stop. He tried to defend *Roots* by describing the many years of research he had put into it. And sometimes he dodged the questions by calling *Roots* "faction."

But his opponents only grew more persistent. Two novelists took Haley to court, accusing him of lifting long passages of their work. Haley settled one case, paying out $650,000. Even worse than the plagiarism, however, was the emerging realization that the genealogical bonds at the heart of the book didn't hold up to scrutiny. An expert on African oral history tracked down the *griot* Haley had met and concluded that he had no way of knowing the details of an eighteenth-century Kinte boy's life. The *griot* had simply told Haley what he wanted to hear. Professional genealogists presented a catalog of errors, cherry-picking, and wishful thinking. They concluded there was no evidence that Kunta Kinte was Toby, or that Toby was Alex Haley's ancestor.

Yet the power in the story brought out many defenders. The fact checkers, they argued, were ignoring what the book meant to readers, how it changed their relationship to the past. "Suddenly, white Americans were tuning in to the horrors of a period too many schoolbooks had tried to sugarcoat," said Clarence Page, an African American journalist. "Suddenly, black Americans were asking their elders relentless questions about a past too many elders had been reluctant to talk about and that too many of us, their children, were reluctant to hear."

To the film critic Eugenia Collier, these outcomes didn't matter. She still felt betrayed. "I believe that Haley sold out," she said in 1979. She accused Haley of getting rich off of a painful absence at the heart of African American life. "I think," Collier said, "that I would give almost anything I own to know who my African ancestors were."

Roots drove another boom in genealogy, not just among African Americans but in its white audience, too. At first these new genealogists could only riffle through the same old library folders, the same parish records, and the same census forms as the genealogists who came before them. But by the end of the twentieth century, the Internet had become a powerful new tool for their searches. Governments and churches put their records online. Genealogists shared their research with each other on online forums and through new companies. By one estimate, genealogy has now become the second-most-popular search topic on the Internet. It is outranked only by porn.

Before the online age, my own family tree looked like a split elm, half killed by a blight. While my mother could trace Goodspeeds and other ancestors back to the Puritan colonies and to England, we knew little about my father's line. But the Internet held a wealth of details about his side of the family. My relatives have documented how my great-grandfather Jacob Zimmer traveled from the Ukraine to Newark in 1892. We learned that some of Jacob's brothers also came to America, while other Zimmers stayed behind. My brother, Ben, who has inherited the genealogy allele from my mother, discovered some pictures of their village on the United States Holocaust Memorial Museum's website. The photos are of heaps of bodies, either freshly shot or excavated just after the war. The Zimmers did not have to be herded into concentration camps to be murdered. The Nazis brought the slaughter to them.

As powerful as genealogy has become, it still gives us only abstract assurances of a biological connection. My birth certificate makes me pretty certain my parents did indeed pass down their genes to me. But babies can be switched, stolen, removed. Fathers can deny paternity. Paperwork can be lost or botched. Online, faulty information can propagate across the planet, their falsehoods infecting one database after another. The only inescapable proof for our biological ancestry is what we inherit in our cells.

———

J udges had to grapple with genealogy's uncertainties for centuries before biologists could offer any help. When faced with paternity disputes, Roman courts relied on the principle of *pater est quem nuptiae demonstrant*: The father is the one whom marriage points out. A married woman's children should always be treated as her husband's children, even if she gave birth a year after his death. In later centuries, judges sometimes followed this principle far beyond what nature could allow. In 1304, a husband who had been away from England for three years came home to find a new child in his house. He went to court to deny being the father. But the judge rejected his case, declaring "the privity between a man and his wife cannot be known."

Over time, judges developed another guiding principle that came to be called "bald eagle evidence." If something looks like a bald eagle, in other words, it probably had bald eagles for parents. "I have always considered likeness as an argument of a child being the son of a parent," a British judge said in 1769. "For in everything there is a resemblance, as of features, size, attitude, and action."

Judges were still deciding if children looked like their fathers well into the twentieth century. But the rise of genetics and molecular biology prompted some scientists to wonder if it might be possible to categorically establish kinship, to see the very atoms of heredity that tie families together.

One of the first attempts to bring this science to court was made by the actor Charlie Chaplin. In 1942, Chaplin began an affair with an aspiring young actress from Brooklyn named Joan Barry. Chaplin treated her like a toy to be discarded. But when he eventually abandoned Barry, she did not go away quietly. Instead, she smashed the windows of his mansion and broke in one night, armed with a gun, demanding he take her back. By then, Chaplin had already moved on to another affair, this time with a teenager named Oona O'Neill. Barry responded by telling a Hollywood gossip columnist that Chaplin had seduced her and left her pregnant. In June 1943, well into Barry's pregnancy, her mother filed a civil paternity suit against Chaplin on behalf of her unborn grandchild. She demanded $2,500 a month, plus $10,000 in prenatal costs.

Soon, Chaplin was facing not just a civil suit but a criminal one as well. J. Edgar Hoover, the director of the FBI, had always found Chaplin a suspicious character; his anti-Nazism seemed to Hoover no different than Communism. Now he relished the opportunity to find some dirt on the actor. In February 1944, Chaplin was charged with violating the Mann Act by transporting Barry across state lines for immoral purposes while she was still a minor. He was also charged with conspiring with Los Angeles police to put Barry in jail for vagrancy.

Gawkers and reporters packed a Los Angeles courthouse for the criminal trial, which dredged up lurid details about Chaplin and Barry's affair. While Chaplin admitted to sleeping with Barry, other men testified that they had been with her during the same period. The jury acquitted Chaplin of all the charges, prompting cheers from around the courthouse.

Next came the civil case over Chaplin's paternity. Between the two trials, Barry had given birth to a girl she named Carol Ann. Chaplin's lawyers came into court ready to raise the prospect that Carol Ann was the daughter of one of Barry's lovers who had testified in the criminal case. And then they would present evidence that Carol Ann could not be Chaplin's daughter, because she had not inherited his genes.

Chaplin's lawyers could not actually read Carol Ann's genes. In the 1940s, scientists still weren't sure what genes were even made of. The best they could manage was to trace the effects of those genes through pedigrees. Sometimes those effects took the form of hereditary diseases, such as PKU. But there was one inherited trait that could be traced through just about everyone: their blood type.

Blood types were first discovered in 1900, and eight years later a Polish serologist named Ludwik Hirszfeld demonstrated that they followed Mendel's Law. A gene called ABO encodes a protein that sits on the surface of red blood cells. The most common variants are A, B, and O. Both A and B are dominant over O—in other words, if you inherit an A variant from your mother and an O from your father, you're A. Only if you inherit two O's do you have O blood type. Inheriting A and B leads to blood type AB.

Hirszfeld realized that these hereditary rules made it impossible for families to have certain combinations of blood types. If a child is blood type A, then one of its parents must carry the A variant. It's simply impossible for a type O mother and a type B father to have a type A son, for example. Writing in the *Lancet* in 1919 Hirszfeld and his wife, Hanka, predicted that under certain circumstances this discovery would make it possible "to find the real father of a child."

In 1926, a court in Germany used blood types to resolve a paternity dispute for the first time. Gradually the practice gained more attention, although many remained skeptical of its accuracy. In the months leading up to Chaplin's civil trial, his lawyers negotiated a deal with Barry's team. In exchange for $25,000, Barry would agree to have herself and her baby tested for their blood types. If the rules of heredity eliminated Chaplin, she would drop her suit.

The tests turned out exactly as Chaplin had hoped. Barry had type A and Carol Ann had type B. Those findings pointed to an inescapable conclusion: Carol Ann's father, whoever he might be, had to have type B blood. Chaplin was type O. Carol Ann had thus inherited nothing from Chaplin.

Yet Barry refused to drop the case. She had gotten a new lawyer, who would not abide by the deal made by her previous ones. Chaplin's lawyers brought the blood test results to the judge to get the case thrown out of court. But blood type tests were still such a novelty in California that the state offered no legal guidance about their reliability. The judge allowed the case to proceed, and in January 1945, Chaplin was back in court.

Throughout the trial, fifteen-month-old Carol Ann sat on her mother's lap. Barry turned her daughter's face toward the jury to allow them to gather bald eagle evidence, judging whether she looked like Chaplin or not. "Showing none of the temperament of her mother, Plaintiff Joan Berry [sic], who sobbed on her attorney's shoulder, or Defendant Chaplin, who shouted his denials, she quietly amused herself by napping, yawning and gurgling," a reporter for *Life* wrote.

Chaplin's lawyers countered the bald eagle with blood. They called a doctor to the stand to explain the blood-type results "with charts, diagrams, and elaborate explanations," as the Associated Press reported. They introduced a report into evidence that included tests from two other doctors, one appointed by Barry's lawyers and a neutral one. "In accordance with the well accepted laws of heredity," the doctors declared, "the man, Charles Chaplin, cannot be the father of the child."

Once the lawyers had introduced all their evidence—the blood test, the stories of sex with other men around the time that Carol Ann was conceived, the bald eagle evidence of her infant face—they left the jury to decide the matter. Barry's lawyer urged them to recognize Carol Ann's place on the Chaplin family tree. "You'll sleep well the night you give this baby a name," he promised them.

To the jury, Mendel's Law could apparently be stretched like taffy. They told the judge they were deadlocked, with seven jurors convinced that Chaplin was not the father, and five that he was. Barry's lawyers filed a second suit. This time, they won, the jury deciding Chaplin was indeed Carol Ann's father.

The decision set off an uproar. "Unless the verdict is upset," the *Boston Herald* declared, "California has in effect decided that black is white, two and two are five and up is down." Nevertheless, Chaplin was ordered to pay $75 a week to support Carol Ann. All told, he would go on to pay her $82,000. The toll that the case took on his reputation was even greater. No one in Hollywood wanted to work with the little tramp anymore. Chaplin left Hollywood for good.

The court's decision ultimately didn't help Joan Barry much either. Her mental health spiraled downward until 1953, when she was found wandering the streets, holding a child's ring and a pair of baby sandals, repeatedly saying, "This is magic." Barry was taken to a mental hospital for treatment. After her release, she disappeared. Carol Ann was left to Barry's relatives to raise, a child of a vanished mother and a never-known father.

The California legislature was so embarrassed by Chaplin's case that it quickly told state courts to treat blood-type tests as conclusive. The tests would go on to resolve many other paternity disputes, although they also had one profound limitation. They could only rule certain men out. They could not definitively rule men *in*. Carol Ann's test showed that Chaplin could not be her father because her father must have had type B blood. Millions of men were left as possible parents. The problem with a blood-type test was that it was based on a comparison of a few different versions of a single protein. It could not reveal things that could definitively prove a hereditary bond between one child and one parent.

Those things do exist, though. They are the bases of DNA in our genomes. But it would not be until the late twentieth century that scientists invented technologies enabling them to read bits of our genetic material. Even in short fragments, there was enough information to reunite some families—even after they had been dead for decades.

In 1917, Tsar Nicholas II and his wife, Tsarina Alexandra, along with their five children, were captured by the Ural Soviets. The Soviet revolutionaries held them in a house in Yekaterinburg for months, only to execute them along with their servants. In the years after the revolution, investigations into the killings failed to find the bodies. Rumors circulated that one or more of the children escaped the slaughter and slipped out of the Soviet Union. Over the years, more than two hundred people came forward claiming to be a prince or a princess.

In the 1970s, a Yekaterinburg geologist named Alexander Avdonin developed an obsession with the mystery. He snooped through archives for clues to what had happened. After years of research, Avdonin and some friends discovered a shallow pit not far from the house where the family had been held. The pit yielded bones from nine different people. Some of the skulls bore bullet holes; some of the bones had been pierced with bayonets.

Avdonin kept the grave a secret until the fall of the Soviet Union in 1991. When he revealed his discovery, the Russian government launched a forensic investigation of the skeletons. Researchers discovered gold and silver in the teeth, a sign the remains belonged to aristocrats.

As part of the investigation, the Russian government enlisted a British forensic scientist named Peter Gill. Gill was able to extract fragments of DNA from the bones. The fragments contained repeating sequences called short tandem repeats. This sort of genetic material can tell scientists a lot about heredity because it is especially prone to mutations. In some cases, cells accidentally duplicate some of the repeats. In other cases they cut some out. (These mutations aren't harmful because the segments are not involved in making proteins.) Over the generations, families will gain distinctive sets of short tandem repeats. Two people who share a matching set are probably close relatives. Examining the Yekaterinburg bones, Gill found that segments in the children matched either one of the adults or the other. In other words, this was a family.

But which family? No one had preserved any tissues from the Romanovs from which scientists could isolate DNA in order to make a precise match. Living relatives would have to stand in for them.

To make the match, Gill took advantage of a special set of genes that lie beyond our chromosomes. They lurk inside mitochondria, the pouches where our cells generate fuel. Each mitochondrion carries thirty-seven genes of its own, which encode proteins essential for its tasks. Mitochondria also divide on their own, making new copies of their own DNA without any meiosis.

What makes mitochondrial DNA especially attractive to geneticists is the way in which it is passed down from generation to generation. Both eggs and sperm contain mitochondria. But if a sperm manages to make contact with an egg, it produces enzymes that shred its own mitochondrial DNA. The mother's mitochondria, and only the mother's mitochondria, becomes the mitochondria of her child.

This quirk means that mitochondrial DNA can act as a record of our maternal ancestry. Meiosis scrambles chromosomes from one generation to the next. But we inherit a precise replica of our mother's mitochondrial DNA. What's more, your mother got her mitochondrial DNA from your grandmother, who got it from your great-grandmother, and so on back through more generations than even the most stubborn child can ask about. Each time mitochondria duplicate their DNA, there is a minuscule chance that it will mutate. That new mutation will be inherited down the maternal line in future generations. If a woman's female descendant picks up a second mutation, the mitochondrial DNA will now get passed down with both distinctive mutations. Relatives can be joined by this mitochondrial record of their shared ancestry.

Like the Habsburgs before them, the Russian Tsars were tightly bound by marriage to the other royal families of Europe. Tsarina Alexandra, for example, was the daughter of Princess Alice of England who in turn was the daughter of Queen Victoria. Tsarina Alexandra thus inherited Queen Victoria's mitochondrial DNA. And Alexandra passed it on in turn to the Romanov princes and princesses.

Looking over royal pedigrees, Gill realized that there was someone alive who also inherited Victoria's mitochondrial DNA: Prince Philip, the Duke of Edinburgh and the husband of Queen Elizabeth II. (Prince Philip is the great-great-grandson of Queen Victoria, through a line of female ancestors.) Gill contacted Philip, who agreed to provide his DNA for the research.

Gill found that Philip's mitochondrial DNA matched the genetic material in the remains of one of the Yekaterinburg adults, along with all the children. This result indicated that the adult was Alexandra. The remains of the other adult in the pit had a different sequence of mitochondrial DNA. Gill found that it matched genetic material from a relative of Tsar Nicholas.

When Gill and his colleagues published their results in 1994, it seemed to many observers that they had tied the Romanov mystery up in an especially neat bow. In 1998, the skeletons were interred in the Cathedral of Saints Peter and Paul in St. Petersburg. Yet, even after the burial, some skeptics questioned the identity of the bones. They raised the possibility that someone else's DNA had contaminated the equipment used to study the bones. If Avdonin really had found three of the five Romanov children, then what had become of the other two? The skeptics speculated that the bones in the shallow pit belonged to relatives. Given how many Russian aristocrats were slaughtered at the time, it seemed like a plausible alternative explanation.

Archaeologists continued to study the area where Avdonin had found the shallow pit, and in 2007, more bones turned up, 230 feet from the original grave. Russian and American anthropologists inspected forty-four bone fragments and teeth from the second site and concluded that they came from at least two individuals. The shape of the remains indicated that some belonged to a girl in her late teens, while the others probably belonged to a boy between twelve and fifteen years old. The silver fillings in their teeth indicated they were aristocrats.

Once again, Gill examined the remains, this time working with researchers from the US Armed Forces DNA Identification Laboratory. They also extracted more DNA from the bones originally found by Avdonin. Once again, DNA from the original five skeletons showed they were parents and children. And the two new skeletons belonged to the family as well. At last all seven of the Romanovs were reunited, through a genetic genealogy.

———

As scientists learned how to analyze larger pieces of DNA, they could uncover more variants joining people together in lineages. They could go beyond close cousins and join people sharing common ancestors who lived thousands of years ago.

According to the Bible, Aaron became the first Jewish priest some 3,300 years ago, and the designation was passed down from fathers to sons since then. Today, many people with surnames like Cohen and Kahn believe themselves to be descendants of those priests, known as Cohanim. In the 1990s, Michael Hammer, a geneticist at the University of Arizona, set out to search for evidence of the Cohanim by studying the Y chromosome, which fathers pass down to sons. Because the X and Y chromosomes do not cross over like other chromosomes, the Y behaves like a male version of mitochondrial DNA, staying nearly identical from one generation to the next.

Hammer and his colleagues tested the story of the Cohanim by getting cheek swabs from 188 Jewish men, 68 of whom had been told by their parents that they belonged to the priestly line. The scientists extracted DNA from their cells and examined mutation-rich regions in the Y chromosomes. Hammer and his colleagues found a single mutation in significantly higher numbers among the self-identified Cohanim than in other Jewish men. The Cohanim, they concluded, inherited their Y chromosome from a common male ancestor.

In later years, Hammer and his colleagues examined Y chromosomes from more men—both Jews and gentiles. A lot of the Jewish men turned out, once more, to share a close common male ancestor. But others had different mutations that pointed back to other men in the past. In 2009, when Hammer and his colleagues published their new research, they proposed that the priestly line started out just as the ancient stories said. But once the Cohanim tradition emerged, other Jewish men, with different Y chromosomes, somehow became priests as well.

At first, reading DNA was such an expensive undertaking that only experts such as Gill and Hammer could do it, and only for scientific research. But the cost dropped so quickly that it became economical for Hammer and others to launch companies that could provide genetic genealogy on demand. People simply spat into a tube that they mailed to the companies, which extracted the DNA inside and compared it to the growing databases of genetic variations in humans. The companies started off looking at a few regions of mitochondrial DNA and the Y chromosome, reporting back about where on Earth a customer's combination of mutations was most common. In later years, they cast a wider net, scanning genetic markers across all the chromosomes. Taken together, the markers made up less than a thousandth of the entire human genome. But they varied so much from person to person that they could reveal clues about people's ancestry or even link them, like the Romanovs, to their relatives.

Some customers took the tests in the hope of finding cousins or connecting themselves back to famous ancestors. Others hoped to reach back across the oceans to places where they came from. Europeans searched for their Viking ancestors. African Americans could leap beyond slavery's void. In 2016, a remake of *Roots* aired on the History Channel. LeVar

Burton, who had played Kunta Kinte in the original version, now served as an executive producer. To promote the new version of *Roots*, Burton took a DNA test from 23andMe, along with Malachi Kirby, the actor playing Kinte in the remake.

"I've always felt there was a piece of me missing," Burton said in a video. As he studied his results on an iPad, he looked deeply moved. "Who I am did not just begin here," Burton said. "To have the proof in my hands is just powerful."

I had heard much the same thing from other people, both friends who got themselves tested and people who have read some of the articles I've written about genetics. One reader told me how she had spent years using historical records to trace her ancestry to Jamaica and Ghana. She then told me that a genetic test showed she descended from two peoples in particular, the Akan and Guan. "In my family, I am tall and my siblings and parents are short," she told me. "It was when I received DNA and folk history that showed a blending of Akan and Guan ancestors that I understood the appearance of tall and short in our family."

I wondered what I might find in my own DNA. Was I carrying a molecular version of *History of the Goodspeed Family*? Would a Zimmer version hold more surprises? I wondered what collection of genetic variants I had inherited from those ancestors, how they had influenced my fate. I thought back to the visit Grace and I had had with a genetics counselor when we had floundered our way through a family history. At the time, the first human genome project was still under way, at a cost of $3 billion. Fifteen years later, my friends were giving ancestry tests as birthday presents. In the fifteen years since my last visit with a genetics counselor, the gene BRCA1 had become a medical celebrity. Certain mutations to the gene drastically raise the odds that a woman will develop breast and ovarian cancer. Those mutations are especially common in Ashkenazi people. I knew of at least one woman on my father's side of the family who had had breast cancer. A friend of mine who had a BRCA1 mutation was dying of breast cancer at age forty-eight. Had my daughters inherited that fate from me? If I found out, how would I tell them?

These questions were humming in my head when an e-mail turned up in my inbox. A geneticist named Robert Green invited me to a meeting. "This should be an extraordinary and very select learning experience if you can make it," he promised me.

The meeting would be on the future of genomes in medicine. Green and other scientists would give talks about how they were using genomes in their research today, and how they might use them in the future. People coming to the meeting could opt to pay to have their genomes sequenced. For $2,700, a gene-sequencing company called Illumina would determine all 3.2 billion base pairs in a person's DNA. Clinical geneticists would then look at the variants people had, searching for mutations linked to 1,200 diseases—some familiar, such as lung cancer, and others obscure, such as cherubism. (Don't be fooled: Cherubism doesn't make you look like an angel. It fills your jaw with cysts.)

I signed up. I was attracted not by the prospect of a medical report; I wanted to get my hands on the raw data itself and find some scientists to help me explore it. Before Green's invitation, I could only guess about my DNA from the stories my relatives told. Now I'd be able to read my genetic heredity, down to the letter.

CHAPTER 7

Individual Z

R OBERT GREEN and I stood inches apart. His eyes scanned across
my face, from ear to ear, from forehead to chin.

"What I'm doing," he murmured, "is looking for any facial features that would suggest an underlying genetic illness." He looked me over as if I were a horse he was thinking about buying. "The shape of your eyes, whether your ears are low set or not," he said. "The complexity of your ears."

Getting my genome was turning out to be a lot more complicated than I had expected. I could not simply spit into a tube and mail it off to a company like 23andMe. In 2007, 23andMe began providing reports on DNA directly to consumers. For $999, they would identify the variants at half a million sites in a person's genome, analyze them for clues to their ancestry, and even supply a report about how the variants influenced risks for disorders ranging from diabetes to Alzheimer's disease. Their service was a profound leap from conventional genetic tests. They had to be approved by the FDA and ordered by doctors. Now 23andMe was delivering information straight to customers. In 2013, the FDA told 23andMe to stop selling unvalidated tests or face the consequences. In response, the company cut back their reports to ancestry and nothing more.

Other companies, such as Illumina, took notice. To get my genome from them, I would have to get a doctor to order it for me as a medical test.

Green, who had originally invited me to get my genome sequenced, also agreed to sign for the test. First, however, he would put me through a thorough, old-fashioned genetic exam—the kind that Lionel Penrose might have given in the 1950s.

"Future clinicians may judge this to be unnecessarily cautious," Green told me. "But there is no standard for how we do whole genome sequencing. So this is how I've decided to do it."

I had taken the train to Boston and made my way to Brigham and Women's Hospital for the exam. I first sat down with a genetic counselor named Sheila Sutti, who took out a form entitled "Family History." She began asking about my relatives. As we spoke, she filled the page with circles and squares, slashing some of them with the diagonal of death. She noted allergies and surgeries. Question marks recorded the many times I shrugged my shoulders in ignorance. Sutti drew a network of symptoms and uncertainty. When I looked at the form, I could not see any signal of heredity.

Green arrived just as Sutti was finishing up. A looming, silent medical student trailed him. Green peered at my face through his narrow frameless glasses. He was taking advantage of the fact that genes play many different roles in our bodies. A hereditary disease that causes hidden damage to the nervous system may also disrupt the development of the face, leaving behind clues that a geneticist can spot with the naked eye. Green then asked me to walk back and forth from wall to wall. He crossed the arms of his white lab coat as he looked down at my feet, sizing up my gait.

Green told me these conventional exams didn't reveal any signs that required a test for a specific disease. He signed the request for my genome, and Sutti led me to another wing of the hospital, where a phlebotomist slid a needle into my arm. I watched blood glide like scarlet motor oil out of my arm and into three tubes.

The tubes were shipped across the country to San Diego, where Illumina's technicians cracked open my white blood cells and pulled out my DNA. They blasted the molecules with ultrasound, shattering them into fragments, and then made many copies of each one. Adding chemicals to the fragments, they were able to determine their sequence.

Now they had to assemble these fragments together like the pieces of a jigsaw puzzle. Just as a puzzle solver can use the picture on the box lid as a guide, the Illumina team consulted a reference human genome to figure out where each of my fragments had come from. Some fragments were too enigmatic to locate, but overall, Illumina was able to rebuild over 90 percent of my genome.

From one person to the next, human genomes are mostly identical. But in a genome stretching over three billion base pairs, the tiny fraction of DNA that varies adds up to millions of differences. Most of these variations are harmless. But some can give rise to a disorder such as PKU. Others raise the risk of more common conditions like cancer or depression. Illumina's clinical geneticists searched my own collection of variants for any especially worrying ones. A few weeks after my visit to Brigham and Women's, Sutti called me with the results.

"The reason we're doing this over the phone and not in person is that we didn't find anything of clinical importance," she said. "You had a very benign report, Carl."

Sutti told me that I didn't have any dominant mutations known to cause diseases with just a single copy. Nor had I inherited two copies of a dangerous recessive mutation. I did find out a few useful things about my health, though. The sequencing revealed variants that could affect the way I respond to certain medicines. If I ever get hepatitis, I know I shouldn't get treated with a combination of interferon and ribavirin.

And, like all humans, I'm also a carrier. That is, I carry single copies of recessive variants. If my children inherited the same variants from both me and Grace, they might develop genetic diseases. In the early 2000s, when Grace and I became parents, DNA sequencing technology was far too crude for me to get a full catalog of my carrier variants. The best we could hope for was to have our daughters tested for a few diseases, such as PKU.

It turned out I'm a carrier for two genetic disorders I never heard of: one called mannose-binding lectin protein deficiency, the other familial Mediterranean fever. I had to do a little research to understand this particular inheritance. I learned that mannose-binding lectin protein deficiency weakens the immune system, leaving babies to develop disorders such as

diarrhea and meningitis. Familial Mediterranean fever, the result of mutations to a gene called MEFV, causes people to suffer from painful bouts of
inflammation in their abdomen, lungs, and joints.

I don't know which of my parents I inherited those mutations from, but
I'd bet that I got my faulty MEFV gene from my father. It is most common
among people of Armenian, Arab, Turkish, or Jewish descent. It's far rarer
in other ethnic groups, like the Irish—from whom Grace descends. It would
be extraordinarily unlikely that she would have a faulty MEFV gene, too. At
worst, my daughters are carriers like me.

And that was all. After more than a century of advances in genetics, I
got a glimpse at my genome, something that had been impossible until recently, and there wasn't much for Sutti and me to talk about. A week after
our phone call, I took the train to Boston to attend the "Understand Your
Genome" meeting. At lunch, an Illumina representative logged me into a
secure web page that elegantly displayed my results. I could compare my
own genome to the reference genome, displayed as two rows of colored letters. Where my DNA differed, the colors were brightly mismatched. Along
with disease-related variants, Illumina revealed a few more associated with
physical traits. They meant little to me. "Your odds of developing male pattern baldness are increased if you are Caucasian," Illumina told me. You
could call me Caucasian, but I have a thick thatch of hair. "Your muscle fibers are built for power," the website lied.

The whole experience was charming but dull. I certainly didn't want the
excitement that comes from discovering you have cherubism. But getting to
see my own genome shouldn't have been boring. I was pretty sure that if I
could dig deeper—or, rather, if I could enlist the help of some scientists to
dig deeper—I'd be able to learn much more about heredity.

After a few weeks of wrangling and paperwork, I managed to get all the
raw data from Illumina. It showed up at my door one January afternoon in
a white cardboard box. Inside was a shroud of green bubble wrap, inside of
which was a kidney-shaped black pouch, inside of which was a slim brushed-
metal hard drive. It contained seventy gigabytes of data—the equivalent of
more than four hundred high-definition movies.

To make sense of that data, I took my genome on the road. On one trip,

I drove down I-95 to the Yale campus and walked up Science Hill to reach the office of Mark Gerstein. Gerstein's office was heaped with scientific bric-a-brac: Galileo thermometers, Klein-bottle coffee mugs, blinking lights that feed off the electric current in your skin. Gerstein's conversation was packed as well, pinging so quickly between genomes and cloud computing and open-access scientific publishing that I sometimes had to look back at my notebook to remember the question I had just asked him.

The idea of telling me about my own genome intrigued Gerstein to no end. Over the course of his career, he has analyzed thousands of genomes— he helped lead a study called the 1000 Genomes Project, for starters—but he'd almost never looked straight in the face of the person from whom one of those genomes came. As I handed him the drive so that he could copy my genome to his computer, he confessed to a vicarious thrill.

"I'd never have the courage to do this—I'm just too timid," he said, laughing. "I'm a worrier. Every time there would be a new finding, I'd look in my genome to see if I had it."

———

While Gerstein and his team got to work, I went to the New York Genome Center, where a group of scientists were building a genealogical database they called DNA.Land. They created a website where anyone could upload their genetic data for scientific research. In exchange, they would analyze people's DNA and share whatever genealogical clues they could find.

My brother, Ben, had gotten his DNA sequenced by a company called Ancestry.com—not his whole genome, of course, but 682,549 genetic markers. I asked him to upload his file to DNA.Land in order to compare our genes.

Thanks to meiosis, Ben and I are not genetically identical. Our genomes are made up of different selections from our parents' chromosomes. Yet, despite our differences, we still have many long stretches of identical DNA in common. DNA.Land could confidently recognize 112 identical segments, each one stretching 100 million bases or more. While we are far from clones, there's no one on Earth who's more genetically similar to me than my brother.

If I were to compare myself to one of my first cousins, I'd find fewer identical segments. We share a pair of grandparents, but they also inherited some DNA from their two other grandparents. The segments we share are also smaller, because there have been two generations between us and our grandparents for meiosis to chop our inherited chromosomes into more pieces.

DNA.Land found 45 other people among its 46,675 volunteers who had enough stretches of identical DNA to suggest they might be my cousins. It was also possible they were not closely related at all, our identical DNA a persistent legacy of ancestors who lived centuries ago. I looked at the names of these possible kin and recognized none of their names. Ben was inspired to do some digging and discovered that one of them—a possible fourth cousin named Elias Gottesman—had a harrowing story.

As a child, he had been sent to Auschwitz with his family, and there the camp's doctor, Josef Mengele, did experiments on him and his twin brother, Jeno. Mengele was especially taken with twins because he believed he could discover the genetic roots of diseases by examining them—sometimes even dissecting them alive. By the end of the war, Gottesman lost his entire family and even lost his name. Only decades later, as an old man in Israel, did he begin searching for them again. A genetic match to cousins in the United States revealed his birth name; his cousins even sent him a picture of his lost parents.

The DNA Gottesman and I inherited from an ancestor might mean we were close kin. But I didn't contact him, or any of the other possible matches that DNA.Land sent me. A genetic connection did not join our lives together. In fact, if I were to compare my genome to those of my fourth cousins, I'd find that I don't even share any DNA with some of them. That may sound impossible, but only because in modern Western culture we've made the mistake of equating DNA with kinship. That's not actually how heredity works.

The more distantly a cousin is related to you, the more generations back you have to go to find your common ancestors. It also means that over those generations, the DNA from those ancestors got cut into ever smaller pieces and was mixed with the DNA from ancestors you and your cousin do not share. It's purely a matter of chance which copy of a DNA segment ends up in an egg or a sperm. And so, in time, one ancestor's genes may disappear altogether. In 2014, Graham Coop, a geneticist at the University of California, Davis, determined that if you brought together 100 pairs of third cousins, one of those pairs would share no identical segments of DNA at all. If you brought together 100 pairs of fourth cousins, 25 would lack this genetic connection.

The same holds true for our ancestors. If I were to compare my genome to those of my grandparents, I'd be able to find large chunks of identical DNA from all four of them, each totaling roughly 25 percent of my genome. In the next generation back, I have eight great-grandparents, contributing more chunks—but smaller ones. With every generation back, my number of ancestors doubles. Roger Goodspeed is among 1,024 ancestors of mine ten generations back. But according to Coop's estimates, I inherited only 628 chunks of DNA from that entire generation. There's only so much room in my genome, and so a lot of their DNA did not finish the journey from my tenth-generation ancestry to me. For any particular ancestor from Roger Goodspeed's generation, there's a 46 percent chance I didn't inherit any of their DNA. I grew up imagining Roger Goodspeed as some kind of American Adam to my family, bestowing Goodspeed genes on all his descendants. But it's pretty much a coin toss whether I have any of his DNA at all. And even if I did, Coop's calculations show I'd be able to trace only about 0.3 percent of my DNA to him.

As you move further back in genealogical time, an even bigger paradox looms into view. We think of genealogy as a simple forking tree, our two parents the product of four grandparents, who are descended from eight great-grandparents, and so on. But such a tree eventually explodes into impossibility. By the time you get back to the time of, say, Charlemagne, you have to draw over a trillion forks. In other words, your ancestors from that generation alone far outnumber all the humans who ever lived. The only way out of that paradox is to join some of those forks back together. In other words, your ancestors must have all been related to each other, either closely or distantly.

The geometry of this heredity has long fascinated mathematicians, and in 1999 a Yale mathematician named Joseph Chang created the first statistical model of it. He found that it has an astonishing property. If you go back far enough in the history of a human population, you reach a point in time when all the individuals who have *any* descendants among living people are ancestors of *all* living people.

To appreciate how weird this is, think again about Charlemagne. We know for a fact that Charlemagne has some living descendants, thanks to the genealogies proudly drawn by the Order of the Crown. But that fact, according to Chang's model, means that every European alive today is a descendant of Charlemagne. The order is hardly an exclusive club.

When Chang developed his model in 1999, geneticists couldn't compare it to reality. They didn't know enough about the human genome to even guess. By 2013, they had gained the technology they needed. Coop and his colleague Peter Ralph, a statistician at the University of Southern California, set out to estimate how living Europeans are related to people who lived on the continent hundreds or thousands of years ago. They looked at a database of genetic variants collected across Europe from 2,257 living people. They were able to match identical stretches of DNA in different people's genomes, which they inherited from a common ancestor.

Ralph and Coop identified 1.9 million chunks shared by at least two of the 2,257 people. Some of the chunks were long, meaning they came from recent common ancestors. Others were short, coming from deeper in the past. By analyzing the chunks, Coop and Ralph confirmed Chang's study, but they also enriched it. They found, for example, that people in Turkey and England shared many fairly big chunks of DNA that they must have inherited from a common ancestor who lived less than a thousand years ago. It was statistically impossible for a single ancestor to have provided them all with all those chunks. Instead, living Europeans must have gotten them from many ancestors. In fact, the only way to account for all the shared chunks Coop and Ralph found was with Chang's model. Everyone alive a thousand years ago who has any descendants today is an ancestor of every living person of European descent.

Even further back in time, Chang and his colleagues have found, the bigger the ancestral circle becomes. Everyone who was alive five thousand years ago who has any living descendants is an ancestor of *everyone* alive today. The Order of the Crown may be big, but an early pharaoh of Egypt might be able to get a club seven billion strong.

———————

I asked the scientists at the New York Genome Center to look beyond my cousins and use my genome to tell me something about my ancestry. They started with the simplest pieces of DNA to interpret: the mitochondrial DNA I inherited from my mother, and the Y chromosome I inherited from my father. By 2015, geneticists had built massive databases of both types of DNA, with sequences of hundreds of thousands of people. They organized the sequences in much the same way a taxonomist might classify insects, dividing them into classes, dividing those classes into orders, and so on. Large groups of men across the world have certain Y-chromosome mutations in common—known as haplogroups. I belong to haplogroup E, I learned. Its ranks are made up mainly of African men, but they also include some men from Europe and the Near East. Within that haplogroup, I belong to a smaller one known as E1, and within that, E1b—and so on all the way down to the haplogroup E1b1b1c1.

That particular haplogroup includes some Jewish men. While that certainly jibed with my experiences with my father's side of the family, the snug fit began wiggling loose when I looked into the haplogroup further. Only a few percent of Jewish men carry E1b1b1c1. Many men who are not Jewish carry it as well; it's found across a range stretching from Portugal to the Horn of Africa to Armenia. When Napoleon died, one of his followers tucked a few hairs from his beard in a reliquary. In 2011, French researchers managed to extract some of his Y chromosome from them. They found that he belonged to the E1b1b1c1 haplogroup, too. The highest percentage of men with E1b1b1c1 yet found don't live in Israel. They live in the Jordanian city of Amman. The second-highest percentage can be found among the Amhara, an ethnic group that lives in the highlands of Ethiopia.

The high percentage of men in Jordan with E1b1b1c1 suggests that it first emerged somewhere in the Near East, perhaps as long ago as ten thousand years—long before the Jewish people existed. Thousands of years later, Arabs, Jews, and other peoples of the Near East spread into Africa and Europe, spreading the haplogroup with them. On its own, my E1b1b1c1 haplogroup cannot let me trace its path back through that ancestry (although I'm pretty sure Napoleon isn't my great-great-great-great-grandfather). All I can know is that there was probably an ordinary Near Eastern farmer some ten thousand years ago who acquired a harmless mutation in his Y chromosome that distinguished a new haplogroup, one that he unknowingly passed down to his son. But even among my male ancestors, that farmer holds no special place. He just happened to be the one from whom I inherited my Y chromosome.

On my mother's side, I discovered that I have a mitochondrial haplotype called H1ag1. It's found throughout much of western Europe, and has been found there for quite a while. When a genome sequencing center was built in Hinxton, England, the construction workers dug up across a 2,300-year-old skeleton. It turned out to have some bits of DNA in its bones. The Hinxton skeleton carried H1ag1, just like me. As for the original Ms. H1ag1, however, I can't say that she lived in Hinxton. I can't even say she lived in England.

People carrying the H1ag1 haplotype can be found today across northern Europe. I know I am their kin along the maternal line, but I can't know where our common ancestor lived. Scientists have drawn a tree of all of humanity's known mitochondrial DNA, and on it my H1ag1 branch sprouts next to other branches common in Europe. The European branches split off from branches common in Asia and the New World. The deepest branches on the tree are found in living Africans. By tracing the mutations along all the branches, scientists can estimate the age of the woman who carried the mitochondrial DNA that gave rise to all haplogroups today. That woman lived in Africa about 157,000 years ago.

The first clues that living humans get their mitochondria from a single woman in Africa first emerged in 1987, thanks to research in the lab of Allan Wilson, a geneticist at the University of California at Berkeley. Reporters swiftly nicknamed this unknown woman Mitochondrial Eve. The name stuck like superglue. *Newsweek* ran a cover story about the research, illustrating their cover with a brown-skinned Adam and Eve.

It would take years for scientists to trace back the Y chromosome of all living men. According to the latest research, he lived in Africa 190,000 years ago, at the dawn of our entire species. Soon enough that man was christened Y-chromosome Adam. He now enjoys a Wikipedia page of his own. It's easy to imagine Mitochondrial Eve and Y-chromosome Adam as the parents of all humanity, dropped down into a Pleistocene Garden of Eden. The fact that Eve didn't show up in the garden until thirty thousand years after Adam died is one of those minor scientific details that cannot undermine a seductive metaphor.

———

It took a couple of weeks for Mark Gerstein to work over my genome. He and his students wanted to analyze the short fragments of DNA with their own software and create their own map. Once they had pinned down the location of the vast majority of Illumina's fragments, they could then determine which variants I carried. And they could try to figure out what those variants meant to me. When I paid my second visit to Gerstein, I was surprised that he wasn't leading me back to his office. Instead, he led me to a conference room down the hall.

Eight of Gerstein's graduate students and postdoctoral researchers were waiting for us, flanking two sides of a long table, all with laptops and wireless keyboards at the ready. They had me sit down at the head of the table so that they could show me slides on a giant monitor on the wall in front of me.

The first slide was labeled "Individual Z Overview."

It was only then that I realized why so many of the scientists I contacted for my little project were proving to be so strangely helpful. To them, I am Individual Z. It was as if I was a frog that had hopped into an anatomy class with my own dissecting scalpel, asking the students to take a look inside.

For the next two hours, Gerstein's team picked over my genome, showing me broken genes and duplicated genes and genes with mutations that altered how my proteins worked. But what struck me most of all was what they found when they compared my genome to two other people's—a pair of anonymous volunteers who agreed years beforehand to have their DNA sequenced and made publicly available. One of them was from Nigeria and the other from China.

Gerstein's team identified a total of 3,559,137 bases in my genome that were different from the human reference genome. These variants are known as single-nucleotide polymorphisms, or SNPs for short. They include the variants that make me a carrier for things like familial Mediterranean fever, as well as ones that influence traits that have nothing to do with disease, like my skin color, and ones that have no effect on my biology at all.

The Nigerian and the Chinese had a similar number of single-nucleotide polymorphisms. But those variants did not distinguish the three of us in any clear way. Sushant Kumar, a postdoctoral researcher in Gerstein's lab, made me a Venn diagram to drive the point home. All three of us have 1.4 million single-nucleotide polymorphisms in common. There were another 530,000 that I shared only with the Chinese person but not with the Nigerian. And there were 440,000 single-nucleotide polymorphisms that I shared with the Nigerian alone. All told, 83 percent of my variants were present in at least one of their genomes.

We were three people of African, Asian, and European descent, from three corners of the world. Three races, some might say. And yet we shared far more than what set us apart.

———

The concept of race is not like the moon or hydrogen. It is not a feature of the natural world beyond our social experience. Up until the Middle Ages, writers never used the word *race* in the sense that it would later take on—referring to a sharply defined biological group of people whose members were bound together by heredity. Ancient writers certainly recognized differences among peoples from different parts of the world. But they didn't explain them with taxonomy.

The word *race* seems to have first taken on a modern complexion during the Habsburg rule of Spain. The country was filled with people of different ancestries—Christian Celts, Romans, Jews, Africans. When the persecution of the Jews began, other Spanish people began to think of themselves as belonging to a particular group—Old Christians. To prove they were Old Christians, noble Spanish families had to demonstrate that they had no Jewish ancestry. In other words, that they didn't have a single drop of Jewish blood. Noble families struggled to prove their ancestry had been pure since time immemorial.

When Spain established an empire in the New World, it now had another group of people to distinguish itself from. The Spanish conquistadors, the conquered Indians, and the imported African slaves now shared the same countries. The governments came up with a legal hierarchy with the Spanish on top, Africans in the middle, and Indians at the bottom.

But the people of the New World would not respect those boundaries. Through marriage or rape, people from different races had children together. The colonial governments needed to invent new categories, with new names. In Mexico, the viceroy sliced his subjects into fine distinctions:

1. Spaniard and Indian beget *mestizo*
2. *Mestizo* and Spanish woman beget *castizo*
3. *Castizo* woman and Spaniard beget Spaniard
4. Spanish woman and Negro beget *mulato*
5. Spaniard and *mulato* woman beget *morisco*
6. *Morisco* woman and Spaniard beget *albino*
7. Spaniard and *albino* woman beget *torno atrás*
8. Indian and *torno atrás* woman beget *lobo*
9. *Lobo* and Indian woman beget *zambaigo*
10. *Zambaigo* and Indian woman beget *cambujo*
11. *Cambujo* and *mulato* woman beget *albarazado*

12. *Albarazado* and *mulato* woman beget *barcino*

13. *Barcino* and *mulato* woman beget *coyote*

14. *Coyote* woman and Indian beget *chamiso*

15. *Chamiso* woman and *mestizo* beget *coyote mestizo*

16. *Coyote mestizo* and *mulato* woman beget *ahí te estás*

To the north, England brought Africans to their own colonies in the 1600s to work their fields. The Africans worked at first alongside European servants, subject to the same laws, but over the course of decades the colonial governments gradually singled out the people from Africa for harsh treatment. By the early 1700s, free Negroes had lost the right to vote or bear arms, while those still enslaved were recognized by the law as slaves for life, and their children inherited their bondage.

Ham's curse grew wildly popular in the British colonies as a moral justification for these laws. Ministers proclaimed Noah's divine prophecy in sermons, and pamphlets circulated in the American South explaining how God turned the skin of Ham's children dark as a sign of their sin. Africans inherited their enslavement as surely as they inherited their color. Over the course of the 1700s, Ham's curse became downright biological. Slavery's defenders now began drawing up catalogs of essential differences between the white and Negro races.

"The Blacks born here, to the third and fourth generation, are not at all different in colour from those Negroes who are brought directly from Africa," a Jamaican plantation owner named Edward Long observed in 1774. Instead of hair, Long claimed, his slaves had "a covering of wool, like the bestial fleece." When Long turned to the minds of slaves, the differences from Europeans seemed even more profound. "They have no plan or system of morality among them," Long declared. "They are represented by all authors as the vilest of the human kind."

By the late 1700s, slaveholders could apply a scientific veneer to these beliefs. Naturalists argued that, just as animal and plant species could be divided into varieties, so, too, could *Homo sapiens*. Carl Linnaeus defined four races: *Americanus* ("reddish, choleric . . . paints himself with fine red lines; regulated by customs"), *Asiaticus* ("sallow, melancholy . . . haughty, avaricious . . . ruled by opinions"), *Africanus* ("black . . . women without shame . . . indolent . . . governed by caprice"), and *Europeaus* ("white . . . inventive . . . governed by laws").

A few decades after Linnaeus, the German anthropologist Johann Friedrich Blumenbach proposed a new system of five races instead of four. His races were Caucasian, Mongolian, Ethiopian, American, and Malay. Blumenbach came up with the label Caucasian after studying a skull in his collection from a woman who lived in the Caucasus Mountains. It was, he later said, the most beautiful skull he ever laid eyes on. She belonged to the same race as people who lived across Europe, Blumenbach believed. He thought the reason that Caucasians had such beautiful skulls was that they were the first people created by God. They retained humanity's original glory, while other people degenerated, producing the other four races.

Blumenbach's system became popular over the nineteenth century, but many of the nuances of his ideas were lost along the way. Blumenbach argued that there was no sharp geographical divide between the races, for example, with each race blending insensibly into neighboring ones. Later anthropologists tried instead to pinpoint fixed anatomical differences. Some even went so far as to reject the idea that humans had a single origin. They argued that every human race had been separately created and forever locked into its place in the divine hierarchy. There was never any question as to how that order was stacked. At the top, one 1852 American textbook explained, was "the white race, who is distinguished above them all: the most perfect type of humanity."

This racial hierarchy had to remain intact and legally clear-cut, no matter how confusing reality was. For all the imaginary walls that were erected between the races, sex forever threatened to bring them down. Early on in the American colonies, black and white indentured servants would sometimes marry and have children. By the end of the seventeenth century, colonial governments had laws in place to stop that practice. The Virginia House of Burgesses labeled the children of black and white parents as an "abominable mixture and spurious issue." These interracial children were deemed Negroes as well, and thus slaves. The words that described them had legal weight, even if they were scientifically absurd. Colonial governments were pretending that the flow of heredity from white parent to Negro child could be arbitrarily severed.

Despite all the laws, more interracial children were born—not just to Negro slaves but to free Negroes as well. Some stayed in Negro communities, where their own children ended up inheriting more African ancestry. Others wound up with so much European ancestry that they sometimes chose to "pass" as white. Like the Spanish governors before them, southern states developed a vocabulary to bring some order to their human property. But if they looked closely at their words, they grew uncertain. Was African blood so potent, so poisonous, lawmakers wondered, that inheriting even a drop would overwhelm a much greater portion of white blood? In 1848, a judge in South Carolina tried to answer the question and failed. "When the mulatto ceases, and a party bearing some slight taint of the African blood, ranks as white, is a question for the solution of a jury," he concluded.

Frederick Douglass took pleasure in forcing his fellow Americans to recognize how badly their racial classifications failed to align with reality. "My father was a white man, or nearly white," he wrote in his autobiography. "It was sometimes whispered that my master was my father."

Douglass's mother was a Maryland slave named Harriet Bailey, who worked as a field hand. His biographers consider it likely that her owner, Aaron Anthony, raped her along with a number of his other female slaves, and then used these children of his for slave labor. Although Douglass may have inherited Anthony's DNA, he did not inherit the legal status that came with it. Instead, Douglass grew up as a slave, driving cows to grazing fields and keeping them out of his father's garden. Anthony loaned Douglass out at age eight to his son-in-law's brother in Baltimore. There Douglass did an assortment of jobs until 1838, when he used false papers to slip aboard a northbound train.

Over the next few years, Douglass started a newspaper and began lecturing across the country in favor of abolition. In 1848, when he traveled aboard a steamboat across Lake Erie to a convention in Buffalo, his fellow passengers recognized him and pleaded for him to give an impromptu speech. Douglass stood up and delivered his case against slavery. "During my remarks, I convicted the slaveholder of theft and robbery," he reported back to his newspaper.

An actual slaveholder, it turned out, was aboard the ship that evening. The man stood up, "with a most contemptuous sneer on his face," Douglass recalled, declaring, "'It was not to be supposed that any white man would condescend to discuss this question with a *nigger*.'"

Douglass decided to reply with "a somewhat facetious account of my genealogy." He told the slaveholder "that he was much mistaken in supposing me to be a *nigger*."

Instead, Douglass declared, "I was but a half negro—that my *Dear father* was as white as himself, and if he could not condescend to reply to negro blood, to reply to the European blood."

The slaveholder could not. He stamped away, astonished, Douglass recalled, that "such sentiments and impudence as he had heard from my lips, could be tolerated and applauded by white men in any part of this Union."

Two decades later, America's slaves would be emancipated. The former Confederate states kept searching for a way to oppress them, and to do so they needed a reliable way to identify different races. Even a drop of black blood became enough to exclude a person from whiteness. In 1924, the state of Virginia enshrined this practice as law by passing the Racial Integrity Act, which barred interracial marriages. The law defined whites much like the Spanish had three hundred years earlier: white people were those "whose blood is entirely white, having no known, demonstrable or ascertainable admixture of the blood of another race."

There was just one problem with this "one-drop rule." The Virginia law defined whiteness as the absence not only of black blood but of Indian blood, too. Ever since the days of John Randolph, many prominent white Virginians had boasted of being direct descendants of Pocahontas. The Racial Integrity Act would have rendered them no longer white. That would simply not do, and so the state legislature tacked on a so-called Pocahontas exception. Even if Virginians were up to one-sixteenth Native American, the revised law held, they would still be considered white. People who were one-sixteenth black, on the other hand, were still black.

It might be comforting to dismiss the Racial Integrity Act as a monstrosity from a vanished racist past. But when the law was passed in 1924, genetics had already been around for almost a quarter of a century. And some of its most prominent figures gave their support to the law. Many eugenicists not only wanted to stop inferior white people from having children; they also wanted to keep the white race genetically pure.

Racism had been a fundamental feature of eugenics ever since Francis Galton coined the word. When Galton studied the heredity of talent, he compared it in different races. Without any reliable way to actually make such a measurement, he simply used his intuitions. Thinking back on his travels through southern Africa, he concluded that he and his fellow white explorers were far more talented than the Africans they encountered. "The mistakes the negroes made in their own matters, were so childish, stupid, and simpleton-like, as frequently to make me ashamed of my own species," Galton wrote.

Africans inherited that childishness, Galton believed, in the same way that they inherited their curly hair or dark skin. The great talents of northern Europeans were just as hereditary, he believed. When Galton promoted eugenics, he promised that careful breeding would make northern Europeans even more talented, and the benefits would redound to all the inferior races, too. Throughout their global empires, Galton believed, northern Europeans ought to use eugenics to improve those lower races as much as their heredity would allow.

Galton wrote about race with the cool abstraction of an English gentleman who spent most of his time in London clubs and meetings of scientific societies. For some white American scientists, the question of race was far more urgent and intimate. In the Jim Crow years after the Civil War, millions of blacks boarded trains headed out of the South, to cities like New York and Chicago. Those cities were also taking in immigrants from abroad at the same time—no longer just northern Europeans but huge numbers of Italians, Poles, Russians, and Jews, along with Chinese and Latin Americans. Some white scientists responded to this sudden mixture by trying to put old-fashioned racism on a new scientific footing.

A scientist named Harvey Jordan experienced a fairly typical anxiety for his time. Jordan grew up in the late 1800s in rural Pennsylvania, where, he later wrote, he "was impressed with the importance of heredity, while playing about the barns." Rather than become a farmer, though, Jordan went to college and became an expert on anatomy, studying at Cornell, Columbia, and Princeton. He spent the summer of 1907 in Cold Spring, New York, where he met Charles Davenport. Davenport taught him the new science of genetics, and, from Cold Spring, Jordan headed straight to the University of Virginia to become an anatomy professor and help modernize its medical school. In all that work, heredity was first and foremost on his mind.

Jordan was appalled to discover in Virginia "the distressing racial conditions in our colored population in the South." But he didn't see these conditions as being caused by social forces. Instead, biology was to blame. Its solution must therefore be eugenics—a state-run program of control over who got to have children. Jordan thought it would be especially useful to encourage mulattoes to have children with full-blooded blacks, in order to spread white genes among their children. They would act like yeast in bread dough, he thought, "as a leaven in lifting the colored race to a higher level of innate mental and moral capacity."

Before such a program could be put in place, Jordan believed it would be necessary to uncover the genetic foundation of the races. He would need the guidance of his eugenic gurus to carry out the task. "I have been wondering if I could be of service in this great work," Jordan wrote to Davenport in 1910, "perhaps in gathering statistics at close range."

Just as Davenport had assigned Henry Goddard to study the heredity of feeblemindedness, he tutored Jordan on how to investigate the heredity of race. They agreed that Jordan would start by studying the most obvious feature that appeared to set the races apart: skin color.

Jordan found four families of mulattoes. To measure their skin color, he brought with him a colored top. The top, a child's toy made by the Milton Bradley Company, had become popular among anthropologists as a way to measure skin color. It had wedges of yellow, black, red, and white. If the top was set spinning fast enough, the colors blurred together into a single hue. By adjusting the size of the wedges, the scientists could change the blurred color. Jordan would have his mulatto subjects hold out their arm and spin his top next to it. He would keep adjusting the colors on the top until he reached a matching shade. Then he would write down the size of the different wedges that produced the match.

Jordan sent his color numbers to Davenport, along with pedigrees of his mulatto families. Their data suggested that the color of children was not simply a blend of their parents' skin. In a single mulatto family, the children might range from light to dark. The way in which they inherited their color, Davenport realized, hinted that the trait followed Mendel's Law, passed down through the generations by hidden factors.

Davenport wanted to publish the data, but he was worried that the results might not hold up. If it turned out that the children were illegitimate, Jordan's pedigrees would be rendered useless. When Davenport told Jordan about his concerns, Jordan assured him he had nothing to worry about. "There isn't the least doubt, I think, about the legitimacy of the children," Jordan wrote. "One man is a minister, one principal of the colored school, one a thriving merchant and one a barber, and all seem considerably above the grade of morality and intelligence of the ordinary stupid and irresponsible negro."

Davenport and his wife, Gertrude, combined Jordan's data with other pedigree studies and published all the results in the *American Naturalist*. For the most part, they wrote with clinical detachment about skin color. It would be hard to tell whether they were discussing humans or pea plants. "Skin color in negro x white crosses is not a typical 'blend' as conceived by those who oppose the modern direction of research in heredity," they declared.

In their private correspondences, though, Davenport and Jordan were frank about their ambitions for a greater study of racial heredity. Skin color was just the start. Jordan went on to publish a study in which he claimed blacks are more prone to tuberculosis than whites. In 1913, he amassed an entire catalog of "unit characters" inherited by Negroes, including physical strength, capacity for routine, and "melodic endowment." Intelligence was not on the list, for "the negro cannot undergo mental development beyond a certain definite maximum," Jordan said.

Davenport shared Jordan's faith in fundamental differences in the mental capacities of blacks and whites. In 1917, Davenport laid out his views in an essay called "The Effects of Race Intermingling." Mixed-race children would suffer because the biology of their parents would be mismatched within them. "One often sees in mulattoes an ambition and push combined with intellectual inadequacy which makes the unhappy hybrid dissatisfied with his lot and a nuisance to others," Davenport wrote.

When Virginia lawmakers began to draft the Racial Integrity Act, Davenport and Jordan pitched in to make it law. Davenport sent advice to the bill's architects, while Jordan worked through Virginia's Anglo-Saxon Club—whose name speaks for itself—to lobby for the bill's passage. The law would stand until 1967, when an interracial couple named Mildred and Richard Loving were convicted of breaking it. The Supreme Court ruled in their favor and struck down the law. By the time the Lovings won their case, many scientists had already decided that race—in the sense of the word as it was used by biologists like Jordan in early twentieth-century America—did not exist.

———

As Davenport and Jordan were spinning their color tops and drawing their racial pedigrees, other researchers were drawing a different image of humanity. They saw the variations in our species as too complex, and too interwoven with historical events, to reduce to simplistic racial caricatures. Starting in 1897, the sociologist and activist W. E. B. Du Bois led a massive study on the Negro residents of Atlanta. His team measured their weight, height, skull size, infant mortality rates, and a host of other vital signs. Du Bois combined the survey results with a synthesis of worldwide anthropological research in his 1906 book *The Health and Physique of the Negro American.*

Du Bois did not present the Negro American as a uniform type of human being. Negro Americans were a population, within which individuals varied tremendously in every regard. In turn, the Negro population itself was intimately connected to other human populations. "The human species so shade and mingle with each other," Du Bois wrote, "that not only indeed is it impossible to draw a color line between black and other races, but in all physical characteristics the Negro race cannot be set off by itself as absolutely different."

Like anthropologists before him, Du Bois studied the outward features of humans. But in the early 1900s, other scientists began observing our inner variability. The Polish serologist Ludwik Hirszfeld proved that blood types were inherited according to Mendel's Law. World War I forced him to put that research on hold, and yet it ultimately provided him with an unprecedented chance to see how blood types varied across human populations.

In 1917, Ludwik and his wife, Hanka, traveled to the Macedonian city of Salonica to work as doctors, treating the thousands of Allied soldiers who were finding refuge in the city. Surrounded by a German cordon, Salonica became "the most crowded and cosmopolitan spot in the universe," one observer later said.

The Hirszfelds saw an opportunity to get a global view of blood types for the first time. Up until then, they had studied the blood types only of Germans, with little idea of how they compared to people from other parts of the world. In Salonica, they were living alongside soldiers from as far away as Senegal, Madagascar, and Russia. The Hirszfelds began asking soldiers and refugees if they'd give some blood. Eventually the couple ended up with samples from 8,400 people, representing sixteen ethnic groups. If the Hirszfelds had tried to gather that much blood in peacetime, their travels might have taken a decade.

The patterns they discovered did not fit any simple division between races. The four known blood types—A, B, AB, and O—turned up in every country they surveyed. The only distinguishing feature was the proportion of types. In England, 43.4 percent had type A, and 7.2 type B. In India, it was type B that was more common, at 41.2 percent; only 19 percent had type A.

The Hirszfelds calculated a "biochemical race-index" for each country, dividing the frequency of Type A by Type B. The index was highest in northwestern Europe and tapered away to the south and east. The Hirszfelds then grouped these "national types" into three regions: the European type, the Intermediate Type, and the Asio-African type. The Hirszfelds were well aware that the types they were constructing would confuse traditionally minded scientists. How, for example, could Asians and Africans be put in one group? "Our biochemical index in no way corresponds to race in the usual sense of the word," the Hirszfelds warned.

The complexity that W. E. B. Du Bois saw in the Negroes of Atlanta, that the Hirszfelds saw in the blood of warring nations, demanded a richer view of heredity—one in which genetic variations were liberally spread across populations and had freedom to flow from one population to another. But in the early 1900s, short of bringing thousands of people together in a besieged city, it was impossible to map the genetic geography of our species. Instead, some of the most important early lessons about race came from other species, such as a little brown fly that lived on the west side of North America.

———

The fly, known as *Drosophila pseudoobscura*, was studied by a Soviet émigré named Theodosius Dobzhansky. Dobzhansky spent his childhood catching butterflies and became a published expert on beetles at age eighteen. His childhood insect hunts gave him a deep appreciation for nature's rich complexity. Looking at the markings and colors of his specimens, he could see the enormous variation that a single species could contain. He could spot differences from one insect to another, and he could also observe differences between populations. Biologists sometimes called these recognizable populations subspecies. Sometimes they called them races.

As a young scientist, Dobzhansky learned of Thomas Hunt Morgan's work on flies. It was a revelation for him. Morgan was tying the visible features of insects that Dobzhansky could see—their wings, their halteres, their spots—to the inner workings of their genes. In 1927, Dobzhansky got a fellowship to spend a year with Morgan in New York. The Soviet Union let Dobzhansky go, assuming he would return home when the fellowship ended. But Dobzhansky cherished his escape from Soviet tyranny and embraced the liberal democracy he found in the United States. He would never set foot in the Soviet Union again.

In 1928, Morgan headed west to the California Institute of Technology, and Dobzhansky went with him to the orange-scented hills of Pasadena. Once Dobzhansky had settled into his new Western home, he drew up a plan to study how genetic variations were spread out over the range of a wild species. He knew he couldn't study Morgan's favorite, *Drosophila melanogaster*. It was a garbage-feeding camp follower. Instead, Dobzhansky picked *Drosophila pseuodoobscura*, a truly wild animal that lived across a range stretching from Guatemala to British Columbia. Dobzhansky bought a Model A Ford and started driving into remote mountain ranges to catch flies from isolated populations. Back in Pasadena, he bred the flies and inspected their chromosomes under a microscope.

Comparing one fly to another, Dobzhansky sometimes spotted a section of a chromosome that was flipped. These so-called inversions acted like a crude genetic marker. Dobzhansky would find many of the same inversions in different parts of North America. Just as with blood types, the inversions marked no sharp geographical divisions between populations of flies. At best, they were more common or less so from place to place.

As Dobzhansky surveyed his flies, his thoughts turned to his fellow humans. The rise of the Nazis in the 1930s disgusted him intensely. He found the way they used a biological definition of race to persecute Jews both vicious and antiscientific. While Dobzhansky dearly loved his adopted country, he also recognized the racism that still infested it, including among many of the older American geneticists he met.

Dobzhansky confronted America's race obsession for himself on a visit to Cold Spring in 1936. He met Edward East, a geneticist who had declared a few years earlier that the Negro race possessed undesirable traits that justified "not only a line but a wide gulf to be fixed permanently between it and the white race." On meeting Dobzhansky, East assured him that, as a brilliant scientist, he could not possibly be a genetically inferior Russian. East was confident Dobzhansky must belong to the small population of Nordics who lived in Russia.

Starting in the late 1930s, Dobzhansky began declaring publicly that popular notions of human races and white superiority "had no basis in biology." In bestselling books, he explained how populations of any animal were a mix of genetic variants. It might be possible to tell one population from another with statistics, but that was a far cry from claiming that all the animals in one population were alike. In fact, the animals with a single population could be tremendously different, genetically speaking. "The idea of a pure race is not even a legitimate abstraction," Dobzhansky wrote. "It is a subterfuge to cloak one's ignorance."

What was true for flies must be true for humans, Dobzhansky asserted. "The laws of heredity are the most universally valid ones among biological regularities yet discovered," he declared. Dobzhansky granted that humans certainly varied, and that some of that variation was spread out geographically. But if human races were sharply defined, then you'd expect to find sharp boundaries between them. And that was almost never possible. While it might be possible to tell an Australian Aborigine from a Belgian by a trait like skin color, another trait—like the prevalence of type B blood— might unite them.

Dobzhansky didn't want to do away with the concept of races completely. He wanted people to see them for just how modest and blurry they really were. Dobzhansky defined races as nothing more than "populations which differ in the frequencies of some gene or genes."

After World War II, a number of other geneticists and anthropologists joined Dobzhansky's campaign. Their efforts culminated in an official statement from the United Nations condemning scientific racism as baseless. But Dobzhansky's new allies pushed the attack further than he had. They demanded scientists give up the term race altogether. It was so fraught with dangerous assumptions that it had to be discarded. The anthropologist Ashley Montagu, for example, switched to using the term *ethnic groups*. But one of Dobzhansky's strongest challenges came from one of his own protégés.

In 1951 a young New Yorker named Richard Lewontin came to Dobzhansky's lab at Columbia to study flies. Dobzhansky was the sort of strong-willed professor who steamrolled his graduate students, pushing them to do the experiments he wanted done and to draw the conclusions he had already reached. But Lewontin pushed right back. He was committed to investigating his own scientific questions. What was most important to Lewontin was finding a new way to measure the genetic diversity in *Drosophila pseudoobscura*, Dobzhansky's favorite fly.

In his own work, Dobzhansky had only managed to get a crude measure of the fly's genetic diversity. He inspected the cells of insects for any that had major changes to their chromosomes. Some flies, for example, had long stretches of DNA that were flipped into reverse order. Lewontin, working with John Lee Hubby at the University of Chicago, developed a new way to look for genetic diversity—one that could detect differences that were invisible under Dobzhansky's microscope.

Lewontin and Hubby would grind up fly larvae and extract proteins from them. They would then put the proteins in a slab of electrified gelatin. The electric field dragged the proteins across the slab, pulling lighter proteins farther than heavier ones. In some cases, the scientists found that all the flies made proteins of the same weight. In other cases, however, some flies had lighter versions and others had heavier ones. And in still other cases, a single fly made both heavy and light versions of a protein.

The different weights of the proteins were the result of variations in the genes that encoded them. Lewontin and Hubby compared the weight of proteins in six populations of *Drosophila pseudoobscura* from Arizona, California, and Colombia. Looking at eighteen kinds of proteins, they found that 30 percent existed in different forms within a single population. In other words, these populations were far from genetically uniform. Even individual flies were surprisingly rich in variations: on average, 12 percent of the proteins in a single fly existed in two forms.

Lewontin then applied this same approach to humans. In the early 1900s, scientists knew of only a single protein that varied from person to person: the blood-type protein that determines people's ABO blood type. By the 1960s, however, scientists had found a number of other kinds of proteins on the surface of blood cells. And these proteins also varied from person to person. A protein called Rh, for example, is present on some people's cells and missing from others'. Doctors have to make sure the Rh factor is the same in a donor and a patient before transfusing blood. Lewontin reviewed studies on these proteins carried out in England. People there had a surprisingly high level of genetic diversity: A third of the proteins varied from person to person.

These results gave Lewontin the confidence to broaden his research and take on the great question of race. He embarked on a new study to see how well racial groups aligned with the actual genetic diversity of humans. If races were indeed biologically significant, Lewontin argued, each race should have a starkly distinctive combination of genetic variants. Most of the genetic diversity should exist between the races rather than between individuals of the same race.

Lewontin gathered measurements of seventeen different proteins in a wide range of human populations, from the Chippewa to the Zulu, from the Dutch to the people of Easter Island. When he sorted people according to their race, he found that the genetic differences between races accounted for only 6.3 percent of the total genetic diversity in humans. The genetic diversity *within* populations, such as the Zulu or the Dutch, contained a staggering 85.4 percent.

In 1972, Lewontin published these results in a profoundly influential paper entitled "The Apportionment of Human Diversity." He concluded that racial classifications had become entrenched in Western society thanks to optical illusions. People defined races based on features "to which human perceptions are most finely tuned (nose, lip and eye shapes, skin color, hair form and quantity)." But these features were influenced by only a small number of genes. It was wrong to assume that all the other genes people carried followed the same patterns.

Given his findings—and given all the suffering that had been justified by racial classifications—Lewontin urged that society set them aside. "Human racial classification is of no social value and is positively destructive of social and human relations," he declared. "Since such racial classification is now seen to be of virtually no genetic or taxonomic significance either, no justification can be offered for its continuance."

It was a sweeping statement to make based on fairly little data. But in the years since, younger generations of scientists have revisited Lewontin's question with better tools. Instead of proteins, they've examined DNA. They've surveyed more people, from more populations. In 2015, for example, three scientists—Keith Hunley and Jeffrey Long of the University of New Mexico and Graciela Cabana of the University of Tennessee—studied DNA from 1,037 people belonging to fifty-two different populations around the world. In each person, they sequenced the same 645 segments of DNA. They looked for the differences in these segments from person to person, calculating their genetic diversity.

Hunley and his colleagues confirmed, like others had before them, that most human genetic diversity can be found within populations rather than between the so-called races. And thanks to the huge scale of their study, they could measure human diversity with far greater precision. The people who live in African populations tend to be more genetically diverse from one another than people who live on other continents, for example. The population with the lowest genetic diversity was a small Amazon tribe called the Suruí. Yet even the Suruí—who number only about 1,120 people— possess about 59 percent of all the genetic diversity in our entire species. If you wiped out everyone on Earth except the Suruí, in other words, nearly two-thirds of humanity's genetic variation would survive.

"In sum," Hunley and his colleagues said, "we concur with Lewontin's conclusion that Western-based racial classifications have no taxonomic significance."

The Venn diagram that Sushant Kumar made for me—showing me all the SNPs that are sprinkled over me, a Nigerian, and a Chinese person—felt like a personal emblem of how badly the concept of race explains human genetic diversity. I'd call myself white, and yet 83 percent of my 3.5 million single-nucleotide polymorphisms are shared by either an African or an East Asian. We may inherit some of those shared variants from common ancestors who lived hundreds of thousands of years ago. Some variants may have arisen later, thanks to a new mutation. They then spread from population to population as people mixed their genes the way people always do. All three of us—me and my pair of anonymous far-flung cousins—got showered in the same genealogical glitter.

———

R ace may not be a meaningful biological concept, but it does exist: It has a powerful existence as a tradition of putting people in social categories. Those categories, then, had profound influences on people's lives. Racial categories served as a legal justification to enslave groups of people and declare their children slaves from birth. Race helped turn other people into scapegoats for economic disasters, justifying their slaughter by the millions. Other people were classified into races judged incompetent to make use of their own land, justifying pushing them off it. And racial categories also gave some people the luxury of enjoying those lands and the profits of slave-based economies without having to learn much about their history. Even after racist institutions and laws were abandoned, their effects have endured for generations, extending race's power.

Because race is a shared experience, it can join people together who aren't closely related. American blacks gained their collective identity only when they came together as cargo on slave ships bound for the colonies. Slave traders roamed up and down the coasts of Africa to capture people separated by thousands of years of history, in Senegal, Nigeria, Angola, even Madagascar. Richard Simson, a surgeon who traveled to South America in 1689 on an English privateer, observed that throwing strangers together was a crucial step in making slavery a profitable business.

The way "to keep Negros quiet," Simson wrote, "is to choose them from several parts of the Country, of different Languages, so that they find they cannot act jointly."

Leaning on the biological concept of race like a crutch has led doctors into some embarrassing blunders in their studies of diseases. "There is no race which is so subject to diabetes as the Jews," declared W. H. Thomas, a New York doctor, in 1904. As late as the early 1900s, Jews were considered a distinct race, with its own diseases. To guide their immigration policies, the United States Congress compiled a book called *Dictionary of Races or Peoples*. The book treated the evidence of the Jewish race as plain to see. "The 'Jewish nose,' and to a less degree other facial characteristics, are found well-nigh everywhere throughout the race," the report declared. Such racial classifications led doctors to look for diseases that were characteristic of each race. Jews, doctors came to agree, had diabetes.

The seed of this notion sprouted in 1870, when a doctor in Vienna named Joseph Seegen observed that a quarter of his patients were diabetic. Other physicians later concluded that Jews died from diabetes at a far higher rate than other groups. German doctors started referring to diabetes as the *Judenkrankheit*: the Jewish disease.

Between 1889 and 1910, New York saw its rate of diabetes triple. To J. G. Wilson, a physician with the US Public Health Service, the cause was clear: the influx of Jewish immigrants. Jews had "some hereditary defect," Wilson said, that made them vulnerable. William Osler, the most important clinical doctor of the early 1900s, blamed the vulnerability of Jews to diabetes on their "neurotic temperament," along with "their racial tendency to corpulence."

And then, in the middle of the twentieth century, the universally recognized fact that diabetes was a disease of the Jewish race simply disappeared. Historians can't definitively say why. It's true that a few scientists questioned the statistical evidence behind the Jewish disease. But no one ever published a definitive takedown. Maybe after Nazis peddled myths that Jews were a naturally disease-ridden race, American doctors quietly decided to retire their own misconceptions.

Myths like Jewish diabetes do not detract from the fact that some people who identify themselves with certain labels—black, Hispanic, Irish, Jewish—have relatively high rates of certain diseases. Ashkenazi Jews have a higher rate of Tay-Sachs disease than other groups, for example. African Americans have a higher rate of sickle cell anemia than European Americans. Hispanics are 60 percent more likely to visit the hospital for asthma than non-Hispanic whites. Researchers have also found significant associations between the race of patients and how their bodies respond to drugs. Chinese people tend to be more sensitive to the blood-thinning drug warfarin than whites, indicating they should get a lower dose.

In some cases, these patterns are the result of the genes people inherited from their ancestors. But sometimes they aren't.

When Richard Cooper went to medical school at the University of Arkansas in the late 1960s, he was stunned at how many of his black patients were suffering from high blood pressure. He would encounter people in their forties and fifties felled by strokes that left them institutionalized. When Cooper did some research on the problem, he learned that American doctors had first noted the high rate of hypertension in American blacks decades earlier. Cardiologists concluded it must be the result of genetic differences between blacks and whites. Paul Dudley White, the preeminent American cardiologist of the early 1900s, called it a "racial predisposition," speculating that the relatives of American blacks in West Africa must suffer from high blood pressure as well.

Cooper went on to become a cardiologist himself, conducting a series of epidemiological studies on heart disease. In the 1990s, he finally got the opportunity to put the racial predisposition hypothesis to the test. Collaborating with an international network of doctors, Cooper measured the blood pressure of eleven thousand people. Paul Dudley White, it turned out, was wrong.

Farmers in rural Nigeria and Cameroon actually had substantially lower blood pressure than American blacks, Cooper found. In fact, they had lower blood pressure than white Americans, too. Most surprisingly of all, Cooper found that people in Finland, Germany, and Spain had higher blood pressure than American blacks.

Cooper's findings don't challenge the fact that genetic variants can increase people's risk of developing high blood pressure. In fact, Cooper himself has helped run studies that have revealed some variants in African Americans and Nigerians that can raise that risk. But this genetic inheritance does not, on its own, explain the experiences of African and European Americans. To understand their differences, doctors need to examine the experiences of blacks and whites in the United States—the stress of life in high-crime neighborhoods and the difficulty of getting good health care, for example. These are powerful inheritances, too, but they're not inscribed in DNA. For scientists carrying out the hard work of disentangling these influences, an outmoded biological concept of race offers no help. In the words of the geneticists Noah Rosenberg and Michael Edge, it has become "a sideshow and a distraction."

To many people, Rosenberg and Edge may sound as if they're ignoring the evidence staring them in the face. While I may share millions of single-nucleotide polymorphisms with a Nigerian, no one would mistake me for someone whose family goes back centuries in Lagos. I once went to Beijing, and never on my trip did someone walk up to me and ask for directions in Mandarin. It is true that humans have physical differences, and some of those differences are spread geographically across the planet. But clinging to old notions about race won't help us understand the nature of those differences—both the ones we can see and the ones we can't.

What matters is ancestry. A small band of hominins in Africa evolved into *Homo sapiens* around 300,000 years ago, after which they expanded across that continent and then across the world. Those journeys shaped the genomes that people inherited from their ancestors. And today, if we look at our own genomes, we can reconstruct some of that history, even back to ancestors who weren't exactly human.

CHAPTER 8

Mongrels

T HE TAITA THRUSH is cloaked in black feathers and tipped by a
vermilion beak. It can be found only in the cloud forests of the
Taita Hills of southern Kenya. Some species of birds fly far across
wide ranges, but the Taita thrush is a homebody. It limits its movements to
a small territory of the forest floor, where it hops about in search of fruit and
insects. This way of life left the bird exquisitely vulnerable to modern change.
Most of the Taita forests were cleared from the hills for farming and pine
tree plantations, leaving behind just a few islands of trees at the summits.
By the end of the twentieth century, only three populations of Taita thrushes
survived. Each numbered just a few hundred.

The isolation of the birds left them especially threatened by extinction.
Before the deforestation, their genes flowed across the landscape as the
birds mated with their neighbors. Now the genes of the Taita thrush were
trapped on hilltop islands. As the years passed, each new generation ran a
greater risk of inheriting two recessive alleles and developing a genetic
disorder—one that might cut a bird's life short or make it infertile.

Hoping to save the species, conservation biologists climbed the hills
and captured 155 thrushes from all three forests. They drew blood from the
birds, and later isolated short segments of DNA from them. They studied
this genetic material to gauge how much diversity was left.

In 1998, a geneticist at the University of Oxford named Jonathan

Pritchard asked the scientists if he could look at the sequences. Pritchard sorted them into three groups, based on their genetic similarities alone. He then asked the conservation biologists where each bird lived. Each of the groups he created perfectly matched each forest.

To sort the Taita thrushes, Pritchard had used a computer program he had recently written with his advisor, Peter Donnelly, and a fellow postdoctoral researcher, Matthew Stephens. They had named the program STRUCTURE.

Sorting 155 birds by DNA alone was a daunting task. At many positions, their genes were identical. Many of the variants shared by only some birds could be found in more than one forest. But Pritchard and his colleagues recognized that certain combinations were more common in each group than others—a signature of their origins. There was a signal buried in all the genetic noise.

When the three forests became isolated, their gene pools got cut off from each other, too. In each pool, some variants were common and some were rare. Without birds traveling between the forests, each generation passed down those variants to their descendants. After many generations of isolation, this pattern still held true. The birds in each forest tended to have some common variants, and it was unusual for them to have rare variants.

Pritchard used STRUCTURE to take advantage of these patterns to sort the birds into groups. He found that three groups worked best. The birds in each of the three groups had a clearer genetic connection to each other than if he had tried sorting them into two groups, or four, or five. STRUCTURE was so good at this sorting that Pritchard could pick out a single thrush, look at its DNA, guess which forest it came from, and almost always get the right answer.

What made this success even more impressive was the similarity of the Taita thrushes. The birds had become isolated from one another only a century beforehand. These were not distinctive subspecies, in other words. From forest to forest, the birds look pretty much identical. They eat the same food. In every forest, males and females form monogamous bonds. The subtle genetic differences that Pritchard used to trace the birds to their homes meant little to the birds themselves.

Pritchard did not invent STRUCTURE only to identify the homes of Taita thrushes. He wanted to build a program that could automatically sort individuals from any species into meaningful groups. He especially wanted to apply it to *Homo sapiens*. In the 1990s, it had become clear that mapping the genetic structure of humanity would be crucial to finding genes associated with diseases.

Scientists had begun searching for these genes by looking for variants that were unusually common in people with a particular disease. But they could end up with misleading results if they didn't take into account people's ancestry. This danger came to be known as the chopstick effect, after a fable spun in 1994 by the geneticists Eric Lander and Nicholas Schork.

Imagine, Lander and Schork said, that a team of researchers in San Francisco decided they would find the genetic cause for why some people in the city ate with chopsticks and others did not. They took blood samples from a random selection of people and scanned their DNA. Lo and behold, the scientists discovered an allele for an immune system gene that was far more common among chopstick users than among people who did not use them. Therefore, the geneticists concluded, inheriting that allele caused people to be more likely to use chopsticks.

They were wrong. The allele was more common in chopstick users for an entirely different reason: because it was more common in Asian Americans than people of European descent. Asian Americans were also more likely to use chopsticks than European Americans. The immune system, in other words, has nothing to do with chopsticks.

A real example of the chopstick effect came to light in the 1980s among the Pima Indians of the southwestern United States. They suffer from Type II diabetes at a catastrophic rate: About half of all adults in the community develop the disease. Diabetes began to wreak havoc on the Pima only in the 1900s, after they lost their land and their sophisticated farming system. Suddenly they had to survive on carbohydrate-rich government-supplied food. That diet could put anyone at greater risk of diabetes, but the Pima proved to be especially vulnerable. Geneticists suspected that their higher risk was due to genetic variants they shared.

William Knowler, a researcher with the National Institute of Diabetes and Digestive and Kidney Diseases, led one of the first studies on Pima DNA. He studied 4,920 subjects on the Pima reservation in Arizona. He discovered that about 6 out of every 100 Pima carried a variant in a gene called Gm, which encodes a type of antibody. The Gm variant seemed to protect the Pima against diabetes. Among those who carried it, only 8 percent developed the disease. Among the Pima who lacked the Gm variant, 29 percent developed diabetes.

Knowler might have stopped there and declared victory. But he was well aware that the Pima he studied did not have a simple history. Native Americans arrived in the Western Hemisphere some fifteen thousand years ago. The Pima probably settled in the Southwest by two thousand years ago, and five centuries ago they came into contact with people of European ancestry: first Spanish explorers, and then Mexican farmers. By the mid-1900s, Pima Indians and Mexican migrant laborers were working together on Arizona cotton farms. Some Pima started families with people outside the tribe. As a result, some of the Pima whom Knowler studied had a fair bit of European ancestry.

To take ancestry into account, Knowler split his subjects into two groups: those with some European background and those with none. When he looked at the Gm variant within each group, the evidence for its defense against diabetes disappeared. Among people with 100 percent Pima ancestry, having the Gm variant didn't lower the risk of diabetes. It also didn't make a difference when Knowler compared the Pimas with some European ancestry to each other.

Knowler had been initially fooled by the Gm variant, he realized, because it was much more common among Pima with some European ancestry. It served as a genetic marker, in other words, rather than as a direct defense against diabetes. Knowler concluded that European versions of certain genes might lower the odds of developing diabetes on a diet high in simple carbohydrates. But he couldn't say from his data which genes those might be. What he did know was that the Gm variant merely came along for the ride.

K nowler managed to overcome the chopstick effect by asking the Pima about their ancestors. Their European forebears had lived recently enough that the Pima could give Knowler a reliable genealogy. He was also fortunate to be studying a small, relatively isolated community. Other scientists who study broader populations of people with mixtures of ancestries and fuzzy family memories do not enjoy Knowler's advantages.

Pritchard and his colleagues, collaborating with Noah Rosenberg at Stanford University, found that they could use STRUCTURE to overcome the chopstick effect, even when they didn't have any information about people's family trees. The geneticists could identify clusters of people based on their DNA alone. To adapt STRUCTURE to the task, the scientists had to reckon with the fact that people are not Taita thrushes. They do not live in a few forests in a small patch of Africa; they span the globe. And rather than living in isolation, humans have migrated over thousands of years, mixing their DNA in their living descendants.

The scientists created a version of STRUCTURE that let them scan the genetic variation in people and assign each individual's DNA to one or more groups of ancestors. Pritchard and his colleagues could then look at how well they could account for the genetic variation in people with different numbers of groups.

In 2002, Pritchard and his colleagues tried STRUCTURE out on people. They looked at genetic variations in 1,056 people from around the planet. Just as in other studies of human diversity, they found that the overwhelming amount of genetic diversity was between individuals. The genetic differences between major groups accounted for only 3 to 5 percent. And yet, with the help of STRUCTURE, the researchers used some of those variants to sort people into genetic clusters. When the scientists allowed people to descend from five different groups, for example, they clustered mostly according to the continents they lived on. People in Africa could trace much of their ancestry to one group, while people in Eurasia were linked to a second one. East Asians traced much of their ancestry to a third, Pacific Islanders to a fourth, and people in the Americas to a fifth.

Much to the chagrin of Pritchard and his colleagues, some people mistakenly took these results as evidence for a biological concept of race. But any resemblance between genetic clusters of people and racial categories concocted before genetics existed can have no deep meaning. It would make just as little sense to say that Aristotle's classifications of animals have been vindicated by comparing the DNA of different species. Aristotle put species into categories based on whether they had blood, whether they had hair, and so on. The genes of animals with hair—mammals—show that they do indeed belong to a group. But Aristotle also threw together species into other categories that have no close evolutionary link. It would be a disaster for biology if scientists cast off two thousand years of progress and followed Aristotle's example. The same is true for race.

Those who claim that STRUCTURE proves the existence of human races also have to ignore how Pritchard and his colleagues actually used it to study human variation. The clusters that the five ancestral groups produced didn't have sharp boundaries. Where two clusters met on a map of the world, the researchers found people who had some DNA that linked them to one group, and some that linked them to the other. What's more, STRUCTURE allows scientists to try out different numbers of ancestral groups to see what sort of clusters emerge. After trying out five ancestral groups, Pritchard and his colleagues decided to see what would happen if they ran their program with six. The results were pretty much the same, with one telling exception: A single population broke off from the Eurasian cluster and formed a cluster of its own.

That population is known as the Kalash, a few thousand people who live in the Hindu Kush mountains of Pakistan. Their separation in Pritchard's study may tell us something important about the history of the Kalash—perhaps a long isolation from other tribes in Pakistan, allowing them to accumulate a small number of genetic variations that set them off from much larger clusters of people. But it doesn't mean that the Kalash are biologically a race of their own.

Pritchard and his colleagues were also able to use STRUCTURE to search for clusters within clusters. For their study, the researchers had picked out five populations of people in the Americas, including the Pima in Arizona and the Suruí of Brazil. When they created a model of those people based on five ancestral groups, they were able to identify people's tribes based on their DNA alone.

In the years since the 2002 paper came out, scientists have been improving on STRUCTURE, developing more powerful statistical tools for tracing people's ancestry. They've also been accumulating more DNA from more parts of the world, to get a more accurate map of the human genetic landscape. Before long, it became possible for genealogy companies to analyze customers' DNA and produce a rough breakdown of their ancestry. It was this approach that allowed LeVar Burton to learn that three-quarters of his ancestry came from sub-Saharan Africa, for example.

One of Pritchard's students, Joe Pickrell, ended up at the New York Genome Center. He and his colleagues there used his own update on STRUCTURE to compare people's DNA and estimate their ancestry. When Pickrell ran my DNA through his computational pipeline, he quickly discovered—not surprisingly—that my ancestry is entirely European. He and his colleagues then examined stretches of my DNA to see if they could trace them to smaller populations within Europe. To find variants that pointed to my northwestern European ancestors, for example, Pickrell and his colleagues looked at the DNA of people from Iceland, Scotland, England, the Orkney Islands, and Norway.

The one group they looked at that didn't have a clear geographical location was the Ashkenazi Jews. While Ashkenazi Jews lived for generations across much of eastern Europe, they remained a culturally closed group, mainly sharing their variants among themselves. They thus became recognizably distinct from neighboring Christians.

After a few weeks, Pickrell and his colleagues sent me a pie chart of my ancestry:

43% Ashkenazi Jewish,
25% northwestern Europe,
23% south-central Europe (Italy, in other words),
6% southwestern Europe (Spain, Portugal, and southwestern France),

2.2% northern Slavic (which means the region running from the
Ukraine to Estonia), and

1.3% that remained too ambiguous to put on the map.

As I pored over the numbers, I grew unsettled. Thinking about all the stories I had told myself about my ancestry since I was young, I realized how often they had let me down.

Names have let me down particularly hard. If your name was Carl Zimmer, you might assume you were German. I certainly did. In school, friends would sometimes greet me with *Guten Tag, Herr Zimmer!* But when my genealogically minded relatives traced our Zimmer ancestors back to my great-great-grandfather Wolf Zimmer, it turned out he didn't live anywhere near Germany. He lived instead in Galicia, a region in what is now the Ukraine.

If we ever manage to reach further back in the Zimmer line, we will probably discover that it vanishes within a few generations. Before the late 1700s, many eastern European Jews did not use family names. The Austro-Hungarian Empire—of which Galicia was then a part—ordered that all Jews take a name so that they could be more readily taxed. Since Yiddish was banished to private life, the Jews chose names that Austrian officials would approve. It's likely that only then did my ancestors become Zimmer. My name is a convenient fiction.

Goodspeed, my mother's name, led me to see England as the other important country of my origins. Reading Shakespeare or Sherlock Holmes tales felt like learning about where I had come from. Genealogy certainly does trace the Goodspeed name back to England. But for me, the Goodspeed name marks only a single branch among many. Pickrell and his colleagues could trace those other branches across many other parts of Europe, perhaps as far away as Spain and Italy—places that my mother's research has never led her to.

After I got these results, I paid Pickrell and his colleagues a visit to pester them with questions. If my father was Jewish, how could I be only 43 percent Ashkenazi? Did that mean that my father was only 86 percent?

Pickrell warned me that their analysis was accurate enough to unsettle my family, but not to give me the final word on my genetic inheritance. "You should treat those numbers as an approximation of reality," Pickrell said.

Those numbers were the best he could manage at the moment, with the genomes and the methods at hand. That might change if I came back to Pickrell in ten years. By then, he expected, geneticists would be able to compare millions of human genomes. Instead of relying on variants that are fairly common in at least one population, they may be using rare variants that arose in individuals just a few generations ago and are shared only by their direct descendants.

"Now it's simply a question of a match—do you have the genetic variant or not?" Pickrell explained to me. "Everyone who shares that genetic variant descends from the same common ancestor who lived two hundred years ago. That must make life so much easier."

Pickrell also warned me that his method could only take me back a few centuries into the past. The groups of people that existed at that time did not necessarily exist a few centuries earlier. *Ashkenazi* is the name for a particular group of people who lived in a particular place at a particular time. Before AD 1000, the Ashkenazi people did not exist. Their ancestors went by other names.

To excavate deeper into my ancestry, I'd need a different genetic shovel.

———

To examine the Ashkenazi-linked DNA in my genome, Dina Zielinski and Nathaniel Pearson of the New York Genome Center used another piece of software, known as RFMix. Developed by scientists at Stanford in 2013, it searches for tiny segments of matching DNA in different people's genomes. Those segments, chopped into little pieces by many generations of meiosis, can reveal ancient kinships. RFMix can match different segments to people from different parts of the world.

"It's a quilt," Pearson told me, "made up of a segment from one ancestor attached to a segment from another ancestor. And we're trying to figure out where those segments came from."

Pearson and Zielinski tested my DNA against two possibilities that historians have raised for where Ashkenazi Jews came from. According to one hypothesis, they descend mainly from a kingdom in present-day southern Russia, on the northwestern banks of the Caspian Sea. These people, called the Khazars, converted to Judaism perhaps a thousand years ago. They then migrated north and west into Europe.

Many historians have dismissed the Khazar hypothesis, arguing instead that Jews were already living in Italy and France at the time when the Ashkenazi ancestors supposedly converted to Judaism far to the east. These scholars argue instead that the ancestors of Ashkenazi Jews originated in Israel and other parts of the Levant. These people traveled in waves to Italy in the age of the Roman Empire, and from there, they expanded into other parts of southern Europe. Later, when Jews became increasingly persecuted across Europe, some of them came together in Poland to seek refuge.

Zielinski and Pearson tested these possibilities by comparing my genome to the genomes of people who would have deep kinship with me. They used genomes from people in France and Italy to look for ancestors in southern and western Europe. They also included a Russian genome to represent eastern Europe. The Khazar kingdom is long gone, and so Zielinski and Pearson used genomes from an ethnic group from the region, called the Adygei. To look for an ancestry in the Near East, they added Palestinians and Druze to the analysis.

The scientists inspected million-base-long segments of my DNA and compared them to the same segments in other people's DNA. They used RFMix to find the closest match to each piece in the genomes of the other people in their study. When they were done, they generated a color-coded map of my chromosomes for me.

Most of my chromosomes matched the genomes of southwestern Europe or the Near East. A few segments showed a Russian ancestry, and even fewer resembled the Adygei. My genome offers no support for the Khazar theory of the Jews.

Zielinski and Pearson carried out only a small-scale study on my DNA—an act of scientific generosity, really. Pearson warned me not to look at their results as the last word on my ancestry. "We have to have tons of grains of salt on the table," he said.

Salt notwithstanding, Pearson and Zielinski's results jibed nicely with a much bigger study carried out in 2016 by Shai Carmi of Hebrew University in Jerusalem and his colleagues. They looked at 252,358 single-nucleotide polymorphisms in the DNA of 2,540 Ashkenazi Jews, 543 Europeans, and 293 people from the Near East. Carmi and his colleagues couldn't study each genome as deeply as Zielinski and Pearson had. But they could compare many more people, hailing from many more regions.

Using RFMix and other software, they concluded that Ashkenazi Jews can trace roughly half their ancestry to the Near East, while the other half comes from Europe. The researchers found hints of two separate pulses of mixing. The first occurred in southern Europe—Italy looks like a strong possibility. The second occurred more recently, bringing together Ashkenazim with northern or eastern Europeans.

While there are a lot of uncertainties in Carmi's study, it also aligns with historical evidence that the Ashkenazi people emerged through a long migration, with plenty of mingling along the way. My parents are part of an ancient tradition.

————

While some of my father's ancestors may have come from the Near East into Europe a thousand years ago, my mother's ancestors were probably there long beforehand. Genetic genealogy can't take me back very far into that history, nor can it lead me to the Stone Age villages where my European ancestors lived. But I can be confident that my European roots run deep. In other words—to resort to the language of censuses—I'm white.

White makes sense as the name of a cultural group, but as a biological label, it's just as dubious as terms like black and Hispanic. We tend to think of whites as the pale-skinned people of Europe and their descendants, a group of humans joined together on one continent, sharing the same uniform heredity that reaches back for tens of thousands of years. The people who lived in Europe twenty thousand years ago might be different in the ways they lived, hunting woolly rhinos instead of posting pictures on Instagram. But we still think of them as white. As scientists have examined the DNA of Europeans—both people who live on the continent today and those who lived there tens of thousands of years ago—they've demonstrated just how wrongheaded those notions are.

In the early 1980s, a graduate student at Uppsala University in Sweden named Svante Pääbo wondered if he could extract DNA from ancient remains. In 1985, he managed to isolate a few thousand bases from a 2,400-year-old mummy of an Egyptian child. He went on to extract DNA out of far older fossils, pioneering a new field called paleogenetics. Pääbo later became the director of the Max Planck Institute for Evolutionary Anthropology, where he gathered a flock of scientists and graduate students to help him fish for more ancient genes. Other scientists built paleogenetics labs in places like Oxford, Harvard, and Copenhagen.

For years, their research was hit-or-miss. Sometimes fossils turned out to have no DNA at all, because they had fossilized in an unforgiving environment. Other fossils had too much DNA—not from humans, but from bacteria and fungi that invaded the bones after death. And even when the geneticists did find human DNA, it often turned out to belong to a technician or some other living person, from whom a flake of skin or a droplet of sweat had wafted into the lab equipment.

Pääbo and other researchers spent years improving paleogenetics. They figured out how to distinguish new, contaminating DNA from ancient material. They learned not just how to grab one particular piece of DNA from a fossil but how to grab it all, sequence it, and assemble it into an entire genome. They even got better at picking which bones to drill for DNA. At first, geneticists had simply cut out chunks from whichever bone a museum curator considered expendable. But in the early 2010s, Ron Pinhasi, an archaeologist at University College Dublin, discovered that one kind of bone was far better than the rest. For some reason, the hard bony case surrounding the inner ear was often rich with DNA, even when none could be found elsewhere in a skeleton.

In 2015, paleogeneticists—especially David Reich's team at Harvard University—began publishing dozens, sometimes even hundreds, of ancient European genomes at a time. The results create a kind of genetic transect. Scientists could trace changes in DNA in Europe over more than forty thousand years, mapping them from Spain to Russia. And because this transect was made from whole genomes, each skeleton could tell scientists about thousands of its own ancestors.

The oldest fossils of modern humans in Europe, dating back forty-five thousand years, look much like the bones of living Europeans. But their DNA doesn't give any indication that any living Europeans inherited their genes. Genetically speaking, they look as if they came from a different continent altogether. It's hard to say what became of them. Their particular combination of genetic variants apparently vanished about thirty-seven thousand years ago.

At a thirty-five-thousand-year-old site in Belgium, paleoanthropologists managed to get DNA out of another skeleton. The skeleton belongs to a culture known as the Aurignacian, which existed across all of Europe that wasn't buried below the glaciers of the last Ice Age. They made tools of stone and bone, painted caves with pictures of woolly rhinoceroses, and carved lion-headed figurines. The DNA from the Belgian skeleton had a distinct genetic signature of its own, different from the oldest Europeans.

About twenty-seven thousand years ago, the Aurignacian culture disappeared from the archaeological record, replaced by a new one called the Gravettian. The Gravettian people used spears to hunt mammoths, and nets to trap small game. Reich's team got DNA from Gravettian skeletons and discovered that they, too, were a distinct people with no direct genetic link to the Aurignacians who came before. For thousands of years, the Gravettian genetic lineage was the only one to be found throughout Europe.

And then, remarkably, Aurignacian DNA made a comeback. A Spanish skeleton dating back nineteen thousand years contains a mixture of Gravettian and Aurignacian DNA. Nobody can say yet where the Aurignacian people disappeared to during the vast intervening time, or how they ended up in Spain, or how people from two such profoundly different cultures ended up having children together. All we know is that for the next few thousand years, everyone in Europe now had genomes blended from these two sources of DNA.

About fourteen thousand years ago, this long stasis was broken. The skeletons from this age now include a third ingredient in their genome. This extra DNA shares some hallmarks with that of people who live today in the Near East. Archaeologists have found that people in the Near East were living as hunter-gatherers at the time. It's possible that as the Ice Age glaciers retreated north, people from the Near East expanded into Europe and began to interbreed with the Gravettian-Aurignacian people. And then, once more, Europe settled into a new genetic equilibrium. For another five thousand years, Europeans inherited their DNA from the same combination of ancestors.

The next wave arrived about nine thousand years ago, and these people came with some important baggage. They were not hunter-gatherers but farmers, bringing crops like wheat and barley with them, along with sheep and goats. These immigrants descended from the first farmers, who had domesticated plants and animals some two thousand years earlier in the Near East. They were not just separated by their culture from the European hunter-gatherers; a wide gulf of ancestry separated them, too. Their common ancestors might have split apart fifty thousand years earlier.

The farmers expanded from the Near East into Turkey and then westward along the southern edges of Europe. As the farmers cleared land, planted crops, and grazed livestock, some of the hunter-gatherers retreated to less fertile lands, while others interbred, mixing their DNA with that of the newcomers. Over centuries, the isolated populations of hunter-gatherers winked out, while the farmers—now carrying a mix of Near Eastern and European hunter-gatherer genes—settled the entire continent.

One more major wave arrived in Europe 4,500 years ago. The DNA of skeletons from that age have many variants in common with a vanished people from the Russian steppes. Known as the Yamnaya, they tended vast herds of sheep grazing on the grasslands, following them with horses and wagons. This nomadic way of life was hugely successful for them. They grew so wealthy, they could build enormous funeral mounds for their dead, which they filled with jewelry, weapons, and even entire chariots.

The DNA of Bronze Age Europeans living 4,500 years ago reveals that the Yamnaya or another closely related people moved into Europe from the Russian steppes. They arrived first in Poland and Germany, building walled cities where they carried on their distinct culture. Within a few centuries, the genetic signature of the steppe people had jumped the English Channel into Britain. At first the steppe people stayed genetically distinct from the farmers and hunter-gatherers around them. But by the end of the Bronze Age, the barriers around the steppe people and the rest of Europe collapsed, as they had so often before. Skeletons younger than 4,500 years contain melanges of ancestry: steppe people, Near Eastern farmers, Near Eastern hunter-gatherers, Gravettians, and Aurignacians. After this last major merger, Europe remained a continent of many cultures. But now its people had taken on a genetic profile from which I and other people of European descent draw their ancestry.

Ancient DNA demonstrated that white people do not share some deep, pure genetic bond reaching back to the earliest days of the human occupation of Europe. The earliest *Homo sapiens* to arrive in Europe have no direct connection to living Europeans at all. Living people of European descent can trace their ancestry to the people who came to the continent in a series of waves, separated by thousands of years. These groups were no more closely related to one another than Laplanders are to Indonesians. But in Europe they encountered each other and mixed their genes. Today's Europeans are fairly uniform, genetically speaking. But that uniformity came out of a biological blender.

Ancient DNA doesn't simply debunk the notion of white purity. It debunks the very name *white*.

From the beginning, skin colors were crucial to Western racial categories. The black skin of Africans was merely the outward mark of an inward curse. Along with whites in Europe and blacks in Africa, there were yellow

Chinese and red Indians. People whose skin was too light or too dark for their race aroused suspicion about their ancestry.

Yet skin color is not a timeless hallmark of human races. It has changed in different places and at different times, probably thanks to natural selection in some cases and thanks to the movement of people in others. Skin-color alleles have traveled the world, working their way into different populations. And the range of light skin tones we call white is only a recent development in this history.

People get their particular color from pigment-producing cells in their skin, called melanocytes. Each melanocyte is stuffed with pouches of pigment called melanosomes. One type of pigment is a yellowish red, and the other is a blackish brown. The amount and the balance of these pigments together set a person's skin color. Since there are many ways to make these adjustments—add more pigment to each melanosome, for example, or increase the number of melanosomes—mutations to a number of genes can produce similar colors.

Today, humans range in color from freckly pale to jet-black. The geography of skin color is complicated. Dark skin is hardly unique to Africa; people in Australia, New Guinea, and parts of southern India are just as dark. Nor is Africa itself uniform: While the Dinka of East Africa are among the darkest people on Earth, the San hunter-gatherers of southern Africa are tan.

Since hominin skin doesn't fossilize, we can't say for sure what skin color our ancestors had four million years ago. But if our closest living primate relatives—gorillas and chimpanzees—are any guide, they likely had light skin. At some point, perhaps two million years ago, our ancestors began adapting to life on the African savanna and lost much of their body hair. Once their skin was directly exposed to sunlight, their skin probably started to change color. That's because the ultraviolet rays in sunlight could now more easily strike skin cells. The damage they caused could lead to skin cancer, and could also destroy an essential molecule in the skin called folate. Mutations that added more pigment to the skin could shield our distant ancestors from this harm.

In 2017, Sarah Tishkoff, a geneticist at the University of Pennsylvania, led a study to uncover some clues about the early evolution of human skin. She and her colleagues measured the skin reflectance of 1,570 people in Ethiopia, Tanzania, and Botswana. They then scanned the DNA of their volunteers for variants that were common in people with lighter or darker skin. They found eight variants that were strongly associated with skin color.

The researchers searched worldwide DNA databases and discovered that these variants were also present in some populations scattered across the world. By comparing the DNA surrounding these variants, the scientists could estimate how long ago they arose in common ancestors. To their surprise, the researchers found that all eight variants were hundreds of thousands of years old—older than our entire species, in other words.

On its own, this result can't let us know what color the first members of *Homo sapiens* were. Some of the ancient variants darken skin, but others lighten it. It's possible that they were all present together in early humans, giving them a medium color. Or perhaps there were early humans in Africa who were very dark and others who were light-skinned.

Within Africa, these variants experienced strong natural selection in different places. Close to the equator, wearing few clothes, the Dinka and related people evolved dark skin. In southern Africa, the sunlight was less intense. For the San, dark skin might have actually been a liability. While too much ultraviolet radiation is harmful, too little can cause trouble as well. When the rays strike the skin, they supply the energy our cells need to make vitamin D. Dark skin may have interfered with its production among the San, leading them to evolve their tan skin.

Somewhere between fifty thousand and eighty thousand years ago, a small group of humans expanded out of Africa. Tishkoff and her colleagues discovered that the dark-skinned people of southern India, Australia, and New Guinea all carry the same dark genetic variants she and her colleagues found in Africa. It's possible that one wave of migrants carried the genes for dark skin on a journey across the southern edge of Asia and into the Pacific.

Some of the ancient variants for light skin made their way into light-skinned populations in Asia and Europe. Along with these African inheritances, Asians and Europeans also gained new mutations that altered their skin color more. One mutation, to a gene called SLC24A5, drastically reduced the pigment that melanocytes make. All living Europeans have it, and so do a substantial fraction of Asian populations.

The discovery of ancient DNA in early humans is allowing researchers to get a better sense of how these newer variants arose. In 2014, researchers studied a hunter-gatherer who lived in Spain seven thousand years ago. They found that he carried mutations for blue eyes, but he lacked the mutations to genes such as SLC24A5 that are known to make the skin of living Europeans light. Scientists therefore suspect that this seven-thousand-year-old hunter from Spain was a dark-skinned, blue-eyed man.

Of course, this was just one man. But when scientists looked at the DNA from other skeletons across Europe, they found that it fit a broader pattern. Across western Europe—in Spain, France, Germany, and Croatia—Europe's hunter-gatherers lacked the skin-lightening mutations found in Europeans today. Farther east, in places like Sweden and the Baltics, scientists have found light-skin mutations in skeletons of other hunter-gatherer groups dating back almost eight thousand years. Meanwhile, to the south, the farmers who expanded into Europe from Turkey had only one light variant, perhaps meaning that they had olive skin. Only about four thousand years ago, as the different populations in Europe started merging, did their skin start to become uniform.

This long lag is puzzling. If all that mattered to skin color was ultraviolet rays, then the early humans who came to Europe should have swiftly evolved to become light-skinned and then stayed that way. Nevertheless, the fact remains that more than forty thousand years passed before people in Europe began to perceive themselves as joined by the paleness of their skin.

The patterns of inheritance that ancient DNA revealed in Europe—of thousands of years of stability disturbed by sudden mergings of deeply different peoples—have also come to light in other parts of the world. In

India today, for example, just about everyone has inherited a mix of DNA from two distinct groups of ancestors. One of those groups was most closely related to Europeans, Central Asians, and people from the Near East. The other group, a more enigmatic one, was closely related to the people who live on a tiny archipelago in the Indian Ocean, called the Andaman Islands. The two groups came together to produce the Indian population (as we know it today) within the last four thousand years.

Africa today is home to more than 1.2 billion people, and it is also where most of the fossil record of our species and our more distant hominin ancestors can be found. But that does not mean that living Africans are ancient relics of humanity's past. The genetic profile of many Africans today is profoundly different from that of the people who lived in the same places just a few thousand years ago. While the history of humans in Africa may be far longer than in other parts of the world, Africans are the product of their own turbulent migrations and mixings.

Much of the evidence for this turbulence comes from ancient DNA. The fact that scientists managed to get ancient DNA from African skeletons came as a huge surprise to scientists themselves. They assumed that the hot climate of Africa would wipe out the DNA in skeletons. But it turns out that the mountains of countries like Malawi, Kenya, and Ethiopia are cool enough to preserve genetic material. In 2017, David Reich and his colleagues published details on ancient DNA from sixteen different Africans who lived as long ago as eight thousand years.

Reich's team found that the branches of living sub-Saharan Africans sprouted early in human history. They first began splitting somewhere between 200,000 and 300,000 years ago, not long after the origin of our species. Over hundreds of generations, the hunter-gatherers in the south, east, and west of the continent gained distinctive genetic profiles. But despite those differences, they did not become entirely cut off from each other. Some genes managed to flow for thousands of miles through a network of small bands.

The ancient DNA Reich and his colleagues studied suggests that at some point the eastern hunter-gatherer population expanded both to the west and east. To the west, Western Africans inherited a substantial amount of their DNA. To the east, these Africans moved out of Africa altogether; their descendants settled across Europe, Asia, and beyond.

But the flow of genes also traveled back into Africa. When Reich and his colleagues studied a three-thousand-year-old girl from a tribe of Tanzanian cattle herders, they found that a third of her ancestry belonged not in Africa but in the Near East, among the first farmers. Younger fossils in Africa revealed that this Near Eastern DNA flowed all the way into South Africa, where it can be found in many living South Africans. These immigrants also appear to have brought the variant in the SLC24A5 gene that makes European skin light—altering the skin of Africans who inherited it.

It is possible that these migrants brought something else: crops and livestock. Meanwhile, elsewhere on the continent, other Africans were also domesticating native plants like yams and bananas. About four thousand years ago a group of farmers and herders, known as the Bantu, began expanding from the present-day border of Cameroon and Nigeria.

Over the next two millennia they expanded east and south, bringing with them iron tools and a distinctive language. The ancient DNA that Reich and his colleagues extracted from fossils shows that in places like Malawi, the Bantu entirely replaced the earlier hunter-gatherers. In East Africa, they remained distinct for generations before merging with the hunter-gatherers there. Only a handful of tiny tribes in East Africa have a strong genetic link to the people who lived there three thousand years ago without any Bantu ancestry. And in southern Africa, only a few small groups of hunter-gatherers still carry a genetic legacy that once stretched throughout the region.

The people of Madagascar, lying off the east coast of Africa, inherited an even more far-flung combination of genes. Half their genetic ancestry comes from East Africans, and the other half from Southeast Asia. It's likely that a small group of people sailing in the Indian Ocean got swept all the way across. A 2016 study was able to trace the Asian ancestors of Madagascar all the way back to a single village in Borneo.

As scientists sequence more DNA from living people and find more in ancient skeletons, they will probably find even more movements and mixings. The further back in time we look, the harder it is to make out the outlines of history, but scientists have already discovered the vestiges of more drastic interbreeding: ancient encounters that introduced DNA into our gene pool from Neanderthals and other extinct humans. To find out about my own inheritance, I took my genome to the place where the study of human heredity had gotten off to such a dubious start: Cold Spring Harbor.

———

O ne sunny late-winter day, I drove down Bungtown Road, along the southern edge of the Long Island Sound, and then climbed a high hill to reach the lab. I parked my car and used a map to navigate my way past a bell tower. The tower housed a staircase in the shape of a DNA double helix. The letters of the four bases, *a*, *c*, *g*, and *t*, were engraved high on its four walls.

I shuffled down broad steps to a cluster of research buildings. Inside one of them I found the office of a young scientist named Adam Siepel. He welcomed me in and had me sit at a table below a giant monitor bolted to a wall. Siepel had a high forehead and kept his hair cropped to near-baldness. He kept a miniature rock garden on a shelf by his desk, a rivulet running through the center and making an endless burble. Near his window, Siepel had set a picture of his young son and daughter. Next to the photo was a peculiar skull, with a brow ridge sloped like a trough. It was a cast of a Neanderthal head.

Ancestor and descendants, I thought.

A century beforehand, the scientists who worked at Cold Spring Harbor would not have taken kindly to the pride of place Siepel gave a Neanderthal. To Charles Davenport and his fellow eugenicists, Neanderthals were nothing more than a brutish victim of humanity's progress, a race below all other races.

Davenport would sometimes travel the forty miles from Cold Spring Harbor to New York City. There he would attend the meetings of the Galton

Society at the American Museum of Natural History. Davenport had helped found the society, along with Henry Fairfield Osborn, the president of the museum. It was made up of scientists and wealthy businessmen dedicated to putting eugenics to work in order to save American society. At their meetings, the Galton Society would grouse about Negroes, immigrants from the wrong parts of Europe, and the feebleminded.

Invitations, Davenport once said, were "confined to native Americans." He did not mean Cherokees.

Osborn had made his scientific mark as a paleontologist who studied mammal evolution. But by the early 1900s, eugenics became his overriding mission. "Heredity and racial predisposition are stronger and more stable than environment and education," he declared. Osborn could not explain genetics to the public as effectively as Davenport or other members of the Galton Society. But he could supply eugenics with its evolutionary back-story. Osborn promoted a eugenic picture of mankind in bestselling books. He even used his museum for the cause, designing its first exhibit about human evolution.

Examining the fossil record of the early 1900s, Osborn argued that central Asia was a nursery for the evolution of new kinds of mammals. Once they evolved there they expanded outward to other continents in a series of waves. Apes and humans were no different, Osborn believed: Their new forms also emerged from the Asian nursery. Each new wave was more sophisticated than the previous one, often eradicating them when they met.

One of the earliest waves to emerge from Asia, Osborn claimed, was the Neanderthals. In 1856, quarry workers in Germany had found the first fragments of Neanderthal fossils. They suggested heavyset humans with thick brow ridges. By the early 1900s, more Neanderthal fossils had emerged across Europe. Looking at their bones, Osborn pictured a lumbering brute: "an enormous head placed upon a short and thick trunk, with limbs very short and thick-set, and very robust; the shoulders broad and stooping." Even the hands of Neanderthals seemed huge and clumsy to Osborn, who said they were "without the delicate play between the thumb and fingers characteristic of modern races."

On tours of Europe, Osborn would visit caves to acquaint himself with Neanderthal remains. He could see that Neanderthals had been capable of hunting big game like horse and bison. But their stone tools were primitive compared to more recent ones. And no one could find any trace of Neanderthal art. This evidence—or the lack of it—only strengthened Osborn's conviction that Neanderthals had subhuman minds.

Not so agreeable were their giant braincases. If Neanderthals were extinct subhuman brutes, they shouldn't have had brains as big as living humans. Osborn dodged this quandary by ignoring the size of their brains and making much of their shape instead. The Neanderthal brain lacked "the superior organization of the brain in recent man," Osborn declared, especially in the prefrontal cortex, "which is the seat of the higher faculties."

When Osborn established the Hall of the Age of Man in his museum, he had murals and busts of Neanderthals put on display. He ordered them to be depicted as dark-skinned, hairy, and brutish. "The Neanderthals represent a side branch of the human race which became wholly extinct in western Europe," he said.

The Neanderthals didn't just quietly disappear, however. They were annihilated by the Cro-Magnons, Osborn believed, a race which had evolved in Asia and which was in no way connected by any ancestral links with the Neanderthals.

Osborn believed that Cro-Magnons were far superior, "a race with a brain capable of ideas, of reasoning, of imagination, and more highly endowed with artistic sense and ability than any uncivilized race which has ever been discovered." Their superiority, in fact, allowed the Cro-Magnons to wrest control of Europe. "They were armed with weapons which, with their superior intelligence and physique, would have given them a very great advantage in contests with the Neanderthals," Osborn said. Judging from this superiority, Osborn believed Cro-Magnons "probably belonged to the Caucasian stock."

Along with Caucasians, Osborn recognized Mongoloids and Negroids as "three absolutely distinct stocks, which in zoology would be given the rank of species, if not of genera."

Osborn was never very clear about the order in which the three stocks arose, but he was sure that Negroids emerged first. For proof, he pointed to how they fared on Henry Goddard's intelligence tests. "The standard of intelligence of the average adult Negro is similar to that of the eleven-year-old youth of the species *Homo sapiens*," he said. Making matters worse, the Negroids had expanded into the tropics, where food was easy to get and intelligence was thus not favored by evolution. "Here we have the environmental conditions which have kept many branches of the Negroid race in a state of arrested development," Osborn said.

To Osborn, the history of white people did not end with the Cro-Magnons. The Nordic race of Caucasians arose twelve thousand years ago and swept into Europe. They had the strongest "race plasm," as Osborn liked to call it, and thus they had produced the greatest men in history, from Columbus to Leonardo da Vinci to Cervantes (never mind that these luminaries were not Nordics, but were born rather in Italy and Spain). To maintain the vigor of the Nordic race, Osborn believed, eugenicists would have to make sure they didn't taint their race plasm by marrying people of the lower races.

In 1935, Osborn died of a heart attack while sitting in his study in his upstate New York mansion. He collapsed as he was "engaged in writing a 1,250,000-word treatise on the evolution of the elephant," according to the *New York Times*. Osborn's death occasioned many obituaries and public memorials, which mostly dwelled on his paleontological achievements and his leadership at the museum. They carefully avoided any mention of the fondness he had developed for Nazi Germany, or the visit he paid there a year before his death to accept an honorary degree. By the time Osborn died, eugenics was losing much of its luster.

Osborn's cherished theory of Asian human origins would eventually be proven wrong. By the 1960s, it was becoming abundantly clear that humans originated in Africa, Osborn's dismal tropical dead end. The oldest human-like fossil yet found, *Sahelanthropus tchadensis*, lived about seven million years ago in what is now Chad. For the next five million years our ancient relatives, known as hominins, lived as small-brained bipedal apes in eastern and southern Africa. Starting about two million years ago, waves of hominins began expanding out of Africa to populate Europe and Asia. But the center of hominin evolution remained in Africa.

By 600,000 years ago, the hominins on our own branch of the family tree had evolved to our own height and brain size. They were making sophisticated tools beyond the skill of earlier hominins. They were not just hominins anymore; they could rightly be called humans. Some early humans stayed in Africa and evolved into our species, *Homo sapiens*. Others expanded their range beyond Africa, gradually adapting to life on other continents. That wandering population became the Neanderthals.

Research on Neanderthals since Osborn's death has dramatically improved their reputation. Paleoanthropologists have found the remains of Neanderthals well outside of Europe, as far away as the Near East and Siberia. For more than 300,000 years, they had a range the size of Australia, encompassing mountains, grasslands, and forests. Neanderthals were versatile at finding food. In addition to catching big game, Neanderthals who lived on the coast would fish, kill dolphins, and harvest mussels. Neanderthals knew how to cook the pitch of birch trees into a glue for fixing stone blades to wooden spear handles. They marked their bodies red with ochre and wore jewelry made from the claws of eagles. They arranged stalagmites into enormous circles deep inside caves, perhaps as sites for subterranean worship.

Despite all their newly appreciated adaptations, Neanderthals still disappeared. The youngest Neanderthal sites are forty thousand years old. Our own species may well have played a part in their demise when some Africans expanded into Europe and Asia. Sometimes this expansion took modern humans into Neanderthal territory, where the two kinds of people overlapped for thousands of years.

In 1995, a technician at the Rheinisches Landesmuseum in Bonn, Germany, did something unheard-of: He switched on a sterile electric saw and pushed it into the fossil of a Neanderthal's arm. This was not just any Neanderthal fossil: It belonged to the first cache of the extinct humans ever discovered, unearthed in 1856 by quarry workers clearing a cave. Now, 139 years later, the technician carved out a C-shaped chunk from the arm fossil. Under the whirring saw blade, it gave off a whiff of burnt bone.

The museum had decided it was time to have a paleogenetics expert look for DNA in some of their fossils, and they chose Svante Pääbo. When they shipped him the Neanderthal bone, Pääbo assigned one of his graduate students, Matthias Krings, to work on it. Krings managed to extract short stretches of mitochondrial DNA from the fossil. One day, he fed the genetic material into a sequencing machine to read it. Krings could only hope that he had not contaminated the sample with the DNA of a living animal, including himself.

The fragment of DNA Krings extracted measured only 379 bases long. He compared it to the same stretch in the mitochondrial DNA of more than two thousand living humans. Most of the fossil DNA matched the human stretches perfectly. But here and there, it had mutations not found in any living person. It differed from every human sequence Krings studied on average by 28 bases.

Late in the night, Krings picked up a laboratory phone to give Pääbo an urgent update.

"It's not human," he said.

It was the first time anyone had found DNA from an extinct, humanlike fossil. The discovery led Pääbo to undertake a scientific expedition unlike anything that had come before. He and his colleagues persuaded other museums to let them drill other Neanderthal fossils. They enlisted paleoanthropologists to send them fresh material from their digs. Pääbo's group built up a genetic portrait of Neanderthals across their entire range, gathering not just mitochondrial DNA but DNA from their chromosomes as well. Their research revealed that Neanderthals varied from one another, but they had much less variation than living humans do. It became clear that they represented a separate branch of humans that lived in small groups with little genetic diversity.

As new kinds of DNA-sequencing technology became available, Pääbo and his colleagues would snatch them and adapt them to their research. They extracted more DNA from each fossil they studied, and their reconstructions grew more accurate. In 2010, they managed to build a rough draft of about 60 percent of the entire Neanderthal genome. To understand how it fit into human evolution, Pääbo enlisted David Reich and his colleagues. Base by base, they compared the Neanderthal DNA to that of a chimpanzee and of humans from different parts of the world.

The scientists found that Neanderthals shared many genetic variants in common with modern humans, variants that aren't shared by chimpanzees. These shared variants must have arisen after our hominin ancestors split away from the ancestors of other living apes—but before the split between modern humans and Neanderthals. Pääbo and his colleagues also cataloged a number of variants that arose only in Neanderthals, and others only in the modern human lineage.

But they were still left with some variants that didn't quite fit in any of these categories. These variants were sprinkled in the DNA of some Europeans and Asians they studied. But they couldn't find any in living Africans.

The strongest explanation for the pattern was one that would have appalled Henry Fairfield Osborn. Neanderthals must have interbred with the modern humans who expanded out of Africa and would go on to settle across the Old World (that included the Europeans whom Osborn believed to be his beloved, pure Nordic race). Pääbo and his colleagues estimated that living non-Africans could trace 1 to 4 percent of their genetic ancestry to Neanderthals. There is thus more Neanderthal DNA on Earth today than when Neanderthals existed.

———

B y 2010, when Pääbo and his colleagues published the first evidence for Neanderthal interbreeding, genetic genealogy was a thriving industry. It was ready to seize such a sensational finding and make the most of it. 23andMe quickly put together a test that they claimed could tell customers just how much of their genome was Neanderthal. When I told people about my reporting about Neanderthals, some of them would eagerly let me know about their percentage. The more Neanderthal DNA they carried, the happier they sounded. Judging from comments that customers have left on 23andMe's website, Neanderthal pride is a common thing.

"I am very proud of my 2.8% Neanderthal DNA," someone named Gayle wrote in 2011. "Neanderthals had larger brains than modern humans, cared for the sick and elderly, buried their dead, wore jewelry in the form of painted sea shells, crafted musical instruments, and gave us hybrid vigor."

Replying to Gayle, another commenter named Lee Ann wondered if Neanderthal DNA might explain the look of her family. "I haven't been tested, although I should, out of curiosity," she wrote. "My genealogy has me as the 27th gr granddaughter of William the Conqueror, the Saxon that made England Anglo Saxon. Every generation in the modern family of the past 200 years has one or two tall, heavy boned people, surrounded by average height and weight. My brother was 5'10", 165 pounds. I am 6' tall, big boned along with carrying some extra weight. I am hairier than he is, he cannot grow a beard, I could grow a beard on my legs alone."

Ancient DNA experts had mixed feelings about this new Neanderthal mania. They were happy to see all the enthusiasm over their research, but they also didn't like the pretzels into which it was getting stretched. The 23andMe test, for example, was based only on the 2010 rough draft of the Neanderthal genome. Only later did Pääbo and his colleagues discover that a Neanderthal toe bone from Siberia was packed with DNA—so much DNA that they were able to reconstruct the entire genome with high accuracy. In 2014, the researchers compared this high-quality Neanderthal genome to more than a thousand human genomes. It was like switching a microscope to a far more powerful lens.

Africans turned out to have between 0.08 and 0.34 percent Neanderthal DNA, likely from the migration of people from the Near East to the continent. Non-Africans had ancestry varying from just over 1 percent to 1.4 percent. That was a far narrower window than the 1 to 4 percent in the original estimates. And within each population, the variation from person to person was far smaller. Central Europeans had an average of 1.17 percent Neanderthal DNA, plus or minus only 0.08 percent. When the Neanderthal researchers published their analysis, they went out of their way to take a swipe at the 23andMe test. For the most part, the researchers said, the test only delivered "statistical noise."

Even if some people actually did have twice as much Neanderthal DNA than others, that wouldn't somehow make them more "Neanderthal," as if Neanderthalness were a spice sprinkled into our genomic soup. Living people who carry Neanderthal DNA have thousands of genetic fragments scattered by meiosis throughout their chromosomes. Most of those fragments probably do nothing at all. Our protein-coding genes make up only about 1 percent of the human genome. It's possible that a few percent more is made up of genes that encode important RNA molecules. Some additional DNA may be important to us as millions of tiny genetic switches, where proteins known as transcription factors can attach to turn genes on and off. But it's likely that the vast majority of human DNA has no function. It's just along for the ride. Inheriting a Neanderthal piece of this so-called junk DNA instead of the human version should make no difference to us.

Some of the Neanderthal DNA we inherit can potentially make a difference if it contains important genes or stretches of DNA that help turn genes on and off. But from one person to the next, the fragments of Neanderthal DNA that survive in the genome are different. Whatever the impact of our Neanderthal inheritance may turn out to be, it will depend on the particular genes each of us inherit.

To find out what my Neanderthal DNA meant to me, I enlisted Adam Siepel, who by then had been studying ancient genomes for several years at Cold Spring Harbor. He was intrigued by the request, admitting that he had never been a big fan of the 23andMe Neanderthal test.

"They just give you a number," he said. "They don't tell you *where* you're Neanderthal."

I arranged for Siepel to get my genome, and then he and two of his colleagues, Melissa Jane Hubisz and Ilan Gronau, set about analyzing it. They used a statistical method they had invented a few years beforehand that can detect mixtures of different kinds of inherited DNA that might go overlooked by other methods.

First the scientists chopped up my genome into thousands of sections, each a million bases long. They then compared each of those sections to corresponding ones in people of European, Asian, and African descent. They also compared my sections of DNA to those of Neanderthals and chimpanzees, the closest living species related to humans. Siepel and his colleagues tested out many evolutionary trees to see which accounted best for all these similarities and differences. They drew trees with different sets of branches and also investigated scenarios in which DNA slipped from one branch to another, thanks to interbreeding. "It builds a coherent model that has to explain everything," Siepel said.

It took days for a computer to churn through all the data, explore all the possibilities, and finally produce an answer. To walk me through the results, Hubisz joined Siepel and me in his office while Gronau, calling in from Israel, peered down at us from a video screen.

"You're definitely pushing us into a new area," Siepel said. "It's actually a little addictive once you start."

Together, they unveiled my genealogical tree, reaching back over half a million years. My genome shared a close ancestry with those of other living Europeans. Beyond Europe, they found that Asians were my closest relatives, as a result of the expansion of humans out of Africa. They compared my genome to a Southern African hunter-gatherer's and estimated we could trace our ancestry back to a common ancestral population that existed over 100,000 years ago. On my genealogical tree, I could see Neanderthals on a far more distant branch, splitting off hundreds of thousands of years earlier.

But some of the DNA that Siepel and his colleagues analyzed didn't travel obediently down these branches. Some had jumped from Neanderthals into humans. Each of the humans outside of Africa that Siepel, Gronau, and Hubisz studied had ended up with a different vestige of Neanderthal DNA.

To show me mine, Hubisz opened up a browser on the monitor. Long black bars marked places where one copy of a chromosome carried DNA from Neanderthals. In some regions, I had inherited Neanderthal DNA from both my mother and father. The largest of those doubly inherited regions spans 189,871 bases. When Hubisz tallied the number of stretches that were 10,000 bases or longer, she ended up with over a thousand.

Some of those spans didn't contain a gene or any other stretch of DNA with a known function. But some of them had promise. "I have a list of interesting regions," Siepel said, pulling out a piece of paper. "There's a lot I don't know about these, but here are some that I've flagged."

One segment contained a gene called DSCF5, for example, that has been linked to coronary artery disease. "We can click on a few others that I found," Siepel said. He threw out names of other genes—CEP350, GPATCH1, and PLOD2.

"Catchy name," he muttered.

Siepel might be able to name my Neanderthal genes, but he couldn't say which Neanderthal I inherited them from. He couldn't say if it was a male or female Neanderthal, or when that Neanderthal lived, or where. He and other researchers were still just trying to make out the broad outlines of Neanderthal interbreeding. As best they could tell, Neanderthals and humans interbred many times, over a period that may have stretched over 200,000 years.

The earliest hints of interbreeding came to light in 2017. Researchers studying the DNA of European Neanderthals determined that their mitochondria came from an ancient human woman who lived more than 270,000 years ago. It's possible that early members of *Homo sapiens* made their way from northern Africa to southern Europe, where they interbred with the Neanderthals there. If this encounter did happen, those early humans must have vanished, leaving behind only their mitochondria in later generations of Neanderthals.

Siepel and his colleagues have found more evidence of human DNA in Neanderthals more than 100,000 years ago. The fossil record offers hints of where this encounter took place: in the Near East. On the coast of Israel, there is a place called Mount Carmel. Inside its caves, scientists have found the fossils of both Neanderthals and modern humans. The Neanderthals lived in the region from at least 200,000 years ago. Modern humans make a brief appearance at Mount Carmel about 100,000 years ago, and then the

Neanderthals return for another 50,000 years before giving way for good to modern humans. It's possible that the 100,000-year-old humans of Mount Carmel belonged to a brief expansion out of Africa. Before they vanished, they may have contributed their own DNA to Neanderthals.

The DNA of living humans documents more recent encounters, after humans expanded successfully out of Africa sometime between 50,000 and 80,000 years ago. In a 2016 study, Joshua Akey and his colleagues at the University of Washington found different patterns of Neanderthal DNA in different groups of people, suggesting that the interbreeding happened in at least three separate episodes.

The first took place not long after modern humans came back to the Near East. This was before non-Africans split into today's major lineages, and so the DNA from this first mingling can be found in all non-African populations. The ancestors of people who live in Australia and New Guinea then split from the rest of non-Africans and moved east along the coast of Asia. The second interbreeding with Neanderthals happened after that split, and today the DNA from that contact can be found in Europeans and East Asians but not in New Guinea or Australia. Finally, after Europeans split from the ancestors of East Asians, Neanderthals bred a third time, with the ancestors of East Asians.

Of course, a study like Akey's has its limits. It can't tell us about other interbreeding that didn't leave any DNA behind in living humans. Nor can it offer all the cinematic details about how the mating took place. Were Neanderthal males having sex with modern human females, or was it the reverse? Did people willingly cross into new societies to raise their children, or were they slaves? For now, we can only craft the different stories we might tell, depending on what the answers to those questions would be.

But the laws of heredity can let us know some things about our Neanderthal inheritance. The children of Neanderthals and humans got half of their DNA from each kind of human. At least some of them must have been welcomed into modern human groups. They must have been cared for and nurtured. They must have gotten the opportunity to have children of their own. Our own DNA is proof.

If those hybrids mated with modern humans, their children would have inherited a quarter of their DNA from a Neanderthal grandparent. Their Neanderthal DNA would be chopped up and shuffled with the DNA of their other grandparents. Over future generations, meiosis split up the Neanderthal DNA into still smaller segments.

Some ancient human fossils turn out to have as much as 6 to 9 percent Neanderthal DNA. Over time, the average amount dwindled. One plausible explanation for that decline is that most Neanderthal DNA is bad for our health. Inheriting a Neanderthal version of a gene may cause people to have fewer children—either because they're less likely to survive to childbearing age, or because they become less fertile. It may be no coincidence that Neanderthal genes that play a role in reproduction are especially rare in humans today. As future generations of humans inherit Neanderthal DNA, some of it may continue to disappear.

But some Neanderthal DNA seems to have endured for tens of thousands of years because it gave our ancestors some benefit. I've discovered that my genome, for example, includes some Neanderthal versions of genes that help fight infections. I'm hardly alone: Neanderthal immune genes are more common in living humans than many other classes of genes.

Once they slipped into the modern human gene pool, these genes appear to have become more widespread with time. Our immune system genes are some of the fastest-evolving parts of our genome, because they need to keep up with the rapid evolution of the parasites that are trying to evade our defenses. People who live in malaria-prone regions have evolved new defenses against the parasites in just the past few thousand years. When early Africans moved onto other continents, they may have encountered a number of diseases for the first time. Neanderthals, on the other hand, had been adapting to those medical challenges for hundreds of thousands of years. Borrowing immune system genes from Neanderthals could have been a quick way to get better odds of surviving in their new home.

One of the most startling revelations about my own heredity came up toward the end of my visit to Cold Spring Harbor. After running through all his results, Gronau said something so casually that I almost missed it.

"There is some Denisovan gene flow in Carl's genome," he said.

I sat up straight. "What?"

"You have a tiny bit, which is more than I see in the other genomes that I've tried," Gronau said.

In 2009, a Russian researcher sent Svante Pääbo a nondescript chip of bone from a pinky. It had come to light during a dig in a Siberian cave called Denisova. There was no reason to expect anything interesting from the chip, but Pääbo's student Johannes Krause discovered that it was actually packed with DNA. While much of that DNA belonged to bacteria that had invaded the bone long after death, there was also a lot that looked humanlike. Krause assumed it would either be Neanderthal or modern human DNA. But when he inspected it closely, it turned out to be neither. It belonged to another extinct human. In honor of the cave where it was found, Pääbo and his colleagues named these phantom people the Denisovans.

In the next few years, Pääbo and his colleagues tested out other fossils for Denisovan DNA and managed to find some more from molars dug up in the same Siberian cave. The scientists cannot look at Denisovan skeletons to compare their anatomy to ours. They're left in the strange situation of knowing a group of extinct human relatives almost entirely from their DNA.

That DNA reveals that Denisovans were most closely related to Neanderthals, not modern humans. They split from a common ancestor as long as 470,000 years ago. In 2017, Pääbo and his colleagues examined the mutations in four Denisovan individuals and estimated that the oldest individual who lived in the Denisova cave was there over 100,000 years ago. The youngest known Denisovan lived there about 50,000 years ago.

Like Neanderthals, Denisovans have left their genetic mark on living humans. The people who carry the highest amounts of Denisovan DNA—in some cases more than 5 percent—live in Australia, New Guinea, and nearby Pacific Islands. Scientists have also found tiny amounts of Denisovan DNA in the genomes of East Asians and Native Americans. Paradoxically, the people who live today near the Denisova cave have hardly any Denisovan DNA at all in their genes.

These clues all point in the same direction: We should think of Denisovans as the eastern Neanderthals. When the common ancestors of Neanderthals and Denisovans spread out across Eurasia, they diverged into two populations. The Neanderthals headed west into Europe, while the Denisovans went the other way. It's possible they reached Southeast Asia, where they later encountered modern humans on their way to the Pacific.

For the most part, Denisovan DNA was as ill-suited to modern humans as the DNA of Neanderthals. But a few Denisovan genes may have provided benefits of their own. One of the best candidates for a beneficial Denisovan gene is called EPAS1. It regulates how many red blood cells our bodies make and how they transport oxygen. Tibetans carry a variant of EPAS1 that protects them from the dangers of life at high altitudes. In 2014, Emilia Huerta-Sanchez and Rasmus Nielsen at the University of California, Berkeley, discovered that the Tibetan form of EPAS1 came from Denisovans. We can't know if Denisovans were adapted for living at high elevations; it's entirely possible that their variant of EPAS1 helped them in some other way, and just happened to prove useful when the ancestors of Tibetans moved toward the sky.

To hear Gronau say that he found some Denisovan DNA in me didn't make sense. It had been unsettling enough for geneticists to say I might be Italian. Did my Denisovan DNA mean that I had some hidden ancestry in the highlands of New Guinea?

"Yeah," Hubisz said offhandedly, poring over her own laptop. "I found it, too."

With a grin, Siepel turned to me. "How are you at high altitudes?" he asked.

Siepel cautioned me that the risk of errors goes up when scientists try to look tens of thousands of years into our genealogical past. Some of my DNA might look Denisovan to Siepel's computer because scientists have yet to find the same sequence in Neanderthal fossils. And even if I do carry some Denisovan DNA, that may not mean that my modern human ancestors made direct contact with Denisovans. It's entirely possible that Denisovans and Neanderthals mated. Later, some Neanderthal descendant passed on that Denisovan DNA to my own ancestors.

"Still, you can't exclude the possibility that you have some Denisovan DNA," said Siepel.

By midday, we were done inspecting my genome and hungry for lunch. Gronau signed off from Israel, and Hubisz, Siepel, and I stood up to stretch our legs. "I wish there was more of a punch line," Hubisz said. "We found a lot of data."

"Well," Siepel said, looking on the bright side, "we found that he was part alien."

CHAPTER 9

Nine Foot High Complete

I N THE LATE 1990S, Joel Hirschhorn became a pediatric endocrinologist at Boston Children's Hospital. As an expert on hormones, he saw a lot of children with diabetes. But he saw almost as many children who were short. "You have parents coming in with their child, and they're worried because their child is not growing quickly—or as quickly as their friends," Hirschhorn told me when I paid him a visit in his office.

Being very short is sometimes a sign of a serious medical problem—an inability to make growth hormone, for example. Mostly, though, Hirschhorn spent his visits with short children calming their parents. "You would end up saying there's probably nothing wrong," he said. "And I would say a good majority of the time, one or both of the parents themselves are short. And so you would just explain how height is inherited."

Hirschhorn would tell the parents about their genes, and how they had given some of them to their children. "You have some versions that make you a little on the shorter side, and you passed some of those to your child, so now they're probably going to end up a little on the shorter side," he said.

The parents sometimes asked Hirschhorn about those genes. He would say that he could be sure there were genes involved, but he couldn't name them. Nobody knew their identity. "And about the twentieth time," Hirschhorn said, "I thought to myself: We could find out what those genes are."

In addition to his work as a doctor, Hirschhorn was doing research on the side. Working at the Whitehead Institute nearby, he developed new methods for pinpointing genetic variants that caused medical conditions such as diabetes. Compared to such disorders, height seemed like an easy thing to study. Diabetes, for instance, takes years to develop, depending in part on what people eat. It's also possible that there are different sets of genes that can make different groups of people at risk for it. Height, by contrast, is simple: It's easy to measure, and you can measure it in anyone. Hirschhorn thought he could just compare tall and short people, look at their DNA, and find the variants that tended to raise or lower their height.

In 2004, Hirschhorn left the Whitehead and moved to the next building over, joining the Broad Institute to continue to study height. And when I visited him at the Broad in 2017, he was still studying height. He had just gotten a new office, which was almost completely bare. He had a phone and a laptop. On a whiteboard, someone had written *Flour* and *Flower*. Hirschhorn looked about as close to average height as a man could be.

During the seventeen years that Hirschhorn had been studying height, he explained, he and his colleagues had made progress. Now his conversations with parents sound a little different. "Instead of saying, 'we don't know what they are,' I usually say, 'we know what some of them are.'"

But if parents came to Hirschhorn with the DNA sequence of their child and asked him how tall their child would become, he still wouldn't be able to tell them. "It's not implausible that we could be there at some point—at least before I retire," he told me.

As I have worked on this book, my daughters have gotten taller. They have entered that phase of life in which they start looking down at their relatives, one by one. There's usually at least one back-to-back comparison at every family gathering. Our girls stand at military attention, the rest of us squinting across the crowns of their heads and patting down their hair. Throughout their growth spurts, Charlotte and Veronica have been good sports. You can tell that they are indulging us, that they don't pay much mind to their increasing height—certainly not compared to an upcoming recital or the insufferable wait for the revival of *The Gilmore Girls*. I can see something of myself when they roll their eyes and smile politely at their height-obsessed elders.

I remember the years when my brother, Ben, and I were barreling upward. The lines that our parents drew on the kitchen doorframe were like the hands of a clock, tracking family time. The jump from one line to the next made clear that both Ben and I were going to outgrow our mother, then our father. As I reached six foot, and Ben six foot one, our heights became a marvel among our shorter relatives. They would tilt their heads up to take in our stature. Sensing an unaccustomed crane, they'd ask, "Where did you *get* that from?" They would vaguely recall a towering great-grandfather, or try to remember the story a great-aunt told about a tall cousin. They searched our genealogy for *someone* from whom we might have inherited our own height. They talked about height as if it were a diamond that an ancestor could have stored away in a safe-deposit box, where it could sit for a century until Ben and I brought it out into the light again.

Sometimes heredity does act with a diamond-like simplicity. Two defective copies of the PAH gene will cause PKU. But heredity's influence is usually much harder to decipher. It's hidden in clouds of complexity, a complexity generated both within our genes and outside of our bodies. It's hard to imagine anything simpler than height. It's nothing but a number, one that can be obtained with a hardware-store tape measure at that. And yet the heredity of height can be as baffling as quantum physics. Light can be at once a particle and waves. Height can be at once shaped by heredity and governed by our experience. Height was among the first puzzles that early scientists of heredity tried to take apart, and yet they haven't finished solving it yet.

———

All of written history is laced with stories of giants and dwarves. The Bible describes races of giants who lived before the Flood. Og, king of Bashan, slept in an iron bed measuring nine cubits (thirteen and a half feet). In other stories from around the Near East, Og and his height also appear. According to one tale, he escaped the Great Flood by wading alongside the ark, the oceans lapping around his knees. In another, one of his bones was laid across a river to serve as a bridge.

The ancient Greeks and Romans would sometimes unearth dinosaur bones and house them in temples, believing them to belong to the skeletons of humanlike giants. They also marveled at the true giants who walked among them, exaggerating their stature with each retelling. According to Pliny the Elder, two men who stood ten feet tall settled in Rome during the reign of Augustus. Similar reports of extraordinarily tall people popped up from time to time all the way into the Renaissance. A seventeenth-century physician named Platerus claimed he once met a man in Luxembourg "nine foot high complete."

By the 1700s, tall people had become professional attractions. In 1782, an Irishman named Charles Byrne, standing eight foot two, dazzled London society. He had been born a normal-size baby, but had quickly begun to grow far faster than other boys. The people in his village said that he grew so tall because his parents had conceived him atop a haystack. As a teenager, Byrne toured fairs around Ireland before traveling to England to make his fortune.

"This truly amazing phenomenon is indisputably the most extraordinary production of the human species ever beheld since the days of Goliath," ran one London newspaper ad. Clad in a frock coat, knee breeches, silk stockings, and frilled cuffs, "the Irish Giant" received paying visitors twice a day, six days a week, in a handsome apartment. Byrne earned more than seven hundred pounds before dying at age twenty-two, reportedly from excessive drinking. "The whole tribe of surgeons put in a claim for the poor departed Irish Giant," one newspaper reported, "and surrounded his house just as Greenland harpooners would an enormous whale."

At the other end of height's spectrum, people with dwarfism were also singled out, sometimes for reverence but more often for great cruelty. In ancient Egypt, dwarves served Pharaohs as sacred dancers, jewelers, textile makers, and priests. The chiefs of some West African tribes appointed dwarves as their attendants, perceiving a connection between them and the gods. The Romans had a brutal fascination, watching dwarf gladiators fight to the death or keeping male and female dwarves around their houses like pets. Often the dwarves simply wandered the houses of their masters, naked except for jewels around their necks. In the 1500s, an Italian noblewoman named Isabella d'Este built miniature marble-lined apartments in her enormous palace to house a colony of dwarves. They would entertain her by performing somersaults, or pretending to be priests, or pissing drunkenly on the floor.

In time, some dwarves were accorded more dignity in European society, although they were no less fetishized. At the royal courts of countries such as England and Russia, dwarves served as royal painters, nurses, and diplomats. Dwarves competed with giants for audiences in the eighteenth century. In 1719, a man named Robert Skinner, who reportedly stood just over two feet high, met an equally short woman, Judith, while they were traveling from exhibit to exhibit. They fell in love, got married, and retired from touring. Their fourteen children all grew to normal height. Somehow, the Skinners' stature had failed to imprint itself on their family.

Twenty-three years after they met, the Skinners ran out of money and went back to London in 1742 to earn some more. This time, they displayed not only themselves but their towering children. The mismatch so astonished London society that the Skinners made a small fortune over the course of two years and were able to retire for good. They spent their retirement traveling around St. James's Park in a custom-built carriage pulled by two dogs and driven by a twelve-year-old boy clad in purple-and-yellow livery.

The Skinners stood out, even within their own family. But there also were rumors since ancient times of entire races of miniature people. According to some stories, they lived in India—or maybe it was Africa—and rode miniature horses into battle against cranes. The forests of northern Europe were reputed to be rife with dwarves and gnomes. Off the coast of England, one of the Hebrides islands was known as the Isle of Pigmies; underneath a chapel, it was told, several miniature human bones had once been dug up.

In 1699, a British anatomist named Edward Tyson tried to dash these stories. Having performed the first dissection of a chimpanzee, Tyson declared that "the pygmies of the Ancients were a sort of Apes, and not of Humane Race." It wasn't until the mid-1800s that European explorers in Africa encountered groups of humans, such as the Baka and Mbuti, who typically never grow taller than five feet. It turned out there were slivers of truth embedded in the old fantasies.

In other parts of the world, European explorers sent back reports of towering peoples. Ferdinand Magellan rounded the southern tip of South America in 1520 and, in the words of his chronicler Antonio Pigafetta, spotted "a giant who was on the shore, quite naked, and who danced, leaped, and sang, and while he sang he threw sand and dust on his head." Magellan claimed that the giant's tribe—which came to be known as the Patagonians—stood ten feet tall. A century later, Sir Francis Drake visited the Patagonians and dismissed that measurement as a lie. The Patagonians were clearly just seven feet tall, Drake said.

With more time, the giants of the world shrank to more realistic heights. Nevertheless, the fact remained that people in some countries were taller than others. In 1826, the British ethnologist James Cowles Prichard observed that the Irish, although not especially tall on average, produced a remarkable number of giants like Charles Byrne. "We can hardly avoid the conclusion that there must be some peculiarity in Ireland which gives rise to these phenomena," he said.

Prichard believed the peculiarity had something to do with the land of Ireland rather than its people. He subscribed to an idea that dated back at least to Hippocrates. "Such as dwell in places which are low-lying, abounding in meadows and ill ventilated, and who have a larger proportion of hot than of cold winds, and who make use of warm waters—these are not likely to be of large stature," Hippocrates explained. Tall people, Hippocrates said, were "such as inhabit a high country, and one that is level, windy, and well-watered."

Hippocrates was well aware that his patients, who all lived in Greece, grew to different heights. He ascribed their differences to the changing weather, which could disrupt the concentration of a man's semen, altering his child's development. "This process cannot be the same in summer as in winter, nor in rainy as in dry weather," Hippocrates declared.

Ancient Greeks seem to have thought about height only in rough figures. "In about five years, in the case of human beings at any rate, the body seems to gain half the height that is gained in all the rest of life," Aristotle wrote. Even in the Renaissance, scholars didn't see the need for precision. In 1559, the Italian physician Pavisi declared that "the growth of infants and children is quite swift, and often in two or three years they add two or three cubits." Three cubits would be four and a half feet. Either Pavisi didn't pay very close attention to infants, or he lived among giants.

The Enlightenment brought a new rigor to measuring height. In 1708, Great Britain enacted the Recruiting Act, requiring that army conscripts be at least five foot five. In 1724, one Reverend Mr. Wasse wrote to the Royal Society to warn that measuring height might be harder than the army realized. Reverend Wasse fixed a nail above a chair high enough that he could just barely touch it with his fingertips. He then pushed a garden roller for half an hour. When the reverend sat down again, a half-inch gap lay between his hand and the nail. Apparently, he had shrunk during his exercise. Reverend Wasse reported to the Royal Society that he also measured the height of "a great many sedentary People and Day-Labourers." He found that people could grow taller and shorter over the course of a day—by as much as an inch in some cases.

"I mentioned it to an Officer," Reverend Wasse reported, "and thereby kept some Persons from being turn'd out of the Service."

The military's cherishing of height helped endow it with a moral value. Tallness turned into a sign of virtue and nobility. The fact that highborn boys ended up taller than lowborns seemed no coincidence. In England, the gap was staggering: At the end of the eighteenth century, the wealthy sixteen-year-old boys who went to the military school at Sandhurst were nearly nine inches taller than the poor boys of the same age entering the Marine Society.

At the same time, the natural philosophers of the Enlightenment began to track the growth of children, with a precision that previous generations hadn't bothered with. The first data came from a French nobleman named Philippe Guéneau de Montbeillard. In 1759, he laid his newborn son on a table and measured him from head to toe. Every six months, except for a few gaps, he would make a new measurement of his son, switching from horizontal to vertical once the boy could stand. Montbeillard saw more than just a series of numbers in these records. They revealed an upward velocity, one that accelerated during growth spurts and later dwindled to zero.

When Montbeillard's work was published, it inspired others to make similar recordings of the height of children in schools and hospitals. As they looked at the multiplying curves, they started to see broad patterns. Children tended to grow at similar velocities even if they were of different heights. A few children broke the rule: late bloomers experienced last-minute surges, while the growth of sick children slowed down drastically, leaving them short for life.

In the early 1800s, a French physician named Louis-René Villermé realized that the height of a group of people could tell him something about their well-being. Serving as a surgical assistant in the Napoleonic Wars, Villermé observed how food shortages afflicted both soldiers and civilians. Children suffered most of all, their growth permanently stunted. When Villermé left the army and began working as a physician, he could see how peacetime ravaged the poor. Traveling across France to study textile workers, child laborers, and prisoners, Villermé became convinced that social reforms were "absolutely demanded by conscience and humanity." Thanks in part to his efforts, France passed a law in 1841 forbidding children between eight and twelve from working over eight hours a day, or doing any night work. School became mandatory till age twelve.

Villermé succeeded because he made his case with data. He determined the rate at which poor people died, which was gruesomely higher than the rate among the wealthy. He also tracked people's heights, measuring the stunting power of poverty. Conscripts from poor regions were shorter than ones from rich regions. In Paris, Villermé documented that the people in wealthy neighborhoods where families owned their homes were taller than in poor neighborhoods where people could only rent.

"Human height," Villermé concluded, "becomes greater and growth takes place more rapidly, other things being equal, in proportion as the country is richer, comfort more general, houses, clothes, and nourishment better, and labour, fatigue and privation during infancy and youth less."

It was controversial to say such things in the early 1800s. Many of Villermé's fellow doctors still followed Hippocrates, believing height was set by air and water, not economics. To advance his cause, Villermé gathered allies. One of his most important converts was a wandering astronomer named Adolphe Quetelet.

In 1823, the twenty-seven-year-old Quetelet came to Paris from Belgium to inspect the city's telescopes. He was in charge of building Belgium's first observatory, and he wanted to see how the French did it. While in Paris, Quetelet met with the greatest mathematicians of the age, people who were developing equations to track the heavens, who were finding hidden order in randomness. Quetelet enjoyed meeting Villermé and learning of his ideas about society, but Quetelet's ambitions were pointed in an entirely different direction. As soon as his observatory was finished, Quetelet would make Newton-grade discoveries about the universe. He once scribbled his motto in the margin of a book: *Mundum numeri regunt,* "Numbers rule the world."

But just as Quetelet was finishing his grand telescopic tour and preparing to go home, Belgium fell into a revolution. Rebels moved into his unfinished observatory, and Quetelet realized that his path to fame wasn't going to run through astronomy after all.

He decided to follow Villermé's example instead. Quetelet turned his attention to people, hoping to find an order in the chaos that had upended his life and his country. He began building a science he called social physics. Like Villermé, he chose to study the statistics of height. Gathering large numbers of measurements of children, he searched for equations that could predict their growth velocity. As Quetelet examined his results, he was startled to see a familiar pattern. Most children were close to average height, and tall and short children were rarer. Plotted on a graph, their heights formed a curving hill, its peak centered on the average.

Quetelet had already seen this hill—known as a bell curve—in the heavens. To calculate the speed of a planet, astronomers would watch it travel across a glass etched with two parallel lines, timing how long it took to move from one line to the other. If two astronomers observed the same planet, they often ended up with different figures for its speed. One astronomer might be slow to check his pocket watch, the other too quick. If the measurements of many astronomers were plotted on a single graph, they formed a bell curve as well.

On his trip to Paris, Quetelet had met mathematicians who had derived an astonishing proof about astronomical bell curves. Even if most astronomers were wrong in their measurements, the average of all their observations ended up being close to the true value. Quetelet came to see a special power in the peaks of bell curves. And when he saw his height measurements form a bell curve as well, he decided that the average height was humanity's ideal. Anyone shorter or taller than average was flawed. He extended this same importance to every other trait in the human body, from weight to the shape of the face. If there was one person who combined all the qualities of "the average man," Quetelet said in 1835, that individual would "represent all which is grand, beautiful, and excellent."

W ord of Quetelet's research spread across Europe. The theory he applied to height—known as the law of error—could also bring order to many other kinds of statistics, be they crime records or weather patterns. Francis Galton saw the law of error as a revolutionary advance for all of science. "It reigns with serenity and in complete self-effacement amidst the wildest confusion," he said. "The huger the mob, and the greater the apparent anarchy, the more perfect is its sway. It is the supreme law of Unreason."

Galton set about measuring British heights. He invented a purpose-built device for the task, complete with a sliding vertical board, pulleys, and counterweights. He had it manufactured and then sent to teachers across England, along with instructions for how to use it on their students. When they sent back their measurements to Galton, he ended up with a bell curve much like Quetelet's.

To Galton, these two curves looked like evidence that height was inherited. Only heredity, he believed, could account for the fact that he drew a bell curve of height a generation after Quetelet drew his. Galton couldn't say how heredity was re-creating the same curve in each generation, though. He also recognized a massive paradox he didn't yet know how to solve. "The large do not always beget the large, nor the small the small," he noted, "and yet the observed proportions between the large and the small in each degree of size and in every quality, hardly varies from one generation to another."

To tackle this paradox, Galton pioneered a new way of studying heredity. While Mendel was tracing isolated, all-or-nothing traits from one generation to the next, Galton set out to study a trait that graded smoothly from one extreme to the other. His work on this paradox would be the most important of his career. Long after his calls for eugenics became a source of shame, his work on height remains part of the foundation of today's research on heredity.

For his new project, Galton needed more than just a bell curve of height. He needed a way to compare the height of one generation to its descendants. "I had to collect all my data for myself," he later recalled, "as nothing existed, so far as I know, that would satisfy even my primary requirement."

When Galton described his project to Darwin and others, they urged him to start simple. Rather than study human height, he should raise peas and measure their diameter. If he had to study animals, it would be better to study the wingspan of moths. Galton gave the sweet peas a go, taking over Darwin's garden to grow enough plants for his research. The initial measurements he got were promising. But Galton grew impatient waiting for the plants to develop, and decided it would actually be faster for him to collect data on human height—"to say nothing of its being more interesting by far than one of sweet peas or moths," he added.

Galton posted a newspaper advertisement, asking for family records and promising prize money for the best entries. He sent cards to his friends, requesting they ask brothers for their heights. In the 1880s, he gathered more data by turning his research into something of a carnival attraction, setting up a public laboratory at the 1884 International Health Exhibition in London. He had handbills printed up and passed around, describing the lab as being "for the use of those who desire to be accurately measured in many ways, either to obtain timely warning of remediable faults in development, or to learn their powers." Over the course of a year, Galton's staff measured 9,337 people at the exhibition. In 1888, he set up a similar lab at the Science Galleries of the South Kensington Museum and examined thousands more. Galton had their height measured, along with many other traits, from their hearing to their hand strength.

A "computer"—a woman who could carry out fast, accurate calculations by hand—worked her way through Galton's thousands of height records, organizing them on a grid. Each column represented the combined average height of the parents (including an adjustment for the shorter height of the mothers). The rows represented the height of the children. The computer put a number in each square to show the number of families with each combination of heights.

Galton would often stare at this grid, trying to make sense of it. In some regions, the grid was blank. Some squares had only one family marked inside them. Others had dozens. Finally, staring at the grid one day as he waited for a train, it came to him. The numbers formed a football-shaped cloud. They clustered around an invisible straight line that extended from the lower left corner to the upper right. The taller parents were, the taller their children tended to be. Some parents had children who were shorter or taller than they were. Very short parents had children who grew taller than they were, and vice versa, drawing their children closer to the average.

Like Mendel, Galton had discovered a profound pattern of heredity. But he was no more clear about what it meant. Galton tried to explain his results by arguing that each child inherited less than half of each trait from each parent. They somehow inherited the remainder from even older ancestors. That extra inheritance, Galton claimed, pulled children back away from the extremes toward the ancestral average. While Galton's "ancestral inheritance" would eventually be proven wrong, his discovery of heredity's signature remains a tremendous accomplishment.

In the 1890s, a young colleague of Galton's named Karl Pearson recognized the importance of his work and gave it a proper mathematical makeover. Pearson invented a formula that let him put a number on how closely children resembled their parents. He could use the same formula to compare siblings as well. To try out his equation on real children, Pearson enlisted his own squadron of teachers to measure the height of their students (along with other traits like the circumference of their heads and the span of their arms). He found that the traits were correlated. In other words, pairs of brothers would tend to have similar traits, presumably due to heredity.

Right around the time that Pearson was developing these new mathematical techniques, Mendel came back to light. A coalition of geneticists, the Mendelians, dismissed the measurements that Galton and Pearson were making. It was more important to them to study heredity the way Mendel did, by tracking recessive and dominant traits. Pearson gathered allies of his own. His coalition—known as the biometricians—accused the Mendelians of being time wasters who were obsessing over the few oddball traits that happen to fall in line with Mendel's simple law. A trait like height was not either/or. People were not either tall or short as Mendel's peas might be smooth or wrinkled. Pearson called for a more powerful explanation for heredity to account for this sort of smooth variation.

In 1918, a British statistician named Ronald Fisher brokered a peace between the Mendelians and the biometricians. He demonstrated that the two kinds of heredity were opposite sides of the same coin. The variation in a trait could be influenced by one gene, or a few, or many. The difference between a wrinkled pea and a smooth one that Mendel studied would turn out to be controlled by variants of a single gene. But a trait like height, with a smooth distribution from short to tall, was likely the result of variations in many genes. People could inherit a vast number of different possible combinations of variants, and for most people, the combined effects of all those variants would leave them close to average. Fewer people ended up very tall or short. The result would be Quetelet's bell curve.

Fisher also found an elegantly mathematical way to take into account the fact that genes do not have sole control over traits such as height. Along with nature, nurture might have a part to play. Fisher argued that the overall variation in a trait could be the result of both genetic variation as well as variations in the environment. Genetic variation might be strong for some traits, and environmental variation might be more important for others. The fraction caused by genetic variation—in other words, the variation that could be inherited through genes—came to be known as heritability. If genetic variation has no influence over the variation in a trait, then its heritability is zero. If the environment has no influence, then the heritability is 100 percent.

Heritability is one of the trickiest concepts in modern biology. It describes variations only across an entire population. If the heritability of a trait in a group of people is 50 percent, that doesn't mean that in any given person, genes and environment are each responsible for half of it. And if a trait has a heritability of zero, that doesn't mean that genes have nothing to do with it. The heritability of the number of eyes is zero, because children are virtually all born with a pair of them. When we walk down the street, we don't pass someone with five eyes, another with eight, and another with thirty-one. If someone has only one eye, it's probably because they lost the other one in an accident or from an infection. Yet we all inherit a genetic program that guides the development of eyes.

As tricky as heritability may be to grasp, it's been a powerful tool for making sense of heredity. Our well-being depends on it, in fact. To a large extent, heritability feeds the world.

How much food farmers can harvest from a given acre of land depends largely on the traits of the crops they plant. A plant that winds up short may produce a low yield. Taller is better, but only up to a point. If plants have to dedicate a lot of resources to reaching a great height, they'll have little left over to produce the seeds or fruit we want to eat. They may also run a greater risk of toppling to the ground and leaving a farmer with no harvest at all.

If the height of a crop were entirely heritable, that would mean that the differences in the plants' height were entirely due to the differences they inherited from their ancestors. Short plants would always produce short plants, tall plants tall. If the heritability equaled zero, on the other hand, the genes of different plants would have no effect on their variation whatsoever. All the differences would arise from their environment. A field of plants growing with the same rainfall, the same rhythm of heat and cold, the same pests and blights, would all grow to roughly the same height.

To measure the heritability of a trait in a crop, scientists can raise plants under carefully controlled conditions and observe how differently they turn out. They grow genetically identical seeds in precisely controlled greenhouses. They can pot them in identical soil, spray them with identical fertilizer, and measure their growth each day of their existence, down to the millimeter. These studies show that height is strongly heritable in some species, and only moderately so in others. This knowledge has helped plant breeders transform crops through artificial selection. It led to the production of "semi-dwarf" breeds of wheat and rice that produce a better yield than taller varieties, because the wind can't flatten them.

Scientists who study human heritability, on the other hand, don't raise babies in laboratories. They don't measure the mashed peas parents feed their toddlers, down to the microliter. Instead, scientists have to hunt for volunteers to study. They can only gather stray fragments of information about their subjects' lives. Errors can thus creep into estimates of human heritability. If children grow to be as tall as their tall parents, that doesn't necessarily mean that the genes they share are the reason. Instead, they might have spent their childhood in the same growth-favoring environment that their parents did.

When Galton first began studying the inheritance of height and other traits, he recognized how hard it would be to reach firm conclusions. But Galton had an inspired idea: Scientists could take advantage of a natural experiment in human heredity. They could study twins.

Galton had no way of understanding the genetic links that twins share, but he had an intuition that they must share a strong common inheritance.

He loved to share stories about the eerie coincidences in the lives of twins. A pair of twins simultaneously came down with the same kind of eye irritation, even though one was in Paris at the time and the other in Vienna. Another set developed the same crook on the same finger of the same hand. Yet another pair decided to buy surprise gifts for each other. They each chose precisely the same set of champagne glasses.

While Galton acknowledged that the experiences of twins might also influence how they turned out, he considered heredity to be paramount. The shared heredity of twins drove their lives along the same path. Galton believed twins demonstrated that everyone's inborn nature had an overriding influence. It guided people through life the way sticks thrown in a stream travel with the current.

"The one element that varies in different individuals, but is constant in each of them, is the natural tendency," Galton said. "It corresponds to the current in the stream, and inevitably asserts itself."

Other scientists soon began investigating twins more rigorously for clues to heredity. But it wasn't until the 1920s that a German dermatologist named Hermann Werner Siemens tapped their full power. By then, scientists had come to recognize that fraternal twins and identical twins are genetically different. Fraternal twins develop from two eggs, each fertilized by a separate sperm. Identical twins arise from a single fertilized egg that splits into two embryos. Fraternal twins are thus no more genetically similar to each other than any other pair of siblings, having on average 50 percent of their variants in common. Identical twins, on the other hand, are essentially clones.

Siemens realized that these two kinds of twins were an opportunity to study heritability. Twins grow up in similar environments, from the womb onward. But the genetic closeness of identical twins would make them more similar in highly heritable traits. By comparing the similarities in both kinds of twins, Siemens could estimate the heritability of a trait.

As a dermatologist, Siemens was most interested in skin diseases. Did people develop them simply due to bad luck, he wanted to know, or because of bad genes? He counted up the moles on the skin of twins and discovered that identical twins didn't develop identical constellations of moles. Those differences told Siemens that the environment had a hand in their development.

But while their moles might not be identical, they did correlate. An identical twin with a lot of moles tended to have a twin sibling with a lot as well. If a twin had only few moles, it was a safe bet the other didn't have many. The moles on fraternal twins were also correlated—but only with half the strength as in identical ones. Siemens concluded that genetic variations played an important part in developing moles, although the environment mattered, too.

Siemens's remarkable study inspired other scientists to use his method to study height. A British researcher named Percy Stocks searched for twins in London's schools and had teachers report on how tall they were. He found that fraternal twins tended to be fairly close in height. But identical twins were closer. The difference between them made it possible for scientists to put a number on the heritability of height. As the studies grew larger, that estimate grew more precise. In 2003, a Finnish researcher named Karri Silventoinen studied the height of 30,111 pairs of twins. He estimated that height was strongly heritable: 70 to 94 percent in men, and 68 to 93 percent in women.

Even a study as sprawling as Silventoinen's rested on a big assumption: that the environmental influences shared by a pair of fraternal twins are no different from those of identical twins. If a trait is more similar in identical twins than fraternal twins, genes can be the only explanation. Scientists can't know that for sure, however, because twins grow up in the wilds of real life, not in a terrarium. Some critics raised the possibility that parents treat identical twins differently from fraternal ones. Since fraternal twins look different, parents might treat them more like ordinary siblings.

Scientists developed twin studies as a way to study human DNA in an age when it was impossible to examine it directly. Once it became possible to read genetic markers in people's genomes, new ways emerged to measure heritability. Peter Visscher and his colleagues found that pairs of siblings can vary tremendously in their genetic similarity, sharing as little as 30 percent of their genetic variants in common to as much as 64 percent. If a trait is highly heritable, Visscher reasoned, then it should be more similar in siblings who have more DNA in common.

In 2007, Visscher and his colleagues examined the height of 11,214 pairs of regular siblings. They found that "twin-like" siblings—those who shared more than half of their DNA—tended to grow to more similar heights. Siblings with less genetic similarity were not so similar. The scientists used these correlations to calculate the heritability of height. They ended up with an estimate of 86 percent.

That's an exceptionally high figure. Nicotine dependence has a heritability of 60 percent. The age at which women go into menopause is 47 percent. Left-handedness is at a mere 26 percent. In the world of heritability, height stands tall.

———

E ven a trait as strongly heritable as height, however, can also be drastically shaped by the environment. In his own research, Louis-René Villermé watched the average height change over the course of a few years. During the Napoleonic War, the average height of young French soldiers declined—the result of wartime food shortages, he guessed. After the war's end, the army's average height rebounded a little—thanks, Villermé said, to "a decrease, however slight, in misery."

Villermé's insight went neglected for the next 150 years, until a small group of economists led by the Nobel Prize winner Robert Fogel started charting height in different countries over the course of decades. They made a compelling case that height could serve as an economic barometer, recording the well-being of societies. They were the first researchers to discover the huge gap between rich and poor boys in late eighteenth-century England, for example.

Their research also gave statistical heft to the stories that Frederick Douglass and other former slaves told about growing up in the antebellum South. Douglass recounted how the sole piece of clothing he was given at age six was a coarse linen shirt. His diet was gruel, served to slave children with as much dignity as slop to pigs.

This cruelty was based on cold economic reasoning: Since slave children were too young to earn money in the fields, their masters chose not to invest in them. When Fogel's followers analyzed plantation records, they found that enslaved American children were much shorter than free ones. But those records also showed that slaves experienced an extraordinary growth spurt in adolescence. That rapid rise was likely the result of the extra food slave owners gave their slaves once they were old enough to turn a profit.

After some small-scale studies in the 1970s, Fogel and his fellow economists widened their spotlight, carrying out a systematic survey of height through history. They looked at military records of conscripts, prison archives, and any other historical data they could get their hands on. They moved from one country to another and pushed back deeper than before into history. When written records failed the researchers, they measured bones from ancient skeletons.

The longest record of height can be found in Europe, where it stretches back thirty thousand years to the Gravettian culture. Gravettian men stood on average six feet tall. When agriculture arrived in Europe some eight thousand years ago, people experienced a tremendous drop in stature. Men lost eight inches of height. The drop was likely the result of Europeans switching to a grain-rich diet much lower in protein. For the next seven thousand years, European stature hardly changed, wavering just an inch or two from century to century. In the eighteenth century, the average European man stood just five foot five.

But they were not locked in at that height. When English people emigrated to the American colonies, men swiftly climbed to five foot eight, becoming the tallest men in the world. By the end of the eighteenth century, American apprentices at age sixteen stood almost five inches taller than poor sixteen-year-olds in London.

In both the United States and Europe, the average height dipped in the first half of the nineteenth century. But then, starting around 1870—at the time Galton began puzzling over height—people in both Europe and the United States started getting taller. Over the next century, Americans grew about three extra inches on average, hitting a plateau in the 1990s. In

Europe, the boom was even more dramatic. With each succeeding decade, Europeans added about half an inch of average height, and kept growing that way into the twenty-first century. Northern and central European countries were the first to begin this ascent, but the southern regions started catching up by the mid-1900s. Today, Latvian women have become the tallest women in the world, jumping from about five foot one to five foot seven. Dutch men rose from five foot seven in 1860 to just over six feet tall, making them the tallest men on Earth.

In 2016, an international network of researchers extended this survey to the world. Over the past century, they found, some countries outside of Europe experienced equally impressive gains. South Korean women experienced the biggest gain, growing eight inches in one hundred years. Among men, Iranians grew the most, now standing six and a half inches taller than they did in the early 1900s. Some people barely grew at all: Pakistani men gained just half an inch. And some countries in Africa, such as Niger and Rwanda, shot up in the first half of the twentieth century only to lose an inch or two after 1960.

Overall, though, the world has gotten much taller. It may be hard to believe that Guatemalan women today—standing only four foot eleven—could have been any shorter in the past. In fact, they have gained four inches since the early 1900s.

Three million years ago, our ancestors in East Africa stood only about a yard high. By 1.5 million years ago, *Homo erectus* grew as tall as five foot seven. Natural selection may have favored genes for greater height because they gave our ancestors long legs that could carry them for long distances across the savanna. Our ancestors kept evolving to greater heights; by 700,000 years ago, they had evolved to our modern stature.

But in some places, natural selection has worked in the opposite direction, making people shorter. African pygmies—to be more accurate, African ethnic groups such as Baka and Mbuti—evolved a new growth velocity. As children, they grow fast, but then they stop early. Some studies suggest this pattern evolved because Baka and Mbuti children faced a higher chance of dying. If children reached sexual maturity faster, they were more likely to have children of their own.

The height boom that started in the late 1800s was too swift to be a product of evolution. If natural selection had been responsible, people with genes for greater height would have had more children than shorter people. The difference would have been stark. Gert Stulp of the University of Groningen and Louise Barrett of the University of Lethbridge estimate that the height boom in the Netherlands would require that a third of short people in every generation of Dutch people have no children at all.

Nothing of the sort actually happened in the Netherlands, and that leaves only one explanation: The environment stretched people out.

How tall children grow depends intimately on their health and diet. A child's growing body demands fuel both to stay alive and to build new tissues. A healthy diet—especially one rich in protein—can meet both demands. If the diet falls short, the body sacrifices growth for survival. Diseases can also stunt a child's growth, because the immune system needs extra resources to fight off infections. Diarrheal diseases are especially brutal, because they also rob children of the nutrients in their food. This fate can get locked in tightly in infancy. As a result, the height of children at age three correlates well with their height in adulthood.

Before the nineteenth century, Europe's rich and powerful families enjoyed the best food and health on offer and got close to their full potential height. The poor were left stunted. Europeans who traveled to the American colonies escaped this growth trap. They moved to a place where they could grow plenty of food for themselves, but where the population was sparse enough that they didn't suffer all the outbreaks that struck Europe's crowded cities.

When the Industrial Revolution came to the United States in the 1800s, these height-favoring conditions faded, and Americans grew shorter. Europeans shrank as well. People who got jobs in factories earned more money than their ancestors, but they had to crowd into cities for the work. Even though the cities were still surrounded by productive farms, the technology did not yet exist to get affordable milk and meat to their residents. As a result, the per capita consumption of meat in the United States dropped by a third in the middle and lower classes. Americans got 2 to 4 percent fewer calories, and they consumed 8 to 10 percent less protein. Making matters even worse, the Industrial Revolution took place decades before the discovery of the germ theory of disease. On the crowded streets of American and European cities, outbreaks flared up and doctors had little idea how to stop them.

By the end of the nineteenth century, things had gotten much better, and people's height reflected the improvement. Clean water and sewer systems helped children stay healthy. Railroad networks brought high-protein food into cities at affordable prices. At the same time, the size of families shrank, making it possible for parents to provide more care to fewer children. Now the scales of the Industrial Revolution tipped in height's favor. Americans started to grow. Europeans started out short at the beginning of the nineteenth century, and they got shorter with the Industrial Revolution. But the balance tipped in the late 1800s, and they sprang up even faster than the Americans.

Similar stories have played out in many other countries. After the Korean War, South Korea's economy rapidly grew to be the eleventh largest in the world, and the country established a universal health care system in 1977. North Korea, meanwhile, stagnated, channeling its income into nuclear weapons and its military while its population starved. South Koreans are now over an inch taller than North Koreans.

No one knows how much taller people in developed countries can become, but in developing countries there's plenty of room for growth. In a 2016 study, researchers at Harvard estimated that 36 percent of all two-year-olds in developing countries were stunted. Improved sanitation, medicine, and nutrition would get rid of much of that deficit and produce much taller people in the future.

But the gains the world has achieved could be easily wiped away. In the late 1900s, shifting economics left many countries in Africa struggling to feed themselves, with the result that children became stunted and their average height declined. The economy of the United States, the biggest in the world, has not protected it from a height stagnation. Height experts have argued that the country's economic inequality is partly to blame. Medical care is so expensive that millions go without insurance and many people don't get proper medical care. Many American women go without prenatal care during pregnancy, while expectant mothers in the Netherlands get free house calls from nurses. Making matters worse, Americans have shifted to a diet loaded in sugar and to sedentary habits. Instead of growing tall, we're growing obese.

———

W hen Jaime Guevara-Aguirre was growing up in a small town in Ecuador, he would sometimes notice grown-ups who stood as tall as a first grader. Otherwise they were like everyone else, with a normal intelligence and life span. Guevara-Aguirre learned to call them *pigmeitos.*

When he grew up, Guevara-Aguirre went to medical school and became an endocrinologist in Quito, where he studied how hormones control people's growth. He wondered about the *pigmeitos* back home in the province of Loja. Sometimes he would get a chance to examine one of them in his office, and he noticed that they all had certain traits missing from other people born with dwarfism. The whites of their eyes had a blue cast, for example. They had trouble extending their elbows. Their voices were high. Blood tests allowed Guevara-Aguirre to make a formal diagnosis: All the *pigmeitos* shared the same condition, known as Laron syndrome.

Before Guevara-Aguirre and his colleagues published their discovery in 1990, only a few people had been identified with Laron syndrome anywhere on Earth. The inherited disorder traveled down through a few families, likely caused by a rare recessive mutation. In Spain, doctors had previously recorded a few cases of Laron syndrome, leading Guevara-Aguirre to suspect that a Spanish immigrant brought the mutation to Loja. In the isolated villages of the province, the mutation managed to become unusually common, and some carriers had children together, creating a cluster of *pigmeitos*. Guevara-Aguirre carried out the first systematic survey of Loja for the condition, traveling back roads from village to village. By the time he was done, he had found one hundred people with Laron syndrome.

At his Quito clinic, Guevara-Aguirre began providing long-term medical care to *pigmeitos* while studying them closely to understand how exactly they ended up so short. They produced growth hormone, he found, but somehow it didn't cause them to reach normal heights. In his research, Guevara-Aguirre also noticed something extraordinary: *Pigmeitos* almost never got cancer or diabetes. Whatever was arresting their growth was also shielding them from diseases that arise as our bodies get old.

After Guevara-Aguirre and his colleagues described the people of Loja, they set out to find the genetic basis of their condition. The scientists drew blood from thirty-eight *pigmeitos* in Ecuador and shipped it to Stanford University. They also sent blood from other members of the *pigmieto* families who were of normal height. At Stanford, a geneticist named Uta Francke and her colleagues pulled immune cells from the blood, and extracted their DNA.

Comparing the *pigmeitos* to their tall relatives, the scientists found one crucial genetic difference. Out of the thirty-eight *pigmeitos*, thirty-seven shared the same mutation on the same gene, a mutation missing from the other subjects. In 1992, the scientists reported that the mutation struck a gene called GHR. GHR encodes a protein that sits on the surface of cells, where it can grab growth hormone molecules. Each time the GHR protein snags one, it sends a signal to the interior of the cell, causing it to turn on a network of growth genes.

Charles Byrne, the "Irish Giant," has provided some clues about how heredity can push people to the opposite extreme from the *pigmeitos*. This was certainly not what Byrne would have wished for. As he lay dying, Byrne grew terrified that grave-robbing anatomists—"resurrectionists," as they were known—would dig his body out of the ground. He begged his friends to bury him at sea. After Byrne died, they put him in a massive iron coffin. The coffin was dumped into the English Channel, but later it turned out that the coffin contained only stones. Somehow—possibly by means of a bribe to an undertaker—a physician named John Hunter ended up with Byrne's skeleton. Shortly after Byrne's death, Hunter posed for a portrait, seated at a table covered with a bell jar and anatomical books. In the upper righthand corner of the painting dangle the foot bones of the Irish Giant.

Yet Hunter appears to have never carefully studied Byrne's skeleton. Instead, the bones were stored in the Hunterian Museum, where they remained until the museum was bombed in World War II. Today, Byrne's skeleton looms on display in the Royal College of Surgeons. A bust of John Hunter sits on a shelf above him, the surgeon pursuing the giant long after their deaths.

In 1909, two doctors, named Harvey Cushing and Arthur Keith, first gave Byrne's skeleton a close look. They thought his bones might have some clues about how humans grow. In the early 1900s, endocrinologists began deciphering the language of hormones that give commands to our bodies. The pituitary gland, located at the base of the brain, releases growth hormone, which stimulates bones and other tissues to get bigger. When Cushing and Keith opened up Byrne's skull, they found a large pit where his pituitary gland had once been. They hypothesized that Byrne developed a tumor in his gland, causing it to produce extra growth hormone, and to keep producing it long after it normally would have shut off. Decades later, other scientists took X-rays of some of Byrne's bones and confirmed Cushing and Keith's suspicion. When Byrne died at age twenty-two, his bones were growing at the rate you'd expect in a seventeen-year-old boy.

Byrne's condition is now known as acromegaly. About sixty people per million suffer from it. While the hormone-producing tumor itself isn't fatal, it can nevertheless cause an early death by spurring runaway growth throughout the body. Doctors now treat acromegaly by surgically removing the tumor, blasting it with radiation, or giving patients drugs that can counteract the extra growth hormone circulating through their blood. When geneticists studied acromegaly, it seemed to fall in a hereditary gray zone. It didn't run in families as starkly as PKU or Huntington's disease. But sometimes a person with acromegaly turns out to have a cousin with it, too.

In 2008, Márta Korbonits of the William Harvey Research Institute in London and her colleagues identified a mutation that was common in families with acromegaly. It affected a gene called AIP, which encodes a protein whose role scientists still don't understand very well. About one in five people who inherit the AIP mutation develop a tumor and may go on to grow to a tremendous height. It's likely that the mutation only triggers its dramatic effects in people who happen to inherit mutations in other genes still to be discovered.

Korbonits's team found that different mutations of AIP could produce acromegaly. But they were surprised to find an identical AIP mutation in four families in Northern Ireland, not far from the village where Charles Byrne had been born. Their clustering suggested that they might have inherited it from a distant common ancestor.

The scientists arranged with the Hunterian Museum to drill into two of Byrne's teeth. More than 220 years after his death, they were able to extract his DNA. Byrne turned out to have a mutation in the same spot in his AIP gene as the living Irish people Korbonits and her colleagues studied; they also found that the DNA flanking the AIP gene was identical. They estimated that this mutation arose in Ireland roughly 2,500 years ago. James Cowles Prichard may have been onto something when he speculated that there was some "peculiarity in Ireland" that produced its giants. It may have been nestled in the DNA of some of its residents, passed down through a hundred generations.

The genes behind Laron syndrome and acromegaly supplied some important clues about human height. By studying people with these conditions, scientists could observe what happens when growth hormones dry up or surge like a river full of snowmelt. But for Joel Hirschhorn, these mutations, limited to a few villages in Ireland and Ecuador, didn't help him understand the height of his own patients. He wanted to find variants that accounted for the heritability of height among billions of people.

Hirschhorn suspected there would be many genes, but he couldn't say how many. To find people to study, he launched collaborations with researchers who were already running studies on the genetics of other conditions, such as diabetes and heart disease. In their exams, the researchers measured height as one of many vital statistics. The data were just waiting for someone like Hirschhorn to take a closer look.

Hirschhorn gathered records on 2,327 people from 483 families, hailing from Canada, Finland, and Sweden. In each subject's DNA, the researchers had sequenced a few hundred genetic markers scattered across their genomes, separated from each other by several million base pairs. Hirschhorn and his colleagues compared the families in each country to see if the children who inherited particular markers tended to grow taller or shorter than the others. They found four regions of the human genome that showed a strong association.

When Hirschhorn and his colleagues published their study in 2001, it was one of the first times that anyone had found a clue about common variants that influence height. But it was a very modest start. Hirschhorn had been able to identify only long stretches of DNA where a genetic variant seemed to be lurking. The variants might reside in one of hundreds of genes in those regions. It was even possible that Hirschhorn's results were a fluke that had nothing to do with height. A number of tall people might have a version of a particular marker thanks simply to chance.

Hirschhorn was not alone in his frustration. Many other scientists were trying to trace traits—especially hereditary risks for certain diseases—to specific genes. At first they enjoyed some high-profile successes, finding links to conditions like diabetes and bipolar disorder. But very often, the links would melt away when other scientists looked at larger groups of people. Soon scientists worried that they were stuck in a dead end. "Has the genetic study of complex disorders reached its limits?" two scientists asked in a 1996 article in the journal *Science*.

Those two scientists, Neil Risch of Stanford University and Kathleen Merikangas of Yale, argued that the answer was no. But to uncover the variants that raise the risk of common diseases, scientists would have to build new tools. Risch and Merikangas predicted that most variants would not be powerful, as in the case of Laron syndrome and acromegaly. Instead, the variants behind many diseases would be weak and numerous.

Risch and Merikangas sketched out a new way to carry out this search. Geneticists needed to step away from their beloved pedigrees. Instead, they needed to look at the DNA of hundreds of people, with no regard to their families. They could search for variants that were unusually common in people with a disease, compared to those who did not suffer from it. Risch and Merikangas dubbed their hypothetical method a genome-wide association study.

It took until 2005 for genome-wide association studies to get their first hit. Josephine Hoh, a geneticist at Yale University, wanted to find genes involved in the leading cause of blindness, a disease called age-related macular degeneration (AMD for short) that ravages the center of the retina. Hoh knew that having a relative with AMD raised the odds that people would develop it in their own life. But studies on families with AMD had failed to reveal a gene associated with the disease.

Hoh and her colleagues gathered DNA from ninety-six people who had AMD, as well as from fifty people who didn't. They scanned the genetic markers and noticed an unusually common one among people with AMD located on chromosome 1. Closely examining that region, they came across a variant in a gene for a protein made by immune cells, called complement factor H. They found that having two copies of the variant drastically raised a person's odds of developing AMD.

The job of complement factor H is to stick to pathogens, triggering inflammation to fight them. Hoh's research indicated that mutant forms of the protein stick instead to retinal cells, causing the immune system to attack the eye. Hoh's findings were later confirmed in other studies. But with such a small group of people in her study, she might very well have missed complement factor H if its effects had been any weaker. She was right, and she was also lucky.

To use genome-wide association studies to find subtler variants, scientists recognized they would have to study thousands or even millions of people. In 2007, a consortium of laboratories working through the

Wellcome Trust in England published the first such large-scale study. Examining fourteen thousand people, they identified twenty-four genes with variants that raised the risk of diseases such as diabetes and arthritis.

After his own frustrating experience studying the height of families, Hirschhorn also turned to genome-wide association studies. He and his colleagues used some of the data from the Wellcome Trust study, adding to it people who had been part of a diabetes study in Sweden. All told, nearly five thousand people became part of the study. The technology for sequencing genetic markers had improved drastically since Hirschhorn had started investigating height. Now, instead of looking at a few hundred markers, he could look at a few hundred thousand of them. The denser spread of genetic markers made it possible to zero in on smaller regions containing fewer genes.

This time, Hirschhorn got a solid hit. One variant, located in a gene called HMGA2, was significantly more common in tall people than in short ones—so common, in fact, that it couldn't be dismissed as a fluke. Hirschhorn and his colleagues tested the association by looking at HMGA2 in more than twenty-nine thousand other people. In the bigger group, taller people once again were much more likely to carry the same variant of HMGA2.

Yet Hirschhorn couldn't say how precisely HMGA2 influenced people's height. A few experiments carried out over the years offered a handful of clues. In experiments with mice, some mutations to HMGA2 could turn the animals into dwarves. Others turned them into giants (by mouse standards).

The evidence about HMGA2's function in humans was even scarcer. In 2005, geneticists at Harvard Medical School published a case report on an eight-year-old boy who had a mutation clipping his HMGA2 gene short. He seemed normal at birth, but at three months he sprouted his first tooth. By the time the boy was eight years old, he was over five foot five, the average height for a fifteen-year-old. His legs and fingers grew crookedly, and he developed lumps of fat and blood vessels under some parts of his skin.

These studies suggest that HMGA2 normally acts like a brake, slowing down our growth-spurring genes. A mutation that shuts down HMGA2 entirely may cause runaway growth. The common variant in HMGA2 that increases height may lift the genetic foot off the brake just enough to make people grow a bit taller—but not enough to lead to deformities or tumors.

The discovery of HMGA2 was like a quarter-carat sapphire: solid, glittering, and tiny. It marked the first time that scientists found a common variant strongly associated with height. Later, when other scientists studied even larger groups of people, they confirmed the link. But the HMGA2 variant accounts for a vanishingly small amount of the variation in the human population. When I got my genome sequenced, I found that I carry one copy of the height-raising form. On average, people with one copy are about an eighth of an inch taller than if they didn't have one. That's the equivalent of putting on a warm pair of wool socks. If I had two copies, it would be like putting on a second pair. And when scientists look at the full range of variation in height, they find that this variant in HMGA2 explains very little—only about 0.2 percent.

Hirschhorn's 2007 study also uncovered some tantalizing clues about many other genes. They contained variants that were more common in tall people than in short ones, or vice versa. But the differences weren't as stark as HMGA2, leaving open the possibility they were the result of chance. To rule out randomness, Hirschhorn would need to measure more people's heights.

Hirschhorn and his colleagues created a new network of hundreds of research groups around the world. They called their consortium the Genetic Investigation of ANthropometric Traits—GIANT for short. The GIANT team examined the height of tens of thousands of people, then hundreds of thousands, and the bigger numbers allowed them to pick out more genetic variants, first dozens, then hundreds. Most of the genes they discovered had a smaller influence than HMGA2. But they also found a number of genes that had a far bigger one. If people carry two variants of a gene called STC2, for example, those alleles will lift them up about an inch and a half. These powerful genes had gone overlooked in earlier studies of height because they were too rare, found in less than 5 percent

of the population. In 2017, a decade after the first genome-wide association study of height, GIANT published a study on more than 700,000 people, bringing the total number of genes influencing height to almost eight
hundred.

——————

To some observers, however, such results seemed like a colossal disappointment. The combined effect of GIANT's eight hundred–odd genes
accounted for just over 27 percent of the heritability of height. The rest remained missing.

Height was not unusual in this regard. Missing heritability dogged
many studies of other traits and diseases, too, even after scientists could
study thousands of people. The shortfall was all the more glaring because of
all the money that had gone into making genome-wide association studies
possible. "The reason for spending so much money was that the bulk of
the heritability would be discovered," the geneticist Joseph Nadeau told a
journalist.

Some critics saw missing heritability as much more than an annoyance.
To them, it was a symptom of a scientific disease. In 2015, two French researchers, Emmanuelle Génin and Françoise Clerget-Darpoux, argued that
missing heritability revealed the futility of genome-wide association studies. Génin and Clerget-Darpoux describe the research as "Garbage-In
Garbage-Out Syndrome." The scientists running the studies were trying to
use brute force to discover the deepest secrets of biology. Yet their repeated
failures simply led them to redouble their efforts, and journal editors to
publish more of their papers. To Génin and Clerget-Darpoux, it seemed as
if geneticists had become trapped in a game they couldn't stop playing.
"Unfortunately, genetics is a clear loser," they concluded.

Other critics say that missing heritability reveals our profound ignorance about heritability itself. Some attacked twin studies, claiming they
lead to estimates of heritability that are much too high. Others argued that
heritability studies miss the way some mutations make the effects of other
mutations stronger. One plus one, in the world of heredity, may be far more

than two. Some critics went even further, arguing that missing heritability is hiding beyond genes, in some other form of heredity scientists have yet to grasp.

W hen I asked Hirschhorn if missing heritability was giving him existential doubts, he shrugged the problem off. "I think a lot of it is just hidden," he told me. "If we had all six billion people on Earth in a genetics study, we would actually get to most of the heritability."

Part of Hirschhorn's confidence came from his own experience over the previous twenty years. The more people he and his colleagues measured, the more heritability they could explain. Some of the genes they found were common but weak, while others were strong but rare. If he could study more people in the future, he expected to find more of both kinds.

Hirschhorn also drew confidence from the work of Peter Visscher, who has given geneticists a new way to study human heritability. Visscher came to research on humans after years of work on livestock. Animal breeders study the heritability of cows to figure out how to get them to make more milk, of pigs to put on more pork. In the 1900s, they used elaborate pedigrees to track the influence of genes on these traits. But at the end of the century, breeders got their hands on technology for reading genetic markers in their animals.

At first, they searched for candidate genes that might have a big effect on their own. Soon it became clear that a trait like milk output was controlled by many genes, each with a tiny effect. Animal breeders found that they could improve their livestock by comparing all their genetic markers in different animals. Animals that were genetically similar overall tended to have similar traits. Breeders could choose which animals to breed based on these so-called genomic predictions.

When Visscher switched from animals to humans in the early 2000s, he realized that he could use genomic predictions on people, too. Visscher and his colleagues took the method out of the barnyard and adapted it to human genetics, dubbing their method Genome-wide Complex Trait

Analysis. To see how well it worked, they unleashed it on the best-studied complex trait of all, human height.

The researchers delved into the data from earlier genome-wide association studies and looked at the genetic markers from thousands of people. They came up with genetic-similarity scores between each pair of people. Heredity turned out to work a lot in humans as it does in chickens. Pairs of people with high scores tended to have similar heights. That tendency reflects the heritability of a trait. The stronger the tendency, the greater the heritability.

When Visscher and his colleagues estimated the heritability of human height from genetic similarity, they ended up with a number close to what had been estimated in earlier studies on families and twins. In 2015, when they published these results in the journal *Nature Genetics*, they declared the missing heritability of height to be "negligible."

Toward the end of my visit with Hirschhorn, I noticed his eyes drifting to the clock on his desk phone. He had a conference call coming up soon with a lot of his collaborators. They were about to take another leap, from 800,000 people to perhaps two million. But before I left, Hirschhorn explained that the years of work he had put into the inheritance of height were not simply to create a catalog of genes. He wanted to use the catalog to understand the mysteries of height. If you stop and think through what it means to grow, the process is astonishing. Each part of the body has to change its shape and size to match every other part. There's no central blueprint for the construction of an adult human. Each cell has to decide for itself, using nothing more than chemical signals and its own network of genes, RNA molecules, and proteins.

As Hirschhorn's list of genes has grown, he and his colleagues have searched them for patterns. They turn out not to be a random assortment. "Most of the action is at the growth plate," Hirschhorn said.

Growth plates are thin layers of cells located near the ends of limb bones. In children, some of the cells in the plates produce signals, which then trigger neighboring cartilage cells to multiply. As the cells divide, the bones get longer. Eventually the cartilage cells change, producing bone instead. They finally commit suicide, tearing themselves open to dump out chemicals that make the surrounding bone even harder.

Hirschhorn and his colleagues found that many of the genes on their list are unusually active in growth plate cells. Obviously, other parts of the body have to grow as well in order for people to become taller. But it's possible that the growth plates lead the parade. Mutations to the genes used by growth plate cells speed up or slow down the increase in limb bones. The rest of the parade has to adjust its speed to follow the leader.

Yet Hirschhorn knew that he would have to find other stories to tell about height. HMGA2, the first gene he and his colleagues discovered influencing height, remained the strongest common variant. It's active in embryonic cells, not in growth plates in children. And despite a lot of research by Hirschhorn and his graduate students, he still couldn't say why it's so important. "That one still boggles my mind," Hirschhorn admitted.

It's possible that Hirschhorn will have to become a Scheherazade of the genome to tell all the stories about how the genes we inherit influence our height. In 2017, Jonathan Pritchard, the scientist who invented STRUCTURE, tried to predict how many genes scientists would ultimately find linked to height. Would Hirschhorn reach a thousand genes and be able to close down his shop? Pritchard thinks the answer is a definite no.

For their study, Pritchard and his colleagues took a closer look at a genome-wide association study that Hirschhorn and his colleagues published in 2014. In that study, Hirschhorn's team scanned 2.4 million genetic markers in a quarter of a million people. They looked for variants at each of the markers with a very strong link to height—so strong that they could confidently reject the possibility that the links were just coincidences.

That study gave Hirschhorn and his colleagues a list of about seven hundred strongly supported genes. But they also found many other ambiguous variants that didn't quite meet their strict standards. Those variants might have a weak effect on height, or they might simply have turned up in Hirschhorn's study by chance. Pritchard used new statistical techniques on those ambiguous variants, to see if he could separate the genetic wheat from the chaff.

He and his colleagues looked for people who carried two copies of each variant and checked their height. Then they looked at the height of people with only one copy of the variant, and that of people with no copies. In many cases, this comparison revealed a small but measurable effect. Two copies of a variant might make people shorter than average, while one copy made them a little taller, and no copies made them taller still. Pritchard and his colleagues then turned to an entirely different group of twenty thousand people to test these results. They found the same effects from the same variants.

What made this study startling was just how many of these variants Pritchard and his colleagues found. At 77 percent of the markers they studied—almost two million spots in people's DNA, in other words—they could detect an influence on height. The markers were not clumped around a few genes on one chromosome. They were instead spread out across all the chromosomes, encompassing the entire human genome.

These variants likely altered the sequence of many genes, changing the structure of their proteins. But they probably also changed the regions of DNA that act like switches to turn the genes on and off. Each of the nearly two million variants had, on average, an exquisitely tiny effect—adding or subtracting the width of a human hair. But collectively, this vast army of weak variants accounted for much more variation in height than the strongest genes that Hirschhorn and his colleagues put together in their catalog.

Traditionally, geneticists have called height polygenic—meaning "many genes." Pritchard thinks a new word is called for: *omnigenic*.

If height really is omnigenic, as Pritchard believes, we may need to rethink the way our cells work. There may be a core group of genes lurking in growth plates that take the lead in determining how tall we get. But some of those genes also have other jobs. They work with other genes in other kinds of cells. You can think of our genes as a set of networks. There's a network of genes that work together in growth plate cells. And you can draw a line from some of those genes to other networks. Thanks to the way these networks are organized, it may take only a few steps to go from any given gene to any other gene in the human genome. With all these connections, a mutation to a single gene can have wide-ranging effects. It can alter a gene that has nothing directly to do with height, but its influence can reach across the networks to affect the ones that do. In science's hunt for how we inherit height, scientists may have to expand their search to the entire genome.

CHAPTER 10

Ed and Fred

I N 1864, when Francis Galton was forty-two, he posed for a photograph. He was now middle-aged and had grown a beard girdling his jaw. His forehead rose into a high, hairless dome. Galton leaned his left hand on a bookshelf next to a globe, the icon of the geographer. A chair stood next to him, a top hat sitting on the seat, the brim turned up like an open pot. The scrolled back of the chair reached almost to the level of Galton's hip. It served as accidental measuring tape, documenting his generous height. The photograph, in other words, is a typical picture of a tall Victorian gentleman: a mass of about 37 trillion nineteenth-century cells nurtured through years of divisions by a wealthy British childhood.

Galton inherited that wealth, but not through his genes. His great-great-grandfather Joseph Farmer opened a small smithy in Birmingham in the early 1700s, where he made a modest living producing sword blades and gun parts. In 1717, Farmer took a big gamble that paid off for generations. Traveling to the American colonies, he set up forges and furnaces in Maryland, where he could smelt the iron from nearby mines. He shipped the metal back to his factories in Birmingham, where his workers could then craft them into more expensive goods. Thanks to the efforts of businessmen like Farmer, Maryland became one of the world's main iron suppliers in the eighteenth century. Farmer would brag about his "plantation" iron—a name it earned for a very simple and lucrative reason: Maryland's ironworks relied heavily on the labor of African slaves.

When Farmer died in 1741, his son James inherited the business, which by then specialized in gunlock springs and musket barrels. His family invested some of the profits in slave-trade companies in Lisbon, bringing them even greater riches. Five years later, James's sister married Francis Galton's great-grandfather Samuel. Samuel Galton had been a draper of modest means, but his new brother-in-law hired him as an assistant. It wasn't long before Samuel became a partner in the firm.

Guns and slavery grew even more intertwined in the Galton family fortune. By the 1750s, the Galtons were delivering more than twenty-five thousand guns a year to European traders, who sold the weapons to African states engaged in increasingly bloody battles. The warring states captured prisoners in the fights, and then sold them to European slave traders. Before long, they demanded to be paid for the slaves with more guns instead of gold.

Samuel Galton took sole control of the firm and began to supply arms to the British government, which used his muskets against American rebels. When Samuel Galton's son, Samuel John Galton, came of age, he joined the firm, and together the two Samuels grew their business for a few decades. When he died, Samuel had amassed 139,000 pounds. "His fortune had been the fruit of God's blessing on his industry," Galton's granddaughter later said.

The Galtons were a pious family of Quakers, but by the end of the 1700s, the wealth they made from war and slavery had largely turned the Society of Friends against them. In 1790, a faction of Quakers tried to bar the Galtons from their monthly meetings. Delegations of wealthy Quakers tried to persuade the Galtons to get into a different line of work. Samuel the elder agreed to stop taking profits from the family's gun business. But Samuel the younger refused. He wouldn't even admit he was doing anything wrong. In a letter read to the monthly meeting in Birmingham in 1796, he cast himself as a helpless prisoner of heredity.

"The Trade devolved upon me as if it were an inheritance," he declared. "My Engagements in the Business were not a matter of choice."

The Quakers didn't buy that excuse. They barred him from their meetings for life. Eight years later, perhaps out of some delayed remorse, Samuel Galton abandoned the gun business to his son, Francis's father, and busied himself with opening a new bank. In 1815, Samuel Tertius Galton closed down the gun business for good. The Industrial Revolution had arrived in Birmingham, and the family bank's investments in factories and canals were proving profitable. By the time Francis Galton was born in 1822, the family fortune had swelled to 300,000 pounds.

As a child, Francis proved to be a prodigy, reciting passages of Shakespeare from memory and discussing the finer points of *The Iliad*. Despite their wealth, the Galtons always felt like outsiders, in part because no one in their family had completed a university education. They loaded their hopes for legitimacy on Francis's small shoulders. At age four, when Tertius asked his son what he hoped for most of all, Francis replied, "Why, university honors, to be sure."

They never came. At age eighteen, when Galton went to Cambridge, his father stocked his rooms with everything a young gentleman at university needed, from silver teaspoons to a steady supply of wine. Above his fireplace, Francis mounted crossed foils and pistols. In a small room next to his bedroom lived his three servants. Once he was settled in, Galton set out to study mathematics, aiming to take the honors examination known as the Tripos. To improve his concentration, he bought a "Gumption-Reviver," a contraption that dripped water on his head and had to be refilled by a servant every fifteen minutes. He hired a tutor with a reputation for brilliance to teach him math.

Despite all his promise and spending, Galton received third-class honors on his first-year exams—the equivalent of a gentleman's C. In the hopes of improving his scores, Galton hired an even better math tutor, who accompanied him and four fellow students on a "reading party" in the Lake District. When it came time to take his first major exam, nicknamed Little Go, Galton made only second class.

In a letter to his father, Galton made light about the score, boasting about "going into the Little Go when I had not read over half my subjects and coming out unplucked." In truth, he was bitterly disappointed to watch his friends—who had studied with the same tutors at the same reading parties—get first honors. One of Galton's tutors urged him to give up his childhood hope. He should simply finish up Cambridge as most students did, and take a so-called poll degree.

Galton refused. Poll degrees were for the mediocre. Instead, he hired a new math tutor and went to Scotland for another reading party. This time the stress of studying gave Galton a nervous breakdown. "A mill seemed to be working inside my head," he later recalled. Looking back at the crisis he went through in the fall of 1842, Galton concluded he had pushed his brain too far. "It was as though I had tried to make a steam-engine perform more work than it was constructed for."

For a few more months, Galton kept up the illusion of a top student. He wrangled a "certificate of degradation" from one of his tutors, which allowed him to put off his final honors examination for another year. He distracted himself from heart palpitations and dizzy spells with wine parties, poetry, and hockey. It was all a facade, one that collapsed when his father suddenly died. Galton left Cambridge with a poll degree and inherited his father's fortune. He was mediocre, yet rich beyond compare.

Galton's failure at Cambridge left him forever insecure about his own standing in the scientific world, always seeking to bask in the genius of others. He would later look back at his Cambridge years with gratitude that he had spent time with "the highest intellects of their age."

Their high intellects may well have inspired Galton's obsession with heredity. He was struck by "the many obvious cases of heredity among the Cambridge men who were at the University about my own time." The students who got the highest honors at Cambridge were exquisitely rare, and yet they generally seemed to have a father, a brother, or some other male relative who had gotten high honors as well. Galton didn't think that was a coincidence.

In later years, this observation mushroomed into a fervent conviction. In Galton's 1869 book, *Hereditary Genius*, he declared that human intellectual abilities "are derived by inheritance under exactly the same limitations as are the form and physical features of the whole organic world."

Galton believed intelligence, like height, was deeply rooted in biology—so deep that it could be inherited. To persuade his readers, he needed a way to measure intelligence in relatives. But in the 1860s, no one knew how to do that. For a crude approximation, Galton got his hands on the scores of seventy-three boys who took the admission test to the Royal Military Academy at Sandhurst.

The scores, he was pleased to discover, roughly followed a bell curve, much like the one he found for height. Most of the boys scored close to average, while the curve tapered off in either direction—to what Galton would call stupidity and genius. He lingered lovingly over the scores of Cambridge students who earned honors in mathematics, drawing up a table in which fewer and fewer students managed to reach higher and higher scores. Yet Galton considered even the lowest-scoring Cambridge honors students to be brilliant in comparison to the majority of English people. "The average mental grasp of even of what is called a well-educated audience, will be found to be ludicrously small when rigorously tested," Galton declared. He never mentioned where he himself fit into that Cambridge continuum.

Galton then gathered evidence for heredity. He followed up on his intuition about his bright college mates at Cambridge, researching their genealogies and building a pedigree of the mind. Galton claimed his data showed that high-scoring students had high-scoring kin. He looked for other examples from history, investigating presidents and scientists and composers, more than a thousand men of talent all told. (Women barely counted in his argument.)

Height and intelligence would remain the twin guideposts for Galton. When thousands of visitors came to his Anthropometric Laboratory, he not only recorded their height but also timed their reactions and recorded the circumference of their heads—two traits Galton suspected were related to intelligence.

But when Galton developed eugenics, height and intelligence would take on very different roles in his thinking. When he dreamed about his hereditary utopia, it was intelligence that he wanted to breed. He pictured a nation of geniuses, not giants.

G alton's disciple Karl Pearson pursued more research on intelligence at the same time as he studied height. He asked teachers at hundreds of London schools to describe their students, picking the most apt words for each one from a list of adjectives such as *slow* and *quick*. When Pearson tallied up the replies and ranked them, he ended up with a bell curve.

To see if heredity played a part in the intelligence of the students, Pearson compared siblings. He found that the ability of siblings was correlated— the siblings of low-scoring students tended to score low as well; the quick ones tended to have quicker siblings. Pearson was impressed that the correlation of intelligence was much like the correlation for physical traits. We inherit the mental abilities of our parents, Pearson declared, "even as we inherit their stature, forearm and span."

Yet Pearson's argument for the inheritance of intelligence suffered from a fundamental weakness: For his measurements, he had to rely on the gut instincts of teachers. In the 1910s, Henry Goddard and other American psychologists replaced those subjective scores with Binet-based test results. And instead of studying hundreds of people, they tested millions.

To Goddard's collaborator Lewis Terman, the army tests confirmed that intelligence was primarily the result of heredity. Among army recruits, immigrants scored lower on average than native-born soldiers. "The immigrants who have recently come to us in such large numbers from Southern and Southeastern Europe are distinctly inferior mentally to the Nordic and Alpine Strains which we received from Scandinavia, Germany, Great Britain, and France," Terman said. Intelligence, the tests made clear, was "chiefly a matter of native endowment," and so "these are differences which the highest arts of pedagogy are powerless to neutralize."

Terman was so convinced of the inheritance of intelligence that he ignored his own data. His test results showed that the longer immigrants lived in the United States, the higher they scored on intelligence tests. Terman and his colleagues built their tests from questions steeped in everyday American life, which required familiarity as well as intellect. Recruits were shown a picture of a tennis game, to see if they noticed the net was missing. They were quizzed about the color of sapphires. They had to complete sentences such as "The Percheron is a kind of . . ." (Answer: Horse).

As it became clear that intelligence test scores could be influenced by people's cultural backgrounds, some psychologists tried to strip that background away. One psychologist, named Stanley Porteus, decided to avoid language altogether by testing people with mazes. He designed mazes of different levels of complexity and had them printed. Traveling across Australia, Asia, and Africa, he searched for people with little contact with the West who he could examine. Porteus found that the so-called Bushmen of the Kalahari scored a mental age of seven. Yet his subjects were navigating their way through Porteus's printed mazes in the middle of a vast desert that they could navigate without a map, finding all the food and shelter they required.

When he presented his results in 1937, Porteus recognized that even a wordless maze test might be skewed by culture. "The Maze is by itself far from being a satisfactory measure of intelligence," he said. In fact, the experience left him wondering what his tests were measuring. "All we can say of it is that the complex of qualities needed for its performance seem to be valuable in making adjustments to our kind of society," Porteus said.

Other researchers have argued that intelligence isn't merely what it takes to survive in one society. They maintain it's a deep-seated feature of the human brain. The neuroscientist Richard Haier, for example, has defined intelligence as "a catch-all word that means the mental abilities most related to responding to everyday problems and navigating the environment."

These abilities are not a random, disconnected collection of skills, tests show. When scientists give people tests on different abilities, their scores are correlated. If people are very good at recalling information from stories, for example, they also tend to do well at recalling words from lists. Different tests for logical reasoning correlate with each other as well. In turn, these broad abilities—such as reasoning, memory, spatial ability, processing speed, and vocabulary—correlate with each other. Psychologists can measure this underlying correlation with a single factor known as *g*, short for general intelligence.

It may seem strange that the speed at which people correctly hit a button can roughly predict whether they can recognize a word like *defenestrate*. Yet the deep connections revealed by intelligence researchers are among the best-replicated findings in all of psychology.

Intelligence is also a surprisingly durable trait. On June 1, 1932, the government of Scotland tested almost every eleven-year-old in the country—87,498 all told—with a seventy-one-question exam. The students decoded ciphers, made analogies, did arithmetic. The Scottish Council for Research in Education scored the tests and analyzed the results to get an objective picture of the intelligence of Scottish children. Scotland carried out only one more nationwide exam, in 1947. Over the next couple of decades, the council analyzed the data and published monographs before their work slipped away into oblivion.

In 1997, an expert on intelligence named Ian Deary stumbled across a mention of the Scottish Mental Survey in a book. Given his line of work—and given that he worked at the University of Edinburgh—he was startled that he had never heard of it before. The book Deary was reading only mentioned the survey in passing, but it was enough to inspire him to find out more about it. Intelligence testing was so time-consuming that researchers typically could manage to examine small groups of people. Here was a test of an entire population. And all the eleven-year-old test takers who were still alive would now be seventy-six. At the time, psychologists were still debating how much intelligence tests taken in childhood said about people's later lives. If Deary could find some of them, he could give them the test once more and get an unprecedented measurement of this influence.

Deary's colleague, Lawrence Whalley, dug into the Scottish Mental Survey reports. Eventually his search led him to a basement stacked with boxes and files containing the original tests. He called Deary with the news. "This will change our lives," Deary replied.

Deary, Whalley, and their colleagues moved the 87,498 tests from ledgers onto computers. They then investigated what had become of the test takers. Their ranks included soldiers who died in World War II, along with a bus driver, a tomato grower, a bottle labeler, a manager of a tropical fish shop, a member of an Antarctic expedition, a cardiologist, a restaurant owner, and an assistant in a doll hospital.

The researchers decided to track down all the surviving test takers in a single city, Aberdeen. They were slowed down by the misspelled names and erroneous birth dates. Many of the Aberdeen examinees had died by the late 1990s. Others had moved to other parts of the world. And still others were just unreachable. But on June 1, 1998, 101 elderly people assembled at the Aberdeen Music Hall, exactly sixty-six years after they had gathered there as eleven-year-olds to take the original test. Deary had just broken both his arms in a bicycling accident, but he would not miss the historic event. He rode a train 120 miles from Edinburgh to Aberdeen, up to his elbows in plaster, to witness them taking their second test.

Back in Edinburgh, Deary and his colleagues scored the tests. Deary pushed a button on his computer to calculate the correlation between their scores as children and as senior citizens. The computer spat back a result of 73 percent. In other words, the people who had gotten relatively low scores in 1932 tended to get relatively low scores in 1998, while the high-scoring children tended to score high in old age. If you had looked at the score of one of the eleven-year-olds in 1933, you'd have been able to make a pretty good prediction of their score almost seven decades later.

Deary's research prompted other scientists to look for other predictions they could make from childhood intelligence test scores. They do fairly well at predicting how long people stay in school, and how highly they will be rated at work. The US Air Force found that the variation in g among its pilots could predict virtually all the variation in tests of their work performance. While intelligence test scores don't predict how likely people are to take up smoking, they do predict how likely they are to quit. In a study of one million people in Sweden, scientists found that people with lower intelligence test scores were more likely to get into accidents.

The long reach of intelligence suggests that it may have some deep biological foundations. Some scientists have proposed that, in one way or another, different intelligence tests all probe how efficiently the brain processes information. Some of the most compelling evidence for this theory comes from a simple exam in which a shape flashes on a computer screen. The shape is made up of two vertical lines and a horizontal one sitting on top, like a sketch of the standing stones at Stonehenge. Each time the shape appears on the screen, one of the vertical lines hangs down lower than the other. Volunteers have to indicate which one is longer.

If the shape flashes too briefly, people guess at random. But if the shape lingers for long enough on the monitor, they can give the right answer most of the time. On average, people can perceive a shape correctly if they can see it for around a tenth of a second. But that sliver of time varies a little from person to person. In one study, scientists found some people needed just 0.02 seconds, while others needed 0.136 seconds.

Time and again, researchers have found that there's a correlation between intelligence and inspection time. People with lower intelligence scores tend to need more time in order to recognize the shape. It's not an iron law (the correlation is about 50 percent), but the link is strong enough to lead scientists to wonder if there's something in common lurking under inspection time and intelligence.

Even the simplest mental operations require our brains to fire neurons in a network of regions scattered throughout our heads. Regions of the middle and back of the brain gather perceptions and organize them. They then feed their own signals through long fibers—known as white matter tracts—to the front of the brain. There, we have regions that specialize in problem-solving and decision-making. The frontal regions then talk back to the others so that they can fine-tune their gathering of perceptions.

But Deary's research raises the possibility that the roots of intelligence dig even deeper. When he and his colleagues started examining Scottish test takers in the late 1990s, many had already died. Studying the records of 2,230 of the students, they found that the ones who had died by 1997 had on average a lower test score than the ones who were still alive. About 70 percent of the women who scored in the top quarter were still alive, while only 45 percent of the women in the bottom quarter were. Men had a simi-lar split.

Children who scored higher, in other words, tended to live longer. Each extra fifteen IQ points, researchers have since found, translates into a 24 percent drop in the risk of death.

In a 2017 study, Deary and his colleagues drilled further into this effect. This time around, the researchers took advantage of the second mental sur-vey that the Scottish government carried out on eleven-year-olds, in 1947. This younger cohort was too young for World War II, so more of them were able to survive into old age. Deary and his colleagues combed over records for more than sixty-five thousand of the test takers. They noted not only who died but *how* they died.

As before, the researchers found that lower intelligence test scores raised people's risk of death. But when they broke down the deaths into the major causes, they found the same rule held true across the board. The people who scored in the top 10 percent were two-thirds less likely to have died from respiratory disease than those in the bottom 10 percent. They were half as likely to have died from heart disease, stroke, and digestive diseases.

It's possible that intelligence test scores measure how well people can take care of themselves. As adults, they may tend to make somewhat more money, which they can spend on their health. Or they may be slightly better able to understand the information that their doctors give them. But the influence of intelligence on longevity is so broad that Deary has proposed a deeper con-nection. Scores on intelligence tests may gauge some broad feature of human biology, in the same way a thermometer or a blood pressure reading does. The efficiency in the brain may have something in common with how well other parts of the body run. And this "system integrity," as Deary calls it, may help determine how long the whole system runs before falling apart.

———

Early intelligence researchers were firmly convinced that heredity had overwhelming control over intelligence. "A person can no more be trained to have it in higher degree than he can be trained to be taller," the

English psychologist Charles Spearman once said. Yet the evidence they offered for these claims was as firm as a soft-boiled egg. Simply running through the annals of great English men could never deliver the proof Galton hoped for. The class prejudices of the early 1900s made it possible for an extravagant fiction like *The Kallikak Family* to be taken seriously for years.

By the 1920s, however, the science of heredity had matured far enough that researchers could begin to study intelligence in a meaningful way. This was the time that twins were emerging as a tool for studying heritability, and intelligence researchers followed the example set by those studying height. Three Chicago scientists, named Frank Freeman, Karl Holzinger, and Horatio Newman, gave intelligence tests to fifty identical and fifty fraternal twins. They found that the identical twins had closer scores to each other than the fraternal twins—suggesting that intelligence was indeed heritable.

The Chicago researchers realized there was another way they could use twins to study intelligence. Rather than compare twins that had grown up together, they could look at how strongly nature influenced twins who had been raised apart. The scientists put out advertisements for adult twins who had been separated as children, typically adopted by different families. Nineteen pairs responded.

One pair, whom the scientists referred to only as Ed and Fred, had grown up in different states. One day, someone walked up to Ed and said, "Hello, Fred, how's tricks?" Ed had dim memories of a long-lost brother, and so he decided it was time to track down this mysterious Fred. When the twins reunited, they were shocked to find they had both dropped out of high school and become electricians. When the scientists gave the twins an IQ test, Ed scored 91, Fred 90.

When the scientists studied other separated twins, they found similarly close scores. Yet Freeman and his colleagues were cautious about the lessons they would draw from their research. "We shall be satisfied if we have succeeded in tracing a few of the threads in the tangled web which constitutes the organism we call man," they said at the end of their 1937 book, *Twins: A Study of Heredity and the Environment.* Despite the similarity of

Ed and Fred, they still agreed with the dictum "What heredity can do, environment can do also."

At around the same time in London, a British psychologist named Cyril Burt was studying the intelligence of twins as well. Burt had been set on a course into psychology as a boy. His father, a doctor, sometimes let him tag along on house calls, and on one trip he met Francis Galton. After talking with Galton, Burt bought one of his books, and his fate was sealed. Burt studied at Oxford and became a teacher, carrying out psychology research on the side. In 1912, he was appointed the London City Council's first psychologist, where he used intelligence tests to identify low-scoring children in need of special education.

Burt wanted to understand to what extent intelligence was "a thing inborn and not acquired." Inspired by Galton's proposal, he searched among his students for twins who separated early in life. In 1955, Burt published a study on twenty-one pairs. Their intelligence test scores were more similar than those of siblings raised in the same house.

Eleven years later, Burt published an even bigger study on fifty-three pairs of twins. The results were the same. Burt used the test scores to estimate intelligence's heritability as being 80 percent. Where Freeman and his colleagues had caviled about nature and nurture, Burt brought down his psychological gavel with a loud crack. Heredity explained most of the differences between people's intelligence test scores, he declared.

Among the people who read Burt's 1966 paper was a Princeton psychologist named Leon Kamin. "Within ten minutes of starting to read Burt," Kamin recalled later, "I knew in my gut that something was so fishy here that it just *had* to be fake."

The results were too pat. They didn't seem to have anything to do with "the messy nature of the real world," as Kamin put it. Kamin dug into Burt's research and found suggestions of fraud. In Burt's 1955 and 1966 studies, twenty of the correlations were identical. In both studies, the correlation between identical twins raised apart were 0.771. That three-digit coincidence alone would have been highly unlikely. Twenty coincidences were astronomically improbable. Elsewhere in Burt's work, Kamin found more signs that Burt had made up much of his twin results. Burt even published papers under fake names to create the illusion that other scientists supported his findings.

In 2007, a Rutgers University psychologist named William Tucker offered an explanation for Burt's long con: He was a eugenicist from start to finish. In 1909, Burt published a study showing that upper-class schoolboys scored higher on intelligence tests than lower-class ones. Their different upbringing could play only a small part in that difference, Burt declared. "The superior proficiency at Intelligence tests on the part of the boys of superior parentage was inborn," he wrote.

Burt's scandal stained all of twin research, leading many to dismiss it as bad science. Yet just because a field attracts a fraudster doesn't make all the discipline's findings wrong. Hundreds of well-designed twin studies have come to the same conclusion: Identical twins have closer intelligence test scores than fraternal twins. Even when identical twins are raised apart, their intelligence test scores stay more similar than siblings raised together. These studies have led scientists to estimate the heritability of intelligence test scores as roughly 50 percent. That's substantially lower than Burt's claim of 80 percent, but it still indicates that heredity has an important role to play in intelligence that should not be dismissed.

As these more rigorous studies piled up, they drew criticisms of their own. Some researchers complained that they relied on the assumption that the only difference between fraternal and identical twins is their genes. Some research suggested that this might not be the case. Because identical twins look alike, they may be treated as interchangeable. Fraternal twins may look different enough from each other that they have an experience more akin to their ordinary siblings. In a 2015 study, a group of researchers investigated twins who experienced bullying, sexual abuse, and other kinds of trauma. They found that identical twins had more similar experiences than fraternal twins. If one identical twin was abused, it was more likely that the other one was, too.

But researchers who have looked closely at the experiences of twins have concluded these effects are weak or nonexistent. One such study was even carried out by a skeptic of twin studies, a Princeton sociologist named Dalton Conley. Conley realized he could investigate the experiences of twins by studying the surprising number of twins who get misclassified.

Some identical twins are recorded as fraternal at birth, and fraternal ones as identical. A genetic test can easily reveal the true nature of newborn twins, but doctors apparently don't bother with it much. In a 2004 study in Japan, researchers found that hospitals misclassify as many as 30 percent of twins. In the Netherlands, researchers tested the DNA of 327 pairs of twins and then asked their parents what kind of twins they were. Nineteen percent of the parents gave the wrong answer.

If genetic differences weren't important, then fraternal twins misclassified as identical should end up more similar to each other. You'd also expect that identical twins would be robbed of this powerful experience if their parents and teachers and everyone else around them treated them like fraternal twins. But Conley and his colleagues discovered no such thing. The cases of mistaken identity had no effect on how the twins turned out. Identical twins ended up more like each other in a range of traits, even when they didn't know they were identical. They reached more similar heights. Their risk of depression was closer. Their grades in high school were more alike. The only explanation for these similarities was heredity.

———

E very behavior scientists have studied turns out to be partly heritable, from smoking to divorce rates to watching television. At this point, it would be astonishing if intelligence *weren't* heritable. But twin studies on intelligence could not say what exactly was being inherited—which genetic variants, in other words, influence people's scores on intelligence tests.

To hunt for those variants, scientists followed the trail blazed by the researchers who studied height. Scientists first linked genes to height by studying people with growth disorders like Laron syndrome. The first genes tied to intelligence also turned up in studies of intellectual disorders such as PKU. Those early discoveries brought tremendous benefits to children.

They made it possible to test for a growing list of disorders and search for ways to treat them—a special diet for some, a special education program at school for others.

But when it came to the heritability of intelligence, those genes were practically meaningless. The severe mutations that lead to intellectual developmental disorders are very rare. PKU affects only one in ten thousand people, for example. For the population as a whole, those variants say nothing about why some people score higher on intelligence tests than others.

At the start of the twenty-first century, behavioral geneticists were full of hope that DNA sequencing technology and a map of the human genome would let them quickly find more genes that influence intelligence. "In a few years, many areas of psychology will be awash in specific genes responsible for the widespread influence of genetics on behavior," Robert Plomin and John Crabbe predicted in 2000.

At first, it looked as if the deluge was on its way. Researchers identified genes that seemed like good candidates for influencing intelligence and studied them in ordinary people. One of those genes, called COMT, encodes an enzyme in the brain. The enzyme keeps the neurotransmitter dopamine in check. It does so by finding the dopamine molecules and chopping them to pieces. One variant of COMT produces an enzyme that chops slowly, allowing the brain's level of dopamine to rise. The slow-chopping variant is fairly common. (Checking my genome, I discovered that I carry one copy of it.) Many scientists suspected that the different versions of COMT might have some effect on intelligence test scores, because dopamine is crucial for memory, decision-making, and other mental tasks. The slow-chopping variant was letting more dopamine build up in people's brains, improving their performance.

In 2001, Michael Egan, a researcher at the National Institute of Mental Health, led a study to test this idea. He and his colleagues gave 449 people an exam known as the Wisconsin Card Sorting Test. It is really just a simple game. The scientists showed the volunteers cards with circles, squares, crosses, and stars. The shapes came in different numbers and colors. The object of the test was to find sets of cards that matched each other according to a rule—a rule that Egan's team didn't tell the volunteers. Through trial and error, the volunteers eventually figured out the rule, but Egan would then switch it, leaving the volunteers to figure out the new rule. The scientists measured how quickly they made this discovery.

Egan and his colleagues found that people with the slow-chopping variant of COMT performed slightly better at the game. Their success prompted other scientists to decide to see about COMT for themselves. A number of them also found a link to intelligence.

It was an exciting discovery, but before long, the excitement curdled into disappointment. Later studies on more people failed to find any effect from the slow-chopping version of COMT. Other researchers tested out other candidate genes for effects on intelligence, only to watch their promising associations collapse.

In hindsight, searching for candidate genes was a strategy pretty much guaranteed to fail. Our brains use 84 percent of our twenty-thousand-odd protein-coding genes. Each type of neuron uses a distinctive combination of those genes, and it turns out the brain is made up of hundreds of cell types—so many that scientists will not finish its catalog for a long time. To think we could just reach into this jumble and pluck out a single gene that had a clear-cut role in intelligence was to pretend we know more about the brain than we really do.

As candidate genes failed to reveal genes linked to intelligence, scientists turned to genome-wide association studies. By searching through genetic markers scattered across the genome, scientists would let the genes speak for themselves.

Ian Deary led the first genome-wide association study on intelligence. As part of his research on the Scottish Mental Survey, he and his colleagues sequenced DNA from some of the test takers. Adding DNA from people who had volunteered for other studies, they analyzed 3,511 people all told. The researchers scanned half a million genetic markers to see if any of them were correlated with high or low levels of intelligence. Nothing on that scale had ever been attempted before. And yet, as Deary and his colleagues reported in 2011, they failed to find even one gene with any clear effect on people's intelligence test scores.

The experiences Joel Hirschhorn and his colleagues had with height had prepared Deary for this kind of disappointment. Complex traits can be influenced by hundreds or even thousands of genes. Their common variants may be so weak that small studies will fail to reveal them. Making matters worse, intelligence is not an obvious trait that can be accurately measured with a simple tape measure. Psychologists may use different intelligence tests depending on who they're studying, what aspect of intelligence they want to study, or how much time they have to examine each person. To try to amass a lot of subjects for their research, scientists often merge smaller studies that used different intelligence tests. The mismatch among the tests can spread a blanket of fog over the influence of genes.

Despite these challenges, genes continued to send signals from their hiding places. Peter Visscher's test for genetic similarity, which he had previously used on height, also confirmed that intelligence test scores were heritable. In fact, he could account for much of the missing heritability of intelligence. The precise amount was different depending on the age of the people the scientists studied. When Visscher and his colleagues examined twelve-year-old children, they could account for a staggering 94 percent of the heritability of intelligence.

The first genes associated with intelligence came to light in a roundabout way. Medical surveys often ask how long people stayed in school, and it turns out that educational attainment is a modestly heritable trait. The correlation between identical twins is stronger than for fraternal twins; full siblings raised together are more similar in their schooling than half siblings. By some estimates, about 20 percent of the variation in how long people stay in school is explained by the variation in people's genes.

In 2013, a team of scientists headquartered at Erasmus University Rotterdam brought together data from dozens of medical studies. They looked for variants in the DNA from more than 100,000 people that correlated with their educational attainment. They found dozens of variants that were more common in people who finished more school than in those who left early.

How long people stay in school depends on a lot of factors, such as motivation and attention. But intelligence plays a part, too: A small amount of the variation in the years people spend in school is explained by their intelligence test scores. The Rotterdam team suspected that a few of the genetic variants that influence educational attainment also influence intelligence. They picked out sixty-nine variants from their educational attainment study and investigated whether any of them had a link by examining 25,000 people who had provided DNA and had taken intelligence tests. In 2015, they reported three variants. Each of the three variants could lift a person's IQ score only three-tenths of a single point. They did not explode like fireworks. They merely popped like champagne bubbles.

These successes spurred other researchers to merge their studies, hoping to find more variants in larger groups of people. In a 2017 study, an international team analyzed nearly 80,000 people. They found fifty-two genes, which they were then able to confirm by turning to other groups of people and finding them once more. Altogether, the genes still account for only a small percent of the variation in people's test scores. And when the scientists looked at each gene's function, no compelling story of biology emerged. A few of the genes control the development of cells throughout the body. A few others are responsible in particular for various tasks inside neurons. Some may work in hidden pathways that scientists have yet to uncover.

If intelligence, like height, turns out to be omnigenic, the fifty-two genes will be only the start of a long list that will keep growing for years. Perhaps there will be a core of genes that shapes the brain in ways that influence intelligence test scores. But the search will take scientists out to rings after rings of more distant networks of genes. And even if scientists gain all this knowledge, they will still be a long way from a complete understanding of intelligence itself.

———

To Galton, Pearson, and their fellow hard-core hereditarians, intelligence seemed a case in which nature trounced nurture. Henry Goddard even convinced himself that all feeblemindedness could be explained by a single Mendelian mutation. In the extreme form of this view, intelligence was like blood types. Your blood type has nothing to do with whether your parents told you to turn off the television, or whether you ate three decent meals a day, or if you got chicken pox in grade school. Your type was fixed as soon as your parents' genes came together to form a new genome.

Intelligence is far from blood types. While test scores are unquestionably heritable, their heritability is not 100 percent. It sits instead somewhere near the middle of the range of possibilities. While identical twins often end up with similar test scores, sometimes they don't. If you get average scores on intelligence tests, it's entirely possible your children may turn out to be geniuses. And if you're a genius, you should be smart enough to recognize your children may not follow suit. Intelligence is not a thing to will to your descendants like a crown.

As hard as it has been for scientists to tease out the genes involved in intelligence, mapping the influence of the environment is even harder. It requires venturing into a daunting wilderness that lies outside the mathematical tranquility of genome-wide association studies. Psychologists who want to figure out the environment's contribution to intelligence have to take into account kindness and trauma, the biochemistry of the womb and the impact of stress on the brain. The influences of the environment cannot be snapped apart into distinct chunks the way genetic variants can. They ramify each other, forming the mycelium of experience.

One reason for this complexity is that intelligence, like height, develops. In an embryo, it does not yet exist. Children need a few years of growth and experience before they can get a meaningful, predictive score on an intelligence test. All along that path, experiences can influence how intelligence develops, and different experiences can lead to different intelligence test scores. While the environment can probably influence intelligence in many subtle ways, scientists understand a few strong effects best.

If a mother drinks heavily during pregnancy, the alcohol can interfere with the growth of neurons, leading to fetal alcohol syndrome. After birth, a child's brain continues growing swiftly, and along the way it stays vulnerable to toxins such as lead paint. Sometimes the enemies of intelligence work together to wreak havoc. In 1999, Brenda Eskenazi and her colleagues at the University of California, Berkeley, went to the farming communities of the Salinas Valley to see how intelligence is influenced by the pesticides sprayed on the fields. They followed 601 women through their pregnancies and then tracked the development of their children. The children of mothers with the highest levels of pesticide in their blood scored low on intelligence tests they took at age seven. And Eskenazi also found that poverty, abuse, and other kinds of adversity worsened the effects of the pesticides.

The environment's power isn't limited to lowering intelligence test scores, however. It can—under certain circumstances—lift them up. One of the simplest ways to do so, it turns out, is to give people iodine.

Iodine is essential for making hormones in the thyroid gland. A lack of iodine can lead to a number of diseases, including a neck swelling called goiter. It can also lead to cretinism, which leads to both dwarfism and severe intellectual disability. Normally, a pregnant mother's thyroid hormones travel into the brain of her fetus, where they help neurons crawl to their proper location in the brain. If she has a deficiency of iodine, she makes fewer hormones, leaving the fetal brain to fail to develop properly.

To keep our iodine levels high, we depend on our food. Seafood is a good source of iodine, because the element is abundant at sea. Meat and vegetables and milk can be good sources, too, but only if they come from places where the soil is rich in iodine. A third of the world's population lives in places that put them at risk of iodine deficiency. Adding it to salt is all that's necessary to give people healthy levels of iodine. When the United States and other countries established this policy in the early 1900s, both goiter and cretinism started to disappear.

Another century would pass before scientists began uncovering evidence that iodine deficiency may have a much wider impact on intelligence. Sarah Bath of the University of Surrey and her colleagues documented this effect in a survey of children growing up in southwestern England. England has never required iodine be added to salt, in the belief that people could get enough of it in milk. That turns out to have been wrong. Bath and her colleagues found that two-thirds of the pregnant women they studied had a mild iodine deficiency. And the children of these women, Bath found, got significantly lower verbal IQ scores at age eight and scored lower at age nine on tests for reading accuracy and comprehension.

The growing appreciation for iodine's importance for intelligence led James Feyrer, an economist at Dartmouth College, to take a fresh look at its history. He took advantage of the fact that the introduction of iodine in the United States fell squarely between the two world wars. Millions of young American men who served in World War I lacked the benefit of iodized salt. Thanks to their iodine deficiency, twelve thousand recruits had goiter, a third of whom couldn't button a military tunic around their neck and were judged unfit for service. But by the time the military inspected recruits for World War II, the rate of goiters had dropped 60 percent.

Feyrer wondered if this shift also affected the intelligence of the recruits. He was not allowed to look at their individual IQ scores, but he and his colleagues found a way to infer them: The highest-scoring recruits were put into the air force instead of the ground forces. Reviewing the records of two million recruits, Feyrer and his colleagues also checked the natural iodine levels in their hometowns. Nationwide, the researchers found, the introduction of iodine raised the average IQ by an estimated 3.5 points. And in the parts of the country where natural iodine levels were lowest, Feyrer and his colleagues estimated that scores leaped 15 points.

It may be hard to believe that such a straightforward change in people's diets could have such a tremendous effect on intelligence. But as public health workers continue to bring iodine to more of the world, the same jumps happen. In 1990, Robert DeLong, an expert on iodine at Duke University, traveled to the Taklamakan Desert in western China. The region has extremely low levels of iodine in the soil, and the people in the region have resisted attempts to introduce iodized salt. It didn't help that the people of the region, the Uyghurs, distrusted the government in Beijing. Rumors spread that government-issued iodized salt had contraceptives in it, as a way to wipe out the community.

DeLong and his Chinese medical colleagues approached local officials with a different idea: They would put iodine in the irrigation canals. Crops would absorb it in their water, and people in the Taklamakan region would eat it in their food. The officials agreed to the plan, and when DeLong later gave children from the region IQ tests, their average score jumped 16 points.

Chemically altering people's brains isn't the only way to change their scores on intelligence tests. James Flynn, a social scientist at the University of Otago in New Zealand, has discovered that across the world, IQ test scores have been steadily increasing. Flynn's first inkling of this shift came in 1984. He had asked a Dutch colleague to send him results of IQ tests administered to eighteen-year-olds in the Netherlands. When the scores arrived in the mail, he settled down to peruse them. A puzzling discrepancy jumped out: Dutch students in the 1980s did substantially better than students in the 1950s.

Flynn found a similar trend in nearly thirty developed countries. In Britain and the United States, for example, test scores increased 0.3 points a year. If the average score in 2000 was 100, then it would have been 70 in 1900. "We are driven to the absurd conclusion that a majority of our ancestors were mentally retarded," Flynn wrote in his 2007 book, *What Is Intelligence?*

Yet this trend—now known as the Flynn effect—has been confirmed many times over. As we've gotten taller, we've gotten smarter. Now the challenge is to figure out what's driving this increase.

As in the case of height, the Flynn effect has been too big and quick to pin on genetic change. For that to be the case, people who scored high on intelligence tests would have to have much bigger families than everyone else to spread their genes, and that hasn't happened. It's possible that what's been happening to intelligence test scores is similar to what's been happening to height. The global height boom has been brought about in part by better food, sanitation, medicine, and—in some places—greater economic equality. Some of the same factors may be at play in the Flynn effect. Better childhood health and nutrition makes the body grow quickly and the brain develop well.

Government regulations have also helped. Feyrer has argued that the push to give people iodine played a part in the worldwide Flynn effect. Exposure to lead can be toxic for the brain, and up until the 1970s, American children were exposed to high levels of lead in paint and gasoline. In 2014, Alan Kaufman, an intelligence expert at Yale, and his colleagues published a study on intelligence tests they gave to hundreds of Americans who were exposed to high lead levels before the 1970s and to hundreds more Americans who were born afterward. They estimated that lowering lead levels in children gave them a boost of 4 to 5 IQ points.

But scientists are also investigating other possible causes, because they're keenly aware that intelligence isn't just affected by molecules that flow through the brain. Our behaviors are shaped by our experiences, particularly the ones we have with other people. As our parents talk to us, they help build our vocabularies, for example. The world's fertility rate dropped drastically over the past century. In 1950, it was 5 children per woman; in 2010, it was 2.5. In a smaller family, the children have the opportunity to listen more to their parents.

Going to school can also raise intelligence test scores. To measure schooling's effect, two statisticians, Christian Brinch and Taryn Ann Galloway, took advantage of reforms that Norway put into place starting in the 1950s. By reorganizing their school system, the Norwegians increased the time students had to spend in school from seven years to nine. Different towns made the switch at different times between 1955 and 1972. Brinch and Galloway looked at how the extra schooling affected the IQ tests that nineteen-year-old men took as part of Norway's universal draft. In 2012, they reported an extra year of education raised scores by 3.7 IQ points.

This natural experiment takes on greater importance when you consider how much more schooling children get now than in previous centuries. In the United States, the enrollment rate in the early 1900s was 50 percent. By 1960, it reached 90 percent. American students went from an average schooling of 6.5 years to 12.

To Flynn himself, the Flynn effect doesn't mean that people in the nineteenth century were intellectually disabled, nor does it mean that people today have neurons that fire signals to each other in a fundamentally new way. Our forerunners relied on ways of thinking suited to their age. In the early 1900s, intelligence tests included questions like "What do dogs and rabbits have in common?" The answer that the test givers wanted was that they were both mammals. But often the answer they got instead was, "You use dogs to hunt rabbits." To people who spent their time hunting rather than learning taxonomy, that fact was the one that mattered.

Twentieth-century schooling began to train students more in thinking in terms of classification, logic, and hypotheses. To get a job, people had to understand how to operate machines, and then computers. Rather than hunting rabbits with dogs, children today are more likely to pass their free time on a smartphone. Flynn's argument is also bolstered by the way the Flynn effect spread over the world. It started in the United States and Europe, but as developing countries became more modernized, their intelligence test scores started their own upward trend.

As in the case of height, intelligence forces us to hold two contradictory ideas in our heads at once. Over the past century the world has gotten taller and smarter, but these increases were not brought about by a shift in our genetic variants. The change is so dramatic that it can be hard to see how genes matter at all. And yet heredity has not stopped mattering. Height was a strongly heritable trait in the early 1900s, when scientists first began to measure it. Intelligence was as well. Today, both remain heritable. Under similar conditions, people will grow to different heights and get different scores on intelligence tests in part because of the genes they inherited.

It's also becoming clear that we can't treat genes and the environment as two distinct forces that act independently of each other. Each one influences the other. In 2003, Eric Turkheimer of the University of Virginia and his colleagues gave a twist to the standard studies on twins. To calculate the heritability of intelligence, they decided not to just look at the typical middle-class families who were the subject of earlier studies. They looked for twins from poorer families, too. Turkheimer and his colleagues found that the socioeconomic class determined how heritable intelligence was. Among children who grew up in affluent families, the heritability was about

60 percent. But twins from poorer families showed no greater correlation than other siblings. Their heritability was close to zero.

It may seem weird that the environment itself can change heritability. We tend to think of genes as rigid purveyors of destiny, the inescapable agents of heredity. But biologists have always known that the two are intimately linked together. If you raise corn in uniformly healthy soil, with the same level of abundant sunlight and water, the variation in their height will largely be the product of the variation in their genes. But if you plant them in a bad soil, where they may or may not get enough of some vital nutrient, the environment will be responsible for more of their differences.

Turkheimer's study hints that something similar happens to intelligence. By focusing their research on affluent families—or on countries such as Norway, where people get universal health care—intelligence researchers may end up giving too much credit to heredity. Poverty may be powerful enough to swamp the influence of variants in our DNA.

In the years since Turkheimer's study, some researchers have gotten the same results, although others have not. It's possible that the effect is milder than once thought. A 2016 study pointed to another possibility, however. It showed that poverty reduced the heritability of intelligence in the United States, but not in Europe. Perhaps Europe just doesn't impoverish the soil of its children enough to see the effect.

Yet there's another paradox in the relationship between genes and the environment. Over time, genes can mold the environment in which our intelligence develops. In 2010, Robert Plomin led a study on eleven thousand twins from four countries, measuring their heritability at different ages. At age nine, the heritability of intelligence was 42 percent. By age twelve, it had risen to 54 percent. And at age seventeen, it was at 68 percent. In other words, the genetic variants we inherit assert themselves more strongly as we get older.

Plomin has argued that this shift happens as people with different variants end up in different environments. A child who has trouble reading due to inherited variants may shy away from books, and not get the benefits that come from reading them. A child who learns quickly how to do math may get encouraged by teachers to do more. As children get older, they gain more power to choose those environments, which can shape their intelligence even more. We think of the environment as our physical surroundings—of heat and cold, of chemicals and food. But as humans, we also build an environment around ourselves of words and numbers.

———————

W hen Galton and Pearson investigated the inheritance of height, they presented their results in cool prose, letting the statistics speak for themselves. When they turned their attention to intelligence, however, their lectures and papers became sermons. Galton promised a galaxy of genius through eugenics. Pearson, writing in 1904, drew a darker conclusion. "For the last forty years," he warned, "the intellectual classes of the nation, enervated by wealth or by love of pleasure, or following an erroneous standard of life, have ceased to give us in due proportion the men we want to carry on the ever-growing work of our empire."

American eugenicists saw the same moral contrast between height and intelligence. They didn't see any need to promote the height of the nation. But to protect American intelligence from feeblemindedness, they were ready to sterilize, to institutionalize, to bar immigrants.

Nazis made much of intelligence as well, their hereditary courts making their decisions based on test scores. Yet those scores had to fit into Nazism's racist ideology, not vice versa.

When German psychiatrists called for the sterilization of even the mildly feebleminded, their proposal was shot down. Many of Hitler's young Brownshirts would have fallen into that category—not to mention 10 percent of the German army. Friedrich Bartels, the deputy leader of the Reich doctors, rejected intelligence tests because they could condemn decent young German peasants. He said it was wrong to judge the value of a Nazi Party member based on how much trivia they knew, such as the year Columbus was born. Such a man might very well have spent his life until then working in the fields rather than taking classes. "It is quite possible that he has never had the chance to learn these things," Bartels complained.

As early as the 1920s, some psychologists were challenging the fatalism of people like Galton and Pearson. Helen Barrett and Helen Koch of the University of Chicago studied a group of children who were moved from an orphanage into preschool, where they were no longer neglected. After six months, Barrett and Koch claimed, their test scores jumped far beyond those of the children left behind in the orphanage. Intelligence was not simply the result of heredity, they argued, but also the quality of homes and schools.

In the 1930s, psychologists in Iowa ran a bigger study that came to a similar conclusion. In 1938, one of the researchers, George Stoddard, went to a conference in New York to describe the results, and a reporter from *Time* delivered the astonishing news. "One of the few fixed stars in the creed of orthodox psychologists is a belief that people are born with a certain degree of intelligence and are doomed to go through life with the same I.Q.," the *Time* reporter explained. But Stoddard—"moonfaced, enthusiastic"—proved "that an individual's I.Q. can be changed."

Stoddard and his team tracked 275 children who were put into foster care. Their parents were poor, badly educated, and scored below average on intelligence tests. After being placed "in better than average homes," *Time* reported, they scored an average IQ of 116—"equal to the average for children of university professors."

The Iowa studies led Stoddard to reject the eugenicist claim that heredity trumped all. "Dull parents are as likely to produce potentially bright children as are clever parents," Stoddard declared. "The way to improve a child's intelligence is to give him security, encourage him in habits of experiencing, inquiring, relating, symbolizing."

Hereditarians attacked Stoddard's work, pointing out its many statistical weaknesses. His critics maintained that intelligence tests measured something fixed in people, and Stoddard's calls for a national network of preschools went ignored. When World War II came, it focused the country's attention abroad, and the prosperity of the 1950s left many Americans unaware of the poverty that others endured in the country.

After World War II, most Americans no longer followed Lewis Terman's example, claiming that Southern and Eastern Europeans inherited low intelligence. But some still maintained that heredity was to blame for the gap between whites and blacks. Henry Garrett, the prominent psychologist who kept the Kallikaks alive in his textbooks, claimed that blacks were as intelligent on average as a white person after a lobotomy.

Garrett was an ardent supporter of segregation, and he brought his considerable credentials—former president of the American Psychological Association and professor at Columbia University—to the fight against racial equality. He served as a star witness for the segregationist defense in *Brown v. Board of Education* and acted as an FBI informant, reporting on the "Communistic theories" his fellow Columbia University professors were spreading about the equality of the races.

In 1955, Garrett retired from Columbia and moved back to the South to fight full-time. He testified to Congress in 1967 against a civil rights bill, lecturing the politicians about the evolutionary "immaturity" of the Negro. He became a director of the Pioneer Fund, a eugenics organization founded in 1937 to promote "the conservation of the best racial stocks." Garrett also became a prolific pamphleteer. Pro-segregation groups distributed over half a million of his pamphlets to American public school teachers, free of charge. Today, neo-Nazi groups continue to sell them.

In his pamphlets, Garrett railed against mixing blacks and whites—both in schools and in marriages. Such a disaster would drag down Western civilization, he warned. Pointing to studies that showed a 15- to 20-point gap between American whites and blacks on IQ tests, Garrett claimed that the differences were fixed by heredity. The notion that blacks and whites were equal was, Garrett declared, "the scientific hoax of the century." He pinned the blame for the hoax, unsurprisingly, on the Jews.

Garrett faced stiff opposition from many of his fellow American psychologists. They argued that poverty had a tremendous power over young minds. Experiments on animals were revealing just how crucial experiences could be to the early development of the brain. If kittens had their eyes sewn shut for just a few crucial days, they were left blind for the rest of their lives. A growing number of psychologists argued that children had a crucial window in their own development. If their early years were deprived, their intelligence would suffer just as their growth would be stunted. In 1965, Lyndon Johnson launched the preschool network that the Iowa psychologists had called for three decades earlier. The Head Start program began enrolling hundreds of thousands of poor children.

Robert E. Cooke, a Johns Hopkins pediatrician and the chair of Head Start's original planning committee, would later describe the program as a rejection of heredity's power. "The fundamental theoretical basis of Head Start was the concept that intellect is, to a large extent, a product of experience, not inheritance," he said.

In later years, social scientists would document many benefits to the Head Start program. It raised the graduation rate of children by more than 5 percent, for instance; among children with mothers who didn't finish high school, the rate rose by more than 10 percent. But it did not help children's intelligence test scores in the long run. At three and four years of age, their scores improved, only to drift back down by first grade.

Critics jumped on these findings as evidence that blacks scored lower than whites on intelligence tests due to their genes. In a 1967 lecture, an educational psychologist named Arthur Jensen declared that the lower average scores that blacks got on intelligence tests "do indeed reflect innate, genetically determined aspects of intellectual ability." In the decades since, others have made similar claims from time to time. For the most part, though, psychologists and geneticists alike have rejected them. It's clear that intelligence test scores are heritable. But just because two groups differ in a heritable trait does not mean that the difference between them is genetic.

The study of height provides a clear and uncontroversial demonstration of this rule. South Koreans are more than an inch taller on average than North Koreans. Height is even more heritable than intelligence. But these two facts don't add up to the conclusion that South Koreans have height-boosting alleles that are missing from North Koreans. In fact, we can be pretty confident they don't. Koreans only became two groups in the 1950s. Only afterward, as South Korea prospered and North Korea fell into the twilight of dictatorship, did their heights diverge.

Nor does the failure of Head Start serve as proof that the 1960s gap in intelligence test scores was some unalterable fact of heredity. The Flynn effect did not leave behind American blacks, for example. In fact, their intelligence test scores have risen dramatically, while American whites had a more modest improvement. The gap between the two groups narrowed by more than 40 percent between 1980 and 2012, according to one estimate.

Studies like these have led some critics to argue that research on the hereditary basis of intelligence is irrelevant at best and toxic at worst. If our goal is to improve the intellect of children, then we have plenty of obvious— but hard—work to do. We should fix the fumbling bureaucracies that run schools, stop programs that don't work, put ones into effect that do, and fix the causes of inequalities in education. We can also look beyond schools, to take on the harmful stress of poverty or the continuing threat of lead in drinking water. "Such efforts need no genetic information—or even I.Q. testing—and are likely to be hindered by the hereditary concept of intelligence," writes Dorothy Roberts, a University of Pennsylvania law professor.

Geneticists have fought back, calling these attacks caricatures of the modern study of intelligence. They are not using weak science to justify the status quo, or to argue for the superiority of one race over another. Nor are they claiming that because a trait is heritable means that interventions are pointless. Some point to the case of eyesight as an analogy. Eyesight is a strongly heritable trait, and yet eyeglasses can overcome the bad vision that children inherit from their parents. It would be absurd to say that there's no point trying to improve bad eyesight, since it's heritable.

In fact, some geneticists argue, understanding how heredity influences intelligence may potentially lead to policies that do a better job of helping children flourish. When education researchers test out new programs, they compare how well students do with it or without. In order for those studies to yield reliable results, the researchers have to make sure the two groups are random samples of students. If one group just so happens to have a lot of the variants known to influence intelligence or educational attainment, the study may deceive the researchers. They may come to believe a program is going to have a powerful effect, only to find out too late that it's just a waste of time and money.

Some researchers have gone further and predicted that DNA scans will make it possible to tailor school programs to each child. Genetic tests can already reveal some severe forms of intellectual disability in newborns and, in some cases, that knowledge is power. A child diagnosed today with PKU doesn't have to face Carol Buck's fate. By inspecting thousands of intelligence-related spots in the DNA of children, it might be possible to make predictions about how they will fare in school. Some variants might turn out to influence *g*, influencing overall intelligence, while others might affect only certain mental abilities. Kathryn Asbury, a lecturer at the University of York, has argued that these genetic tests might allow parents to intervene early with children who are, for example, dyslexic. "If a simple blood test at birth could spot children with a probable risk of struggling in any of these areas," she says, "then imagine the tailored interventions that might nip such risk in the bud or at least reduce its effects."

"Precision education," as this approach has been called, has a sleek, futuristic appeal. But for now, and probably for decades, it's only a placeholder of an idea. In the meantime, less sexy tasks will help children more, such as getting the lead out of drinking water in schools and getting enough textbooks for students. Instead of providing concrete help, research on heredity may end up fueling fallacies about the nature of intelligence. Unfortunately, psychologists have found, our minds have vulnerabilities when it comes to such matters. There's a reason that *The Kallikak Family*—with its simplistic, destructive vision of heredity and society—sold as well as it did.

In 2011 the psychologists Ilan Dar-Nimrod and Steven Heine dubbed this kind of thinking "genetic essentialism." Dar-Nimrod and Heine argued that genetic essentialism arises from how we make sense of the world. Decades of psychological research have shown that our minds instinctively sort things into categories. We ascribe the same essence to everything in the same category. Birds all have birdiness, and fish have fishiness. When psychologists try to get people to describe these essences, words often fail them. Feathers are a manifestation of birdiness, but if a bird gets sick and loses its feathers, we still consider it a bird. We use essentialism to make sense not just of birds but of each other as well. Early in childhood, we learn to view people as having essential characteristics, which we perceive as being present at birth and enduring throughout life.

Our built-in essentialism made it easy for us to misunderstand heredity. Genes seem like they belong to our essence. We inherit them from our parents and carry them till we die. It's tempting to conclude that nothing we do in our lives can change what our genes bring about. We are successful because we have genes for success. Races are different because the people in each race share genes not found in any other race.

Genetic essentialism turns out to be stronger in some people than in others. In one study, psychologists measured people's racism by asking them questions such as whether they'd approve of their child marrying a black partner or whether they think blacks have only themselves to blame for not doing well. It turns out that people who ascribe more of the differences between races to genes score higher on tests of racism.

It's even possible to manipulate people's genetic essentialism. In 2014, Dar-Nimrod and his colleagues had 162 college students fill out questionnaires about the foods they liked and their eating habits. Then they all read what looked like a newspaper article, which in fact had been written by the scientists. Some students read an article explaining that obesity is caused by bad genes. Others read an article explaining how people get obese if they're surrounded by friends who eat too much. Another group read an article about food with no mention of obesity. Finally, all the students were led to another room, where there was a big bowl of chocolate chip cookies that were broken into pieces.

The scientists explained they were going to use cookies in another experiment but they wanted to make sure they tasted right. They asked the students to eat some cookies and give their opinion. In fact, the cookies were part of the experiment, too. After the students left, the scientists measured how much they had eaten. The students who had read the genetic article ate almost 52 grams of cookies. The ones who read the article on social networks ate only 33 grams, and the ones who read the article that didn't mention obesity ate 37 grams.

The concept of genes driving people's appetite caused them to lose some control of their own. In a society that practically worships DNA, we are running this experiment on a colossal scale.

PART III

The Pedigree Within

CHAPTER 11

Ex Ovo Omnia

D IVIDE YOUR MIND'S EYE like a split screen. On the left, picture a single bacterium. On the right, a fertilized human egg.

The bacterium grows, duplicates its DNA, and splits itself in two, then four, then eight. The eight bacteria are kin, joined to the original microbe by the bonds of heredity. They inherited copies of its chromosome. Each new generation of bacteria is made of the proteins, RNA molecules, and other ingredients of their mother cell. You could trace their heredity over time by drawing a branching family tree.

The fertilized egg does much the same thing. It grows, copies its DNA, and splits in two, then four, then eight. The human cells may be hundreds of times bigger than the bacteria, and they may divide at a far slower pace. But in a developing embryo, cells share a heredity as well. Each pair of new daughter cells inherits copies of their mother cell's DNA, along with half of its other molecules.

It may seem strange to use the language of heredity for what happens in our own bodies. We tend to think of heredity only as a way to link ourselves biologically to the past and to the future. Yet heredity does not stop when a new life begins. Each of the 37 trillion cells in our bodies resides on a branch of a genealogical tree that runs all the way back to our origin at conception.

Before long, the split screen diverges. As the bacteria divide, they grow into a jumbled colony of identical cells. The egg's descendants, on the other hand, develop into a human form, complete with a head and face, with fingers and toes. And along the way, the cells that make up the new embryo give rise to different kinds of cells. Now its inner heredity shifts to a new style. Each cell in the stomach lining gives rise to two cells of the stomach lining. Now bone cells divide reliably to make more bone rather than pads of fat.

Textbooks say that the human body has about two hundred cell types, but recent studies have rendered that figure a laughable understatement. No one can say how many cell types there are, because the more scientists examine cells, the more they break down into more types. Immune cells may all carry out the same mission to save us from pathogens and cancer, but they are an army with hundreds of divisions. All our cell types are separate branches on the body's genealogical tree, like rival dynasties descended from a first monarch.

This transformation poses the chief question that developmental biologists have asked for centuries and continue to ask: How does a cell with a single set of genes give rise to the complexity of the human body? Heredity supplies the answer, but it does so in different forms. Heredity, in other words, is more than one thing.

———

A ristotle asked himself this question, but he could try to answer it only by cracking open chicken eggs. If he opened an egg the day it was laid by a hen, he saw only the white and the yolk. He observed nothing more on the second day, or the third. But on the fourth day, he could make out a red speck. This he took for the heart. "This point beats and moves as though endowed with life," Aristotle said.

Over the next few days, other things became visible in the egg. Tubes filled with blood sent out branches. The dim mass of a body emerged. Aristotle could eventually make out a head, adorned with a pair of bulging eyes. A chicken embryo at this stage looked a lot like the embryos of other animals Aristotle had studied. But as the days passed, the similarity faded, overtaken by the peculiar features of birds—a beak, feathers, claws, wings. "About the twentieth day, if you open the egg and touch the chick, it moves inside and chirps," Aristotle observed.

Some philosophers of Aristotle's age believed that the parts of a chicken, or any other animal, somehow existed in miniature even before this development. Aristotle would have nothing of it. He saw the development of an embryo as akin to making cheese. A cheese maker added fig juice to milk, setting about a transformation that created something that did not exist beforehand. When chickens mated, the rooster's semen triggered a similar transformation of fluids inside the hen. Organs curdled into existence in an unfolding sequence. The spirit within the semen shaped a heart, which in turn shaped other organs, which in turn shaped more of the chick's body until it was complete.

For two thousand years, Western scholars and physicians hewed closely to Aristotle's vision, but the Scientific Revolution brought a realization that he had gotten some things wrong. William Harvey, the royal physician to King James I and Charles I of England, searched for Aristotle's curdling fluids inside does and hens. He could find none. Like Aristotle, Harvey inspected embryos of chickens. The heart did not form first, he realized: Blood vessels took shape earlier. To account for these differences, Harvey came up with a different vision of life's beginning: It was from eggs that all animals grew. When he published a book about his idea in 1651, he emblazoned it with the Latin motto *Ex Ovo Omnia*.

You can search Harvey's book from cover to cover and still be left baffled by what he meant by *ovo*. While Harvey was sure that mammals had eggs, he could never find any evidence of them. Instead, he speculated that the hypothetical eggs were produced by the female body much as the mind produces thoughts. Semen then acted on the eggs so that they began to develop into embryos.

In this regard, Harvey still remained faithful to his hero Aristotle. The different parts of the body all emerged from a homogenous beginning. He called this unfolding *epigenesis*.

Other scholars in the seventeenth century began promoting a radically different theory for how new generations emerged. Known as preformationists, they argued that all the anatomy of an animal already existed before conception. In the 1670s, a Dutch naturalist named Nicolaas Hartsoeker discovered sperm, using newly invented microscopes. He drew the head of a sperm with a tiny human lodged inside.

Preformationism held sway until the mid-1700s, when new observations exposed its flaws. A German medical student named Caspar Friedrich Wolff studied chicken embryos more carefully than anyone ever before and could find no trace of a miniature bird in their earliest stages. Instead, he saw a blob of unorganized tissue that gradually took on new structures, which only later turned into recognizable parts of a chicken's body.

The 1800s brought more powerful microscopes, which scientists used to make new discoveries about development. Only then, for example, did they finally discover the mammal eggs that Harvey had dreamed up two centuries earlier—first in a dog, and then in a woman. After fertilization with sperm, eggs could then develop. It also became clearer what the eggs were developing into. Under the new microscopes, animal bodies resolved into tiny units. These units looked different depending on what tissue they came from: the blobs in blood, the long fibers in muscles, the brickwork of skin. But researchers realized they were all variations on a theme. "There is one universal principle of development for the elementary parts of organisms," the German zoologist Theodor Schwann declared in 1839. "And this principle is in the formation of cells."

The new cell theory raised new questions of its own, such as how cells came to be. Some naturalists argued that cells spontaneously formed out of biological fluids, the way hard-edged crystals could form out of a featureless soup of chemicals. But a group of German biologists proved that new cells emerged only from old ones. It was a microscopic form of like engendering like. The biologist Rudolf Virchow decided it was time to update Harvey's motto. *Ex ovo omnia* became *Omnis cellula e cellula*—every cell comes from a cell, inheriting its traits from its ancestor.

Cells, it became clear, were not unique to animals. Plants were made of cells, as were fungi. Bacteria and protozoans had bodies made up of a single cell. Different forms of life made new cells in different ways. Bacteria, for example, simply divided in two. A species like yeast used a somewhat different cycle. A mother cell might split in two, but the two daughter cells would remain firmly stuck together. As the division continued, the yeast grew into a mat—not exactly a body like ours, but not a group of isolated organisms either. Animals and plants, on the other hand, developed into giant collectives of cells, which reproduced by making new collectives.

To reproduce, some aquatic animals such as corals and sponges simply break off a packet of cells known as a bud. The bud drifts away, settles on a new spot on the seafloor, and grows into its own full-size body. When animals reproduce by budding, it's hard to draw a clean dividing line between ancestors and descendants. They all belong to the same unbroken line of cell divisions. Even though the new animals have distinct bodies, you can still think of them as overgrowths of their ancestors.

Most animals, humans included, can't reproduce by budding. Cut off your arm, and it will not grow into a second you. We and most other animal species develop from a single fertilized egg, called a zygote. Like other cells, zygotes don't leap into existence out of the void. Each is the fusion of two older cells. The continuity of zygotes with the previous generation led some scientists to propose that children are the overgrowth of their two parents.

Overgrowth sounds like a chaotic, weedy explosion of life. But the development of animal embryos is nothing of the sort. Most animal embryos change from a nondescript ball into a shell, with a clump of cells stuck to its interior wall. The cells making up the shell become the placenta, while the clump becomes the embryo itself. The clump spreads out into a sheet made up of three layers. Those layers came to be known as the ectoderm, the endoderm, and the mesoderm. You started out from these three layers, and so did a grasshopper, and so did a tapeworm. Those layers go on to form different tissues of the body.

When biologists began to look at later stages of developing embryos, they could start making out newly formed tissues. Each type was made up of its own distinctive set of cells. Yet no matter how distinctive they became on the outside, they still remained similar within. A sprawling neuron and a sheet-shaped epithelial cell both had a nucleus at their core, inside of which they held identical chromosomes.

To August Weismann, the man who cut off mouse tails to champion Darwin over Lamarck, this unfolding variety was hard to fathom. "How is it," he asked, "that such a single cell can reproduce the *tout ensemble*"—the total impression—"of the parent with all the faithfulness of a portrait?"

Weismann's many years of looking at animal embryos led him to an answer. When a fertilized egg divided, it bequeathed its nucleus to its offspring. And inside that nucleus was a mysterious thing Weismann called "hereditary tendencies." Those descendant cells passed down the same tendencies when they divided. Weismann reasoned that the only way that cells in an embryo could take on different identities would be to inherit different hereditary tendencies.

When a cell divided, in other words, it had to determine which of its daughters inherited which tendencies. Early on in development, a cell might bequeath the tendency to become ectoderm to one cell, and the tendency to become mesoderm to the other. Each cell could then pass down only its specific tendency to its daughter cells. Later on, an ectodermal cell might divide its hereditary tendencies unequally once more. Some of its descendants might inherit only the tendencies for becoming a skin cell, others a nerve.

For Weismann, in other words, the development of an embryo was a saga of loss. By the time organs like stomachs and thyroid glands emerged, their cells lacked most of the original hereditary tendencies in the fertilized egg. They could only divide into more stomach cells or thyroid cells. They could never produce a new animal, *tout ensemble*.

Thinking about development this way, Weismann turned his attention to how embryos produced their own supply of new eggs or sperm. He had observed this process for himself, and he had been struck by how early they developed, and how they then were set aside while the rest of the embryo continued to grow. Weismann became convinced that this early isolation was vital, because eggs and sperm had to be set aside before they lost too many hereditary tendencies. There must be a profound distinction between sperm and eggs, which Weismann called germ cells, and the rest of the body, which he called somatic cells.

Weismann split heredity in two. One form of heredity joined parents to their children. According to Weismann, parents were custodians of the germ-plasm, a mysterious hereditary substance that could produce an entire human being. Over the generations, the germ-plasm never lost its ability to give rise to new life.

Germ-plasm heredity gave geneticists the concept they needed to make sense of Mendel's experiments, to observe how hereditary factors could hop down through the generations like rocks skipping across a pond. To geneticists, what happened during development didn't matter much. It was just a dead end made of disposable flesh.

But Weismann also recognized another kind of heredity playing out inside of each of us. He illustrated this inner heredity with pictures. In his book *The Germ-Plasm: A Theory of Heredity*, Weismann represented a threadworm's development in the form of a tree—an embryonic pedigree, as it were. At the base of the tree he drew a circle, representing a single fertilized egg. The circle sprouted a pair of branches, to stand for the zygote's division into two daughter cells. One branch led to a white dot, which branched in turn into more white dots. These represented ectodermal cells. The other branch gave rise to other lineages—of endoderm, mesoderm, and germ cells. If you didn't know you were looking at a threadworm, you might think you were looking at the Habsburg dynasty's family tree.

This tree, Weismann was quick to caution, was just a "theoretical illustration." He drew it only to convey his ideas about the crucial division between germ cells and somatic cells. But the picture inspired other biologists to look at the development of real embryos, and draw trees of their own.

One of the first biologists to draw these cellular genealogies was a young American graduate student named Edwin Grant Conklin. Conklin started his own artwork in the summer of 1890, when he traveled to the seaside village of Woods Hole, Massachusetts, to find something to study for his PhD. He ended up scraping slipper limpets off crab shells and harvesting their eggs. The limpet eggs were so big and transparent that Conklin could observe them clearly under a microscope. He drew the portrait of a limpet egg, down to the nucleus and other structures within. He drew another when the egg divided in two. Each time the embryo divided, he made a new portrait in pencil, identifying each new cell according to its ancestor. His portraits changed from tiny batches of cells into large spheres, and then into more complex shapes.

"I followed individual cells through the development, followed them until many people laughed about it," Conklin later recalled. "They called it cellular bookkeeping."

The hours that Conklin spent staring through his microscope made him an object of mockery in the lab. One day a fellow graduate student named Ross Harrison "came behind me while I was anxiously studying some of the cleavage forms under the microscope and hung a crab on my left ear," Conklin said. "That crab pierced the ear lobe and could not be taken off except by some of the other sympathetic people in the laboratory."

Harrison sprinted away, and Conklin bolted in chase. "I ran for half a mile or more without catching him," Conklin said.

Despite these distractions, Conklin managed to draw a prodigious number of images. When he got back to Baltimore, he numbered each cell so that readers could follow their multiplication from stage to stage. Conklin wrote a paper about the limpet's transformation, which he gave to his advisor, William Keith Brooks, to read. A few days later, Brooks brought the manuscript back to Conklin.

"Well, Conklin," he said loudly, so that the other students in the lab could hear, "this University has sometimes given the doctor's degree for counting words; I think maybe it might give one degree for counting cells."

The other students roared with laughter. "I certainly felt pretty small," Conklin said.

The next summer, Conklin went back to Woods Hole and collected more limpets. One day, a professor named Edmund Beecher Wilson walked up to his lab table and said he had been doing a similar study on the larvae of leeches. Conklin and Wilson sat down together and compared their drawings. They were shocked to see how similar their embyros were, even from their earliest stages of development. Wilson became Conklin's mentor, introducing him to other scientists and helping him publish his research in scientific journals. Conklin went on to draw painstakingly detailed embryos of other species, tracing the lineages of cells further than anyone had managed before.

As Conklin uncovered the lineages, he found a new way to tackle the centuries-old debate about how a single egg developed into a complex body. He could trace the division of cells as they produced tissues and organs. He watched generations of cells gradually part ways, committing to an existence as muscle or nerves, or some other tissue. In some cases, the fate of cell lineages was fixed at the start; in other cases, cells seemed to hold on to the capacity to take on a range of different final forms.

Conklin turned cell lineages into an essential part of embryology. Later generations of scientists examined these embryonic pedigrees as they sought to understand *how* the cells reached their final identities, and how their identities became locked for life.

Although genetics was booming at the time, embryologists didn't think it could help them solve this mystery. They thought geneticists, who had yet even to show what genes were made of, were supremely arrogant to think they would be the ones to solve Aristotle's mysteries. "The 'Wanderlust' of geneticists is beginning to urge them in our direction," Conklin's crab-wielding nemesis, Ross Harrison, warned an audience of embryologists in 1937. This "threatened invasion," as Harrison called it, would lead only to nonsense. There was no way that simplistic explanations based on genes and their mutations could make sense of the majestic unfolding of development. Geneticists could busy themselves with finding mutations that changed the color of a fly's eye, Harrison said. He and his fellow embryologists were after bigger game: how the eye itself came to be.

Harrison was certainly right that, in 1937, geneticists didn't know enough to help explain embryos. But even as Harrison rallied his troops to man the university ramparts, a British embryologist was already thinking about how he could let the enemy in. Conrad Waddington carried out experiments at the University of Cambridge, moving bits of tissue around chick embryos to see if he could disturb their development. But he also had a philosopher's detachment. He could rise above the fine details of ectoderm and endoderm, and think in the abstract about how genes might guide development.

Each cell in an embryo, Waddington hypothesized, was a little factory. It used its many genes to produce many proteins, some of which could spread to other cells. Different cells produced different proteins, creating complicated chemical blends that were different from one place to another in the embryo. The particular blend a cell was exposed to could cause it to take on a new identity as it developed.

Waddington shared Weismann's fondness for pictures. To illustrate the development of embryos, he drew a hillside shot through with forking valleys. He imagined a lineage of cells as a ball rolling down this landscape. The slope of the surface could guide it down one valley or another— committing to become a particular type of cell. Waddington had an artistic friend illustrate this landscape with two pictures, one looking down from above, and the other looking up from below. The underside of the landscape was anchored with guy wires, pulling down to create the valleys that drove cells to their final states.

Waddington liked to call this strange terrain the epigenetic landscape, borrowing the old language of Harvey and Aristotle. Waddington used the term, as he explained in a 1956 textbook, "for the theory that development is brought about through a series of cause interactions between the various parts."

Waddington freely admitted that his epigenetic landscape was just an idea, one that was valuable mainly as a way to guide his thoughts. "Although the epigenetic landscape only provides a rough and ready picture of the developing embryo, and cannot be interpreted rigorously," he wrote, "it has certain merits for those who, like myself, find it comforting to have some mental picture, however vague, for what they are trying to think about."

The pictures that Weismann, Conklin, and Waddington drew were like visions from the future. They captured some of the overall truth of how we develop, but they lacked the specifics. The three biologists made mistakes, albeit forgivable ones. Weismann, it turned out, was wrong to say that hereditary tendencies are divvied up between daughter cells. The DNA that encodes someone's genes is replicated in full each time a cell divides. What makes a sweat gland cell different from a taste bud cell is the combination of genes that are active in each of them, as well as the combination that is silent. And that difference can be passed down from mother cells to daughter cells.

It's an inheritance, but not of some particular mutation. It's the inheritance of a state, a configuration of life's network. And the first glimpse of how that network is configured came to a woman whose day job was to prepare for the apocalypse.

I n the 1950s, hydrogen bombs were lighting up the world in test after test. It seemed as if nuclear war might not be far away. Movies distilled the anxiety and projected it onto theater screens. In *Godzilla*, a radiation-induced monster tramples Tokyo. The giant ants of *Them* slaughter people with formic acid. *The Day the Earth Caught Fire* imagines nuclear bombs pushing Earth out of its orbit around the sun.

Nuclear nightmares featured not just incineration but also a gruesome transformation of heredity. The people who did not evaporate in a blast would be pierced with radiation. It could damage cells deep in their bodies, causing radiation sickness and cancer. If an alpha particle crashed into an egg or sperm cell, it could alter the DNA inside, thereby extending nuclear war's ravages to future generations. The survivors might pass down the mutations to their descendants, along with the diseases they caused.

The British government decided it needed a lab to study this sort of damage, and it needed scientists like Mary Lyon to work there. Lyon, a quiet, focused thirty-year-old geneticist, was hired in 1955 to work at the Radiobiological Research Unit of the Medical Research Council.

It was still rare for a woman to hold such a job. When Lyon had gone to the University of Cambridge to read zoology, she was allowed to receive only a "titular" degree, even though she worked as hard as her male colleagues. Nevertheless, she made such a strong impression on her advisors that they helped her get a spot as a graduate student with Ronald Fisher, the geneticist who had combined Mendel and Galton's concepts of heredity into a new form in the 1920s.

Fisher turned out to be an irascible screamer who regularly threw graduate students out of his lab. But Lyon earned his respect, and he put her in charge of some of the mutant mice he studied. She carried out elegant experiments to see how one mutation could give rise to different traits, such as a splotchy coat and a loss of balance. But she decided Fisher's rage-choked lab was too toxic for her to grow as a scientist. She was a fan of Conrad Waddington's new ideas about epigenetics, and so she went to the University of Edinburgh, where he had become the chairman of the biology department, to finish her PhD.

Lyon thrived scientifically in Edinburgh, staying on after finishing her dissertation to continue her research. Waddington provided his scientists with the latest technology, and they discussed the newest ideas about heredity and development. Though pensive and humble, Lyon nevertheless gained a reputation in Edinburgh for seeing right through scientific problems. She would politely challenge her male elders if she found their reasoning flawed. The friends Lyon made in Edinburgh grew accustomed to her long silences as she composed what she wanted to say next. Although she thrived as a scientist, her parents still couldn't understand why a woman would waste her time fussing over some odd mice.

"They wanted me to get married at one point," she later recalled in an interview.

"What did you think about that idea?" her interviewer asked.

"I didn't like it."

The differences between the sexes came to dominate Lyon's own work. In Edinburgh, Lyon got the chance to study the first mice ever isolated with mutations on the X chromosome. She used them to explore how traits on the X and Y sex chromosomes were passed down. Lyon took her mice with her when the British government transferred her and other Edinburgh biologists to the Radiobiological Research Unit near Oxford. There they were expected to uncover the genetic risks of nuclear warfare. After five years of intellectual ferment in Edinburgh, Lyon found the government bureaucracy in the unit stultifying. As much as she could, she "always tried to stick to the mouse work," she said.

One strain of mice she studied, called mottled, had a particularly intriguing form of heredity. The female mottled mice developed an assortment of colored patches randomly scattered across their coat. The male mice met one of two very different fates. Either they ended up with a uniform coat or they died before birth.

These clues suggested to Lyon that a deadly mutation lurked on the X chromosomes of mottled mice. Males died if they inherited it, because they carried only a single X. Females, with two X's, had better odds of surviving. If one of their X chromosomes lacked the mutation, they would develop normally.

Lyon suspected that the mutation was also responsible for the coat patterns of mottled mice. One normal copy of the X chromosome in males produced hairs with identical colors. Somehow, carrying two copies of X chromosomes gave females a mottled coat. Not only that, but the mottling developed into different patterns from one female mouse to the next.

Poring through earlier studies on the X chromosome, Lyon searched for evidence that could account for all these strange outcomes. She was led from the narrow question of mottled mice to a much deeper question about X and Y chromosomes.

With two copies of an X chromosome, a female ought to make twice as many proteins from their genes than a male. All those extra proteins ought to throw a female's biology into deadly chaos. The big mystery about X chromosomes, Lyon realized, was how females could be healthy with two of them, and males with just one.

Lyon realized that Canadian researchers had discovered a possible answer in the 1940s when they examined the cells of female cats. In every cell, they saw, one of the two X chromosomes was packed down into a dark clump. The other X chromosome remained open, like all the other chromosomes. Perhaps, Lyon thought, females shut down one X in each cell, silencing its genes. As a result, they made only one chromosome's worth of X proteins, just like males.

There was just one problem with this explanation. It couldn't account for Lyon's mottled mice.

If the female mottled mice shut down one of their X chromosomes, they ought to end up with the two fates that the males met. Shutting down their normal X chromosome ought to cause them to die before birth. Shutting down their mutant X chromosome instead ought to give them a uniform coat of fur. Somehow, the females avoided both fates.

Turning over facts like these in her mind, Lyon alighted on an idea that could explain them all. She sat down, typed it out in seven paragraphs, and sent them off to the journal *Nature*.

Lyon proposed that as a female embryo develops, its cells shut down one of their two X chromosomes. But each cell chooses at random between the two. After making this choice, the cell divides and causes its daughter cells to shut down the same X chromosome. They in turn pass down that same selection to all their own descendants. A female mouse's body is made up of lineages of cells, half of which have silenced one X chromosome, and half the other.

This kind of inner heredity could explain mottled mice. Lyon speculated that they had one X chromosome with a mutation that disrupted the development of the skin. A female mottled mouse had skin composed of clusters of cells. All the cells in each cluster shut down the same X chromosome. Some clusters produced normal fur as a result, and others produced altered colors.

Nature published Lyon's short paper in 1961. Other biologists read it and wished they had thought of the idea themselves. Lyon meanwhile went on looking for more evidence. She investigated the fur of cats, finding that tortoiseshells and calicos had coat patterns that fit her model. A number of human diseases seemed to support her hypothesis as well.

As Lyon published her new evidence in further papers, other scientists generally found her idea irresistible. They dubbed it the Lyon hypothesis, or just L.H. The random silencing of X chromosomes came to be known as "lyonization"—although Lyon herself disapproved of the name.

In 1963, when Lyon traveled to New York to speak at a scientific conference, newspapers and magazines lavished praise on her. *Time* marveled that the star of the meeting should turn out to be "a quiet Englishwoman who presented no paper and who is, of all things, editor of the semi-annual *Mouse News Letter.*"

But Lyon also incurred the wrath of a formidable opponent, a German-born geneticist named Hans Grüneberg. Grüneberg fled the Nazis in 1933, finding refuge in England and becoming a professor at University College London. In the mid-1900s, Grüneberg did more than anyone else to turn mice into a model for human heredity. He even wrote the definitive guide to the subject, *The Genetics of the Mouse.*

Grüneberg had examined Lyon's thesis defense in 1950. A decade later, he read her paper in *Nature* and found it ridiculous. "He may not have realized I wasn't a PhD student anymore—that I didn't have to ask him for permission," Lyon later speculated.

While other scientists hailed her work, Grüneberg launched a crusade against her. In his own research, Grüneberg studied mice with a mutation on the X chromosome that produced defective teeth. According to the Lyon hypothesis, a female's teeth should be made of a patchwork of cells, some using the healthy version of the X chromosome and some using the defective one. But when Grüneberg peered into the mouths of the mice, he found that their teeth all looked the same.

When Grüneberg looked over studies of human diseases, he also failed to find compelling evidence for the Lyon hypothesis. "It is concluded," Grüneberg declared with the solemnity of a judge, "that the behaviour of sex-linked genes in man (like that in other mammals) gives no support to the Lyon hypothesis."

Other scientists were appalled by Grüneberg's ruthlessness. Year after year, paper after paper, conference after conference, he kept up the attacks. His colleagues were embarrassed for him, too, because he refused to accept the evidence in favor of lyonization as it continued to pile up. One of the most important studies in favor of L.H. came out in 1963. Ronald Davidson, a geneticist at Johns Hopkins, and his colleagues studied a blood disease called G6PD deficiency. It's caused by a mutation on the X chromosome, causing a defect in proteins called G6PD that makes red blood cells fall apart. Men who inherit the G6PD mutation always suffer the disease. Women, on the other hand, can escape the symptoms if their other copy of the X chromosome has a normal version of the G6PD gene.

Davidson inspected individual skin cells from women who inherited the mutation. He showed that half of the cells silenced the X chromosome with the defective gene, and the other half silenced the working version. Overall, the women's cells produced enough G6PD to keep them healthy.

Grüneberg refused to accept Davidson's evidence. Instead, he started attacking Lyon's supporters, too. For a decade, Lyon later said, Grüneberg made her life difficult and depressing. Yet she maintained her unflappable tact. By the 1970s, scientists stopped asking if lyonization was real. They just wanted to know how it happened.

———

The answer turned out to lie in the many molecules that swarm around our DNA. These molecules—a combination of proteins and RNA molecules—control which genes become active and which remain silent. Some silence genes by winding stretches of DNA up tightly around spools. Others unwind it, allowing gene-reading molecules to reach the exposed DNA. Some proteins clamp down on a gene, shutting it down until they fall off. Since each cell may make many copies of a silencing protein, another will soon take its place. Cells can also shut down genes for the long-term by coating them with durable molecular shields. This shielding—called methylation—lasts beyond the life of a cell. When the cell divides, its two daughter cells build new shields to match the original pattern.

A number of scientists have dedicated their careers to finding the molecules that shut down X chromosomes. Their search has led them to one stretch of DNA on the X chromosome, dubbed Xic, where several crucial genes reside. Early in the development of a female embryo, the two X chromosomes in each cell are guided toward each other, their Xic regions lining up neatly. A flock of molecules descends on the pair of Xic regions, drifting between them in what is essentially a molecular version of eenie-meenie-minie-moe. Eventually they settle on one of the two Xic regions, where they switch on genes that will shut down the entire X chromosome.

One of the genes they switch on is called Xist. The cell uses Xist to manufacture long, snakelike RNA molecules. They slither along the X chromosome, finding a place where they can take hold. While one end of an Xist molecule grips the X chromosome, the other end snags proteins passing by to help it. Together, they twist and coil the X chromosome, until it has shrunk down to a compact nugget of DNA. The other X chromosome meanwhile remains active by keeping its own copy of the Xist gene silent.

Each cell in the early female embryo rolls the genetic dice to pick which X chromosome to silence this way. Its pick is permanent. When the cell divides, it painstakingly unpacks the inactivated X chromosome to make a copy. In the two new daughter cells, the same X chromosome is folded back up again. The chromosome becomes like a box of old kitchen utensils you move from apartment to apartment, without ever making use of anything inside.

We can now see lyonization not only in molecular detail but also across the entire body. In 2014, Jeremy Nathans and his colleagues at Johns Hopkins University figured out a way to make active X chromosomes light up. They inserted a gene into a mouse's X chromosome that could produce a red glowing protein if the scientists exposed the mice to a particular chemical. They engineered another line of mice to produce a green protein instead. With some careful breeding, they were able to produce litters of mice that inherited a green chromosome from one parent and a red one from the other. When they added both chemicals to different parts of the mice's bodies, their cells lit up like Christmas lights. Each cell was either red or green, depending on which chromosome fell quiet.

Neighboring cells typically glowed in different colors. But when Nathans pulled back to look over greater distances, new patterns took shape.

Purely by chance, large swaths of cells might mostly have the father's X turned on, while others had the mother's. This imbalance could affect entire organs. Some mice had brains with one hemisphere mostly red and the other mostly green. Some saw out of their left eye with retinal cells mostly using the father's X, while the mother's X gazed out of the right. The variations even included entire mice. In some animals, almost all the X chromosomes from one parent were shut through the whole body. In others, the opposite was true.

Much of the research scientists have carried out on X chromosomes has focused on their special capacity to make us sick. For men, carrying a single copy of the X chromosome means they can't hope to be rescued from a mutant gene by a working backup. As a result, most X-linked hereditary diseases almost exclusively strike men. Muscles require a protein called dystrophin to work properly, for example, and it just so happens the dystrophin gene sits on the X chromosome. Duchenne muscular dystrophy—a disease that causes muscles in many parts of the body to turn to jelly— almost always strikes boys. They inherit it from their unknowing mothers, who don't suffer the disease because some of their muscle cells make enough dystrophin to keep up their strength. Women, meanwhile, face their own troubles if silenced X chromosomes become active, throwing off their balances of proteins.

But Nathans and his colleagues suspect that lyonization might have an upside, too. It may expand the scope of heredity for women. In the brain, some neurons may inherit an active X chromosome that guides them to sprout branches in one pattern, while other neurons branch in another. The power of the human brain comes from its diversity—from different kinds of neurons, from different kinds of circuits, from different types of chemicals for communication. Lyonization may make women's brains inherently more diverse.

On Christmas Day 2014, Lyon enjoyed a holiday lunch and had a glass of sherry before taking a nap. She had been long retired at that point, with many laurels. In 1998, Cambridge had held a special ceremony to give her an official degree to replace her titular one. A Medical Research Council building was named in her honor. The Genetics Society of America launched the annual Mary Lyon Award to recognize an outstanding geneticist. The biologist James Opitz complained that it was all "too small an honor for one, whom many I have known deemed ready for the Nobel Prize." During her Christmas nap, Lyon died. Opitz only wished that in her final moments, she had her tortoiseshell cat, Cindy—a living demonstration of lyonization—on her lap to keep her company.

M ary Lyon did far more than reveal how women manage life with two X chromosomes. She opened up a new way of thinking about our inner heredity. Her hypothesis offered an example of how cells could commit themselves and their descendants to using some genes and not others. It turns out that similar commitments allow cells in the early embryo to turn into different tissues and organs. In the decades that have followed Lyon's pioneering work, other scientists have documented more steps on the journey. It's a journey that starts at conception, continues through development, and lingers on for the rest of our lives.

At the moment of fertilization, when a sperm cell fuses to an egg and unloads its delivery of chromosomes and other molecules, a distinctive set of genes switches on. This special combination makes zygotes totipotent, meaning that they have total potency. A single zygotic cell has the potential to become any type of cell in the body, or even a cell in the placenta. When the zygote divides, it produces two new totipotent cells, and then four. If a doctor were to pluck any one of these totipotent cells and rear it in a petri dish, it could multiply into a complete embryo along with a placenta.

These cells, in other words, inherit not only DNA from their mother cells but their totipotency as well. This state endures from one generation of cells to the next, thanks to the molecules that hover around their DNA, determining which genes the cells can use and which stay silenced. A few master genes create powerful proteins, each of which can keep hundreds of other genes turned on and off. Those master genes also sustain each other in feedback loops. One gene promotes another, which then turns on yet another, which turns on the first. When a totipotent cell divides, its daughter cells inherit this same balanced network of proteins. The molecules go right back to controlling the DNA in the two new cells, so that the new cell inherits the totipotency of its ancestor.

Totipotent cells can maintain this delicate balance for a few divisions. But then each new cell loses its totipotency, its future possibilities narrowed. The cells forming the outer shell of the embryo commit to becoming the placenta. The other cells, forming a clump of cells inside the shell, can only become part of the embryo itself. Instead of totipotent, these cells are now just "pluripotent," meaning they still have several potential destinies.

The cells change their identities because their networks of genes and proteins rearrange themselves. When a totipotent cell makes proteins from its master genes, it doesn't produce them on a smooth assembly line. Sometimes its molecule machinery stalls and its supply of a protein runs low. Sometimes it races forward and produces a burst of molecules.

These fluctuations can throw the cell's feedback loops out of whack. One of the master genes in a totipotent cell, called Nanog, keeps many genes shut down. If a cell doesn't make enough Nanog proteins, the muzzled genes can spring into action—silencing Nanog itself. Once these gene networks flip, they can't flip back. The cell turns from totipotent to pluripotent.

Pluripotent cells get pushed farther across Waddington's landscape, falling into even deeper ravines and committing themselves to even narrower possibilities. Their random bursts of proteins continue to help drive them forward, along with signals that the cells get from their neighbors. The pluripotent cells end up in one of the three germ layers. Once a cell has become a mesoderm cell, it has surrendered its chance to become one of the other germ layers, to help build an eye or a lung. And with each new commitment, the methylation of genes—the long-term shielding of DNA—becomes more widespread. Cells begin to silence many of their genes so firmly that they can't be roused again. The networks of genes that maintain their identity as bone or muscle or gut become stronger, able to withstand errant bursts of proteins. When they divide, they reliably produce more of their kind, with the same methylation, the same coils to spool their DNA.

When cells divide, one of the most obvious things their daughter cells can inherit from them is their shape. In an embryo's nervous system, many neurons become long and spindly, with two slender branches extending out from a tiny cell body containing DNA. And when they divide, their daughter cells end up long and spindly as well.

These cells are the sensory neurons, which let our bodies feel. Buried in the skin of your thumb, for example, are feathery nerve endings connected to a sensory neuron that reaches from the thumb to the base of the hand, bends around the elbow and heads up to the shoulder before finally reaching a swelling of neurons around the spinal cord. The pain of scraping your thumb on a thorn is relayed along the two branches to the spinal cord, where it gets passed on to other neurons headed for the brain.

Leila Boubakar, a neuroscientist at the University of Lyon, and her colleagues wondered how it was that sensory neurons inherited their two-branch shape from their two-branched precursors, known as neural crest cells. Carefully observing neural crest cells divide under a microscope, they saw something remarkable happen: The neural crest cells got rid of their two branches before dividing, leaving behind only their blob-shaped cell bodies. But as soon as a neural crest cell divided, its daughter cells sprouted two new branches from the same sides as the ones that had sprouted from their mother cell.

To figure out how this was happening, the scientists added glowing tags to some of the proteins inside the neural crest cells. They discovered that the cells laid down what Boubakar and her colleagues call a "molecular memory" of their two-branch shape. It is a memory that can be inherited by their daughter cells. Before the neural crest cells start to divide, they move special proteins called septins to the base of their two branches. After the branches die back, the clusters of septins remain, marking where the branches had been.

The neural crest cell then divides into two sensory neurons, each of which inherits a septin mark. At that mark, each of the new neurons sprouts a new branch. Boubakar's experiments suggest that the septins then travel to the opposite side of the new sensory neuron. There the septins form a new cluster, which marks the spot where the second branch on the new neuron will grow.

Boubakar's research shows how heredity can incorporate more than genes. When cells divide, everything inside them is a living legacy to their descendants. It is unquestionably true that sensory neurons inherit genes from their mother cells. But that genetic inheritance alone does not explain why like engenders like inside our nervous system. Sensory neurons do not inherit their shape from their mother cells simply by inheriting genes for septins and other molecules. The mother cell's proteins carefully orchestrate the renewal of its branches in its offspring.

———

B y the time my daughters were born, they had developed sensory neurons throughout their bodies. They had also developed virtually all their other cell types: red skeletal muscle cells and white skeletal muscle cells; white fat cells and brown fat cells; liver lipocytes and Paneth cells of the intestines. But when they were born, my daughters were far from finished growing.

Throughout their childhood, many of their cell types continued to multiply. Most by now were deep in Waddington's canyons, their inner heredity now relentlessly rigid. I'm grateful for this tight epigenetic control. It kept their eyes from turning to kidneys, their fingernails from growing teeth. As I worked on this book, Charlotte and Veronica reached their mature height, determined in part by hundreds or thousands of genetic variants they inherited from Grace and me, combined with the influence of a twenty-first-century American environment rich in sunlight and pizza. As they approached their final height, all their cell lineages slowed down in a harmonious braking. Their full-size lungs now fit snugly in their properly sized rib cages. Their earlobes do not graze the floor.

But some of their cells went on giving rise to new types, and this creative flame will flicker on throughout their lives. Some parts of the human body perpetually renew as old cells die and new ones develop to take their place. In people in their thirties, the average fat cell is only eight years old. Red blood cells survive for only four months. Skin cells last just a month; taste buds, ten days; stomach lining, as little as two.

Scattered through the human body are hidden refuges of stem cells that can replenish these short-lived cells. In our long bones, our pelvis, and our sternum are cavities of bone marrow. The stem cells they harbor can divide into two kinds of cells, called myeloid cells and lymphoid cells. The myeloid cells have their own lineage, which branches into red blood cells as well as platelets and bacteria-gobbling immune cells called macrophages. The lymphoid cells have a different tree: They develop into T cells, which can command infected cells to commit suicide, and B cells, which make antibodies that can precisely attack certain pathogens. The stem cells lurking in the stomach lining rebuild it as old cells slough off. The same renewal happens in our skin.

Some stem cells generate new tissue only in an emergency. So-called satellite cells, nestled in our muscles, will produce new muscle cells to help repair damage. If you cut your hand, stem cells lurking in hair follicles will make new skin cells that crawl to the wound and heal it over.

Stem cells need to hide in their refuges to cling to their special nature. There they can swim in a pool of chemical signals, ensuring that the right network of genes stays turned on. In these refuges, the stem cells perform the same magic trick over and over again. They divide in two: One daughter cell goes on dividing to become mature types of cells, while the other is yet another stem cell. The cells manage this feat by manipulating the way their daughter cells inherit their molecules. Stem cells don't simply split up their molecules fifty-fifty. They move certain proteins and RNA molecules to one side but not the other. When they split in two, one of the new cells inherits a combination of molecules that allow it to stay a stem cell. The other cell flips its network to a new wiring and takes on its new identity.

One of the most important places where new cells develop is also one of the last places where they were discovered: in the brain. In fact, generations of neuroscientists were convinced that the neurons in the brain stopped dividing altogether shortly after birth. In order for us to learn, the neurons in our brains only grew new connections and pruned their old ones. In 1928, the Nobel Prize–winning neuroscientist Santiago Ramon y Cajal put the dogma of the twentieth century into a simple declaration: "Everything may die; nothing may be regenerated."

It wasn't until the late 1900s that this dogma started to crack. Some of the most elegant evidence of adult neurogenesis was made possible by the fact that everyone on Earth is partly made up of nuclear fallout.

Aboveground nuclear testing started in the mid-1950s, and was carried on until the 1963 Partial Test Ban Treaty. Each explosion sent neutrons racing through the atmosphere, sometimes crashing into nitrogen atoms and converting them into carbon-14. By 1963, carbon-14 had reached twice its level before the testing had begun. As plants absorbed carbon dioxide from the air, they incorporated extra carbon-14 into their leaves and stems and roots. The animals that ate those plants also accumulated high levels of the isotope. Plants absorbed it into their leaves and stems, and animals that ate the plants absorbed it into all their own tissues. Those animals included people alive at the time. They used the carbon-14 to build many new molecules. The RNA molecules and proteins sooner or later got torn apart and recycled. But the DNA remained unchanged. Ever since 1963, the carbon-14 level in the atmosphere has been dropping toward the level it was before the Atomic Era.

In the early 2000s, Jonas Frisén, a cell biologist at the Karolinska Institute in Stockholm, realized he could use carbon-14 levels in brain cells to estimate their age to within a couple of years. He and his colleagues began studying people who donated their bodies to science. They clipped bits of tissue from different regions of the brain and measured the levels of carbon-14 in them. By checking the year of their birth, the scientists could determine how old they were when the neurons had formed.

At first, the results confirmed the dogma. Frisén and his colleagues looked at the cerebral cortex, the thick outer layers of the brain that carry out much of our higher-level thinking. The neurons there dated all the way back to people's birth. But then the scientists looked to a small region tucked deep in the brain called the hippocampus. They were curious about it because scientists have long known that the hippocampus is vital for learning and for laying down long-term memories.

Because the hippocampus was so small, their initial tests weren't sensitive enough to measure carbon-14 levels accurately. It wasn't until 2013 that Frisén and his colleagues were finally able to take the measurements. Some of the neurons in the hippocampus turned out to be young. In fact, the scientists calculated, seven hundred new neurons were added to each hippocampus every day.

Adding seven hundred neurons to the eighty billion in an adult human brain is like dumping a tablespoon of water into an Olympic-size swimming pool. Yet some scientists suspect this tiny infusion may make an important difference to how our brains work. When mice are prevented from growing new neurons in their hippocampus, they take longer to learn how to make their way through a maze or how to press on a screen to get a reward of food. New neurons may make it possible to erase old, faulty memories and form new ones. In other words, Weismann's genealogy of our inner heredity may extend from the moment of conception to the last lesson we've learned.

CHAPTER 12

Witches'-Broom

I N MEDIEVAL EUROPE, travelers making their way through forests would sometimes encounter a terrifying tree. A single branch sprouting from the trunk looked as if it belonged to a different plant altogether. It formed a dense bundle of twigs, the sort that people might fashion into a broom to sweep their floors. The Germans called it *Hexenbesen.* The word was later translated into English as witches'-broom. Witches supposedly cast spells on trees to grow brooms, which they used to fly across the night sky. They could summon forth other branches as nests for sleeping. Elves and hobgoblins used the nests, too, as did the evil spirits who traveled about to sit on people's chests and give them nightmares.

In the nineteenth century these terrors faded, and plant breeders began using these rare, strange growths to create entirely new cultivars. Cuttings from monstrous branches could take root and grow into trees of their own, producing seeds that would grow into a new generation of plants with the same monstrous shape. Some of today's most popular landscaping plants got their start as witches'-broom.

Dwarf Alberta spruce, a tree that grows only ten feet high, is a common sight in suburban yards. But it originated from white spruces that grow as tall as ten-story buildings in northern Canada. In 1904, a pair of Boston horticulturalists visiting Lake Laggan noticed that a white spruce there had sprouted a witches'-broom. Seeds had fallen from the freakish branch to the ground, where they had grown into squat little shrubs. The horticulturalists took some of the shrubs home with them and dubbed them *Picea glauca* "Conica," or dwarf Alberta spruce. The only trouble these shrubs cause their owners is that they sometimes reclaim their ancestral glory. Sometimes a branch will jut out from a dwarf Alberta spruce and race upward, taking on the titanic shape of its giant predecessors back at Lake Laggan.

Plant breeders didn't have to go into the north woods to find witches'-broom, however. They could look in their own orchards and gardens. When they spotted an odd branch, they dubbed it a bud sport. In the early 1900s, a Florida farmer found a notable bud sport while inspecting his grove of Walters grapefruit trees. Tree after tree bore white fruit, except one. On that tree, the farmer spotted a branch weighed down with pink fruits. From that single bud sport, all pink grapefruits descend.

To make sense of witches'-broom or bud sports, scientists had to study how plants grow. As plant cells divide, the daughter cells inherited the same hereditary factors that were in the mother cell. In some cases, a cell would change, and its descendants would inherit its new quirk. Those cells might produce a new branch, complete with leaves, fruit, and seeds. But bud sports could alter plants in other ways, too. As a red sunflower bloomed, half of it might grow yellow leaves. Sometimes an ear of corn developed a patch of dark kernels. A pale red apple might develop a wedge-like stripe of green running down one side, right next to a stripe of umber.

Charles Darwin would pore through issues of the *Gardeners' Chronicle* to find new reports of bud sports. He noted branches on cherry trees that bore their fruit two weeks after the rest of the branches. His curiosity was piqued by the story of a French rose that mostly produced flesh-colored flowers but also grew a branch covered by deep-pink blossoms.

As he struggled to make sense of heredity, Darwin believed studying these sports could help. They seemed to contain the same mysterious power of generation as seeds or eggs. Sports were not mere freaks, deformed by a cold snap or a disease. Something triggered a drastic change inside them, Darwin declared, like "the spark which ignites a mass of combustible matter."

Half a century later, it became clear that this combustible matter lay in the chromosomes of the plants. When plant cells divided, they usually produced identical copies of their genetic material. But on rare occasion, one of the new cells would mutate, and its own descendants within the plant would inherit that mutation.

"It appears that a change in the hereditary constitution of the cells has occurred in the soma or body," the biologist T. D. A. Cockrell wrote in 1917, "without having any connection with the process of sexual reproduction." Cockrell called this change a somatic mutation. He coined the term to distinguish it from a germ line mutation—a mutation that germ cells could pass down to the next generation.

When Cockrell investigated somatic mutations, scientists knew so little about genes that it was hard to say exactly how they occurred. One possibility was that newly formed pairs of chromosomes got entangled and swapped parts. The strange stripes on apples—known as twin spotting—might occur because a cell had two copies of a gene for color. One copy might be a light variant, the other dark. When the cell divided, it accidentally bequeathed two dark variants to one daughter cell, and two light ones to the other. When those cells multiplied, their daughters would inherit those new combinations. And since they grew next to each other, the result would be dark and light stripes.

As geneticists studied these peculiar plants more carefully, they gave them a new name: mosaics. The name hearkened back to the ancient artworks composed of thousands of tiny colored tiles. Nature created its mosaics from cells instead of tiles, in a rainbow of different genetic profiles.

Plants first brought mosaicism to our attention, but in the early 1900s, scientists started to appreciate that animals can be mosaics, too. Their attention might be caught by a parakeet with a splash of dark plumage across one wing, a rabbit with a peculiar white patch of fur.

But modern science was slow to recognize that we humans are mosaics as well. It's not as if human mosaics were invisible. Some were downright impossible to miss. Human mosaics might be born with port-wine stains on their face. Others looked as if a charcoal artist had applied stripes and checkerboards to their skin (a condition that came to be known as the lines of Blaschko, named for the German dermatologist Alfred Blaschko, who first described the condition in 1901). One human mosaic even became a celebrity in Victorian England. He called himself the Elephant Man.

When Joseph Merrick was born in 1862, he seemed heathy and normal. But within a few years, his forehead began to swell forward like a ship's prow. His feet became nightmarishly large, and his skin grew rough, lumpy, and gray like an elephant's. As his appearance altered, his parents became convinced that his deformities were the result of his mother's being knocked over by an elephant at a fair while she was pregnant with him.

Merrick went to school until he was thirteen and then found work rolling cigars in a factory. His deformities continued to worsen, his head broadening out until it was thirty-six inches in circumference. His right arm expanded into a paddle-like shape, forcing him to quit his job. He tried to work as a peddler, but the authorities soon revoked his license because they deemed him too grotesque.

Merrick decided to follow the examples of Charles Byrne, the Irish Giant. He turned himself into an attraction, traveling around England as the Elephant Man. His manager, Tom Norman, would warm up the crowds by warning them about what they were about to see: "Brace yourselves up to witness one who is probably the most remarkable human being ever to draw the breath of life."

In London, Merrick exhibited himself in a shop across the street from the Royal London Hospital. Medical students came to gawk, and eventually a doctor at the hospital, Frederick Treves, followed them over. He was startled by "the most disgusting specimen of humanity I had ever seen," as he later recalled. He persuaded Merrick to visit the hospital and be examined by the hospital doctors. But after a few inspections, Merrick decided he felt like "an animal in a cattle market," and stopped going.

Merrick's business tapered off, prompting him and Norman to try their luck on the continent. Things didn't go much better there, and soon Norman abandoned Merrick, who was then robbed of all his possessions. Destitute and filthy, he managed to make his way back to England in 1886, whereupon Treves set up an apartment for him in the hospital.

When Treves first met Merrick, he'd thought the Elephant Man was intellectually disabled. But in the comfort of his new home, Merrick flourished. He wrote poetry, made cardboard dioramas, and received visits from aristocrats. Alexandra, Princess of Wales, brought him a signed photograph of herself and sent him a Christmas card each year. Merrick enjoyed this happy existence for four years before dying at age twenty-seven in his bed. It is likely he died when his massive head fell back suddenly, severing his spinal cord.

Try as he might, Treves never figured out Merrick's condition. He brought in medical experts, who speculated Merrick might be suffering from a nervous system disorder. Merrick's death did not quench Treves's curiosity: He had plaster casts made of much of Merrick's body, and had his bones bleached and boiled. Treves observed that the growths on Merrick's skeleton were enormous, and yet he could see they were not tumors. No one in Merrick's family had suffered his condition, making it unlikely that it was inherited. And, most puzzling of all, his deformities were scattered in random patches across his body. The other parts of his body were entirely normal.

Merrick's case, along with the lines of Blaschko and port-wine stains, were all dramatic examples of mosaicism, but their true nature remained hidden for decades. Part of the reason for this oversight was the lack of scientific tools, but there were other reasons for the lag. As scientists studied the genetic variations among people, they gave little thought to the genetic variations *within* each one.

It is hard to think of another explanation for how a scientist could correctly realize that cancer is a form of mosaicism in 1902, only to die years later before other researchers proved he was right.

In the late 1800s, Theodor Boveri carried out a series of studies on chromosomes that assured his place in the history of science. His experiments made clear, for example, that chromosomes carry hereditary factors. Boveri did most of this work on sea urchins at a marine biology station in Naples. He would carefully inject sea urchin sperm into eggs and then observe them develop, duplicating their chromosomes with each division.

After a few years of this research, Boveri and his wife, Marcella, got an idea for an experiment. They wondered what would happen if they injected two sperm instead of one into a single sea urchin cell. The result, they discovered, was chaos.

The extra DNA delivered by the two sperm overwhelmed the fertilized egg, leaving it unable to separate all its chromosomes into equal sets. When the egg divided, some of its daughter cells ended up with more chromosomes than others. Some even ended up with no chromosomes at all. The aberrant cells continued to copy their chromosomes and divide. Eventually they broke apart into embryonic fragments, and some of those clumps of cells continued to develop. Some became healthy sea urchin larvae, while others ended up as deformed pieces of tissue.

Observing this chaos, Boveri wondered if it was akin to cancer. In the late 1800s, biologists who studied tumor cells under microscopes noticed their chromosomes had odd shapes. They couldn't see the chromosomes well enough to understand the precise nature of those differences. But they saw enough to speculate that chromosomes had something to do with cancer.

Now looking at sea urchin cells run amok, Boveri had an insight of uncanny brilliance. In order to grow normally, he reasoned, cells needed to inherit the same set of chromosomes as their ancestors. If some disturbance ruined the process, cells might end up with too many chromosomes or too few. Many of these mutant cells would die. Sometimes these cells would multiply at an unnatural rate. Their daughter cells inherited the same abnormal chromosomes, and continued to proliferate. The result would be a tumor.

As soon as Boveri floated his theory, he faced intense opposition. "The skepticism with which my ideas were met when I discussed them with investigators who act as judges in this area induced me to abandon the project," he later said. Boveri set the idea aside for twelve years, only making it public in 1914 in his book *Concerning the Origin of Malignant Tumors*. Even then, he was met with skepticism. Boveri died the following year, never knowing if he was right.

It would take until 1960 for scientists to observe chromosomes carefully enough to test Boveri's theory. David A. Hungerford and Peter Nowell discovered that people with a form of cancer called chronic myelogenous leukemia were missing a substantial chunk of chromosome 22. It turned out a mutation had moved that chunk over to chromosome 9. The altered chromosomes drove cells to become cancerous.

Like Boveri before them, Hungerford and Nowell could observe only the large-scale changes that occurred in chromosomes. Later generations of scientists would gain the technology necessary to study cancer cell DNA at a finer scale, sequencing entire genomes from tumor cells. And when they looked closer, they found that far smaller changes than the ones Hungerford and Nowell had observed could also drive cells toward cancer.

Healthy cells make a number of proteins that guard them against becoming cancerous. Snipping out a short stretch of DNA or misreading a single base in their genes disables these guards and lets the cells run wild. Some genes, for example, make proteins that regulate how quickly cells grow and divide. Shutting down one of these genes may be like disabling the brakes in a car rushing downhill. A succession of mutations can then push the descendants of a cell farther down the path to cancer. They can make precancerous cells invisible to the immune system, which continually searches for new tumors. They can make the cells send out signals that lure blood vessels their way, feeding their wild growth.

Each new generation of cancer cells inherits these dangerous mutations, and by the time they've produced a full-blown tumor, it may harbor thousands of new mutations not shared by healthy cells. These mutations can allow cancer cells to thrive at their host's expense, but they can also damage the cells themselves. Mutations to the DNA in mitochondria, which generate a cell's fuel supply, can leave it without enough energy to grow. Cancer cells can solve this particular dilemma with a bold change to their DNA: They steal mitochondrial genes from healthy cells to replace their own damaged set.

It's hard to think of cancer having anything in common with a pink grapefruit. Yet they are both the product of mosaicism: living lineages of cells set off from the rest of a body by the mutations they inherit from their mother cells. Once scientists finally realized that cancer is a deadly form of mosaicism, they wondered how many other forms it might take.

As scientists looked more closely at how cells divide in the body, simple arithmetic hinted that mosaicism might be everywhere. A single fertilized egg will multiply into roughly 37 trillion cells by the time a person reaches adulthood. Each time one of those cells divides, it must create a new copy of its three billion base pairs of DNA. For the most part, our cells manage this duplication with stunning precision. If they make a mistake, one of their daughter cells will acquire a new mutation that was not present at conception. And if that daughter cell produces an entire lineage, a potentially vast pool of cells will inherit it, too. Based on estimates of the somatic mutation rate, some researchers have estimated that there might be over ten quadrillion new mutations scattered in each of us.

But simple arithmetic on its own could not reveal the precise nature of mosaicism. When a mutation arose in a cell, it might kill it. Our bodies might experience a kind of internal natural selection, favoring cells that retained the genome we started with as fertilized eggs. It was also possible that other mutations were harmless, accumulating without any effect for good or bad. Without technology to inspect DNA, researchers could not find out which possibility was true. They still managed to discover new examples of human mosaics, but only when those examples were impossible to ignore.

On August 5, 1959, for example, a baby was born at New York University Medical Center with both a penis and a vagina, and lacking testicles. The doctors extracted cells from the baby's bone marrow to study their sex chromosomes. Out of twenty cells the doctors looked at, eight of the cells had an arrangement found in boys: one X chromosome and one Y. But twelve of the cells had only a single X chromosome.

The baby had started out as a zygote with an X and Y chromosome, the doctors realized. But at some point during pregnancy, a dividing cell in the embryo accidentally failed to pass on its Y chromosome to one of its daughter cells. Without a Y chromosome, the cell could not produce some of the proteins involved in developing the male anatomy. It divided and passed down its Y-free chromosomes to its descendants, giving rise to some female anatomical parts. The baby became a mosaic of XY and X cells.

As scientists worked out more details of how embryos developed, they recognized that other conditions were mosaicism as well. The lines of Blaschko, for example, were already present when babies were born, suggesting they were the result of some kind of genetic disorder. But geneticists could not trace the lines of Blaschko through family pedigrees, suggesting the mutation was not passed down from parents to children.

In 1983, a team of Israeli geneticists examined the chromosomes of a boy with lines of Blaschko running up and down the right side of his body. They collected epithelial cells that had been shed into his urine, skin cells from his arms, and white blood cells. The skin cells from his right arm had an extra copy of chromosome 18, as did half of his white blood cells. The rest of the cells were normal. The doctors concluded that a chromosomal mistake had arisen early in the boy's development. It marked the start of a new lineage of cells, all of which carried the same extra copy of chromosome 18. Later, that lineage of cells differentiated into various tissues, including immune cells and skin cells. Only in the skin cells did the mutation produce a visible change.

Joseph Merrick proved to be a mosaic, too, but his case was especially hard to solve. For many years after Merrick's death, doctors generally agreed that he suffered from neurofibromatosis, a hereditary condition that makes neurons prone to develop benign tumors. While Merrick did indeed have some of the symptoms of neurofibromatosis, some researchers noted that he had other symptoms that didn't fit the diagnosis. Merrick's feet, for example, developed moccasin-like overgrowths—a symptom not caused by neurofibromatosis.

In 1983, researchers recognized a few other people with Merrick's precise combination of symptoms. Proteus syndrome, as they dubbed the condition, struck fewer than one in a million people. While Merrick's disease now had a name, scientists didn't yet understand its cause. In the early 2000s, Leslie Biesecker, a geneticist at the National Human Genome Research Institute in Bethesda, Maryland, led a search for its genetic basis. He and his colleagues collected samples from six people with Proteus syndrome—from diseased skin, as well as from healthy tissue and blood.

Instead of looking for large changes in chromosomes, the scientists used a newer method, called exome sequencing. They decoded all the protein-coding stretches of their genome—about 37 million bases of DNA per cell. Biesecker and his colleagues found that all six subjects had the same mutation in common. It struck a gene called AKT1, which is known to be important in controlling the growth of cells. But the mutation was present only in some of their cells, and not others. The mixed results suggested that Proteus syndrome was a case of mosaicism.

Biesecker's team then turned to twenty-nine other people with Proteus syndrome. They sequenced the AKT1 gene from cells in a variety of their tissues, too. The scientists found the same mutation in the diseased skin of twenty-six of the subjects. But the scientists couldn't find the mutation in any of the white blood cells they examined.

Beisecker and his colleagues reared some of the cells in flasks to see how the mutation affected them. They found that it didn't shut AKT1 down. Just the opposite: It made the gene even more active, spurring skin and bone to grow more—precisely what you'd expect from a mutation that could produce the Elephant Man. It was the first time scientists used exome sequencing to find the cause of a mosaic disease. And once the researchers knew what gene was responsible for Proteus syndrome, they could search for a drug that could attack it. Biesecker and his colleagues found one, which they began testing with promising results. Now that Joseph Merrick's disease had finally been revealed to be a case of mosaicism, it may one day become curable.

As scientists have pinned down the genetic causes of more mosaic diseases, they are building a chronicle of our inner heredity. A mutation may arise at any stage of development, from the first division that splits a zygote in two, to the last mitosis before death. Depending on when it strikes, a disease may affect a few cells or many. A skin disorder called CHILD strikes early, just as an embryo's cells are dividing the body into its left and right sides. It produces a body that's half-dark, half-light. The lines of Blaschko arise much later, as an embryo's skin starts to develop. Epidermal cells stream in rivers from the body's midline over the surface of the body. If they pick up a mutation to their pigment genes, they will trace lines across the skin.

The timing of development is so powerful that it can cause the same mutation to produce a different kind of mosaicism, depending on when it arises. A condition called Sturge-Weber syndrome causes a cluster of devastating changes to the head. It can trigger an aggressive bloom of blood vessels that push down dangerously hard on the brain. Depending on where the vessels press, they may cause epileptic seizures, paralyze one side of the body, or cause intellectual disability. If the blood vessels push against the eyes instead, they can cause glaucoma. Sturge-Weber syndrome also creates a massive pink birthmark across as much as half the face. It looks like an extravagant version of a port-wine stain.

The resemblance to port-wine stains is so strong that some scientists have wondered if the two conditions are related. In 2013, Jonathan Pevsner of the Kennedy Krieger Institute led a study to find out. They took a sample of pigmented skin from three people with Sturge-Weber syndrome, along with samples of their unpigmented skin and blood. Pevsner and his colleagues extracted the DNA from the different tissues and sequenced their entire genome. In each patient, they discovered that the pigmented skin cells shared the same mutation to the same gene, called GNAQ. Following up with twenty-six other people with Sturge-Weber syndrome, they found twenty-three had the mutation as well in their altered skin.

Having found the genetic basis of Sturge-Weber syndrome, Pevsner turned his attention to port-wine stains. When he and his colleagues examined the stains on thirteen people, they discovered the same mutation to GNAQ in twelve of them. Their study suggests that the two conditions arise from the same mutation but take on different forms depending on when it appears during development. Sturge-Weber syndrome occurs if the mutation takes place early in development. As the mutant cells divide, they can turn into skin, blood vessels, and other tissues. If the mutation arises in

GNAQ later in development, it becomes limited to skin cells, causing only port-wine stains. The two conditions differ only in time.

————

Conditions like port-wine stains and Proteus syndrome brought mosa-icism to the body's surface and made it visible. More recently, scientists have searched for buried mosaicism hidden from view. Annapurna Poduri, a pediatric neurologist at Harvard, investigated a brain disorder called hemimegalencephaly. In people with this condition, one of the brain's hemispheres becomes massively swollen, leading to severe seizures. The fact that the disease affected only half the brain raised the possibility that it was a case of mosaicism.

As plausible as this was as an idea, it would be hard to test. Poduri and her colleagues couldn't simply draw blood from people with hemimegalencephaly or snip off a bit of their skin. The mosaic mutation might be hiding only in the brain.

Poduri and her colleagues took advantage of surgeries that people may get to treat hemimegalencephaly. Surgeons will sometimes remove part of the overgrown hemisphere, or take it out completely. The scientists were able to examine brain tissue taken from eight people. In the first sample they looked at, some of the cells had a lot of extra DNA. It turned out that in those cells, a long stretch of chromosome 1 was duplicated. In other cells from the same patient, chromosome 1 was normal. When the scientists looked at a second patient, they once again found another duplication of DNA in the same region of chromosome 1.

That region contains an intriguing gene called AKT3. Looking back at earlier studies on the gene, Poduri and her colleagues found that a loss of AKT3 sometimes led babies to develop abnormally small brains. Perhaps, they thought, an extra copy of the gene might push brains in the other direction. Poduri and her colleagues sequenced the AKT3 gene in brain tissue from six other people with hemimegalencephaly. One of them had a mutation in AKT3, but only in about a third of his brain cells.

Hemimegalencephaly probably gets its start early in the development of embryos, when neurons are climbing up cellular ropes to build the brain. The neurons divide as they climb, and a mutation arises in the AKT3 gene, or perhaps another gene that helps it. While other neurons eventually stop dividing, the mutant neuron's lineage does not. Its proliferation is not the runaway growth of a tumor. Instead, the extra neurons spread out across a hemisphere, nestling in among normal cells. Even though they make up only a small fraction of the total neurons, they somehow trigger some hemisphere-wide damage.

The genetic differences that mosaicism creates between our cells are far fewer than the differences between two people. If I could compare cells from my left and right hands, they would not be genetically identical, but they would be vastly more similar to each other than to any cell from my brother, Ben. Yet a somatic mutation that alters even a single base can have a profound effect on our health while eluding our best medical tests. To diagnose a standard hereditary disease—one that was already present in a zygote—geneticists can look at the DNA of any cell in a patient. But in a mosaic disease, one cell cannot stand in for all cells.

In 2013, doctors at Lucile Packard Children's Hospital Stanford in Palo Alto, California, discovered how vexing mosaicism can be when a woman named Sici Tsoi gave birth to her third child, a daughter named Astrea. The first clue that Astrea had a problem came in the thirtieth week of pregnancy. Tsoi's obstetrician noticed something peculiar about the baby's heartbeat. "The beat was long and short and long and short," Tsoi explained to me.

It was possible, Tsoi's doctors worried, that Astrea had a hereditary disorder known as long QT syndrome. Normally, the heart beats by releasing regular bursts of electric charge across its muscles, causing them to contract. After each beat, the heart moves charged atoms through tunnels in its cells to build up a new charge. In about one in two thousand births, babies are born with defective tunnels. Some don't develop enough of them; others produce deformed tunnels that can block the flow of charged atoms. These defects can slow down the heart's recharging, creating long lags between beats, and throwing off the heart's precise choreography of electric waves. Left untreated, the chaos caused by long QT syndrome can be fatal.

A definitive diagnosis of long QT syndrome would require putting electrodes directly on Astrea's chest after birth. For the time being, Tsoi's doctors kept tabs on Astrea's fetal development with a twice-weekly echocardiogram, using ultrasound to monitor her heartbeat from a distance. The longer the doctors could extend the pregnancy, the healthier Astrea would be after birth.

In her thirty-sixth week, Tsoi's doctor spotted a suspicious buildup of fluid around Astrea's heart. It might be a sign that she was experiencing heart failure. They decided Tsoi would need to have an emergency caesarean section.

When Tsoi woke up in her hospital room after the delivery, she expected a nurse would bring Astrea to her bedside. Hours passed without a glimpse of her new daughter. Tsoi asked her husband, Edison Li, to go to the neonatal intensive care unit. He came back saying that there were so many doctors surrounding Astrea that he couldn't even see her.

The next day, Tsoi's doctor visited her with forms to sign. "Then I realized it was something serious," Tsoi said. Her doctor explained that Astrea did indeed have a severe form of long QT syndrome and had gone into cardiac arrest shortly after birth. It was hard for Tsoi to make sense of all the medical terminology, but she understood that surgeons were going to have to operate on Astrea's day-old heart to save her life.

After Tsoi and Li signed the forms, the surgeons implanted a cardioverter defibrillator in Astrea's heart. When her heartbeat lurched out of control, the defibrillator delivered an electric shock that reset her heart and established a normal rhythm again.

Astrea's medical team included a pediatric cardiologist named James Priest from Stanford Medicine's Center for Inherited Cardiovascular Disease. Priest sent some of Astrea's blood to a genetic testing company to see if they could find the cause of her long QT syndrome. Rather than look for a single mutation, Priest ordered a so-called panel test that could search for mutations on a number of genes that are firmly tied to long QT syndrome. The panel's results might tell Priest which kind of tunnel was altered in Astrea's heart. Some tunnels pump sodium atoms, while others pump potassium. Different drugs for long QT syndrome work better on different tunnels.

But Priest was keenly aware of the limits of the panel test. For one thing, it was slow. He might have to wait a couple of months to finally get the results back—a vital window during which Astrea might benefit from being put on the right kind of drug. Priest also knew that about 30 percent of patients with long QT syndrome got no genetic diagnosis at all from panel tests. Scientists at the time were still a long way from identifying all the genes that can, when mutated, give rise to long QT syndrome. Thus, nearly a third of patients ended up in what doctors call genetic purgatory.

In 2013, Priest and his colleagues were beginning to sequence the entire genomes of some of their patients to better understand their diseases. Rather than inspect one gene at a time, they wanted to look at all genes at once. When Priest talked about Astrea's case with his fellow scientists, they realized that genome sequencing might be both quicker and more thorough than the standard panel test. But they knew such an experiment would have no guarantees of success.

Priest spoke to Tsoi and Li, explaining what he wanted to do. "Everybody's genome is like a book with 23 chapters," he told them. "You have two copies of each chapter, one from your dad and one from your mom. Whole genome sequencing looks for everything. It looks for missing chapters, missing paragraphs, every misspelled word."

Tsoi and Li gave their consent, and Priest drew some blood from Astrea—now only three days old. He shipped it to Illumina, which rushed the job. Six days later, Priest got all their raw data. He set up a program to assemble the short reads into Astrea's entire genome, and then he searched through it for mutations that might be responsible for her long QT syndrome.

Astrea had millions of variants, of course, but Priest was quickly drawn to one in particular. She carried a rare mutation of one copy of gene called

SCN5A. That particular gene encodes sodium tunnels in the heart, and Priest himself had found that, in another patient, a mutation at precisely that same spot caused long QT syndrome. "It totally hit me over the head," said Priest. "I wasn't going to find anything better."

The next day, Priest informed Tsoi and Li of his discovery. Astrea, now only ten days old, was put on a drug to treat sodium channels. Priest then went back to Astrea's genome to wrap up the case, to confirm his diagnosis before writing up the results.

And that's when his story fell apart.

The Illumina technicians had sequenced Astrea's genome as they had mine and thousands of other people's. They broke open her white blood cells and chopped up the DNA inside. They then made many copies of those fragments—known as reads—and sequenced them all. Priest had used his computer to figure out where each read sat in Astrea's genome. Because the sequencer made so many reads, around forty of them lined up at every spot in her DNA. On average, half of the reads in a gene came from one copy of a gene, and the remaining ones came from the other. Priest found the SCN5A mutation in eight out of thirty-four reads. It wasn't a perfect fifty-fifty split, but it was close enough, Priest decided. He assumed that one copy of her SCN5A gene had the disease-causing mutation.

Priest followed up on the genome sequencing with a more focused exam of Astrea's DNA. He pulled out the SCN5A gene from some of Astrea's white blood cells and made millions of copies of it so he could examine it in fine detail. He expected to find a fifty-fifty split between the normal version and the mutant one. But he found no mutation at all. It was as if he had examined two different babies, one with a lethal mutation and one without it. "I was just flabbergasted," he said.

Priest wondered if there was some unusual heredity in Astrea's family that had tricked him. Neither Tsoi nor Li showed any sign of having long QT syndrome. They never had any problems with their hearts, and Priest found that their EKGs were normal. It was possible that one of them carried an extra broken copy of SCN5A. Sometimes a mutation will trigger the accidental duplication of a gene but in a form that can't make a protein.

Perhaps Astrea had inherited a so-called pseudogene of SCN5A, and perhaps Priest had mistaken it for her working version. If that was the case, then SCN5A would have nothing to do with Astrea's ailing heart, and Priest would be back at square one. He'd have to start a new search for her long QT mutation.

To search for a pseudogene, Priest sequenced DNA from Tsoi and Li. Instead of sequencing their entire genomes, he sequenced only their protein-coding genes. Again, he ended up empty-handed. Neither of Astrea's parents had a pseudogene for SCN5A.

Finally, Priest considered the most extreme possibility: that Astrea was a mosaic. Perhaps the SCN5A mutation was only in some of her cells but not others. To investigate this possibility, Priest brought Astrea's blood to Stephen Quake. Quake, a Stanford scientist, had developed a way to sequence a genome from a single cell. Rather than throwing together DNA from millions of Astrea's cells, he could inspect them one at a time.

Quake and his team inspected thirty-six of Astrea's blood cells. In three of them, they discovered a mutation on one copy of the SCN5A gene. In the other thirty-three cells, both copies of the SCN5A gene were normal.

Quake's test confirmed that Astrea's blood was a mosaic. To get a broader survey of her mosaicism, Priest and his colleagues also examined cells from her saliva and urine. Now they had samples of cells that had developed from the three germ layers. (Blood comes from the mesoderm. The lining of the mouth comes from the ectoderm. And the urinary tract develops from the endoderm.)

In all three tissues, the scientists found the SCN5A mutation in between 7.9 and 14.8 percent of Astrea's cells. She was a mosaic through and through, in other words. And she must have become one before she had developed the three germ layers, when she had been just a ball of cells. One cell in that embryonic ball had mutated, and when it divided, it passed down that mutation to its descendants. The cells that inherited the errant SCN5A gene ended up mixing into all three germ layers.

As Priest and his colleagues were deciphering Astrea's mosaic nature, she recovered well enough from her surgery for Tsoi and Li to take her home. The drugs Priest had recommended kept her long QT syndrome under control, and she enjoyed a happy infancy. One day, when Astrea was seven months old, Tsoi's phone rang.

"I got a call from the doctor, and she asked if Astrea was doing okay," said Tsoi. Astrea was right in front of her, playing with toys, Tsoi said.

It turned out that Astrea's defibrillator had just shocked Astrea's heart. It had sent a wireless message to her doctors to let them know. They needed to get Astrea back into the hospital as quickly as possible. "I couldn't absorb that information fast enough," said Tsoi.

When the Stanford doctors examined Astrea, they discovered that her heart had become dangerously enlarged—another risk posed by SCN5A mutations. Astrea would need a new heart in order to survive. Not long after Astrea came back to the hospital, her heart stopped, and her doctors struggled to bring her back, clamping a mechanical pump to her heart to keep it functioning.

"On the night that she was almost gone," Tsoi said, "I was thinking, 'If it's too hard or it hurts too bad on her, it's okay, just go.'"

Astrea recovered and regained her strength. And a few weeks later, a donated heart became available. Astrea underwent transplantation surgery, and she was back home again after a few days. The first few months at home were rough for the entire family, with Astrea throwing up constantly. But gradually she recovered. Except for having to take anti-rejection drugs three times a day, Astrea got her childhood back. She listened over and over again to songs from the movie *Frozen*. She did cartwheels with her sister.

For Priest, Astrea's heart transplant gave him a chance to find out once and for all if mosaicism had been to blame for her condition. After surgeons removed her heart from her body, they clipped off some pieces of muscle for Priest to study. On the right side of the heart, he and his colleagues found that 5.4 percent of the cells had mutant SCN5A genes. On the left, 11.8 percent did. Little grains of mutant cells were mixed in with the ordinary tissue. Priest and his colleagues built a computer simulation of Astrea's heart with those levels of mutant cells and let it beat. The simulated heart thumped irregularly, in much the same way Astrea's did.

Astrea had lost her mosaic heart, but the rest of her body remained a genetic mix. Yet now her SCN5A mutations could no longer threaten her life. Priest was left wondering how many other cases of long QT syndrome are actually the result of mosaicism like Astrea's. "It's hard to say I'll be involved in such an interesting case for the rest of my life," Priest said.

———

The search for the causes of diseases has uncovered a number of cases of mosaicism. But scientists have also discovered some people in which mosaicism can heal.

A team of Dutch dermatologists and geneticists described the first case of mosaic healing in 1997. They examined a twenty-eight-year-old woman whose skin was so fragile that even a gentle rubbing would raise blisters. This painful condition is caused by a mutation to a gene called COL17A1. Normally, skin cells use this gene to make a type of collagen that makes them stretchy.

Both of the woman's parents were carriers. They each carried a mutation on one copy of their COL17A1 gene. (They had different mutations in different locations—a detail that will turn out to matter tremendously in a little while.) Because each parent also had a normal copy of the COL17A1 gene, they could still make enough collagen to keep their own skin healthy.

The woman had the bad luck to inherit each parent's bad copy of the gene. Those defective copies were present when she was still a fertilized egg. They were passed down to every cell that zygote gave rise to. When she developed skin, her skin cells needed to switch on her COL17A1 gene to make collagen. The gene failed at its job, and she was left with skin that couldn't stretch.

Remarkably, however, the woman's doctors noticed that she had a few patches of normal skin on her arms and hands. They didn't blister when they were rubbed. The woman had been aware of some of the patches for as long as she could remember. Others had emerged more recently and were expanding. When the doctors looked at the molecular makeup of her healthy patches, they found healthy collagen.

Looking closely at the DNA in her cells, the geneticists figured out how these patches had developed. Each arose from a single faulty skin cell. Before it divided, the cell duplicated its DNA. And during that duplication, it mutated in a peculiar way: It swapped a section of the COL17A1 gene between its chromosomes.

When the two daughter cells pulled away from each other, one cell no longer carried the woman's mutation from her mother. It had been replaced by the working portion of her father's COL17A1 gene. Now altered, the cell could make collagen again. And when it divided, its daughter cells inherited a working version of the gene as well. The woman's mosaics had repaired her defective genes.

Since that initial discovery, scientists have found more genetic diseases partially cured by mosaics. Their list now includes hereditary forms of other skin diseases, along with anemia, liver disorders, and muscular dystrophy. The growing inventory of mosaicism—causing diseases or healing them—raised the question of just how mosaic humans are in general. The definitive answer would come from breaking down people into their 37 trillion cells and sequencing every base of DNA in each one. For now, scientists are only carrying out rough surveys. But even these preliminary studies have come to one clear conclusion: We are all mosaics, and we have been so pretty much since our beginnings.

In the first few days of an embryo's existence, over half of its cells end up with the wrong number of chromosomes, either by accidentally duplicating some or losing them. Many of these imbalanced cells either can't divide or do so slowly. From their initial abundance, they dwindle away while normal cells create their own lineages. If the supply of chromosomes is too abnormal—a condition called aneuploidy—then the mother's body will sense trouble and reject the embryo altogether.

But a surprising number of embryos can survive with some variety in their chromosomes. Markus Grompe, a biologist at Oregon Health & Science University, and his colleagues looked at liver cells from children and adults without any liver disease, most of whom had died suddenly, by drowning, strokes, gunshot wounds, and the like. Between a quarter and a half of their liver cells were aneuploids, typically missing one copy of one chromosome.

A trained expert can spot aneuploid cells with a microscope. Finding smaller mutations—such as short deletions, duplications, or single-base changes—has required far more sophisticated technology. In 2017, for example, researchers at the Wellcome Trust Sanger Institute in England sequenced the entire genomes of immune cells they got from 247 women. In each volunteer, the scientists found around 160 somatic mutations, each present in a sizable fraction of her cells.

Because these somatic mutations were so common, the researchers suspected they arose early in development. To test the idea, they sequenced the genomes of cells from other tissues in the women. They could find most of the somatic mutations in a fraction of those other cells, too. Based on their research, the Sanger scientists estimated that an embryo gains two or three new mutations every time its cells double. As those new mutations arise, embryonic cells pass them all down to their descendants as a mosaic legacy.

Christopher Walsh, a geneticist at Harvard who studies mosaicism in the brain, wondered how extensive mosaicism is in our neurons. To find out, he and his colleagues got hold of tissue samples from three people who underwent brain surgery. From each sample, they isolated around a dozen neurons and then sequenced the genomes of each one. They then looked for somatic mutations that set each neuron apart from other cells in the brain, as well as the rest of the body.

Every neuron, Walsh found, was a mosaic. It carried around 1,500 single-nucleotide variants, a unique genetic signature that set each neuron apart from the cells in other parts of the body. These mutations accumulated gradually, through many generations of dividing neurons. Recent mutations were shared by only a few neurons, while older ones were shared by many.

It occurred to Walsh that he could use the mutations to reconstruct the cell lineages of the brain—not to watch the lineages grow forward as Conklin had, but to work his way back up their branches like a genealogist, back to the womb.

To make this trip, Walsh and his colleagues studied a seventeen-year-old boy who had died in a car accident. The boy's family donated his body for scientific research. Walsh got hold of frozen pieces of the boy's brain, and his team plucked 136 neurons from the tissue. They then sequenced the entire genome in each cell. As a point of comparison, they also sequenced DNA from other organs in the boy's body, such as his heart, his liver, and his lungs.

Scanning the trillions of bases they sequenced, the researchers spotted hundreds of somatic mutations in each neuron. Many of the mutations were shared by some of the neurons, but not all of them. Some were found in only a few of the neurons, and some were unique to a single cell. The researchers used this pattern to draw a genealogy of the brain, linking each neuron to its close cousins and its more distant relatives. Walsh and his colleagues found that the cells belonged to five distinct lineages, the cells in each one inheriting the same distinctive mosaic signature.

The shared mutations must have all arisen when the boy was still an embryo, when the neurons in his brain were still multiplying quickly. But Walsh got even deeper insights into the development of the boy's brain when he compared the neurons to cells from his other organs. One lineage of neurons also included cells from the boy's heart. Other lineages included cells from other organs.

Based on these results, Walsh and his colleagues pieced together the biography of the boy's brain. When he was just an embryonic ball, five lineages of cells emerged, each with a distinct set of somatic mutations. Cells from those lineages migrated in different directions, becoming different organs—including the brain.

The cells that joined to become the brain were transformed into neurons. And these new neurons wandered throughout the brain before settling down and dividing a few more times. That's why Walsh and his colleagues could find neurons belonging to different lineages sitting near each other. The boy's brain ended up divided into millions of patches of tiny cellular cousins.

Mosaics were once the stuff of superstitions, of freak shows. Then they gained recognition as diseases, both rare and common. Now we can see them everywhere. A single genome can no longer define us, because our inner heredity toys with DNA, altering just about every piece of genetic material we inherit. Even in our skulls, we grow a witches'-broom.

CHAPTER 13

Chimeras

I N 1779, John Hunter, the British anatomist, sent a letter to the Royal Society. He wanted to describe to them a peculiar sort of cow. If a mother gives birth to twins of the opposite sex, Hunter wrote, "the bull-calf becomes a very proper bull." The cow-calf, however, turns out very improperly. "They are known not to breed: they do not even show the least inclination for the bull, nor does the bull ever take the least notice of them," Hunter explained.

"This cow-calf is called in this country a *free martin*," he wrote, "and this singularity is just as well known among the farmers as either cow or bull."

By 1779, freemartins already had a long history. The Romans had called them *taura*. Farmers knew that a freemartin couldn't provide them money by producing calves or milk. But that didn't mean it was worthless. A freemartin could work almost as hard as an ox, and it brought a good price for its meat. "The flesh of a fatted free martin will fetch a halfpenny a pound more than any cow beef," according to the 1776 book *A Treatise on Cattle*.

A few years before Hunter became famous for dissecting Charles Byrne, the Irish Giant, he had studied freemartins. When he autopsied a freemartin calf, it looked to him like a normal female cow. But when Hunter got the opportunity to inspect a freshly slaughtered adult freemartin, he saw that it had undergone a bizarre change. On the outside, it still resembled an

370

370

ordinary female cow. But it now lacked ovaries. In their place, the freemartin grew what looked to Hunter like testicles. He concluded that freemartins were "unnatural hermaphrodites."

Later generations of anatomists didn't know what to make of freemartins either. Some argued that they developed from the same fertilized eggs as their brothers. Others thought freemartins and their brothers were fraternal twins, growing from two eggs instead of one. Some experts argued the freemartin was a cow that became bullish, or a bull that became cowish.

The true nature of freemartins was far stranger than they could imagine, but it would not emerge until the twentieth century. The discovery would ultimately challenge how we trace heredity from children to parents.

The first step toward deciphering freemartins came in the early 1900s, when a University of Chicago embryologist named Frank Lillie started dissecting cow fetuses supplied to him from the Union Stock Yards a few miles away. As part of his research, Lillie examined fraternal calf twins and discovered a strange feature of their development. The calves grew from two fertilized eggs that implanted themselves at different spots on the wall of their mother's uterus. Each then developed its own placenta, which pushed fingerlike growths into their mother's blood vessels. But Lillie also noticed that some of the placental blood vessels linked the calves together. Blood could flow from the mother into one calf, then out into its placenta, and then into the other calf. When Lillie injected ink into the umbilical cord of one twin, the placentas of both calves ended up dark.

In 1916, Lillie speculated that these hidden networks of vessels were responsible for freemartins. A fetal bull produced male hormones. If its fraternal twin was female, she could receive those hormones through their joined placentas. The chemicals would then bathe her sex organs and masculinize them. "Nature has performed an experiment of surpassing interest," Lillie concluded.

Lillie was right to see blood vessels as part of the solution to the freemartin mystery. But it was not hormones that transformed the female calves. The freemartins actually inherited cells from their brothers, which took hold in their bodies and grew, making them a combination of two different animals in one.

This insight would have to wait for another three decades. It would strike another midwestern biologist, by the name of Ray David Owen. He would look again at freemartins, and he would realize freemartins are cellular mergers.

————

C ows were Owen's life. His father came to the United States from Wales on a cattle boat transporting purebred Guernseys, and in Wisconsin he established a dairy farm. Owen grew up working long hours on the farm, witnessing the births and deaths of cows on a regular basis. School was an afterthought. Owen went to a two-room schoolhouse with a pair of teachers for eight grades. To pass the time while the older children did recitations, he would practice sewing.

When Owen began traveling to the nearest town for high school, his teachers generally assumed that he would go back to his family's farm afterward and tend to his cows. Only his English teacher, Miss Grubb, recognized his potential for something else. When she suggested he take French, Owen's vocational agriculture teacher snapped, "What the hell do you expect him to do, swear at the cows in French?"

A full scholarship from a small college nearby allowed Owen to continue his education, although he still came home from classes each day to do his farm chores. His family expected he'd become a schoolteacher. But as his graduation from college approached, Owen decided to become a biologist instead.

He went to the University of Wisconsin, where he would fill bushel baskets with chicken heads so that he could examine their irises. He inseminated naked pigeons to trace the genes that robbed them of their feathers. He studied how the germ cells of birds burrowed their way into the depths of embryos to find their proper anatomical place. This work ensured that Owen never forgot that development is more than just the multiplication of cells; it is a time of migrations as well.

After getting his PhD, Owen started working in 1941 for a genetics lab that supported itself by performing paternity tests on cows. "It was a kind of bio-business venture," he later said. Farmers from around the country were starting to get their cows inseminated with the sperm from champion bulls. They wanted to make sure their calves were inheriting the expensive pedigree they had paid for, rather than being fathered by some bull that randomly crossed paths with their cows one day.

Not only did the lab make money off the arrangement, but it ended up awash in cow blood. "They'd bleed the whole herd," Owen said.

For Owen and his fellow biologists, the blood was a scientific godsend. Each sample was accompanied by a wealth of information about the animal it came from, along with its relatives. They could measure different kinds of proteins in the blood—not just the proteins that produced the ABO blood type groups, but many others—and learn about how the cows passed down genes to their offspring. They could ask fundamental questions, such as whether complicated traits could be encoded by lots of simple genes, or genes that were linked together somehow. It all worked out very well. Careers were made.

Only one problem got in the way. "There was something funny about twin calves," Owen said.

To be more specific, there was something funny about freemartins. Owen would compare the blood proteins of freemartins to their twin brothers. Since they were fraternal twins, he expected that their proteins would be as different as those of any pair of siblings. Instead, the proteins from the freemartins and their brothers were identical. Despite being the opposite sex, they looked biochemically like identical twins.

Owen didn't know how to account for this result. As he puzzled over freemartins, a Maryland cattle farmer got in touch with Owen for some help. One morning, the farmer had bred a Guernsey cow with a purebred Guernsey bull. Later that day, a white-faced Hereford bull broke through a fence and bred with the Guernsey, too. Nine months later, the cow gave birth to twins.

"They were a remarkable pair," Owen later recalled, "because, while one was a female and looked as a Guernsey should, the other was a bull and had the dominant white-faced marking of the Hereford. It seemed evident, just from looking at this pair, that they were twins from different fathers."

The farmer asked if Owen could sort out the paternity. He sent Owen the blood of the calves, their mother, and both bulls. When Owen looked closely at the proteins in their blood, he discovered something no one had seen before. Both calves carried proteins matching both bulls.

Thinking back to Lillie's research, Owen speculated that each calf had been fathered by a different bull, but then their blood mixed together through their united placentas. He wondered how much they had blended. After all, red blood cells last for only a few months, replaced by cells from bone marrow. Owen decided to keep track of the Maryland calves as they grew older to see if they developed into normal animals.

Owen arranged to get more blood from the calves when they were six months old. Their blood still remained a mixture. Even on their first birthday, Owen was surprised to find, they still had blood proteins of both bulls. Owen realized that it wasn't a blood transfusion that the calves had given each other. They had transplanted their stem cells into each other's bone marrow.

With this discovery, Owen proved just how fragile our notion of heredity really is. We think of ourselves as having inherited our genes from our parents, brought together by a single egg and a single sperm into a single zygote, defined by a single genome. Now Owen had discovered cows whose bodies were made up of cells belonging to different lineages.

You could trace some of the Guernsey purebred calf cells back to its own original cell. But you could also trace some of its stem cells back into its Hereford twin. If embryologists were to draw the pedigree of their cells, they would have to draw two trees, with separate bases and intermingled branches. And if they were to trace the genes in those cells to the previous generation, some would go back to the Guernsey bull, and the rest to the Hereford. Despite breaking the rules of heredity, however, the calves were perfectly healthy. An amalgam of different cells from different parents, a divergent heredity, worked just fine.

Owen wondered if he had merely discovered a rare fluke. He inspected the blood of hundreds of pairs of twin calves. In 90 percent of the cases, he found, their blood was a mix. What made Owen's discovery all the more remarkable was that the immune systems of the twins didn't seem bothered by the blending. By the 1940s, blood transfusions had become standard medical practice, but only because doctors could very carefully avoid giving patients the wrong blood type and triggering deadly immune responses. Perhaps, Owen thought, an early exposure to foreign cells taught the immune system tolerance.

Owen published the story of the freemartins in October 1945, and on the strength of those findings Caltech offered him a job. He and his wife left behind the Wisconsin winters for Southern California, where Owen gave up his research on freemartins. Settling into a conventional lab, he turned his attention to rats instead, stitching together blood vessels from one rodent to another to see if they could also trade stem cells through their shared circulation.

His work on freemartins might have slipped into obscurity if it hadn't caught the attention of a British physician a few years later. Peter Medawar was, at the time, running pioneering experiments on transplantation. He fell into the line of research in World War II, hoping to find a way to treat Royal Air Force pilots covered in burns. Medawar found that if he took healthy skin from a patient's own body and cultured it, he could then graft it successfully to the wound. But if he transplanted tissue from other people, the skin usually died.

Sometimes Medawar would try to graft a second patch of skin from the same donor to the same patient. Now the patient rejected the skin in even less time. Medawar realized that the patient's immune system was attacking the transplant like an invading enemy and launching its assault faster as it grew familiar with the foreign tissue.

That discovery led Medawar to wonder how immune cells could tell the difference between self and nonself. He suspected that in the developing embryo, the immune system learned to recognize genetically encoded proteins on cells as an identity badge. If they later encountered cells without that right badge, they turned hostile. If that were true, then Medawar realized there was a simple way to find out. Identical twins, having the same genes, should accept transplants from each other. Fraternal twins and other siblings would be more likely to reject them.

Medawar and his colleagues traveled to a research farm in Staffordshire to run the test on cows. They punched out bits of skin from the cows' ears and inserted them into the withers of other cows. The experiment proved both a success and a failure. Siblings typically didn't accept transplants, while identical twins did. So far so good. But Medawar was surprised to find that fraternal twins—including freemartins—accepted transplants as well.

The results confused Medawar at first because, unlike Owen, he didn't understand cows very well. That confusion disappeared when he discovered Owen's research. As embryos, Owen had shown, fraternal twin calves trade cells with each other through their joined bloodstream. Medawar realized that their developing immune systems accept both kinds of cells as their own self. When Medawar gave an adult freemartin a skin punch from her brother, her immune system gave a molecular yawn.

Building on Owen's insights, Medawar went on to gain an even deeper understanding of the immune system. Ultimately, his research would open the way to the modern practice of transplanting organs. In 1960, Medawar won the Nobel Prize, but later he sent a letter to Owen complaining they should have shared the honor.

The skin-punch experiments told Medawar important things not only about the immune system but about heredity as well. Freemartins and other fraternal cow twins represented a kind of heredity not documented before, one in which the cells in their bodies belonged to more than one lineage. Medawar thought they deserved a name of their own. He called them chimeras.

The name echoed back thousands of years, to Greek myths that were probably inspired by strange births. Chimera was the name of a monster that had the front of a lion, the back of a snake, and a goat for its middle. For Medawar, the word also had a more recent resonance. Horticulturalists like Luther Burbank would sometimes graft the top of one plant onto the stem of another, creating so-called graft-hybrids. In 1903, a German botanist named Hans Winkler produced an exceptional graft: a tomato plant on one side and nightshade on the other. He chose to call his creation a chimera. Winkler's new name became familiar among botanists, but only for their handiwork.

Owen, Medawar declared, had discovered a natural animal version of Winkler's botanical monster: "a genetical chimaera."

―――――――

When Medawar published his skin-punch experiments in 1951, he didn't dwell much on whether chimeras were limited to cows—thanks to their unusual placentas—or if other animals might become chimeric, too. But two years later, Medawar got a letter from a London scientist named Robert Race announcing what looked like the discovery of the first human chimera.

We know this original human chimera today only as Mrs. McK. In the spring of 1953, Mrs. McK, then twenty-five years old, went to the Sheffield Blood Transfusion Centre in northern England and donated some blood. Before storing it, the center tested it for Mrs. McK's blood type. They added antibodies to the blood that would make type A blood cells clump together. Some of the cells clumped, but more of them didn't. The blood looked like what you'd get if you blended together type A and type O.

Ivor Dunsford, a doctor at the clinic, assumed there had been a mix-up. Maybe Mrs. McK had gotten a blood transfusion recently. Perhaps she was type O and had accidentally been given type A. But when he looked into the matter, he learned that Mrs. McK had never gotten a transfusion in her life.

Dunsford got in touch with the Medical Research Council Blood Group Unit in London for help. Robert Race, the director of the unit, was the country's expert on blood groups and relished puzzling cases like that of Mrs. McK. Dunsford supplied him with Mrs. McK's blood, and Race replicated the analysis. He ended up with the same result, separating her blood into O and A once more.

In all his years studying blood groups, Race had never seen such a thing.

It reminded him of Owen's discovery of twin cows that had traded blood cells. Eight years had passed since Owen had described the first genetical chimeras, but nobody had ever demonstrated whether humans could end up the same way. Race wrote back to Dunsford, instructing him to ask Mrs. McK if she had a twin.

When Dunsford relayed the question to Mrs. McK, she was startled. Indeed, she did have a twin brother. But he had died of pneumonia when he was three months old.

Race was intrigued by the news. "I suppose Mrs. McK is not obviously a freemartin," he mused to Dunsford. "Has she been pregnant?"

Race wondered if cells from Mrs. McK's brother had disturbed her sexual development, in the same way that bull calves render their freemartin sisters sterile. It turned out Mrs. McK had a son, and so her ovaries must be in good working order.

Race wasn't put off by the news. He and his colleagues continued to investigate the possibility that Mrs. McK was a chimera. They took more careful measurements of her blood types, and determined that she had two parts O to one part A.

Race then wrote to Medawar about the case. Fascinated by Mrs. McK, Medawar offered the expertise he had gathered studying chimeric cows. He suspected that Mrs. McK was a chimera, too, and thought up a way to test the idea.

Medawar knew that the gene for blood types—the ABO gene—was active not only in red blood cells but in the salivary glands, for reasons that still aren't clear. Medawar suggested to Race that he collect Mrs. McK's saliva and inspect its ABO proteins. They might offer a clue as to which version of the proteins was original to Mrs. McK, and which she had acquired from her brother.

Race's team discovered that her saliva was type O—the same type that made up about two-thirds of her blood. Race now had his answer: Mrs. McK had inherited type O genes from her parents, and then she had acquired some of her brother's type A stem cells in the womb. His cells established themselves in her bone marrow, where they still contributed to her blood supply.

On July 11, 1953, Dunsford, Race, and their colleagues published "A Human Blood-Group Chimera" in the *British Medical Journal.* "In 1916 Lillie wrote: 'In the case of the free-martin, nature has performed an experiment of surpassing interest,'" they recounted at the end of the report. "No doubt the same could be said of Nature's experiment on Mrs McK, were we able to appreciate more of its implications."

In later years, Dunsford kept Race up-to-date about Mrs. McK. The fraction of her blood cells that came from her brother slowly declined. When Race would think back on the case, he marveled that he had been able to determine the blood group of a boy who had been dead for a quarter of a century. When we say people are dead, it goes without saying that their cells are dead with them. Parents can cheat death in a way by using a few of their cells to create new lineages of cells, known as children. By these lights it was hard to know quite what to call Mrs. McK's brother. His infant heart had stopped beating after a bout of pneumonia. But by then his stem cells had nestled into his sister's bones months before, and decades later they were still casting off new blood cells.

Medawar dedicated several pages to the ghostly boy in an essay he called "The Uniqueness of the Individual":

There is no telling how long Mrs McK will remain a chimera, but she has now been so for twenty-eight years; probably, in the long run, her twin brother's red blood cells will slowly disappear, and so pay back the still outstanding balance of his mortality.

———

Three years after solving the mystery of Mrs. McK, Race was delighted to discover another pair of human chimeras. If he found enough examples, he said, he might be able "to lift the phenomenon out of the 'freak' category."

As years passed, more chimeras came Race's way. He recorded them in new editions of his book, *Blood Groups in Man*, up till the 1970s. In 1983, another researcher at the Blood Unit, named Patricia Tippett, followed up with a list of her own. She counted seventy-five cases of human chimeras in total. Tippett and other researchers suspected there were many more wait-ing to be discovered. At the time, the best clue that people were chimeras was that they carried two different blood types. But blood type tests were so crude in the 1980s that they would come up negative if one type made up less than a few percent of a person's blood.

In the 1990s, Dutch researchers invented a better test. They designed a collection of fluorescent tags, each of which could grab onto the cells of a certain blood type. They could spot the glowing tags even when a blood type was as rare as one in ten thousand cells. The scientists then used this new method to search for chimeras. They asked hundreds of parents of twins if they'd send blood samples to their lab. Using their new test, they discovered that 8 percent of the twins were chimeras. When scientists looked at triplets, 21 percent contained a pair of chimeras, too.

But these new blood type tests had their limits. If a pair of twins both had O blood, a blood type test cannot tell if they've mingled their cells. By the 2000s, scientists in search of chimeras were moving away from blood types, to DNA itself.

In 2001, a thirty-year-old woman in Germany discovered she was a chi-mera while she was trying to get pregnant. For the previous five years, she and her husband had been trying to have a baby. They were fairly certain the problem didn't lie with her biology, because she had gotten pregnant when she was seventeen and had had regular menstrual cycles ever since. A fertility test revealed that her husband had a low level of viable sperm, and so they made plans for IVF.

As a routine check, the woman's doctors took blood samples from her and her husband. They looked at the chromosomes in the couple's cells, to make sure neither would-be parent had an abnormality that would torpedo the IVF procedure. The woman's chromosomes looked normal—if she were a man. In every white blood cell they inspected, they found a Y chromosome.

Given that she had given birth, this was a weird result. And a careful exam revealed that all her reproductive organs were normal. To get a broader picture of the woman's cellular makeup, her doctors took samples of her muscle, ovaries, and skin. Unlike her immune cells, none of the cells from these other tissues had a Y chromosome in them. The researchers then carried out a DNA fingerprinting test on the different tissues, looking at the women's microsatellites—the repeating sequences that can distinguish people from one another. They found that her immune cells belonged to a different person than her other tissues.

It turned out that the woman had had a twin brother who died only four days after birth. Although he was unable to survive on his own, his cells took over his sister's blood and lived on within her.

As scientists learned more about chimeras, they also realized that there's more than one way to become them. In 1960, a girl was born in a Seattle hospital with a clitoris so enlarged it looked like a penis. She grew normally for the next two years, when she underwent surgery to have her clitoris reduced. At the time, doctors were beginning to appreciate how hormones given to pregnant women as medicine could alter the development of fetuses, causing them to become hermaphrodites. But when geneticists at the University of Washington gave the girl an exam, they realized that hers was another story.

A few clues struck them at first. One of her eyes was hazel, the other brown. When the doctors inspected the girl's ovaries, they found that the right one was normal, but the left looked more like a testicle.

The Seattle scientists took tissue samples from the girl's skin, her ovaries, and her clitoris. They carefully examined cells from each part of her body, counting up the chromosomes. Sometimes the cells had two X chromosomes. Sometimes they had an X and a Y. In the ovary, they found, all the cells were double X. But everywhere else, the cells were a mixture of XX and XY. When a blood group expert at the university named Eloise Giblett looked at the child's blood, she found a mixture of two types. But the genes for those two types could only have come from her father, not her mother.

Giblett and her colleagues realized that the girl's father had fertilized two of her mother's eggs with two of his sperm cells. One of his sperm cells carried a Y chromosome, the other an X. They also carried different variants for blood types. The two sperm fertilized two of her mother's eggs, and she became pregnant with a pair of fraternal twins. Under most circumstances, those twins would have gone on to become a brother and sister. But in this particular case, the two embryos fused early on into a single clump of cells. The cells from both twins were still totipotent at this stage, so they could develop into any tissue, given the right signals. Together, they produced a single healthy child.

Today, this girl would be called a tetragametic chimera—meaning that she arose from four sex cells (gametes) rather than the regular two. Tetragametic chimeras pose an even greater challenge to our conventional notions of heredity than Mrs. McK. Mrs. McK could point to another distinct person as the source of some of her cells. Tetragametic twins start out as two embryos with separate genomes, and then merge entirely. Only one child is born, and there's no other human being to point to. All we can do is trace their intimately mingled cell lineages to their separate sources.

If two embryos of the same sex form a tetragametic chimera, it's much easier for them to go undetected. The cells from the two twins seamlessly merge together to produce a girl or a boy with ordinary genitals. Only a close inspection of their DNA will reveal their true heredity. And even when the results are clear, people may refuse to believe them.

———

In 2003, a woman in Washington State named Lydia Fairchild had to get a DNA test. Fairchild, who was then twenty-seven, was pregnant with her fourth child, unemployed, and single. To get welfare benefits, state law required that she prove that her children were genetically related both to herself and to their father, Jamie.

One day, Fairchild got a call from the Department of Social Services to come in immediately. A DNA test had confirmed that Jamie was the father of the three children. But Fairchild was not their mother.

The Social Services workers began interrogating her, suspecting her of a crime. Perhaps she had stolen the children. Perhaps she was running some kind of surrogate-mother scam. In any case, she must be guilty of welfare fraud. Fairchild was told her children would be taken away and she would go to jail.

Fairchild desperately tried to prove that the children were hers. She gathered their birth certificates, showing she had delivered them in local hospitals. She contacted her obstetrician to vouch for her. "I saw them come out," her mother later told ABC News. Six decades earlier, Charlie Chaplin couldn't get courts to look at DNA evidence to decide parenthood. Now DNA was the only evidence the courts would accept, and it was telling Fairchild something that couldn't be true. Fairchild's father confessed that despite his trust in his daughter, the tests gave him doubts. "I have always had faith in DNA," he later said.

Most lawyers did, too, and it took Fairchild a long time to find one willing to ignore the test results and take on her case. He persuaded Fairchild's judge to order two more DNA tests. The new results also ruled out Fairchild as the mother of her children. When Fairchild was rushed to a hospital to deliver her fourth child, a court officer was there to witness the birth. The officer also oversaw a blood draw for a DNA test. The results came back two weeks later. Once again, Fairchild's DNA didn't match her child's. Even though the court officer had witnessed the child's birth, the court still refused to consider any evidence beyond DNA.

It began to look like there was nothing more to be done. The state prepared to put Fairchild's children into foster care and prosecute her for fraud. But then Fairchild's lawyer read about another mother who had been informed that her children were not her own. In Boston, a woman named Karen Keegan had developed kidney disease and needed a transplant. To see if her husband or three sons were a match, her doctors drew blood from the whole family in order to examine a set of immune-system genes called HLA.

A nurse called Keegan with the results. Not only were her sons not suitable as organ donors, but the HLA genes from two of them didn't match hers at all. It was impossible for them to be her children. The hospital went so far as to raise the possibility she had stolen her two sons as babies.

Since Keegan's children were now grown men, she didn't have to face the terrifying prospect of losing her children as Fairchild did. But Keegan's doctors were determined to figure out what was going on. Tests on her husband confirmed he was the father of the boys. Her doctors took blood samples from Keegan's mother and brothers, and collected samples from Keegan's other tissues, including hair and skin. Years earlier, Keegan had had a nodule removed from her thyroid gland, and it turned out that the hospital had saved it ever since. Her doctors also got hold of a bladder biopsy.

Examining all these tissues, Keegan's doctors found that she was made up of two distinct groups of cells. They could trace her body's origins along a pair of pedigrees—not to a single ancestral cell but to a pair. They realized Keegan was a tetragametic chimera, the product of two female fraternal twins.

The cells of one twin gave rise to all her blood. They also helped give rise to other tissues, as well as to some of her eggs. One of her sons developed from an egg that belonged to the same cell lineage as her blood. Her other two children developed from eggs belonging to the lineage that arose from the other twin.

When Lydia Fairchild's lawyer heard about the Keegan case, he immediately demanded that his client get the same test. At first, it looked as if things were going to go against Fairchild yet again. The DNA in her skin, hair, and saliva failed to match her children's. But then researchers looked at a sample taken from a cervical smear she had gotten years before. It matched, proving she was a chimera after all. Fairchild got to keep her children.

The stories of Lydia Fairchild and Karen Keegan both ended happily. But they left the women with haunting questions not only about their families but about themselves. Fairchild's eggs, cervix, and perhaps some other tissues in her body all had a direct genetic link to her children. But what of the rest of her body? Was she partly their aunt, too? As for Keegan, were her sons half brothers to each other, with two sisters for their mothers? We use words like *sister* and *aunt* as if they describe rigid laws of biology. But despite our genetic essentialism, these laws are really only rules of thumb. Under the right conditions, they can be readily broken.

Speaking years later to National Public Radio, Keegan admitted that telling her sons about the test results was the hardest part of the experience. "I felt that part of me hadn't passed on to them," she said. "I thought, 'Oh, I wonder if they'll really feel that I'm not quite their real mother somehow, because the genes that I should've given to them, I didn't give to them.'"

———————

I t might have been some consolation to Keegan to learn that her sons were probably chimeras as well. They probably carried some of Keegan's own cells in them. And she was probably a twofold chimera, carrying some of her own children's cells inside her.

As an embryo's placenta draws in nutrients from its mother, it blocks the cells in her blood with a tight filter. But it's not perfect, and sometimes a mother's cells end up in the embryo. Other times, the traffic goes the other way.

It was in 1889 that a doctor first took note of this traffic. A German pathologist named Christian Georg Schmorl examined the bodies of seventeen pregnant women who had died of seizures. He noticed that their livers contained some "very peculiar" cells. Judging from their size and shape, Schmorl guessed they had originally come from the placentas of their unborn children.

It was easy to dismiss those cells as pathological, dislodged by the women's disorders. But in 1963, Rajendra Desai and William Creger, two doctors at Stanford University, discovered that this traffic might be a regular part of pregnancy. They collected blood from nine pregnant women and spiked it with a drug called Atabrine. Although Atabrine was initially used to prevent malaria, it also proved to be useful to research scientists who wanted to track cells. Certain types of cells will swallow up Atabrine and then glow green under a fluorescent light.

Desai and Creger injected the women's Atabrine-treated blood back into their bodies and waited for them to give birth. The doctors then examined the umbilical cords of the babies. When they smeared the cord blood onto slides and lit them up, six of the nine babies glowed green. The mothers' white blood cells were crawling around the bloodstreams of their children.

Three years later, Desai and colleagues in Boston ran the reverse experiment. They took advantage of blood transfusions that fetuses sometimes get when they develop anemia in the womb. Desai added Atabrine to blood that was to be transfused into seven fetuses. A few hours after the procedure, he drew blood samples from the mothers. In almost every case, Desai found green-glowing white blood cells and platelets that had been injected into the fetuses. The mothers were becoming chimeras, taking on their children's blood.

————

D esai's experiments proved that the placenta was a leakier barrier than scientists had previously thought. But it was hard to know just how significant the migrant cells were in their new homes. Perhaps they simply died out soon after making the crossing. It would take three decades for other scientists to show that these cells can endure, and that mothers can become permanent chimeras with their children.

These results had their origins in a failed attempt to come up with a test for Down syndrome. In the 1970s, the only way to test for Down syndrome was to pierce the amniotic sac surrounding a fetus with a needle and draw off some fluid. The fluid contains some cells shed by the fetus, which geneticists could inspect to look for chromosomal abnormalities. But this test, known as amniocentesis, had many drawbacks. It sometimes falsely indicated a fetus had Down syndrome, and it sometimes failed to find genuine cases. Making matters worse, inserting a needle into the uterus put women at greater risk of a miscarriage.

A Stanford University researcher named Leonard Herzenberg decided to invent a blood test to take the place of amniocentesis. He would draw blood from pregnant women, which—as Desai had shown—contained some cells from the fetus. He could then examine the fetal cells without ever disturbing the woman's pregnancy.

The big challenge of Herzenberg's project was to find a way to separate fetal cells from maternal cells that was both quick and accurate. Herzenberg and his students figured out how to put fluorescent tags on the HLA proteins that sit on the surface of cells. They used only tags that would attach to the HLA proteins a child inherited from its father and not shared by its mother. That step would ensure that only the child's cells would glow.

In 1979, Herzenberg and his students demonstrated that they could use their new method to sort the cells of a fetus out of the bloodstream of its mother. To improve the method even more, one of Herzenberg's students, Diana Bianchi, carried on research at Tufts University. In her own lab, she came up with a new strategy. Herzenberg had tagged lots of different types of fetal cells. Bianchi developed a tag that would mark only the stem cells that give rise to red and white blood cells. In adults, these stem cells are locked away in bone marrow and never slip into circulation. Any stem cells in a pregnant woman's blood would almost certainly have been shed by her fetus.

Bianchi crafted a new set of molecular tags, which she successfully used to fish out fetal stem cells. She was delighted with her success—until some of the pregnant women she had studied started giving birth.

From some of the women, Bianchi had drawn out stem cells with Y chromosomes. This was to be expected from women who were pregnant with boys. But when some of these women gave birth, their babies turned out to be daughters.

Even more startling to Bianchi were the results of a control experiment she ran on women who were not pregnant. Some of those women had Y chromosomes, too. All of them, Bianchi learned, had given birth in the past to sons.

As a search for a blood test, Bianchi's study was a bitter failure. She could not reliably isolate fetal cells from a current pregnancy. But Bianchi got a fabulous consolation prize: She discovered that fetal cells can survive for years in women.

Bianchi decided to keep studying these cells by finding more mothers with sons. The women she selected for her research had never had a blood transfusion or an organ transplant. Out of eight such women, Bianchi found fetal cells with Y chromosomes in six of them. One of the women with Y chromosomes had a twenty-seven-year-old son—meaning that his cells had remained established in her body for over a quarter century.

When Bianchi wrote up her results, they were rejected by three journals. The reviewers complained that it didn't make sense that fetal cells could endure for so long in another person's body. Finally, *Proceedings of the National Academy of Sciences* agreed to publish her results in 1996. "Pregnancy may thus establish a long-term, low-grade chimeric state in the human female," Bianchi and her colleagues wrote.

———

To distinguish this new form of chimerism from other ones, Bianchi coined a new term: *microchimerism*. In the years since her paper came out, other scientists confirmed that most mothers experience microchimerism. Y chromosomes were the easiest markers for this condition. But some researchers also began looking in mothers for other segments of DNA from their children. Their research has revealed that all pregnant women have fetal cells in their bloodstream at thirty-six weeks. After birth, the fraction drops, but up to half of mothers still carry fetal cells in their blood decades after carrying their children.

These microchimeric cells swim against heredity's current, a legacy in reverse. Other forms of chimerism play different games with heredity. Very often, a mother's cells will infiltrate her children's bodies, where they can endure and grow long after her death. According to one estimate, 42 percent of children end up with cells from their mothers.

Chimeras even create side flows in heredity. At the University of Copenhagen, scientists got blood samples from 154 girls ranging in age from ten to fifteen. They cracked open the cells in the blood and searched them for Y chromosomes. In 2016 they reported that twenty-one girls—more than 13 percent—had them. Because the girls did not have sons of their own, the scientists concluded that their Y-chromosome–carrying cells originated in their brothers, were left behind in their mothers after birth, and then made their way into the bodies of the girls while they were still fetuses. It's also possible that the cells came from male fetuses that their mothers miscarried or had aborted.

Charting the full scope of microchimerism is difficult, because foreign cells can work their way remarkably far into the nooks and crannies of the human body. Just because chimeric cells aren't present in blood doesn't mean they're not hiding in some hard-to-reach organ. The best way to hunt for chimeras is to cut open a cadaver.

In 2015 a group of researchers at Leiden University in the Netherlands did just that. They searched Dutch hospitals for tissue samples taken from women who had been pregnant with boys when they died, or who had died within a month of giving birth to sons. The researchers found twenty-six such women and collected samples from their kidneys, livers, spleens, lungs, hearts, and brains. At least some of the women had their son's cells in every organ. Out of seventeen hearts they examined, five were chimeric. Out of nineteen lungs, all nineteen were. The cells of their sons were also in their brains, five for five.

Autopsies of older women have also shown just how long fetal cells can endure in a mother. Lee Nelson, a rheumatologist at the Fred Hutchinson Cancer Research Center, and her colleagues examined the cadavers of fifty-nine women who died, on average, in their seventies. In 63 percent of the women, the scientists found Y chromosomes in their brains.

Fetal cells don't simply migrate around their mothers' bodies. They sense the tissue around them and develop into the same types of cells. In 2010, Gerald Udolph, a biologist in Singapore, and his colleagues documented this transformation with a line of engineered mice. They altered the Y chromosomes in the male mice so that they glowed with the addition of a chemical. Udolph and his colleagues bred the mice, and then later they dissected the brains of the mothers. They found that the fetal cells from their sons reached their brains, sprouted branches, and pumped out neurotransmitters. Their sons helped shape their thoughts.

———

C himeras took much the same scientific path as mosaics: from monster to fluke to fixture. And as scientists came to recognize that a substantial fraction of humanity are amalgams of cells from different individuals, they wondered what effects their divided inheritance had on them.

In 1996, Lee Nelson proposed that microchimerism might make some mothers sick. With half their genetic material coming from their father, fetal cells might be a confusing mix of the foreign and the familiar. Nelson speculated that being exposed to fetal cells for years on end could lead a woman's immune system to attack her own tissues. That confusion might be the reason that women are more vulnerable to autoimmune diseases such as arthritis and scleroderma.

To test this possibility, Nelson and Bianchi collaborated on an experiment. They picked out thrity-three mothers of sons, sixteen of whom were healthy and seventeen of whom suffered from scleroderma. Nelson and Bianchi found that the women with scleroderma had far more fetal cells from their sons than did the healthy women. Other scientists who carried out similar studies got the same results for a number of other diseases. These findings aren't definitive proof that microchimerism made these women sick, however. It was also possible that the diseases came first, and the fetal cells only later flocked to the diseased tissues, where they could multiply.

It's also possible that being a chimera can be good for your health. Bianchi's first clue that chimerism might have an upside came in the late 1990s, when she was searching for fetal cells in various organs. She discovered a mother's thyroid gland packed with fetal cells carrying Y chromosomes. Her gland was badly damaged by goiter, and yet it still managed to secrete normal levels of thyroid hormones. The evidence pointed to a startling conclusion: A fetal cell from her son had wended its way through her body to her diseased thyroid gland. It had sensed the damage there and responded by multiplying into new thyroid cells, regenerating the gland.

In another woman, Bianchi discovered that an entire lobe of her liver was made up of Y-chromosome–bearing cells. Bianchi was even able to trace the paternity of the cells to the woman's boyfriend. She had had an abortion years before, but some of the cells from the fetus still remained inside her. When her liver was damaged later by hepatitis C, Bianchi's research suggested, her son's cells rebuilt it.

It's also possible that fetal cells help mothers fight cancer. In 2013, Peter Geck of Tufts University and his colleagues looked for cells with Y chromosomes in the breast tissue from 114 women who died of breast cancer and 68 women who died of other causes. Fifty-six percent of the healthy samples had male fetal cells in them. Only 20 percent of the cancerous tissue had them. Geck speculated that fetal cells swooped into the niches of breast tissue that are good for proliferating cells. Those may be the same niches cancer cells need to find in order to grow into tumors.

As chimerism rises out of the freak category, it also raises unexpected ethical questions. Somewhere around a thousand children a year are born to surrogate mothers in the United States alone. As Ruth Fischbach and John Loike, two bioethicists at Columbia University, have observed, the rules for surrogacy are based on an old-fashioned notion of pregnancy. They treat people as bundles of genes. As a society, we are comfortable with a woman nourishing another couple's embryo and then parting ways with it, because she does not share the hereditary bond that a biological mother would. If the pregnancy goes smoothly, the surrogate mother is supposed to leave the experience no different than before the procedure.

But Fischbach and Loike observed that a surrogate mother and a baby may end up connected in the most profound way possible. Cells from the fetus may embed themselves throughout her body, perhaps for life. And she may bequeath some of her cells to the child. This is not merely a thought experiment. In 2009, researchers at Harvard did a study on eleven surrogate mothers who carried boys but who never had sons of their own. After the women gave birth, the scientists found Y chromosomes in the bloodstreams of five of them.

Fischbach and Loike don't argue that surrogacy should be banned because of chimerism. But they do think that surrogates-to-be need to give informed consent that's truly informed. It may come as a surprise to them that their own DNA could have a long-term influence on the health of an unrelated child, and that they may end up with some of the child's cells— complete with a separate genome. These women need to know that heredity's tendrils can't be pruned as easily as we might imagine.

The Tasmanian devil couldn't have been dead long.

Elizabeth Murchison got out of her car in a cool, damp gully. It was the summer of 2006, and Murchison had just spent a week hiking the Central Highlands of Tasmania. Now she appreciated the shady break from the heat. She noticed a cloud of flies swirling over a black creature the size of a Jack Russell terrier. A wound on its neck was still trickling blood, presumably from a recent collision with some vehicle. Murchison turned over its body, hoping to see something. And there it was: a pea-size lump of pink flesh swelling from its face.

Murchison had grown up in Tasmania listening to the howls of devils in the night. When it came time for college, she headed to mainland Australia to study genetics, and in 2002 she traveled to Cold Spring Harbor in New York to earn her PhD. Murchison studied microRNAs, molecules made by our cells to silence genes when needed. Not surprisingly, she was the only Tasmanian in Cold Spring Harbor, and so she often found herself answering questions about Tasmanian devils. She would explain how the real animals were nothing like the Looney Tunes version, the drooling cartoon tornado that swept across American television screens. Tasmanian devils were actually the largest living species of marsupial predators, and a tough species at that. They have a habit of biting off pieces of each other's faces, whether in a fight over food or a courtship.

Tough as they might be, though, Tasmanian devils were in trouble. A singular epidemic was sweeping across the island, not quite like anything veterinarians had seen before. A devil would develop a fleshy growth in or around its mouth. In a matter of weeks, the growth would balloon, and within a few months, the animal would starve or suffocate.

These growths were first observed in 1996 in the northeast corner of Tasmania, and over the next few years they spread over most of the island, killing off tens of thousands of devils. By the early 2000s, the species looked like it might become extinct in a matter of decades, killed by a disease scientists didn't understand. It would take them years to realize that these devils were chimeras, and that their cancers descended from the cells of a long-dead animal.

———

Murchison and her fellow graduate students would try to guess at the diagnosis for the devils. On an individual animal, the disease looked like a tumor (hence its name, devil facial tumor disease). But, as Boveri first recognized, a typical tumor is a mosaic. It arises from an animal's own body, thanks to a series of mutations that pile up along a lineage of cells. The reports of new devil facial tumor cases formed a wavelike pattern, as if they were a contagious epidemic.

Some of Murchison's colleagues guessed a virus was to blame. Some viruses do cause cancer by infecting cells and disrupting their biochemistry. But once a virus infects its host, it can take years to produce cancer. Devil facial tumor disease was moving far faster. An even bigger problem with the virus hypothesis arose when scientists closely examined the tumors themselves. The tumor cells were not mosaics, descended from a devil's own zygote. They had an entirely different pattern of bands in their chromosomes, indicating that they came from a different animal altogether. The devils were actually chimeras.

Soon afterward, an Australian geneticist named Kathy Belov led a more detailed study on the disease. She and her colleagues sequenced microsatellites from tumors and healthy tissues taken from a number of devils. The DNA fingerprint from tumor cells didn't match the healthy cells from the same devil's body. Instead, they matched cancer cells from devils who died dozens of miles away. It was as if all the sick devils had gotten a cancer transplant from a single tumor.

When Murchison encountered the roadkill devil in the Tasmanian forests, she decided then and there she would study the tumors. At the time, powerful new technologies for reading genomes were just becoming affordable enough for small groups of scientists to use. In 2009, Murchison brought samples from the roadkill devil to the Wellcome Trust Sanger Institute in England to read its DNA more closely than anyone had read the tumor DNA before.

———————

The genome in the tumors had undergone many changes, but Murchison could trace back its history through earlier generations. She found a pair of X chromosomes in the cells, and no trace of a Y chromosome. So the cancer must have begun in a female devil. Murchison then looked for clues about which kind of cell it started out as. She sequenced the microRNAs from tumor cells. The cells were making a combination of molecules typically only found in one kind of cell. It's a type of nerve known as a Schwann cell, which normally wraps an insulating sleeve around other neurons to help them send their signals.

At some point in the early 1990s, Murchison's research showed, a single

Tasmanian devil in the northeast corner of the island got cancer. The mutations may have initially arisen in a Schwann cell. The descendants of that original cancer cell grew into a tumor. During a fight, another devil bit off the tumor. The cancer cells did not end up digested in the attacker's stomach. Instead, they likely lingered in the devil's mouth, where they were able to burrow into the cheek lining and work their way metastatically into the other tissues in the devil's head.

The cancer cells continued dividing and mutating, until they broke through the skin of the second devil's face. At some point, that new victim also got bit, and its own attacker took in the cells from the original cancer. A single carrier could pass the cancer on to several other devils if it was especially aggressive, and thus help accelerate its spread. Passing through host after host, the tumor cells gained about twenty thousand new mutations.

As strange as it had become, the devil's contagious cancer has not escaped the bonds of heredity. All the tumors that grew in thousands of Tasmanian devils were united along an unbroken line to their female Schwann cell ancestor. But the conventional language we use to describe heredity fails to describe what is happening in Tasmania.

August Weismann had crystallized our conception of heredity by giving germ cells a chance for immortality. Now a batch of somatic cells from a mammal had gained an immortality of its own by moving out of its original body and ending up in new ones. The original cancerous Schwann cell died inside a Tasmanian devil decades ago, as did the subsequent victims of its cancer. But the tumors endured because they were in the right place for another Tasmanian devil to bite off a chunk of them and let them live on in a new home.

Murchison's research showed just how cobbled together heredity is. It's not a cosmic imperative but a process that emerged from biological ingredients and has been modified into new forms. And yet, on their own, devil facial tumors might have ended up as little more than a philosophical curiosity. It was easy to dismiss them as the product of a weird species restricted to a remote corner of the world. In fact, contagious cancers were cosmopolitan, spanning the planet.

M urchison knew that there was at least one other form of contagious cancer: a disease in dogs known as canine transmissible venereal tumor (CTVT for short). It's a particularly ugly disease, causing oozing tumors to grow around the genitals. The disease is found in many countries and is most common in stray dogs. Because it's disgustingly obvious, pet owners quickly take their animals to the vet for treatment.

While devil facial tumors only came to light in the 1990s, the symptoms of CTVT were already known by 1810. In a book called *A Domestic Treatise on the Diseases of Horses and Dogs*, the British veterinarian Delabere Blaine described a "fungous excrescence" that formed around the genitals of dogs. At the time, some physicians believed the growths were caused by bad humors, while others thought they spread like the plague.

It wasn't until 1876 that a Russian veterinarian named Mstislav Novinski carried out the first successful transplant of the cancer. He cut pieces of a canine transmissible venereal tumor from a sick dog and inserted them in the skin of two puppies. After a month, each puppy had a tumor of its own. Novinski then cut off a piece of one of the new tumors and grafted it to another dog. The second-generation tumor grew as well.

Later generations of scientists carried out more detailed versions of Novinski's experiment. In 1934, a veterinarian and a pathologist reported carrying the cancer forward through eleven dogs. But they could succeed only when they used living cancer cells. Dead cells wouldn't do the trick, nor the fluid from tumors. In a few cases, the cancer became metastatic and killed the dogs, but most of the time it grew for a few months before disappearing. "The origin of the cells forming this tumor is unknown," the researchers wrote. They suspected that sometimes dogs could spontaneously develop the cancer and then pass it to other dogs through mating.

Cancer biologists paid CTVT little mind, because it didn't work like standard cancer—which is caused either by random mutations or by certain viruses. But the discovery of contagious cancers in Tasmanian devils led a number of scientists to take another look. In 2006, a British biologist named Robin Weiss and his colleagues collected CTVT tumors from forty dogs on five continents. They sequenced short pieces of DNA from each one, along with healthy cells from other parts of the dogs. Weiss and his colleagues found that the dogs were also chimeras, not mosaics. All the cancer cells shared a common set of genetic variants—variants that couldn't be found in the healthy cells in any of the dogs. Rather than cropping up from time to time in different dogs, the scientists realized, the disease had arisen just once.

M urchison followed up on Weiss's study with one of her own. Instead of looking at snippets of DNA, she and her colleagues sequenced the entire genomes of two tumor cells. One came from a cocker spaniel in Brazil and the other from an Aboriginal camp dog in Australia. Each tumor cell had just over 100,000 unique mutations not shared by the other one. But the two cells shared 1.9 million mutations in common that Murchison didn't find in ordinary dog DNA.

The cancer cells still carry clues about their origins. Murchison suspects that they descend from some kind of immune cells, based on the genes they use and the ones they keep quiet. That signature may be a distant echo from the origin of CTVT: An immune cell in an ancient dog turned cancerous, and eventually the cancer found a way to other dogs.

Murchison and her colleagues also discovered that CTVT is far, far older than devil facial tumor. Mutations accumulate in cancer cells at a roughly clock-like rate. Based on the mutations in different CTVT cells, Murchison and her colleagues estimated that the cancer arose in a single dog about eleven thousand years ago, around the end of the Ice Age.

If it had been an ordinary tumor, CTVT might have endured for a few years, until its host died. Instead, it escaped mortality's trap and endured a thousand times longer than a typical dog. In different parts of the world, the tumor cells acquired new mutations that they passed on to their descendants. Those mutations sketched out a map of the tumor's journey. It began somewhere in the Old World; when Europeans brought dogs on ships to new continents like Australia and North America, they brought CTVT with them. While devil facial tumor may be limited to a corner of the world, humans have spread CTVT everywhere they brought their dogs.

————

Two cases of contagious cancer would have been extraordinary enough. But as Murchison was studying dogs and devils, another case came to light—not on land, but in the sea.

In the 1960s, marine biologists discovered a leukemia-like cancer in some species of clams. Their immune cells multiplied explosively, invading all of the animal's tissues. It typically took just a few weeks for the cancer to claim a clam's life.

The biologists had never seen anything quite like the cancer before. What made it especially baffling was how quickly new cases turned up. Once one clam in a population got sick, it didn't take long before almost all the animals died. This cancer roared up the East Coast until, by the 2000s, it was killing soft-shell clams from New York all the way north to Prince Edward Island in Canada.

Two biologists at Columbia University, Stephen Goff and Michael Metzger, studied the clams to see if a virus was triggering them to form tumors. They found no sign of cancer-causing viruses. And when Metzger and his colleagues looked at DNA from the cancer cells, they were flummoxed to find that the cells in different clams shared a common set of genetic markers. It looked as if the scientists had stumbled across a third contagious cancer.

To test this idea, Metzger brought clams into Goff's lab and put them into tanks. If he put a healthy clam in a tank with a cancerous one, it became cancerous, too.

Surveying cancer cells from along the East Coast of North America, Metzger could find mutations in some of them and not others, allowing him to draw a genealogical tree. Some long-ago clam had developed leukemia, Metzger and Goff concluded, and it expelled some of its cancer cells into the ocean currents. As nearby clams filtered seawater for bits of food, they also sucked in the cancer. Over the years, the outbreak had hopscotched its way hundreds of miles from its origin.

Metzger and his colleagues knew that marine biologists had been finding fast-spreading cancers in some other species of clams as well as other bivalves. In all these cases, scientists had tried to find a cancer-causing virus and had come up empty-handed. Metzger got in touch with colleagues around the world and was able to get hold of three sickened species to study: mussels, cockles, and golden carpet shell clams. All three of their cancers also turned out to be contagious.

The cockles proved to have not one strain of contagious cancer but two. And the cancer in the golden carpet shell clams had particularly weird DNA: It didn't start out as a golden carpet shell clam. Instead, it came from pullet shell clams, another species that lives in the same intertidal beds off the coast of Spain as the golden carpet shell clam. That discovery led the scientists to inspect pullet shell clams, but they failed to find any of the cancer cells in them. They concluded that a strain of contagious cancer must have started in pullet shell clams and then jumped species, infecting golden carpet shell clams. It killed off all the vulnerable pullet shell clams, leaving only resistant ones behind.

Meanwhile, back in Tasmania, biologists were busy trying to protect the devils from extinction. They built colonies of healthy animals at zoos and turned a small island off the coast of Tasmania into a refuge. If the devil facial tumor did kill off all the wild devils of Tasmania, the protected animals could be brought back. But in March 2014, a University of Tasmania graduate student named Ruth Pye made an ominous discovery. She was analyzing cancer cells from a devil caught south of the city of Hobart and found something in them that shouldn't have been there: a Y chromosome. Murchison and her colleagues had already established that the progenitor of the disease had been a female devil. Pye's new cancer looked as if it had started in a male.

A close examination of its entire genome revealed a wealth of unique mutations not found in any of the other tumors. The data pointed to an inescapable conclusion: Pye had discovered a second devil facial tumor. Later she and her colleagues found another seven sick devils in the region with the same form. They dubbed the two lines of contagious cancer DFT1 and DFT2.

While DFT1 arose around 1990, DFT2 has fewer mutations of its own, suggesting it emerged more recently. It's a worrying finding, because it means that the process by which contagious cancers emerge in devils may be more common than previously thought. If conservation biologists manage to save the devils from DFT1 and DFT2, who's to say they won't face DFT3 before long?

From the cancers identified so far, scientists can start developing hypotheses for how they emerge and manage to break heredity's rules. It looks as if the ability to become a contagious cancer is not limited to one particular type of cell. It just has to start down the road to cancer. One of the biggest challenges to a developing tumor is the immune system, which continually patrols the body for suspicious-looking cells. Cancer cells can evolve a range of tricks to evade detection. They can cut back on their production of surface proteins such as HLA that immune cells recognize in order to tell good cells from bad, for example.

No one knows yet what sort of adaptations cancer cells need to survive the first journey from one host to another. Perhaps it all comes down to an accidental mutation that keeps them from desiccating in the dry air. Once they've made the first jump, their immune evasions become even more valuable, because now they are clearly foreigners, rather than traitors from within. And once the contagious cancer cells move into a new host, some of the adaptations they gained in their normal existence as a tumor may turn out to help them once more. They can send out signals to the surrounding tissue, hypnotizing cells into helping them get nourishment.

Contagious cancer is not all that different from an ordinary tumor that becomes metastatic and spreads from one organ to another. The new organ is, in effect, another animal. But unlike ordinary tumors, contagious cancers no longer face an inescapable death. Instead of gaining a few years' worth of mutations, they can gain centuries of them. After eleven thousand years circulating among dogs, for example, CTVT has acquired an impressive arsenal of mutations in genes linked to immune surveillance. And just like ordinary cancer cells, CTVT cells have stolen mitochondria to replace their own. The only difference is that they steal mitochondria from a series of dogs—at least five different dogs over the past two thousand years. From the days of the Roman Empire onward, CTVT has recharged itself like a vampire, with the youth of its canine victims.

————

Contagious cancers were a scientific secret hiding in plain sight for two centuries. Once scientists realized what they were and started to wonder if there were others, they started finding more. It's almost certain that the eight cases identified so far are not the last. No one knows just how many there will be. It's possible that some species will prove to be more prone to them than others. The small population of Tasmanian devils may make them vulnerable, because they have a low level of genetic diversity. In such a species, it's easier for a contagious cancer to hide inside new hosts, because it doesn't look so suspiciously foreign. Contagious cancers also need an easy route from one host to the next. Dogs have long bouts of sex that can last half an hour and leave them with broken skin. That's a promising route for a tumor adapted to growing around the genitals. But we shouldn't underestimate the lengths that contagious cancer cells can go to in order to get into new hosts. After all, now we know they can swim.

Indeed, even if scientists come across more cases of contagious cancers, the true rate may be far higher. Perhaps, like gene drives, contagious cancers are regularly emerging from animals, only to disappear—either because the immune systems of their hosts evolve a potent defense or because they drive their hosts to extinction.

All of which raises the disturbing question: Could we humans develop a plague of contagious cancer? I'm not talking about an epidemic of cancer-causing viruses, such as HPV, which can cause cervical cancer. I'm talking about human cells taking up residence in other humans: tumors that travel across the social planet.

The scientific literature is sprinkled with intriguing cases of cancer cells moving from one person to another. A transplanted kidney turned out to contain a tumor. A surgeon accidentally nicked his hand, allowing his patient's skin cancer to slip into his body. A few pregnant women with leukemia passed some of their cancerous immune cells to their fetuses. In an exceptionally bizarre case, a man in Medellín, Colombia, got infected with a tapeworm from which a cell grew into a tumorlike mass.

Yet all of these cases only document cancer making a single leap. There's no evidence of the same lineage of cancer cells moving on to a third victim. It may be that our immune systems are so strong that cancers never get the chance to evolve into parasites that can leap from host to host.

Why our immune system should react so violently to foreign tissue is puzzling. For the most part, our immune system is exquisitely adapted to particular threats. Our cells can sense invading viruses and commit suicide to stop their infection. We can make antibodies to destroy a strain of bad bacteria while sparing our beneficial bugs. These are the products of evolution, which gave our ancestors better odds of surviving and reproducing. But our ancestors were not constantly transplanting lungs and spleens into each other.

So why should our immune system be so well prepared to respond? One way to make sense of this enigma is the risk of contagious cancers. Perhaps our early animal ancestors 700 million years ago regularly faced invasions of parasitic offshoots of other animals. They had to fight off the heredities of others, or die and lose their own.

PART IV

Other Channels

CHAPTER 14

You, My Friend, Are a Wonderland

WHEN THE DARK of moonless nights arrives, the one-fin flashlight fish emerges from its hiding place.

This fish (scientifically known as *Photoblepharon palpebratus*) lives in the waters off the Banda Islands, a scattered archipelago in Indonesia. It spends the daylight hours resting in caves a hundred feet or more underwater. When the sea turns black, the fish swims out of its caves, up to the surface waters. As it hunts for little invertebrates, its body emits a cream-colored light.

Like any animal, the flashlight fish is actually a collection of organs. Its skin acts as a barrier, protecting it from the surrounding sea. Its gills draw oxygen. Its stomach digests its prey. Each organ is distinct, because its cells produce a distinctive collection of molecules and operate a distinctive network of genes. The light made by the fish comes from a pair of jelly-bean–shaped structures under each eye. To produce their light, the cells in those jelly beans manufacture proteins that glow.

Shining a light might not seem a very smart thing to do when you're a small fish swimming in a sea of predators. But flashlight fish can actually use it to escape their enemies. To flee, they dash straight ahead for a while, their light organs tracing a forward-moving line. They then roll each light organ into a pocket in their head. The fish suddenly go dark and then break away from their straight line, leaving predators barreling forward into empty water.

Their light organs thus help the fish survive long enough to reproduce. The males cast sperm into the water to fertilize female eggs, which develop into larvae and finally into mature fish. *Like engenders like* is as true for flashlight fish as for any other animal. Every new generation developed fins like their ancestors did, along with eyes, jaws, and gills. And light organs.

In 1971, a pair of scientists—Yata Haneda of the Yokosuka City Museum and Frederick Tsuji of the University of Pittsburgh—journeyed to the Banda Islands to investigate the one-fin flashlight fish. In the evenings, they would push off from shore in a canoe. Eventually they would extinguish their lights and gaze into the water, searching for the fish's gleam. When Haneda and Tsuji caught some in a net, they put the animals in jars of seawater and later dissected the light organs. Even after they were carved out of the animals, the organs still glowed. (Banda fishermen use the light organ as a lure, putting it on a hook, where it can glow for hours.)

Haneda and Tsuji inspected the light organ cells under a microscope to understand how they worked. It was possible that the fish produced their light like fireflies. Fireflies carry a gene in their DNA for a protein called luciferin. The insects store luciferin in the cells in their tail. When they want to send a signal to their fellow fireflies, they use other proteins to alter the luciferin, unleashing its stored light. Fireflies inherit the luciferin gene from their parents, along with the rest of their genes.

But the one-fin flashlight fish has no genes for making light, Haneda and Tsuji discovered. The glowing cells in their light organ do not belong to the fish itself. They are bacteria.

They are not just any bacteria, however. If you look in the light organs of any one-fin flashlight fish, the glowing microbes always belong to the same strain, known as *Candidatus* Photodesmus blepharus. And if you want to find *Candidatus* Photodesmus blepharus, the one-fin flashlight fish is the only species on Earth where you'll find it. The same waters around the Banda Islands are also home to a nearly identical species, the two-fin flashlight fish, with its own bacteria-loaded light organ. But the bacteria glowing inside them is different. Each species of flashlight fish inherits its own exquisitely rare microbial partner.

All species of animals and plants are shot through with microbes, ourselves included. By one estimate, each human being contains about 37 trillion human cells and about the same number of bacteria. It's easy to ignore our bacterial half, because human cells are hundreds of times bigger than microbes. Yet that's no reason to ignore them. We have thousands of species of bacteria within us, each carrying thousands of its own genes that are fundamentally unlike our own. In this respect, we are no different from any other animal—any Portuguese man-of-war, any desert scorpion, any elephant seal. We're not even very different from a sugar maple tree or an evening crocus.

The bacteria we're most familiar with are those that cause diseases, chewing through our skin and raging through our guts. Yet even in the best of health, we are still rife with permanent lodgers. Some harmlessly cling to their hosts, scavenging molecular scraps. Others perform tasks on which their hosts depend for their survival. If the flashlight fish had no bacteria, it would have no flashlight. Other microbes carry out tasks that are harder to see but no less important. They synthesize vitamins, they nurture a well-tempered immune system, they form a living barrier against dangerous pathogens. The microbiome, as this collective is known, blurs any simple notion of what it means to be an individual organism. If we turned into true individuals, sterilized of our microbiome, we'd become sick and might well die.

In each species, every new generation acquires a microbiome. In some regards, this cycle of renewal looks a lot like heredity. A new animal does not acquire its own genes out of the blue, synthesized from scratch. Its genes have been duplicated over and over again inside the cells of its ancestors, taking an extraordinary journey to get to each new animal. A man, for example, starts off as a zygote full of genes, which are copied each time that fertilized egg divides. The genes end up in totipotent cells, and then in pluripotent cells, and then in cells destined for different tissues. Some of those cells end up as germ cells. As these cells migrate through the body, they bring their genes to the region where the testicles will later develop. Years later, the descendants of these cells may develop into long-tailed sperm, each containing only one copy of each of the man's genes. Although a man makes billions of sperm over the course of his life, only a tiny fraction of them at most will ever manage to leave his body and enter a woman's reproductive tract, and fewer still will deliver his genes into an egg.

The genes of bacteria can take strikingly similar routes of their own through the generations of their hosts. One of the most remarkable of these journeys takes place thousands of feet underwater, where thick beds of vesicomyid clams thrive around cracks in the seafloor. The clams soak up hydrogen sulfide—the toxic chemical that gives rotten eggs their awful smell—rising out of the cracks. They absorb it into their muscles, and then their circulatory system delivers the compound to their gills. Special cells in the gills—cells that don't exist in other species of clams—split sulfur atoms from the hydrogen sulfide molecules, releasing the energy stored in their bonds. The clam uses this energy to combine carbon, hydrogen, and oxygen into sugar molecules. The clams act much like trees, except that they capture a subterranean chemical energy instead of sunlight.

To be precise, it's not the clams themselves that seize the seafloor's energy. The specialized cells in their gills are actually bacteria. They carry the gene for an enzyme that can break down hydrogen sulfide. In exchange for this service, the clams supply the bacteria with a well-appointed home. Without those bacteria, the clams would starve; without the clams, the bacteria would barely eke out an existence.

This relationship is remarkable for many reasons, not the least of which is geography. Because vesicomyid clams can grow only where hydrogen sulfide seeps out of the seafloor, clam beds may be miles from each other. The clams broadcast their sex cells into the surrounding water, and after fertilization the clam larvae drift through the sea. Most of them will land in the marine desert and die. Only a few will end up at a site where they can grow. They bring with them the bacteria they need to survive, as their ancestors did.

To understand how the clams manage to hold on to their partners is difficult, because it's nearly impossible to rear a deep-sea creature in the comfort of a laboratory. Instead, scientists haul up clams from the seafloor and pick apart their dead bodies for clues. In 1993, S. Craig Cary and Stephen Giovannoni of Oregon State University mapped the bacteria inside clams by searching for their DNA. They found some in the gill cells that house the microbes. But they also found some bacterial DNA in the ovary-like organs where clam eggs develop. Somehow the bacteria were traveling through the clams from the gills to the eggs, which they could invade in order to get into the next generation. The new clams are born infected, inheriting an expanded set of genes—some animal and some bacterial.

It's hard not to wonder what Darwin would have thought of these clams. When he pictured heredity, he saw gemmules streaming from across the body to the germ cells, coming together to carry on the body's traits to the next generation. His theory of pangenesis turned out to be wrong, and biologists set it aside as one of his exceptional blunders. They turned instead to August Weismann's stark division between the germ line and the soma. Now researchers are finding that deep-sea clams use a gemmule-like form of heredity to carry their partners into the future.

If these deep-sea clams were the only species on Earth to inherit a vital trait this way, it might be possible to dismiss them as an oddity, in the same way it was once possible to dismiss contagious cancers as a Tasmanian fluke. But they have company. Many animals ensure that essential bacteria get inside their eggs. And some of them—like cockroaches—are a lot easier to study than deep-sea clams.

Among the microbes that live inside cockroaches is one called *Blattabacterium*. Just like clams, cockroaches develop special cells inside of which *Blattabacterium* can dwell. Instead of feeding on chemical-laden seawater, cockroaches graze on organic matter on land, able to survive on what they find on the floor of a forest or a New York apartment. *Blattabacterium* is essential to their global conquest. As the insects eat, they store away nitrogen in an organ in the cockroach abdomen, called the fat body. Inside the fat body are some cells infected with *Blattabacterium*. The bacteria convert that nitrogen into amino acids and other building blocks that the cockroach needs to grow.

Sometimes the cells housing *Blattabacterium* will take a trip. They crawl out of the fat body and seek out the cockroach's eggs. They attach themselves to the eggs for a few days before ripping themselves open. Their resident bacteria spill out, to be swallowed up by the eggs, so that a new generation of cockroaches can continue to conquer the world.

————

These strictly in-house bacteria—known as endosymbionts—did not always live this way. Their ancestors lived outside of hosts. The free-living cousins of endosymbionts have helped researchers learn about how some bacteria have evolved into such intimate partners. In case after case, the microbes took a gradual slide into a life inside.

This research shows that when the free-living ancestors of endosymbionts came into contact with a host—be it a roach, a clam, or one of millions of other species—they could grow on it by good fortune, or even inside it. By sheer coincidence, these bacteria provided some benefit to their hosts— perhaps casting off a useful amino acid in their waste. If their hosts did well as a result, the bacteria had more opportunity to reproduce in them. Natural selection favored bacteria that could do their hosts more favors because their interests were becoming aligned. Likewise, their hosts evolved to nurture the bacteria. Evolving special cells to shelter the bacteria ensured that animals could enjoy their services.

As the bacteria grew ever more pampered, genes that were once essential in the outside world became useless. Mutations that broke these newly superfluous genes no longer guaranteed extinction. The bacteria became genetically streamlined, their genomes shrinking in size by 90 percent or more. Some endosymbionts have lost the ability to do just about anything except the one thing that their host can't do.

Both the bacteria and their animal hosts became trapped in an evolutionary rabbit hole from which there is no escape. Once they were locked in symbiosis, their evolution began to follow identical paths. When an insect species split in two, its endosymbionts split as well. Their evolutionary trees became mirror images, with identical branches splitting from each other for tens of millions of years.

The story of the one-fin flashlight fish is a lot like those of the clams and the cockroaches. It also builds a special shelter—the light organ—where its bacteria can thrive. Every new generation of flashlight fish inherits a fresh supply of the same species of bacteria. In effect, the fish are expanding their genomes to include light-producing genes. Those genes just so happen to belong to a separate species. As the bacteria have adapted to life inside light organs, they have lost 80 percent of their genome.

There is one important difference, however. A female flashlight fish does not carefully move bacteria inside her body, transferring them from her light organ into her eggs. Her offspring hatch from their eggs lacking the microbes they need to glow. To gain their own flashlight, they have to get infected.

Each day, as an adult one-fin flashlight fish hunkers down in a cave, it sheds some of its bacteria. While *Candidatus* Photodesmus blepharus has lost most of the genes required to live outside an animal, it still clings to a few. Some of the genes enable it to build tails it can whip back and forth to swim through the sea. It also still retains genes for making chemical-sensing proteins, which it likely uses as a molecular nose, sniffing its way to young flashlight fish that it can invade. Ultimately, though, it's up to the fish to let the bacteria into their light organ. They've got a strict admission policy: The same waters also teem with the bacteria that give light to the two-fin flashlight fish, but those microbes can't get in.

This pattern of heredity is looser than the strict transmission of bacteria in clams and cockroaches. And yet it still embodies some of heredity's essential features. In the journey from one generation to the next, the bacteria and their genes aren't neatly bundled together with the host's genes in an egg. But the outcome is the same: A combined genome continues to produce a cream-colored light in the Banda Sea in each new generation, as it has for millions of years.

Our own microbiome is yet another step away from standard heredity. We don't develop a special pouch that exists only to be packed with one species of bacteria. If you give antibiotics to a vesicomyid clam and destroy its sulfur-feeding bacteria, it will die. But there's no single species of bacteria upon which our own life depends. In fact, there's not even a single species of bacteria that we humans all share. We house personalized zoos.

I got an intimate appreciation for our variety a few years ago when I went to a science conference. Wandering from talk to talk, I encountered a biologist named Rob Dunn who waved a Q-tip in front of me. He asked if I'd give him a sample from my belly button for a survey he was carrying out. I am the sort of person who says yes to such requests without missing a beat, and so within a few minutes I was in the nearest men's room, knocking out lint from my navel and swiping it with Dunn's Q-tip, which I dropped into a plastic vial of alcohol.

Dunn and his colleagues collected hundreds of these vials and extracted DNA fragments out of each of them. Most of those fragments were obviously human. But some belonged to bacteria. Dunn and his colleagues searched for matching sequences in online databases to figure out which species they came from. In my belly button, they found fifty-three species of bacteria. When Dunn sent me a spreadsheet with my personal navel catalog, he added a message. "You, my friend, are a wonderland."

Having fifty-three species of bacteria in one's navel is nothing special, I should point out—Dunn and his colleagues have found twice as many in some other people. To get an overall sense of this diversity, the scientists analyzed results from sixty people. All told, they identified 2,368 species. None was present in every person. Eight species were present in at least 70 percent of Dunn's subjects. But 92 percent could be found only in 10 percent or less of the subjects. The majority was found only in a single person. When I looked over my spreadsheet, I could see that seventeen of my species were unique to me. One type, called *Marimonas*, had only been known from the Mariana Trench, the deepest spot in the ocean. Another, called *Georgenia*, lives in the soil. In Japan.

On discovering this, I e-mailed Dunn to let him know I'd never been to Japan.

"It has apparently been to you," he replied.

The weirdness of my spreadsheet stems from our profound ignorance of the microbial world. Microbes are unimaginably diverse, with thousands of species in a single spoonful of soil. Although microbiologists have been naming species of bacteria for well over a century, they've described only a tiny fraction of the Earth's single-celled diversity. *Marimonas* was named for a deep-sea microbe, but the lineage likely includes many other species adapted to other environments, including human skin.

Within the howling complexity of the human microbiome, however, you can still hear heredity's signal. It begins when a mother seeds her children's microbiome. When this seeding starts is still not clear. Although researchers have long held that embryos start off sterile, in a bacteria-free amniotic sac, a few studies have hinted that at least some maternal bacteria may slip into the fetal sanctuary. What is abundantly clear, however, is that once a baby starts moving down through the birth canal, it gets contaminated. The bacteria growing on the canal walls slather the baby in a microbial coat. Some of the bacteria grow across its skin, while some slip into the mouth and make their way to the gut.

A nursing mother can inoculate her baby with even more bacteria. Breasts foster microbes, which they allow to mix into their milk. Small-scale studies suggest that the strains that move most successfully through nursing into the babies are especially good at breaking down milk sugar and converting compounds in milk into vitamins that babies need. Mothers appear to play favorites, promoting certain species of bacteria in their babies and filtering out others. While breast milk contains a lot of nutrients that a baby can absorb, it also contains certain sugars, called oligosaccharides, that are indigestible. Indigestible by humans, to be specific. Certain strains of gut bacteria delight in oligosaccharides, multiplying in the guts of nursing infants. Mothers may thus transmit microbes to future generations in a heredity-like way.

To see how closely our microbes have followed us through the generations, a University of Texas microbiologist named Howard Ochman and his colleagues looked far back in evolutionary time. They compared the microbiomes of humans and those that live in our closest primate relatives: gorillas, chimpanzees, and bonobos. (Bonobos are a species of ape that split from the ancestors of chimpanzees about two million years ago.) They found that many lineages of bacteria that live in the human gut do not exist in the guts of our fellow apes. Instead, those apes have their own related strains of bacteria.

When Ochman and his colleagues compared the evolutionary trees of the hosts and the bacteria, they lined up closely, branching in the same patterns. Microbes in chimpanzees tend to be more closely related to ours than those that live in gorillas—just as chimpanzees themselves are our closest kin. Ochman's study suggests that for more than fifteen million years, our ancestors have been in a tight coevolutionary dance with their microbiome.

As hominins split off from other apes, they adapted to new kinds of diets, and their microbes may have adapted as well. Our ancestors evolved ways to foster only the strains of bacteria that belonged to us and not to other species. The oligosaccharides in human milk are different from those in the milk of other mammals. They may be adapted to foster some of our own strains of bacteria, shutting out others that can grow in other species.

In a few cases, bacteria get passed down so loyally from human parents to children that they can serve as rough genealogical records. A species known as *Helicobacter pylori* adapted long ago to life in the human stomach. Impervious to the digestive juices we make, it guzzles glucose in the food we eat. How a microbe can get from one human stomach into another is a mystery, but epidemiological studies show that infections with *H. pylori* start early in childhood. The bacteria have been found in the plaque on people's teeth, carried there by refluxes into their mouths. It's possible that mothers and other family members infect babies by transmitting the bacteria from their mouths to the children.

Whatever route *H. pylori* takes, it's a tremendously successful one. By some estimates, it lives in the stomachs of over half the people on Earth. Before the advent of antibiotics, that figure might have been closer to 100 percent. A small fraction of people who carry the bacteria will go on to develop ulcers and gastric cancer, but *H. pylori* is, for the most part, our friend. It sends signals to the developing immune system in children, helping it learn how to respond carefully to threats rather than overreacting and harming our own bodies. In billions of people's stomachs, the microbe grows and divides. The mutations that it accumulates along the way have allowed scientists to draw an evolutionary tree of the bacteria.

The history recorded in its branches bears a striking resemblance to the history of our own species. *H. pylori* first colonized humans in Africa more than 100,000 years ago, and people carried it around the world with them. If you want to know something about your ancestry, you can look at your own genes. But you can also get some clues from the *H. pylori* that you inherited from your ancestors.

Children do not inherit all their microbes only from their mothers, or even just their families. They can pick up bacteria from friends' toys they stick in their mouths, from teachers who wipe the dirt off their cheeks, even from the air they breathe. Yet even the bacteria that move freely from stranger to stranger also become intertwined with our own heredity.

To see this intertwining, you first have to think about our microbiomes as a heritable trait, just like our height, intelligence, and risk of getting a heart attack. And you have to study it as such. Julia Goodrich, a microbiologist at Cornell University, and her colleagues did just this, investigating the microbiomes of twins to see how their genetic similarities influenced the species they carried.

The scientists collected stool samples from 1,126 pairs of twins and cataloged the microbial inhabitants. Out of thousands of species of bacteria, they identified twenty that were more strongly correlated in identical twins than in fraternal ones. In other words, if one identical twin carried a particular species, the other twin was more likely to carry it, too. The scientists found that some species were more heritable than others. The most heritable of all was a kind called *Christensenella*. Goodrich and her colleagues estimated its heritability at around 40 percent. That's on par with moderately heritable traits, such as anxiety.

These results suggested that the genes we inherit from our parents help determine which microbes we end up harboring. To investigate this possibility further, Goodrich and her colleagues took a different approach: They scanned people's genomes, looking for people who shared certain variants and certain kinds of bacteria. They discovered that people with one variant have a high population of microbes belonging to a group of species called bifidobacteria.

The nature of that genetic variation hints at why it favors bifidobacteria. It controls a gene for a protein we use to break down a sugar called lactose. Babies make lots of this protein—called lactase—to break down the lactose in breast milk. The majority of children stop making lactase as they shift to eating solid food. But others have a different genetic variant that lets them continue making lactase, allowing them to digest milk sugar into adulthood.

Bifidobacteria thrives on the lactose that doesn't get digested by the time it reaches the large intestine. People who can take it up tend to have fewer bifidobacteria. But those who shut down their lactase wind up feeding a bigger population of microbes.

It's not so clear why *Christensenella,* the most heritable bacteria of Goodrich's study, is heritable. Perhaps that mystery has something to do with the fact that the microbe was only discovered in 2012. Scientists have determined that *Christensenella* breaks down a variety of sugars, and other types of bacteria feed on its by-products.

There are hints that *Christensenella* acts like a gatekeeper, helping to control how much of the energy in our food actually gets to our body instead of to our microbiome. One clue comes from looking at who carries *Christensenella* and who doesn't: Lean people are more likely to carry it than overweight ones. Another clue emerged from an experiment Goodrich and her colleagues ran on mice. They infected baby mice with *Christensenella* and then waited for them to grow to adulthood on a regular diet. The bacteria left the animals slim. Mice without *Christensenella* put on 15 percent more weight and ended up with 25 percent body fat. Mice with *Christensenella* gained only 10 percent more weight and reached 21 percent body fat.

These findings raise the possibility that we have to take the microbiome into account to understand why a trait such as weight is heritable. Some of the genetic variants behind the heritability of weight may not directly influence how our cells store fat. Instead, we inherit a variant that fosters *Christensenella* in our guts. It's the bacteria that take it from there.

———

There is one species of bacteria that has merged snugly into our bodies— even more snugly than the microbes that give the flashlight fish its light. This microbe has actually merged into our heredity, becoming such an intimate part of our existence that for decades many scientists refused to believe it started out as a free-living organism. I'm speaking of mitochondria, the tiny pouches that produce fuel inside our cells.

Mitochondria first came to the attention of biologists in the late 1800s as they developed new chemicals for staining the interior of cells. The stains revealed that the cells of animals were packed with mysterious granules. A German biologist named Richard Altmann published an entire book on these strange objects, filled with loving drawings of extraordinary accuracy. Altmann was astonished by how much the granules looked like bacteria. Not only were they shaped like bacteria, but sometimes Altmann's stains revealed them dividing in two like bacteria. Altmann developed an obsessive conviction that these granules were alive. He called them "elementary organisms." Altmann believed that cells themselves came into existence when these granules assembled into colonies and built a shelter of protoplasm around themselves.

The idea sounded absurd to other biologists. They rejected it so completely that Altmann turned into a bitter recluse. He would slip in and out of his lab through a back door, avoiding all human contact. His colleagues began referring to him as "the ghost." In 1900, Altmann died under mysterious circumstances at age forty-eight.

"Things went from bad to worse," the biologist Edmund Cowdry wrote cryptically in a 1953 history of mitochondrial research, "and the end was tragic and of the sort expected."

Cowdry forgave Altmann his error about mitochondria, "for the similarities between them and bacteria really are remarkable," he said. Ultimately, though, Cowdry and most other researchers judged the similarities only superficial. Mitochondria were simply parts of the cell, their construction encoded by the cell's own genes.

Years of subsequent research revealed that mitochondria performed an essential job: They use oxygen and sugar to create a cell's fuel supply. Researchers also discovered that mitochondria were shared not only by all animals but also by plants, fungi, and protozoans—in other words, by all eukaryotes. Tracing these lineages on the tree of life revealed that mitochondria must have evolved in the common ancestor of eukaryotes, some 1.8 billion years ago.

In the early 1960s, an astonishing fact about mitochondria came to light: They contained more than just proteins. Scientists also discovered they store their own DNA—if only a little. Human mitochondria have only thirty-seven genes, compared to about twenty thousand protein-coding genes in the nucleus. Nevertheless, the discovery of mitochondrial DNA baffled scientists. Our cells have many compartments—lysosomes for breaking down food molecules, for example, and the endoplasmic reticulum for moving proteins around the cell. But of them all, only mitochondria have their own set of genes.

Lynn Margulis, a biologist at the University of Massachusetts, argued that there was only one way to make sense of the discovery: It was time to revisit the old theories of Altmann and other early cell biologists. The evidence pointed to mitochondria starting out as free-living bacteria, and still holding on to a few of their original genes.

Margulis would be proven right. Starting in the 1970s, scientists began sequencing mitochondrial DNA. When they looked for the most similar genes in other species, they found that mitochondria most resembled bacteria. They were even able to narrow the genetic resemblance down to one lineage in particular, a group of species called alphaproteobacteria.

Before gaining their mitochondria, the evidence now suggests, our ancestors were microbes that survived by slurping some kind of molecular debris from their surroundings. About 1.8 billion years ago, a small species of bacteria—an alphaproteobacteria, to be specific—ended up permanently inside of them. Living alphaproteobacteria have given scientists inspiration for ideas about how this merger happened. Some researchers have argued that the alphaproteobacteria slipped into the larger cells as parasites. Their host did their best to destroy the invaders, but the alphaproteobacteria evolved defenses. In time, they stopped spreading from cell to cell. When their host divided, the alphaproteobacteria wound up in both the daughter cells.

Other scientists have proposed that the two microbes lived side by side at first. They traded essential nutrients, helping each other thrive. The closer they were to their partners, the more reliably they could exchange these gifts. Eventually, they merged entirely.

Whichever is the case, gaining mitochondria marked one of the great leaps in the evolution of life. A cell now could harvest the fuel made by its new lodgers. The more mitochondria a cell could house, the more energy it could use. This symbiosis spiraled upward, allowing eukaryote cells to become far bigger, far more complex, than any cell before. Instead of feeding on molecular debris, eukaryotes now had enough fuel to chase after bacteria and engulf them. Later, these single-celled predators began clinging together, evolving into multicellular creatures.

Ensconced in their new home, mitochondria followed the same path that endosymbionts so often do. They abandoned many of the genes they had once needed to live freely on their own. Yet mitochondria never gave up their own form of heredity. Altmann might have been wrong to think that mitochondria were independent life-forms. But he was right to think of bacteria when he saw mitochondria dividing. Within a cell, a mitochondrion will sometimes split in two, and the daughter mitochondria inherit copies of its DNA, just as their free-living ancestors did nearly two billion years ago.

When our own cells divide, their daughter cells inherit a portion of their mitochondria, which keep dividing over the course of our lifetime. Our bodies don't get overrun by mitochondria because our cells sometimes destroy them, keeping their numbers in check. Our deaths bring an end to the lineages of mitochondria in our bodies; the only ones with a chance to escape to the future are those that dwell in women's eggs. A man's mitochondria have no future, because their sperm destroy them during fertilization.

The fact that mitochondria are inherited only down the maternal line makes their DNA a powerful genealogical tool. It allowed some scientists to reunite the family of the Tsar Nicholas. It allowed others to reunite all living humans, tracing our mitochondrial DNA to a single woman in Africa 150,000 years ago. Yet mitochondria's distinctive patterns of heredity have also created deep confusion.

When mitochondria copy their DNA, they can make mistakes and introduce mutations. Some of those mutations can disrupt their fuel-generating assembly line, while others can cause devastating hereditary diseases. They can make eyes go blind, ears go deaf, muscles waste away. Many of these mitochondrial diseases went overlooked for decades by geneticists, because they flouted Mendel's Law. In some families, a disease will only sporadically strike relatives over the generations. In other families, the same disease will reliably occur in every child of a mutation-carrying mother.

It wasn't until the late 1980s that scientists began pinpointing the genetic basis of mitochondrial diseases. Since then, they've identified hundreds of these disorders, which together afflict one in four thousand people. Strangely, though, these people often have relatives who carry the same mutations in their mitochondria but don't suffer the same symptoms.

This confusion dissolves when you bear in mind that mitochondria are our resident bacteria, following their own rules of heredity. If a single mitochondrion mutates, the cell that carries it will continue functioning normally, because it still has hundreds of other healthy ones. When the cell divides, one of its daughter cells inherits that one mutant mitochondrion. As the mutant mitochondrion itself divides, it becomes a bigger burden on cells. When the number of mutant mitochondria rises above a certain threshold, a cell will start to fail.

Mutant mitochondria can continue to become more common from one generation to the next. A woman with low levels of mutant mitochondria may give birth to children who cross the threshold into a full-blown mitochondrial disease. Thanks to chance, some of her children may get sick, while others remain healthy.

Studying mitochondrial diseases may eventually lead scientists to an answer to the biggest question about their heredity: Why does it follow only the maternal line? We all need mitochondria, males and females alike, to stay alive. Sperm need mitochondria to power their swim toward conception. Scientists have discovered a few species in which both parents pass down their mitochondria to their offspring. Ink cap mushrooms are one. Geraniums are another. In mussels, sons inherit mitochondria from both parents, while daughters inherit them only from their mothers. But in the overwhelming majority of species, fathers never pass down their mitochondria.

All these clues hint that there must be some powerful advantage to limiting mitochondria to the maternal line. It's possible that this kind of heredity evolves because mixing together mitochondria from two parents can be a disaster for children. In 2012, Douglas Wallace, an expert on mitochondrial diseases at the University of Pennsylvania, and his colleagues injected mitochondria from one healthy line of mice into the cells of a genetically distinct line. They then used those blended cells to produce mouse embryos. When the animals became adults, they suffered a host of problems, especially in their behavior. The mice became stressed-out, lost their appetite, and did badly at learning their way out of a maze.

Limiting mitochondrial heredity to one parent may help organisms move ahead in the evolutionary race. And once a species restricts mitochondria to eggs, mothers sometimes evolve ways to inspect their eggs, eliminating ones with too many mutations. The bacteria that sometimes infected our ancestors have now become so much a part of our heredity that their quality is the standard by which new human lives can come into existence.

CHAPTER 15

Flowering Monsters

T IM OVER HERE has the original Linnaeus flower."
I had come back to Cold Spring Harbor, this time for its plants.
A transplanted Englishman named Robert Martienssen met me in
front of his lab, and we spent the morning admiring his mats of duckweed
and tall stands of experimental corn. We went to one of the laboratory
greenhouses to meet the farm manager, Tim Mulligan. He brought with
him a black plastic pot with a flower.

Mulligan set it down on a counter made of planks, and I leaned in to
inspect it. The pot contained a single plant, sprouting a dozen or so bright
yellow blossoms. The flowers looked to me like miniature herald trumpets.
The petals wrapped around each other to create a long, closed tube. Each
tube curled out at the end, forming a spiked, five-sided rim.

It was a lovely plant, but if I encountered it on a walk through a meadow,
I might well have crushed it under my boot. To Martienssen, however, it
was one of the most interesting organisms in the world. It represented an
enduring mystery about heredity and the forms it can take.

The flower I was looking at has a clear-cut pedigree. It's a direct ancestor
of a plant that was discovered in 1742 by a Swedish university student
named Magnus Zioberg. Zioberg was hiking on an island near Stockholm
when he happened to notice a trumpet-flowered plant. It confused him,
because—aside from the flowers—it looked like a familiar plant called

toadflax. The flowers of normal toadflax plants have a mirrorlike symmetry. They grow a few small yellow petals, some sprouting off to the left and others to the right, and a spike develops at the base of these flowers, pointing toward the ground. The flowers on the plant that Zioberg stumbled across had a circular symmetry instead.

Zioberg plucked the flower out of the ground, pressed it in a book, and brought it back to Uppsala University to show to his professor Olof Celsius. Celsius was thunderstruck. He immediately brought the flower to his colleague—and one of the most important naturalists in history—Carl Linnaeus.

Linnaeus was working at the time on a new system for classifying all plants and animals. It's the system we still use today. To classify plants, Linnaeus paid particular attention to the shape of their flowers. When he looked at Zioberg's discovery, he thought Celsius was having a joke at his expense. Celsius must have glued flowers from another species onto a toadflax plant to fool him. But Celsius assured Linnaeus it was genuine.

Zioberg had found a monster, Linnaeus decided. But such monstrous flowers were supposed to be sterile, and Linnaeus discovered that Zioberg's specimen was fertile, growing the structures it would need to produce viable seeds. Linnaeus became even more astonished the closer he studied the structure of the flowers. They were unlike anything that Linnaeus—or any botanist before him—had ever seen. He begged Zioberg to go back to the island and bring him back some plants that were still alive.

Zioberg did so, and returned to Uppsala with a living plant that still had intact roots and stems. It was planted in the university's botanical garden, but it languished and died. Linnaeus desperately made the most of the flower's brief existence, writing down a wealth of observations. He produced a long report on that single plant, his surprise radiating off each page.

"This is certainly no less remarkable than if a cow were to give birth to a calf with a wolf's head," Linneaus declared. He considered the trumpet-shaped flower a species of its own. He named it *Peloria*—from the Greek for "monster."

To make sense of this "amazing creation of nature," as he called it,

Linnaeus speculated that it descended from ordinary toadflax. Pollen from another species had fertilized a toadflax plant, somehow triggering a sudden leap into a new form. To say such things in the 1740s—a century before Mendel's and Darwin's work—verged on heresy. Species were supposed to be fixed since creation. Heredity could not abruptly change course and make a new species.

"Your *Peloria* has upset everyone," a bishop wrote in an angry letter to Linnaeus. "At least one should be wary of the dangerous sentence that this species had arisen after the Creation."

In his later years, as he studied other specimens, Linnaeus became less sure of what the plants really were. He discovered that sometimes a single *Peloria* plant grew a mix of monstrous trumpet flowers and ordinary mirrorlike ones. He couldn't decide whether they were indeed a species of their own or some kind of strange variant that defied botany's rules.

Peloria would continue to intrigue later generations of botanists. Reared in botanical gardens across Europe, the plant went on passing down trumpet flowers to later generations. Goethe, who was just as interested in flowers as he was in poetry, made sketches of *Peloria* alongside toadflax flowers. Hugo de Vries thought for a time he might discover proof for his mutation theory in *Peloria*. The monstrous flower must have arisen through a mutation to an ordinary toadflax, he believed, creating a new species in a single jump.

Peloria refused to surrender to such an easy explanation. If a mutation really had produced the plant's trumpet flowers, it would have rewritten a piece of toadflax DNA. Later generations of *Peloria* would have inherited that mutation. Instead, the descendants of the original *Peloria* plants sometimes grew ordinary mirror flowers and sometimes monstrous trumpet ones, displaying no clear pattern that Mendel would have recognized.

In the late 1990s, a group of English scientists turned their attention to *Peloria*, using the tools of molecular biology. Enrico Coen of the John Innes Centre in England and his colleagues examined a gene involved in making flowers, called L-CYC. In order for ordinary toadflax plants to develop flowers, they must switch on the L-CYC gene in the tips of their stems. In *Peloria*, Coen discovered, L-CYC stays silent.

This difference is not due to a mutation that altered the gene for L-CYC in *Peloria*. Coen and his colleagues found that the gene is identical in toad-flax and *Peloria*. The difference between them was not *in* their DNA but *around* it.

Coen found a different pattern of methylation around the L-CYC gene in *Peloria* and in normal toadflax. In *Peloria*, L-CYC had a heavy coating of methyl groups, preventing the flower's gene-reading molecules from reading it. Coen and her colleagues noticed that as they bred new *Peloria* plants, they sometimes produced flowers that looked more like those of regular toadflax. When the scientists inspected the L-CYC gene in these throw-backs, they found that the gene had lost some of its methylation, allowing it to become more active again.

In *Peloria*, it seems, heredity has traveled down two channels. The flower has passed down copies of its genes, which guided the development of toadflax-shaped plants. But these plants also inherited a peculiar pattern of methylation that was not encoded in their genes. At some point before Zio-berg stumbled across it in 1742, a toadflax plant accidentally added on methyl groups to its L-CYC gene. By silencing the gene, this methylation caused the flower to develop into a new shape. This newly altered flower then produced seeds, which inherited the same epigenetic mark. They fell to the ground, sprouted, and produced the same monstrously lovely flowers. Over the centuries that followed, some of their descendants lost the epigenetic mark, blooming into ordinary toadflax flowers once more. But other *Peloria* plants continued to inherit the wolf's head of botany.

When I visited Martienssen, he was starting an experiment on *Peloria*. No one knew exactly how these plants kept inheriting the epigenetic marks for their monstrous flowers for so many generations. Martienssen had an idea for how to find out, but his experiment almost didn't happen. When he asked Coen where he could get a supply of *Peloria*, Coen told him the flower had vanished. As far as Coen could tell, no one in the world had any *Peloria* left.

"They lost it at Kew Gardens," Martienssen told me. "They lost it at Ox-ford Botanic Gardens, where they had it for two hundred years."

After months of searching, Coen finally discovered a cache of the historically important flowers. He found it not in a botanical garden or in a scientific laboratory. A California nursery offered to ship *Peloria* anywhere in the world. Martienssen put in a big order, and once the plants arrived in Cold Spring Harbor, he and Mulligan started building up a supply of their own.

"We're trying our best to make sure we don't lose it," Mulligan said.

I n the late 1800s, Charles Darwin and Francis Galton first turned heredity into a scientific question. Scientists such as Hugo de Vries and William Bateson believed that in genes they had found an answer. They found a way by which living things today could be correlated with their biological past. But in the process, they didn't just look for evidence in favor of genes as a vehicle for heredity. They also sought to refute any other alternative.

When August Weismann argued that the germ line carried heredity, walled off from somatic cells, he singled out Jean-Baptiste Lamarck as his opponent. Chopping off mouse tails was his way of refuting Lamarck's claim that acquired characters could be passed down. "If acquired characters cannot be transmitted, the Lamarckian theory completely collapses and we must entirely abandon the principle by which alone Lamarck sought to explain the transformation of species," Weismann said in 1889.

Weismann cleared a scientific path for geneticists to follow in the early 1900s, and they, like him, fashioned themselves as opponents of Lamarck and the so-called Lamarckians of their own day. In 1925, Thomas Hunt Morgan declared that genetic studies "furnish, in my judgment, convincing disproof of the loose and vague arguments of the Lamarckians." Any Lamarckian who did not abandon those loose and vague arguments in the face of all the evidence had to be confusing wishful thinking for science. "The willingness to listen to every new tale that furnishes evidence of the inheritance of acquired characters arises perhaps from a human longing to pass on to our offspring the fruits of our bodily gains and mental accumulations," Morgan sniffed.

Lamarck has remained an icon of pre-genetic thinking ever since. It's a role that's unfair both to him and to history. The inheritance of acquired traits had been widely accepted for thousands of years before Lamarck was born. In Europe, scholars from the Middle Ages to the Enlightenment treated it as fact. When Lamarck developed his theory of evolution from the inheritance of acquired traits, he felt no need to argue that it was true, because the matter had been settled so long ago. "The law of nature by which new individuals receive all that has been acquired in organization during the lifetime of their parents is so true, so striking, so much attested by the facts," he once said, "that there is no observer who has been unable to convince himself of its reality."

Regardless of whose name should be put on the idea, it continued to fall out of favor over the course of the twentieth century. As genetics explained more and more about life—with the discovery of the structure of DNA, with the fine details of its inheritance charted in thousands of experiments—the evidence for other forms of heredity remained weak: an odd frog or a stray stalk of wheat that seemed to pass down acquired traits. But some scientists continued to fight for conceptual room for more than one form of heredity. If we simply redefine heredity as genetics, they argued, we will never even look for those other channels.

———

Toward the end of the twentieth century, a few cases came to light that looked an awful lot like the inheritance of acquired traits.

In 1984, a Swedish nutrition researcher named Lars Olov Bygren launched a study of people in Överkalix, a remote region of Sweden where he had grown up. For centuries, Bygren's relatives had eked out a difficult existence along the banks of the Kalix River, fishing salmon, raising livestock, and growing barley and rye. Every few years, they suffered devastating crop failures, leaving them with little food to eat during the six-month-long winters. In other years, the weather would swing far in their favor, bringing bumper crops.

Bygren wondered what sort of long-term effects these drastic changes had on the people of Överkalix. He picked ninety-four men to study. Studying church records, he charted their genealogies and discovered a correlation between their own health and the experiences of their grandfathers. Men whose paternal grandfathers lived through a feast season just before puberty died years sooner than the men whose grandfathers had endured a famine at that same point in their life. Women, Bygren found in a later study, also experienced an influence across the generations. If a woman's paternal grandmother was born during or just after a famine, she ended up with a greater risk of dying of heart disease. It had long been known that a woman's health while she was pregnant could influence a fetus, but Bygren's research suggested the effects could stretch even further, to grandchildren or beyond.

Experiments on animals produced some similar results. In the early 2000s, Michael Skinner, a biologist at Washington State University, and his colleagues stumbled across one while they were investigating a fungus-killing chemical called vinclozolin. It's used by farmers to protect fruits and vegetables from mold, despite some evidence it can interfere with sex hormones.

Skinner and his colleagues gave vinclozolin to pregnant rats, and their offspring developed deformed sperm and other kinds of sexual abnormalities. Later, one of Skinner's postdoctoral researchers mistakenly bred these offspring and produced a new generation of rats. That error allowed Skinner to discover something he would not have expected: The grandsons of the poisoned rats also produced defective sperm, despite having no direct exposure to vinclozolin.

Skinner and his colleagues launched a new study to see how far this effect could get passed down. They exposed more female rats to vinclozolin and then bred descendants for several generations. Even after four generations, they found, males kept on developing damaged sperm. Exposures to other chemicals, like DEET and jet fuel, could also alter the rats for generations.

Skinner's work inspired other researchers to look for other kinds of changes that could be inherited. Brian Dias, a postdoctoral researcher at Emory University, wondered if mice might even pass down memories.

Each day, Dias put young male mice in a chamber into which he periodically pumped a chemical called acetophenone. It has an aroma that re-
minds some people of almonds, others of cherries. The mice sniffed the
acetophenone for ten seconds, upon which Dias jolted their feet with a mild
electric shock.

Five training sessions a day for three days was enough for the mice to
associate the almond smell with the shock. When Dias gave the trained
mice a whiff of acetophenone, they tended to freeze in their tracks. Dias
also found that a whiff of acetophenone made the mice more prone to star-
tle at a loud noise. In other trials, Dias would pump an alcohol-like scent
called propanol into the chamber instead, without giving the mice a shock.
They didn't learn to fear that odor.

Ten days after the training ended, researchers from Emory's animal resources department paid Dias a visit. They collected sperm from the trained
mice and headed off to their own lab. There they injected the sperm into
mouse eggs, which they then implanted into females. Later, after the pups
had matured, Dias gave them a behavioral exam, too. Like their fathers, the
new generation of mice was sensitive to acetophenone. Smelling it made
them more likely to get startled by a loud sound, even though he had
not trained the mice to make that association. When Dias allowed this new
generation of mice to mate, the grandchildren of the original frightened
males also turned out to be sensitive to acetophenone.

Dias then examined the nervous systems of these mice, hoping to find
physical traces of the association. When a mouse smells acetophenone, the
signal takes a precise path through its nervous system. The molecule latches
onto only one type of nerve ending in the mouse's nose, and those nerves
then send impulses to one small patch of neurons in the front of the mouse's
brain. When mice learn to fear acetophenone, previous studies had shown,
this patch gets enlarged.

The same patch of brain tissue that was enlarged in the trained mice was
enlarged in their descendants as well. Yet the only link from the frightened
fathers to their children and grandchildren was their sperm. Somehow, those
cells had transmitted more than genes to their descendants. And somehow

the animals passed down information not carried in their genes but gained through experience.

Dias's work raised the possibility that behaviors could be acquired and then inherited. Other researchers have come to a similar conclusion with their own experiments on mice. Stressful experiences when mice are young can change the way they cope with stress as adults. Young mice that are separated from their mothers for hours at a stretch act a lot like depressed people, for example. If they're put in water, they give up swimming quickly and just float helplessly. Male mice can pass down this helplessness to their offspring, and then on to their grandchildren.

The fact that fathers as well as mothers appear to influence future generations is especially intriguing. Unlike females, they have no direct link to developing embryos. In fact, males seem to be able to pass down behavioral traits by in vitro fertilization alone. If these experiments are sound, there must be something inside sperm (and eggs, too, presumably) that can pass down these mysterious marks. And since it can be influenced by experience, it can't be genes.

———

To explain this eccentric heredity, some scientists looked toward the epigenome, that collection of molecules that envelops our genes and controls what they do. By the late 1900s, it had become clear that the epigenome is essential for the proper development of eggs into adults. Our cells coil up their DNA and alter their methylation as they divide. The distinctive combinations of genes they keep switched on help to commit them to becoming muscle, skin, or some other part of the body. These patterns can be remarkably durable, enduring through division after division. That's how little hearts grow into bigger hearts, instead of turning into kidneys.

Yet the epigenome is not simply a rigid program for turning genes on and off in a developing embryo. It is also sensitive to the outside world. Over the course of each day, for example, our epigenome helps drive our bodies through a biological cycle. We get sleepy and wakeful; we warm and cool; our metabolic flame rises and falls. Our cycles stay on track with the twenty-four-hour rotation of our planet, thanks to the changing levels of light that enter our eyes over the course of each day. During the day, certain genes are active, making proteins important for waking life. As darkness falls, a growing number of proteins land around these genes, winding up their DNA and altering their methylation. The genes will stay silent through the night, helpless until the morning's army of molecules wakes them again.

The epigenome can alter the workings of genes not only in response to reliable signals like dawn and dusk but to unpredictable ones as well. When we develop an infection, immune cells bump into the pathogen and go into battle mode. They can start spewing out deadly chemicals or send signals to surrounding blood vessels to swell with inflammation. To undergo these changes, the cells reorganize their DNA, allowing certain genes to start making proteins while silencing others. And as the immune cells multiply, they pass down this battle-ready epigenome to their descendants as a kind of cellular memory.

The memories we store in our brains may also endure thanks in part to changes we make to the epigenome. Starting in the mid-1900s, neuroscientists found that we sculpt the connections between neurons as new memories form. Some of the connections get pruned, while others get strengthened, and these patterns can endure for years. More recently, researchers have found that the formation of new memories is accompanied by some epigenetic changes. The coils of DNA in neurons get rearranged, for example, and new methylation patterns get laid down. These durable changes may ensure that neurons preserving long-term memories keep making the proteins they need to keep their connections strong.

Plants don't have brains, but they have a memory of their own—one that can respond to infections, deadly influxes of salt, or drought. Struggling against these challenges can prime a plant to prepare for more in the future. If a drought-stricken plant enjoys a shower of rain, it will still remember its lack of water. Even a week later, it will respond to drought more strongly than a plant that has never faced such a threat to its existence. And researchers have found that long-term changes to a plant's epigenome are essential for laying down these enduring responses.

The malleability of the epigenome is not an unalloyed good, though. Some studies suggest that stress and other negative influences can alter epigenetic patterns inside our cells, leading to long-term harm.

Some of the strongest evidence for this link has come from the laboratory of Michael Meaney at McGill University. In the 1990s, Meaney and his colleagues started a study to see how rats experience stress. If they put rats in a small plastic box, the animals got anxious, producing hormones that raised their pulse. Some rats reacted more strongly than others to the stress, and, after some searching, Meaney and his colleagues found the source of the difference. It turned out that the rats that made more stress hormones had been licked less as pups by their mothers.

Working with Moshe Szyf, a McGill geneticist, Meaney investigated the physical differences that more licking or less licking produced in the animals. They knew that mammals control their stress response with the help of the hippocampus, that memory-forming region that keeps making new neurons through life. When stress hormones latch onto these neurons, the cells respond by pumping out a protein. Those proteins leave the brain and make their way to the adrenal glands, where they put a brake on the production of stress hormones.

Meaney and Szyf inspected the neurons in the hippocampus, looking closely at their methylation. In rats that get licked a lot, they found relatively little methylation around the gene for the stress-hormone receptor. In rats that get licked a little, the methylation is much greater. Meaney and Szyf proposed that when mothers lick their pups, the experience alters neurons in the hippocampus: Some of the methylation around their receptor gene gets stripped away. Freed from the methylation, the gene becomes more active, and the neurons make more receptors. In the well-licked pups, these neurons thus become more sensitive to stress, and rein it in more effectively. Rats that get little licking develop fewer receptors. They end up stressed-out.

Given that rats and humans are both mammals, it's possible that children may also undergo long-term changes to their stress levels from their upbringing. In one small but provocative study, Meaney and his colleagues looked at brain tissue from human cadavers. They selected twelve who had died of natural causes, twelve people who had committed suicide, and another twelve who had committed suicide after a history of abuse as children. Meaney and his colleagues found that the brains of people who had experienced child abuse had relatively more methyl groups around their receptor gene, just as in the case of the under-licked rats. And just as those rats produced fewer receptors for stress hormones, the neurons of victims of child abuse had fewer receptors as well. It's conceivable that the child abuse led to epigenetic changes that altered emotions in adulthood, snowballing into suicidal tendencies.

Meaney and Szyf's work has inspired many other studies on how epigenetics may link the environment to chronic disorders. But even in the absence of trauma or poverty, the epigenome changes over our lifetime. In fact, a geneticist named Steve Horvath has proposed that our epigenome changes at a steady rate, like the ticking of a biological clock.

The idea of an epigenetic clock came to Horvath in 2011 while he was studying spit. He and his colleagues had collected saliva from sixty-eight people and fished out some cells from the cheek lining that had been shed into the fluid. Initially, Horvath tried to find a difference in the methylation patterns between heterosexuals and homosexuals. But no clear pattern came to light. Hoping to salvage the study, he decided to compare the saliva according to the ages of the subjects.

Horvath and his colleagues found two spots along people's DNA where the methylation pattern tended to be the same in people of the same age. When they looked at other kinds of cells, they found other places where the methylation changed even more reliably as people got older. By 2012, Horvath was able to look at the methylation at sixteen sites in the DNA of nine different cell types. He could use those patterns to predict people's ages, with an accuracy of 96 percent.

When Horvath wrote up his experiments, two journals rejected them. It wasn't that his results were too weak. They were too good. The third time he got rejected, he drank three bottles of beer as fast as he could and wrote a letter back to the editor, objecting to the reviews. It worked, and the paper appeared in October 2013 in the journal *Genome Biology*. When a team of researchers in the Netherlands read the study, they quickly tested out the epigenetic clock with samples of blood they had collected from Dutch soldiers. They could accurately guess the soldiers' ages to within a few months.

As provocative as such studies are, it's still far from clear whether the epigenetic clock matters much. The same uncertainty hovers over studies on how negative experiences can trigger epigenetic changes in the brain and the body. These studies tend to be small, and sometimes when other scientists replicate them, they fail to see the same results. It's even possible that the way scientists search for epigenetic change may trick them into seeing it where none exists. Perhaps the epigenetic clock is not produced by cells changing their epigenetic marks, for example. Perhaps some types of cells become more common as we get older, and those cells have different epigenetic marks than the cells more common in youth.

These uncertainties have not scared off scientists from studying epigenetics, however. The stakes are just too high. By cracking the epigenetic code, researchers may discover a link between nurture and nature. And if we can rewrite that code, we may be able to treat diseases by altering the way our genes work.

———

These studies raised the possibility that the mysterious kinds of heredity that Dias and others were observing were the result of epigenetic changes getting passed down from one generation to the next. Within our bodies, it's clear that a cell can experience a change to its epigenetic pattern, and when it divides, its daughter cells will inherit that change. If those daughter cells happen to be germ cells, perhaps they could pass acquired traits on to later generations.

The prospect of this new kind of heredity made many people giddy. The mystery of missing heritability was solved, they claimed, because heredity was more than genes—it could be epigenetic, too. "If the 20th century belonged to Charles Darwin," the epidemiologist Jay Kaufman declared in a 2014 commentary, "it is looking increasingly as if the 21st century will be handed back to Jean-Baptiste Lamarck, given the explosion of recent developments in epigenetics."

A lot of people started talking about Lamarck again, making him the symbol of a more pliable kind of heredity. When *Nature Neuroscience* published Dias's study on memories of smells, they put a picture of Lamarck on the cover, complete with a thatch of gray hair and a high cravat. *New Scientist* covered the study in the same spirit, describing it in an editorial entitled "Mouse Memory Inheritance May Revitalise Lamarckism."

Transgenerational epigenetic inheritance, as this new flavor of Lamarckism came to be known, inspired giddiness far beyond scientific journals. It implied that our health and even our minds were shaped by an alternate form of heredity. If you let your imagination run wild through the possible implications, it can be hard to get it back on its leash. The fact that vinclozolin and DEET can have transgenerational effects is worrying when you consider that many other chemically similar compounds might as well, including some of the chemicals in plastics. In 2012, 280 million tons of plastics were produced worldwide, and much of it ended up in the environment. It's bad enough to envision their potential to disrupt hormones in people and animals. It's worse to picture a legacy of this pollution enduring through the generations.

Now imagine that poverty, abuse, and other assaults on parents also impress themselves epigenetically on their children—who might then pass down those marks to their own children. Think of all the social ills you might explain with Lamarck. In 2014, a journalist named Scott C. Johnson indulged in this speculation in a feature entitled "The New Theory That Could Explain Crime and Violence in America." He wove the story of a black family in Oakland, California, beset by poverty, addiction, and crime into a scientific history of epigenetics—starting, wrongly of course, with Lamarck—and then running up to recent experiments on mice. "Forget what you've heard about guns and drugs," Johnson exhorted us. "Scientists now believe the roots of crime may lie deep within our biology."

If, on the other hand, you suffered from upper-middle-class anxieties, epigenetics could become your new yoga. A hypnotherapist named Mark Wolynn started running workshops around the United States where he rummaged back in the genealogy of his clients for hidden epigenetic troubles. "The newest research in epigenetics tells us that you and I can inherit gene changes from traumas that our parents and grandparents experienced," Wolynn declared on his website. He promised to deprogram those inherited changes by helping build "new neural pathways in your brain, new experiences in your body, and new vitality in your relationship with yourself and others." All for only $350 per workshop.

Coming out of an epigenetic workshop, your neural pathways rebuilt, you might be shocked to discover just how much skepticism there is in scientific circles about transgenerational epigenetic inheritance. Many critics see no basis for drawing huge lessons from the evidence gathered so far. They are suspicious of the small size of many of the most sensational studies. Results that look like evidence of transgenerational epigenetic inheritance may often be random flukes. In some cases, the results may be genuine, yet the causes may have nothing to do with inherited epigenetic marks.

But some of the most potent attacks on this form of inheritance have been directed at the molecular details. It's hard to see how exactly the experiences of parents can reliably mark the genes of their descendants. While it's true that the methylation pattern in cells can change during people's lifetimes, it's not at all clear that those changes can be inherited.

The trouble with this hypothesis is that it doesn't fit what we know about fertilization. A sperm carries its own payload of DNA, which has its own distinct epigenome as well. For example, sperm have to tightly wind their DNA in order to fit it inside their tiny confines. During fertilization, the sperm's genes enter the egg, where they encounter proteins that attack the father's epigenome. As the embryo starts to grow, the epigenetic drama continues. The totipotent cells strip away much of the remaining methylation on their DNA. And then they reverse course and start putting a fresh batch of methyl groups back on.

This new methylation helps cells in an embryo take on new identities. Some cells commit to becoming the placenta. Others start giving rise to the three germ layers. And when the embryo is around three weeks old, a tiny wedge of cells receives a set of signals that tell them they have been picked for immortality. They will become germ cells. The newly formed primordial germ cells alter their epigenome yet again. They strip off much of the methylation from their DNA.

Many scientists doubt that inherited epigenetic marks can survive all this stripping and resetting. If heredity is a kind of memory, methylation suffers radical amnesia in every generation.

It's concerns like these that led a number of scientists to question Brian Dias's claim that mice can inherit memories. Kevin Mitchell, a neurogeneticist at Trinity College, Dublin, took to Twitter to express his skepticism. He delivered a rant worthy of August Weismann.

"For transgeneration epigenetic transmission of behaviour to occur in mammals," he wrote, "here's what would have to happen:

> *Experience—>Brain state—>Altered gene expression in some specific neurons (so far so good, all systems working normally)—>Transmission of information to germline (how? what signal?)—>Instantiation of epigenetic states in gametes (how?)—>Propagation of state through genomic epigenetic "rebooting," embryogenesis and subsequent brain development (hmm . . .)—>Translation of state into altered gene expression* **in specific neurons** *(ah now, c'mon)—>Altered sensitivity of specific neural circuits, as if the animal had had the same experience itself—>Altered behaviour now reflecting experience of parents, which somehow over-rides plasticity and epigenetic responsiveness of those same circuits to the behaviour of the animal itself (which supposedly kicked off the whole cascade in the first place)*

For scientists like Mitchell, an epigenetic form of heredity suffers from more than just biological gaps. It demands rewriting entire fields of science that researchers already understand very well.

———

In 2014, Robert Martienssen coauthored the definitive cold-water bath for the new Lamarckism. He and Edith Heard, a biologist at the Curie Institute in Paris, looked over all the research to date and published a review in the journal *Cell* entitled "Transgenerational Epigenetic Inheritance: Myths and Mechanisms."

"Might what we eat, the air we breathe, or even the emotions we feel influence not only our genes but those of descendants?" Heard and Martienssen asked. For all the attention that scientists and others had drawn to that question, they saw no reason to answer the question with a yes. "So far there is little support," they wrote.

When I visited Martienssen at Cold Spring Harbor in 2017, he still didn't see any reason to revise his judgment. The research on animals—and people in particular—remained too skimpy to get excited about. He saw no compelling evidence for a mechanism that could carry epigenetic traits across many generations of animals.

Yet Martienssen found it funny that he had gained a reputation as a naysayer. While he finds the evidence weak in the animal kingdom, he spends most of his time in the kingdom of plants. And there the evidence is actually overwhelming. "This sort of thing happens all the time in nature," Martienssen told me.

Plant scientists got their first clues to this extra channel of heredity in the mid-1900s. Corn kernels took on new colors, but their offspring didn't follow Mendel's Law, and after a few generations the ancestral color sometimes returned. A careful inspection of corn DNA showed that these changes to their color were not the result of mutating genes. It was the pattern of methylation that was changing. Each time plant cells divide, they rebuild the same pattern of methylation on the new copies of DNA that they make. But every now and then, plant cells alter the pattern: They add an extra methyl group where none was before, or a methyl group falls away and isn't restored. These changes can silence a gene in a plant or allow it to become active—triggering, among other things, new colors in corn kernels.

This strange inheritance has turned up in other crops as well as in wild plants—including toadflax. Enrico Coen and his colleagues discovered that *Peloria* reliably produced its trumpet flowers because it passed down a distinctive methylation of its L-CYC gene. Other researchers gathered other plants from the wild and found that some of them inherited epigenetic patterns that influenced their size, shape, and tolerance for harsh conditions. In experiments, they stripped off methyl groups from certain segments of DNA in plants and then bred them. The plants could reliably pass down these new epigenetic patterns for twenty generations or more.

It's possible that plants make it easier for transgenerational epigenetic inheritance to occur than do animals. Unlike animals, plants don't set aside germ cells early in development and reset their methylation. A red oak acorn will break open, and its cells will develop into roots and a stem, and over years its cells will multiply into a tree. After it has grown for about a quarter of a century, it will prepare to reproduce, reprogramming some of the cells on the tips of its branches into a botanical version of stem cells.

These cells swiftly divide, forming flowers, some with pollen (the plant equivalent of sperm) and some with ovules (the plant equivalent of eggs). The tree's ovules may get fertilized by pollen from other trees, developing into acorns. The following year, the same oak will produce a new batch of stem cells at its branch tips that will grow into flowers and sex cells. It will keep doing so for centuries. In other words, there's plenty of time—and plenty of cell divisions—for the epigenetic patterns in red oaks to change before their somatic cells turn into germ cells. And since plants don't reset their epigenetic marks in germ cells the way animals do, there's an opportunity for a new red oak tree to inherit new epigenetic marks from its parents.

There's another important difference between animal epigenetics and plant epigenetics. Even though plants cover their genes with the same methyl groups, they use different molecules to apply them. Martienssen and other researchers have discovered that plants do so by producing small RNA molecules, each of which can home in on specific segments of DNA.

Once they reach their target, the RNA molecules draw proteins around them, which add methyl groups to the DNA. When these cells divide, their daughter cells inherit these RNA molecules, which can continue to control how their genes work.

Something similar might have happened in *Peloria*. Now that Martienssen had tracked down the last source in the world, he could see if his hunch was right. His plan was to pull out RNA molecules from the strange flowers. "I'm hoping," he said, "that we can finally close this chapter and explain the monster."

———

The biology of animals may offer less of an opportunity for transgenerational epigenetic inheritance than that of plants. But that does not necessarily slam the door shut on the possibility. To a number of scientists, it remains ajar.

Our understanding of epigenetics depends on how well we can see it. When scientists began mapping the methylation that coats DNA, they could barely see it at all. In the 1990s, Enrico Coen could cut out a single gene and inspect it for methylation. Scientists then developed the tools for mapping the methylation across all the DNA in a cell. But they had to pull the DNA out of millions of cells at once to do so. If those cells belonged to subtly different types, each with a different pattern of methylation, the scientists could see only an epigenetic blur. By the 2010s, scientists were learning how to put cells on a kind of microscopic conveyor belt where they could inspect all the methylation in each cell, one at a time.

As our epigenetic focus has sharpened, old assumptions have turned out to be wrong. In 2015, for example, Azim Surani, a biologist at the Wellcome Institute in England, led one of the first studies on the epigenetics in human embryonic cells. In particular, he and his colleagues examined the cells that were on the path to becoming eggs or sperm. They observed these so-called primordial germ cells stripping away most of their methylation before applying a fresh coat. But a few percent of the methyl groups remained stubbornly stuck in place on the DNA.

A lot of the cells shared the same resistant stretches of DNA that held on to their old epigenetic pattern. These stretches contained virus-like pieces of DNA called retrotransposons. They can coax a cell to duplicate them and insert the new copy somewhere else in the cell's DNA. Methylation can muzzle these genetic parasites.

Retrotransposons typically sit near protein-coding genes, and it is possible that those genes get muzzled, too. Surani and his colleagues found that some of the genes near the stubborn methylation sites have been linked to disorders ranging from obesity to multiple sclerosis to schizophrenia. Based on their experiments, the scientists concluded that these genes are promising candidates for transgenerational epigenetic inheritance.

It is also possible—but, again, not proven—that other molecules may carry out transgenerational epigenetic inheritance. Sperm cells, for example, deliver RNA molecules into the eggs they fertilize, along with their chromosomes. Some of those RNA molecules help orchestrate the earliest stages of an embryo's development. Tracy Bale, a biologist at the University of Pennsylvania, has carried out experiments to see if the RNA molecules in sperm can allow experiences of fathers to influence their offspring.

In particular, Bale and her colleagues investigated the effect of stress that male mice experienced early in life. They found that when these stressed mice matured, they produced sperm with an unusual blend of RNA molecules. The scientists wondered what sort of effect these RNA molecules might have on offspring. They injected an RNA cocktail into the sperm of mice that had not experienced a lot of stress, and then fertilized eggs with the sperm. The pups that these eggs developed into handled stress badly. Bale's research suggests that the RNA in a stressed father's sperm can shut down certain genes in the cells of their offspring. And by silencing these genes, fathers can permanently alter their offspring's behavior.

A few other researchers have also found tantalizing hints of the hereditary power of RNA in animals. Antony Jose, a biologist at the University of Maryland, tracks RNA molecules produced inside the body of a tiny worm called *Caenorhabditis elegans*. RNA molecules created in the worm's brain can make their way across its body and end up inside its sperm, where it turns off a gene. Other researchers have found that RNA molecules in the worms can turn off the same gene in the next generation, and for several generations after that. It appears that the RNA molecules sustain themselves through the generations by spurring young worms to make more copies of themselves.

We are not worms, of course, but a number of experiments have demonstrated that human cells can send RNA molecules to each other on a regular basis. Very often, they are delivered in tiny bubbles, called exosomes. Scientists have observed more and more types of cells releasing exosomes, and more and more taking them up. In some species, embryos may use exosomes to send signals between parts of the body to make sure they all develop in sync. Heart cells may release them after a heart attack to trigger the organ to repair itself. Cancer cells spew out exosomes with exceptional abandon—probably as a way to manipulate surrounding healthy cells into becoming their servants. In 2014, an Italian biologist named Cristina Cossetti observed that exosomes cast off by cancer cells in male mice could deliver their RNA into their sperm cells.

These studies are far from conclusive, but they've been provocative enough to send scientists back to reread *The Variation of Animals and Plants Under Domestication*. Darwin's gemmules certainly don't gather genes from around the body. But perhaps—just perhaps—exosomes are a modern incarnation of gemmules, ferrying the RNA molecules that allow the experiences of one generation to influence the next.

———

But even if there is a link from somatic cells to the germ line and to future generations, it won't be enough to resurrect Lamarck. What made Lamarck's theory so seductive in the nineteenth century was the idea that the acquired traits were *adaptive*. In other words, they helped animals and plants survive, enabling species to fit themselves to their environment. Lamarck believed his version of evolution could explain why species were so well matched to their surroundings. In Lamarck's world, giraffes stretched their necks and ended up with the longer necks they needed to get food.

There is no solid evidence that transgenerational epigenetic inheritance is adaptive in the sense Lamarck intended. The few experiments that come closest to support have been carried out on plants. In one such study, researchers at Cornell University put caterpillars on a small flowering plant called *Arabidopsis thaliana*. The plants responded by making toxic chemicals that slowed down the onslaught. The researchers then bred the plants for two generations and then unleashed a new assault of caterpillars on the third-generation offspring. The plants still made high levels of toxins that stunted the growth of the insects.

Martienssen finds these experiments intriguing but doesn't see them as solid proof of Lamarckism. *Arabidopsis thaliana* is the lab rat of the plant kingdom, for example, having been bred by scientists for many generations in caterpillar-free conditions. Their response to insect enemies may not reflect what happens outside, in the insect-infested world.

"Finding that is still a Holy Grail of epigenetics," Martienssen said. "I mean, there are reports out there, but nothing has really, really stuck."

It's entirely possible that some inherited epigenetic changes are good for plants. But it's also possible that others are bad, and still others indifferent. The flowers of *Peloria* grabbed Linnaeus's attention, but no one has demonstrated that they do the plants any more good than the ordinary toadflax flower. An epigenetic flip simply swapped one flower for another.

Dandelions, scientists have found, can inherit epigenetic patterns that make them sprout early or late. Wild populations of *Arabidopsis* inherit some patterns that make some of their roots grow deep and others shallow. It's possible—although it has yet to be proven—that this overall variety helps out plants. If a drought strikes a prairie, a population of flowers may avoid extinction because of the long roots that a few of them have, thanks to the lingering luck of the epigenetic draw.

Regardless of what transgenerational epigenetic inheritance does for plants, Martienssen told me, he saw them as a legitimate part of how their ancestors influence their descendants.

At one point in our conversation, Martienssen surprised me by asking if I had ever heard of Luther Burbank. I had indeed; just a few weeks before,

I had made my own pilgrimage to his garden in Santa Rosa. But here at Cold Spring Harbor, at a modern shrine to genetics, I didn't expect to hear his name. Martienssen said that Burbank fascinated him. Burbank might not have been a rigorous scientist, but he could perceive patterns that still matter to science today. Martienssen stared off and recited to me a line of Burbank's that he drops whenever he can into his lectures and papers.

"Heredity," Burbank declared, "is only the sum of all past environment."

CHAPTER 16

The Teachable Ape

O NE AFTERNOON I picked up my older daughter, Charlotte, from preschool. As I was pulling her lunch bag and coat from her cubby, I came across a typed sheet of paper. I stopped for a moment to read it. It was from a Yale graduate student who was studying how children learn. He was looking for parents who would volunteer their young.

I got home and immediately e-mailed the student, a young man named Derek Lyons, and asked what he was investigating. Lyons responded later in the day. He explained that his research might give him some clues about what makes the human mind unique, perhaps even about how our species evolved.

I suppose a lot of parents might have been put off by that. But I signed right up.

Lyons later paid a visit to Charlotte's preschool to give her a screening test, and a few days later I drove her to New Haven for more trials. I held her hand as we shuffled down the worn steps in Sheffield-Sterling-Strathcona Hall to reach a basement lab. Lyons, thin, whippet-faced, greeted her like an old friend. Charlotte smiled back and took his hand, following him into another room. I read a book while the experiment took place. When the two of them returned, Charlotte was carrying little plastic animals.

I asked Charlotte how it went as we left the building. She shrugged and said it was fun. We had been talking a lot about princesses and atoms on the ride over, and she was more interested in getting back to those subjects as we drove home. Whatever secrets she had helped unlock would have to remain locked for now.

Lyons was studying young children like Charlotte to follow up on a series of earlier experiments on young chimpanzees. A team of Scottish scientists had shown each chimpanzee a clear plastic box with some fruit inside. The box had an assortment of doors and bolts on it. The scientists demonstrated to the apes how to slide open the bolts and open the doors in order to get to the reward. Sometimes they showed the chimpanzees how to open the box using the minimum number of actions. Other times, they added needless flourishes, like opening doors and shutting them, or using a bolt to tap the sides of the box.

The chimpanzees would patiently watch the demonstration, and then seize the box as soon as they got the chance. They would open it as swiftly as possible, jettisoning the needless extra steps, and grab the fruit. Their own sense of physics trumped any urge they might have to imitate humans.

Lyons's experiments were designed to observe how the battle of physics versus imitation plays out in the mind of a child. A couple of weeks after Charlotte's trial, he invited me back to his lab to take a look at the videos he had filmed. Driving into New Haven, I couldn't help feeling as if she had taken some sort of interspecies SAT test. I hoped she had scored one for *Homo sapiens.*

We sat down in front of Lyons's computer monitor, and he switched on a video. I could see Charlotte sitting on a carpet at school, her legs folded to one side. She looked at Lyons as he told her he was going to give her a box with a prize inside. The box he set down in front of her was made of clear plastic walls fastened together with Velcro. On the top of the box was a bar, and on the front wall was a little transparent door. The box contained a green toy turtle.

Charlotte didn't bother to open the door. She saw a quicker, rougher solution. She ripped the wall from its Velcro and seized the turtle. "I've got it!" she shouted.

"This is an unusual strategy," Lyons said diplomatically. "The important thing is that she totally ignored the bar."

A chimpanzee couldn't have done better, I thought.

Next, Lyons showed me Charlotte's trials at the lab. This time an undergraduate named Jennifer Barnes presented Charlotte with another box, with its own combinations of doors and bolts. Unlike the earlier trials, Charlotte now had to watch Jennifer open it first. Barnes had added some extra, useless steps to the process. Jennifer slid the bar back and forth across the top of the box. She picked up a stick and gave the box three careful taps. Only then did she turn the knob on a door, open it, and pull out the toy.

Charlotte's earlier performance had demonstrated that she understood physics well enough to figure out on her own how to open Lyons's boxes. But when Barnes finished her demonstration and handed Charlotte the stick, the box ripper vanished. In her place was a girl driven to imitate every irrelevant step she had been taught.

I could almost hear the chimpanzees hooting. Barnes showed Charlotte four other boxes, and time after time Charlotte followed her useless lesson to the letter.

When the last video ended, I wasn't sure what to say. "So . . . how did she do?" I asked.

"She's pretty age-typical," Lyons said. He had already studied dozens of children, and it was rare that one of them would not perform every step they were taught. Sometimes, Lyons would shake things up by suddenly telling the children that he had to leave soon. Even then, the children wouldn't skip the extra steps. They just performed them faster.

Charlotte's tapping was not a childish mistake. It was actually a little window through which I could glimpse something profound about human nature. We are well adapted for inheriting culture. We pass genes down through generations, but we also pass down recipes, songs, knowledge, and rituals. Our genetic inheritance endows us with many vital things, from the feet on which we can walk for miles, to the brains we can use to solve problems and think into the future. But if we inherited only our genes, we would not be long for this world. You can't use your brain to reinvent thousands of years of technology and customs on your own.

To say that culture is an important part of our lives doesn't really do the word justice. Culture is not a part of our life. We are a part of it. Lyons's experiment helped me see how adapted we humans are to immersing our-selves in culture. The Harvard anthropologist Joseph Heinrich has combed through history for unplanned experiments that demonstrate just how dan-gerous it is for humans to try to survive outside of it. One of the most striking of these experiments took place in 1861, in a region of deserts, mountains, and swamps in eastern Australia, a place called home by a group of Aborigi-nals known as the Yandruwandha.

The ancestors of the Yandruwandha likely arrived in Australia around sixty-five thousand years ago and then moved into the deep interior over the millennia that followed. Over hundreds of generations, the Yandru-wandha built up knowledge about their part of the continent and transmit-ted it to their children. They came to know where to find the water holes where they could drink and fish. They learned to keep small fires going through the night to survive the outback's winter chill. They gradually fig-ured out how to make bread and porridge from a clover-like fern that they called nardoo that grew in the creeks and swamps of the region.

To dine on nardoo is actually a remarkable feat. Nardoo's cell walls are so tough that you can eat it all day without extracting any nourishment. You will starve even as your stomach feels full and taut. Making matters worse, nardoo contains a toxic enzyme called thiaminase. When it gets into the bloodstream, it destroys the body's supply of thiamine (otherwise known as vitamin B_1). People who lose too much thiamine develop a disease called beriberi, which makes them extremely fatigued and shiver with hypother-mia as their muscles waste away.

Yet the Yandruwandha could make nardoo a major part of their diet, because they learned how to make it safe to eat. They collected the plant's seedlike sporocarps in the morning and immediately started roasting them in the embers of a fire. The heat destroyed some of the thiaminase, and the ashes likely eliminated more of it by altering the nardoo's pH. The Yandru-wandha women then ground the sporocarps between two broad, flat grind-ing stones, periodically adding water. This procedure inactivated more thiaminase and also broke down the plant's cell walls, turning the nardoo into a digestible flour. The women then used the flour to prepare bread or porridge. It's customary to eat the porridge with a mussel shell as a spoon. This probably serves as yet another safety measure. If Yandruwandha were to use a leaf instead to eat the porridge, chemical reactions might take place between the leaf and the nardoo that could make it toxic again.

In the summer of 1861, three European men dressed in ragged clothes and leading a sick camel wandered into Yandruwandha territory. Almost a year beforehand, the men—Robert Burke, William Wills, and John King—had marched triumphantly past thousands of cheering people as they left the city of Melbourne. They were part of an eighteen-man crew setting off to do what no European had ever done before: find their way from the southern coast of Australia, through the interior, all the way to the Gulf of Carpentaria on the north side of the continent. The expedition party left with twenty-six camels, twenty-three horses, a two-year supply of food, and a collection of essential equipment for Victorian gentlemen. They even brought enough oak furniture to fill a dining room.

The expedition party made its way across arid wastelands and treacherous swamps. Burke, King, Wills, and a fourth explorer, Charlie Gray, pulled ahead of the rest of the group and stopped about halfway along their journey at a place called Cooper Creek to wait for the rest of their straggling party to catch up. A month passed and no one arrived. Burke, the leader, decided to push forward and leave orders for the rest of the party to wait for them to return.

As the four men continued north, they slowed down even more. At last they reached an estuary filled with salty water. They realized they must be close to the gulf, but they couldn't find the sea. Instead, they kept struggling through mile after mile of coastal wetlands. Exhaustion eventually forced them to turn around and head south without ever setting eyes on the gulf.

The journey south proved even worse. Their supply of food began running out, and the British explorers had little idea how to hunt in Australia for their meals. Gray shot a python, but when he ate it, he contracted dysentery and died. Burke, King, and Wills struggled on toward Cooper Creek for months. When they arrived back at their camp, they found it abandoned. The rest of the explorers had reached Cooper Creek and waited for three months. When the advance party failed to return, they had headed for home, taking all the food with them. Burke, Wills, and King realized that if they tried to follow the rest of the explorers back south, they would starve. They needed to find help as soon as possible. To the west, they knew there was a cattle station. But to reach it, the three explorers would have to cross an expanse of swamps and then a long stretch of desert. Then they would find the cattle station, near the foot of Mount Hopeless.

Burke, Wills, and King agreed on this dangerous change of plan and headed west. The swamps turned into a maze of creeks, and as they struggled to find the way out, they grew weaker. Their camels suffered even more, dying one by one. Eventually, only one was left. Now their trip across the desert would be doomed, because the sole surviving camel wouldn't be able to carry enough water. It was then, in their greatest desperation, that the explorers met the Yandruwandha.

To Yandruwandha, this place was not a godforsaken wilderness but a place they'd called home for thousands of years. When the Europeans stumbled into the territory, the Yandruwandha assumed they had fallen under a spell that caused them to wander aimlessly, unable to return to the place they had come from. Nevertheless, the Yandruwandha welcomed Burke, King, and Wills. They let the explorers camp by their watering hole and fed them fish and nardoo bread.

Over the next few weeks, the Europeans regained some of their strength. Wills even grew friendly with some of the Yandruwandha. But Burke felt humiliated that they had to accept charity from savages. Surely they shouldn't need help from an inferior race—their superior intelligence alone should have sufficed. This resentment apparently led to conflicts between the travelers and their hosts, causing the Yandruwandha to pack up their camp and slip away.

Burke, King, and Wills were left again to rely only on their intelligence. They tried to fish the local watering holes but couldn't catch anything. It's not clear why they failed where the Yandruwandha succeeded so well. One possibility is that they didn't realize they should use nets like the Yandruwandha. Even if they wanted to use nets, though, they probably had no idea how to fashion them from the local plants.

Without enough fish to eat, the explorers turned to nardoo. They boiled the plants and ate four or five pounds' worth of the stuff every day. But no matter how much they ate, they kept growing more gaunt. "I am weaker than ever although I have a good appetite and relish the nardu much but it seems to give us no nutriment," Wills wrote in his journal.

Within a month after the trio parted ways with the Yandruwandha, Wills was dead. Burke and King buried his body and tried to find a way out of the swampy maze. Before long, Burke collapsed and died, and King was now alone. He grew dangerously ill as well as he wandered the swamps, until he ran into a second group of Yandruwandha. They took him in once more. Nestled in their culture, King recovered.

King spent a month in their company before a rescue party from Melbourne found him. They brought King home, where he recounted his misadventures to the city's reporters. News of the expedition quickly traveled around the world, and for years afterward Australians repeated the story until it became a patriotic fable of heroism. Burke and Wills were celebrated with statues, coins, and stamps. Yet their achievement was to have died in a place where others had thrived for thousands of years. The Yandruwandha got no honors for that.

I n the years after the Burke and Wills expedition, European anthropologists began scientifically studying the culture of the Australian Aboriginals. They also studied cultures on other continents and remote islands. They would contact isolated groups of people and spend years with them, systematically documenting the words in their languages, their creation stories, their rules for marriage. Anthropologists looked for universal patterns, for variations from place to place. As skeletons of ancient humans came to light in caves, they fit their observations of living cultures into a history of humankind that extended back thousands of years. Victorian anthropologists liked to build a simple history that followed a line of progress predictably ending with their own European culture. But in the early 1900s, as anthropology matured, the line was replaced by a branching tree representing the diversification of cultures over time.

The tree of culture took on a striking resemblance to the tree of life. Many anthropologists began to borrow ideas and methods from evolutionary biology in the hopes of coming up with a scientifically precise theory of their own. They wanted to distill culture into mathematical equations, to make predictions about how cultures change. For most of the twentieth century, the work of these evolution-inspired anthropologists didn't draw much notice outside of academic circles. In 1976, however, the British evolutionary biologist Richard Dawkins offered an idea about culture that became a cultural force of its own: the meme.

Dawkins unveiled memes at the end of his first book, *The Selfish Gene*. The book mostly concerns itself with genes and biological evolution. The most important quality of a gene, Dawkins argued, did not lie in the details of its chemistry. What mattered about a gene was that it could be inherited. When parents have children, a new copy of the gene gets replicated, and those genes that do a better job of getting replicated become more common over the generations. In a sense, Dawkins argued, we exist simply as vehicles that genes build to get themselves transported into the future.

At the end of *The Selfish Gene*, Dawkins added a provocative coda. He argued that replicators didn't have to be made from DNA. Astronauts traveling to another world might find life built from some alternate molecules. And we didn't even have to go to space to find other replicators.

"I think a new kind of replicator has recently emerged on this very planet," Dawkins declared. With the emergence of humanity, Earth was overrun with self-replicating pieces of culture: "tunes, ideas, catch-phrases, clothes fashions, ways of making pots or of building arches," Dawkins wrote. These new replicators deserved a name of their own, he decided, one as catchy as *gene*. He dubbed them memes, harkening back to the Greek word for an imitated thing.

"When we die," Dawkins said, "there are two things we can leave behind us: genes and memes."

Dawkins wrote about memes so beguilingly that the concept lodged itself in the heads of many readers. Some researchers tried to build a science out of memes, even launching the *Journal of Memetics*. They used memes to explain religions. They claimed that memes were responsible for the evolution of our big brains. And the appeal of memes reached far beyond academic circles. It became a popular label for trends and slogans. Advertisers saw themselves as meme designers. And when the Internet arose, it proved to be an environment tailor-made for memes. "The Net has effectively become a meme factory," the *Financial Times* declared in 1996.

In 1996, only 2 percent of the world's population was on the Internet. They struggled to reach it from sluggish desktop computers and through screeching modems. In the following decades, the Internet expanded its reach, infiltrating phones and cars and refrigerators. By 2016, almost half the world was prowling its nodes. The early listservs and forums gave way to giant social media platforms. Across this new memetic ecosystem, LOL-cats and the Crazy Nastyass Honey Badger began to roam. A website called Know Your Meme cataloged thousands of digital replicators, to help the befuddled keep up with new memes, and to help the forgetful to recall those of years gone by. The 2016 United States presidential election became a war of memes as operatives looked for the stories and photographs—genuine or doctored—that could spread a political message.

The nature of the Internet itself gave memes extra legitimacy. A gene, Dawkins often observed, is digital. It encodes a protein or an RNA molecule in a string of bases, with only four possible choices for each position. Because genes are digital, they have the potential to be replicated precisely. A computer file, made up of a string of zeros and ones, can be copied just as accurately.

Of course, both kinds of digital replication can fall short of perfection, acquiring mistakes thanks to sloppy enzymes or a dropped server connection. But molecular proofreading and error-correction software can fix most of them. Social media platforms have worked hard to make this replication not merely perfect but easy. You don't have to dig into the HTML code for your favorite political slogan or your favorite clip of an insane Russian driver. You press SHARE. You retweet. It's not just easy to spread memes; it's also easy to track them. Data scientists can track memes with all the numerical precision of a geneticist following an allele for antibiotic resistance in a petri dish.

Forty years after the publication of *The Selfish Gene*, Dawkins wrote an epilogue to an anniversary edition in which he looked back at his idea with satisfaction. "The word meme seems to be turning out to be a good meme," he declared.

It's entirely possible, though, that you have never heard of the Crazy Nastyass Honey Badger. An Internet meme is successful only insofar as it holds people's attention long enough for them to be amused or horrified by it, and to feel moved to transmit it to other people. They don't endure in people's minds very well. They don't combine with other concepts or values to grow into a more complex cultural development. Internet memes get knocked out of the spotlight by the next amusement, the next horror. The Nastyass Honey Badger never taught anyone how to eat nardoo.

Dawkins had broader ambitions for memes, to explain everything from technology to religion. But those ambitions have mostly gone unmet. The *Journal of Memetics* shut down in 2005, and no new journal took its place. Many researchers who study culture decided memes were too superficial to help them dig deeper. In 2003, the Stanford scientists Paul Ehrlich and Marcus Feldman went so far as to declare memes ready for their scientific funeral. "Identifying the basic mechanisms by which our culture evolves will be difficult," they said. "The most recent attempts using a 'meme' approach appear to be a dead end."

———

Instead of dissecting memes, many researchers searched for the biological building blocks that made human culture possible in the first place. The most fundamental ingredient seems to be learning.

But not just any learning will do. Wills and Burke might well have gotten back safely to Melbourne if they had been able to learn for themselves how to prepare nardoo, by dint of their own reasoning. But they simply weren't smart enough—certainly not as smart as the collective experience of hundreds of generations of Yandruwandha. Culture runs on social learning, on the ability of people to learn from other people. Yet it's become clear that social learning evolved long before human culture. Our species has no monopoly on that skill. In 2016, a team of scientists discovered that even bumblebees can learn cultural practices from one another.

Lars Chittka, a biologist at Queen Mary University of London, and his colleagues created an experiment in which bumblebees had to learn how to get sugar out of fake flowers. Each flower was a blue plastic disk with a well at the center holding a dollop of glucose.

Chittka and his colleagues tied a string to each of the flowers and tucked them under a tiny transparent table made of plexiglass. The scientists then put bees into this flower chamber and observed them. The bees could see the flowers through the plexiglass but could not directly reach them. They could eat the sugar only if they first pulled on the string to draw the flower out from under the table. Chittka and his colleagues put 291 bees to this test. Not one could figure out how to get to the sugar.

To make the puzzle easier for the bees to solve, the scientists broke it down into simpler tasks. In the first lesson, the bees could land directly on the flowers and drink from them. After the bees learned that they could get food from the flowers, the scientists presented them with a new challenge. They tied a string to each flower and nudged them only partway under the plexiglass table. The bees were now trained well enough to fly straight to the flower. But to reach the sugar, they had to push their heads against the edge of the table and extend their tongues. Once they learned this new technique, the researchers pushed the flowers completely out of reach underneath the table. Now the bees could get the sugar only if they pulled the string to draw out the flower.

Once their task was broken down into these smaller puzzles, some of the bees could learn how to drink the sugar by pulling on the string. Only twenty-three of forty bees figured out the task completely, though, and only after five hours of lessons. But, after all that hard work, Chittka and his colleagues had twenty-three trained bees. Now the scientists could see if other bees could learn from them.

Chittka and his colleagues built a tiny observation box where untrained bees could sit and watch the trained bees go about pulling out the flowers and drinking sugar. Each untrained bee got to watch ten different bees perform the entire routine. When Chittka then put the observers into the flower chamber, 60 percent of them—fifteen out of twenty-five—headed to the plexiglass, tugged on the string, and pulled out the flower.

That experiment showed that the bumblebees could learn by observing other bees. For his last study, Chittka wanted to see whether the practice of string pulling could spread, meme-like, through a group of insects. Chittka maintains a number of colonies in his lab, each made up of a few dozen bumblebees. From three colonies, Chittka and his colleagues picked out a single bee to train. They returned it to its colony, and then hooked up a tunnel so that any of the bees could walk into the flower chamber if they were so inclined.

The scientists then played the role of doorman, letting two bees at a time crawl down the tunnel of the colony to the flowers on a first-come, first-served basis. They repeated this procedure 150 times with each colony and then tested the bees to see how many knew to get the sugar by pulling the string. In each colony, they found, about half the bees had learned the lesson.

Only some of the bees learned directly from the original trained insect. The rest of them learned secondhand—or even third- or fourthhand. In fact, the original trained bee in one of the colonies died in the middle of the experiment, yet the string pulling continued to spread. Her intellectual legacy extended well beyond her own life.

What makes these experiments especially striking is that scientists have not found any sign that bumblebees in the wild learn from each other. For bumblebees, social learning may be a dormant skill, one that can be awakened only by an unnatural experiment.

Vertebrates, on the other hand, don't need a scientist to help them show off their social learning. They learn from each other on a regular basis in the wild, and their cultural practices can last for years—and perhaps far longer.

One of the first of these animal traditions to be documented arose in 1921 in a small English town called Swaythling. The people of Swaythling were annoyed to find that someone was vandalizing the milk bottles on their doorsteps. The foil caps that covered the mouths of the bottles were pierced, torn, and sometimes ripped clean away.

It turned out that the vandals were birds—in particular, a chickadee-like species called the blue tit. The birds would land on a bottle, pull away the foil, and then sip from the layer of cream atop the milk. Bird-watchers were so taken with the innovation that they looked for it in other villages, and kept looking for it for years. In 1949, two scientists, James Fisher and Robert Hinde, contacted birders and mapped three decades of their observations across much of the country.

These maps revealed that a few blue tits had independently started ripping milk bottle foil. Other members of their species took up the practice as well, either after watching them in action or just discovering a cap torn by another bird. The movement of birds between towns spread the foil-ripping long after the original inventors had died. The practice endured across much of Britain for decades, finally disappearing when milkmen stopped delivering foil-covered bottles of milk each morning.

Yet the birds have not lost their capacity to invent new ways to get food, or to inherit them from other birds. In 2015, Ben Sheldon of Oxford University and his colleagues gave a lesson to birds known as great tits, close relatives of blue tits. They captured wild birds in a forest and took them into their lab, where they were taught how to open a box full of mealworms. Each bird learned one of two ways to open the box: either sliding a red door to the left or a blue door to the right. The scientists set more of the boxes out in the woods, where they released the tutored birds.

Now in the wild, the birds continued to open the boxes as they had been taught in the lab, opening either the red or blue door. And the wild birds around them watched what they did. Three-quarters of the forest's population ultimately learned how to get mealworms out of the boxes. The tradition spread through the social networks of the birds—mostly among relatives or friendly birds that spent a lot of time together. And they reliably passed on the technique they had seen other birds use: Some learned to open the red door, the others the blue.

Sheldon and his colleagues also found that the memory of the birds helped extend their traditions. They left the boxes in the forest for three weeks and then took them away. Over the next nine months, over half of the original birds died. The scientists then brought the boxes back to the forest and waited to see what would happen. The older birds started opening them again, and the new generation of birds learned from their elders once more.

Other researchers have searched for more animal traditions, and they've found examples everywhere from the open ocean to dense jungles. In 1978, a University of Rhode Island biologist named James Hain and his colleagues discovered a striking case among humpback whales in the Gulf of Maine. When the whales chase after schools of fish, they typically dive about sixty feet underwater and use their blowhole to produce nets of bubbles that corral their prey. Once the fish have retreated into a tight clump, the whales lunge at them with an open mouth. In 1978, while observing whales in the Gulf of Maine, Hain spotted a single animal use a new technique.

Before spraying a bubble net, the whale first slapped the water with the underside of its tail. In the years that followed, whale observers recorded other animals in the gulf slapping the water as well. The proportion of whales performing this so-called lobtail feeding increased in fits and starts for decades.

In 2013, Luke Rendell of the University of St. Andrews, Scotland, and his colleagues mapped the records of lobtail feeding onto the social network of the whales. Humpback whales have a loose social organization, coming together into small groups to feed and mate. Rendell found that whales typically didn't just start spontaneously lobtail feeding on their own. Spending time with a proficient lobtailer made them much more likely to try out the method for themselves.

Rendell argued that this new tradition became entrenched in the Gulf of Maine for the same reason that blue tits started drinking cream: We changed their food supply. The whales fed on herring until the stocks collapsed due to human overfishing. They then switched to sand lances for their prey. It's possible that a whale discovered that smacking the water stunned the sand lances, preventing them from escaping whale attacks.

When scientists can't observe a new tradition take hold in the wild—which is most of the time—they can get some clues to its history by simply comparing the populations that make up a species. The chimpanzees that live in the Kibale Forest in Uganda, for example, use sticks to extract honey from logs. In the Budongo Forest, not far away, chimpanzees get honey in a different way: They chew up leaves and then use them as a sponge. Rather than being some instinct shared by all chimpanzees, these behaviors were probably invented by some innovative chimpanzees and then became local traditions.

The traditions of chimpanzees are especially important to understanding human culture because they're our closest living relatives. Primatologists have cataloged dozens of traditions among these apes, many found only in some populations but not others. Any one population may have a unique combination of traditions—for getting food, for using plants as medicine, for grooming each other, for making calls to other chimpanzees, for performing courtship rituals. Andrew Whiten, an expert on animal traditions at the University of St. Andrews, has argued that these bundles of traditions should be treated as cultures. If Whiten is right, that means that our cultural history reaches back at least seven million years, to our common ancestor with chimpanzees.

———

O nly after our ancestors split off from other apes did truly human culture emerge. One way to study its rise is to run experiments in which humans and other species perform the same task, and then see how they do it differently.

At the University of St. Andrews, a graduate student named Lewis Dean designed one such test. He created a "puzzle box" that dispensed rewards with the right combination of actions. When monkeys and chimpanzees tried to solve the box, they were rewarded with fruit; three- and four-year-old children got shiny stickers.

To get the rewards, the subjects had to uncover three hidden chutes. They could reveal the first chute by sliding a door to one side. Once they learned how to do this, they could learn how to push a button on the box that let them slide the door farther, revealing the second chute. And if they then turned a dial in the right direction, they could slide the door farther still, revealing the third chute.

Dean showed the box to two groups of capuchin monkeys. To get them started, he slid the door open to demonstrate how to get the first reward, but he then left them to their own devices after that. All told, each group of monkeys played with the puzzle box for fifty-three hours. And yet, after all that time, only two of the monkeys learned how to get the second reward. None got the third. When Dean put the puzzle box in front of chimpanzees, they fared almost as poorly. After thirty hours, only four chimpanzees got the second reward, and only one got the third.

The children, who were three or four years old, did far better. Dean showed the box to eight groups, each consisting of up to five children. In five of the groups, at least two children figured out how to open up the third chute. Many others managed to get the second one. And they made all this progress in just two and a half hours.

The children did so well because they—unlike the primates—could go beyond simply solving isolated problems. They could accumulate knowledge, and they did so as a group. When some of the children figured out how to reveal the first chute, the other children could learn it from them. And then the other children could use their own insights to add a new step in the procedure, revealing the second chute.

Many anthropologists now argue that this so-called cumulative culture is a hallmark of our species. The cultural practices that chimpanzees carry out are simple, requiring just a few steps to complete. They have never shown any capacity to learn a practice and then build on it. Humans, by contrast, are constantly adding on to the practices they've learned, creating complex new forms of culture. They can build up elaborate recipes for nardoo, modify canoes until they are able to cross the Pacific, and turn wooden guitars into electric ones.

Dean's experiment illuminates a few of the crucial traits that make cumulative culture possible. Friendliness matters, for one thing. When Dean gave the puzzle box to the primates, they competed over it rather than cooperating. The ones that figured out how to get fruit out of a chute always gobbled it up, never sharing it with others. The children, on the other hand, were more comfortable in each other's company. Some of them even spontaneously gave the stickers they got out of the puzzle box to others who didn't. The friendlier children were, Dean found, the better they did on his test.

A number of other studies have come to a similar conclusion: We humans have evolved more tolerance for each other than other species. That gives us more opportunity to learn from one another. And that extra social learning may be crucial for making culture cumulative.

The children in Dean's experiments stood apart from the chimpanzees and monkeys in a second way: They sometimes taught each other. When children figured out how to open a new chute, they sometimes demonstrated to the others how to do it. Dean never witnessed a single lesson taught by the chimpanzees or the capuchins.

Anthropologists were slow to recognize how important teaching has been to our species. In the mid-1900s, Margaret Mead and other experts claimed that teaching was merely a peculiar feature of Western societies, a cultural penchant for herding children into schools. In other cultures, children were supposedly left to their own devices, to learn for themselves. In recent years, though, many anthropologists have changed their minds, in part because they've reconsidered what it really means to teach. Teaching doesn't have to happen in a lecture hall or run on a semester schedule. At its essence, teaching is a behavior that one person uses to help another person learn a skill or acquire a piece of information. And by that definition, teaching appears to be common in human societies.

In the rain forests of central Africa, for example, Barry Hewlett, an anthropologist at the University of British Columbia, found evidence for teaching among the Aka, a group of pygmy hunter-gatherers. The Aka don't run schools, and they don't even have a word for *teach*. And yet Hewlett found that adults are constantly teaching children. The lessons happen in brief bouts, some as short as a few seconds. Sometimes the adults don't utter a word. But they still manage to convey crucial lessons to their young children, such as how to build a fire, how to use a machete, and how to find yams and dig them out of the ground.

Once researchers settled on their new definition of *teaching*, they wondered if other species teach as well. They have found only a few cases where the answer seems to be yes. Meerkats are one. In the deserts of southern Africa, they hunt many species of prey—some of which, like scorpions, are dangerous. Killing and eating a scorpion without getting fatally stung is a difficult skill to learn. Adult meerkats make it easier for pups by letting them practice safely. They start by bringing dead scorpions to their young. When the pups get older, the adults only wound the scorpions, and also make sure to rip off their stingers. Later, when the pups have developed some skill at handling live scorpions, their parents don't wound their prey as badly. At the end of the lessons, the pups are ready to kill live scorpions on their own.

As impressive as this teaching may be, meerkats appear to be an exception among animals rather than the rule. Even among our closest relative, animal teachers appear to be rare. In 2016, researchers from Washington State University reported that they had spotted chimpanzee teachers in a forest in the Republic of Congo. The adult chimpanzees there have a cultural practice of fashioning sticks into tools for fishing termites. Sometimes the adults will hand their tools to a young chimpanzee so that it can give it a try. Given the decades that scientists have been watching chimpanzees in the wild and in zoos, it's remarkable that the Washington State University team made the first observation of any behavior that might be called teaching. Teaching would make a tremendous difference to the well-being of chimpanzees. Cracking nuts with stones is so difficult that young chimpanzees can take as long as four years to master the skill. Yet in all that time, no one has seen an adult chimpanzee offer some guidance.

For humans, by contrast, teaching comes easily—so easily that children will spontaneously teach each other about games and toys. This tendency (which goes by the technical term *natural pedagogy*) may be rare in the animal kingdom because it demands a lot from teachers and students alike. Successful teachers have to be able to gauge what their students do and don't know, which demands an ability to get inside their heads. Teachers also need to communicate information clearly so that their students will be likely to learn something new. If teachers fail at any these challenges, all their efforts are wasted.

What's more, the best teachers in the world can't teach students unable to absorb their lessons. Humans seem to have evolved to be especially good students. One of the most important adaptations we've evolved—and one of the strangest—is what Derek Lyons was testing by studying Charlotte: extreme imitation.

Children are willing to imitate teachers even when they should know better. In a species like ours, in which teachers transmit important lessons, extreme imitation is a smart strategy. Children don't have to struggle to reinvent all of human culture from scratch. Dean's experiment gives a glimpse at imitation's value. The children who tried solving his puzzle box imitated each other fairly often, and the children who imitated the most got the most rewards.

There are certainly risks to imitating someone rather than wondering about each step along the way. If children end up imitating people who don't know what they're doing, they'll reproduce failure. But humans appear to have evolved some defenses against this sort of mistake. If children can choose whom to direct their attention, they'll tend to focus on adults who appear to be trustworthy experts. "The theory of natural pedagogy," the British psychologist Cecilia Heyes says, "suggests that blind trust is at least as important as smart thinking."

These ingredients for cumulative culture must have emerged after our ancestors split off from the ancestors of chimpanzees some seven million years ago. We don't have a lot of evidence to narrow that window and pin down the precise timing of when we started teaching, imitating to the extreme, and the like. We have to rely mostly on the material products of cumulative culture.

You can tell that living humans have cumulative culture by their stuff: If you haul together all the man-made objects in an Aka camp, they'd be different from what you'd collect from a Greenland Inuit settlement or a middle-class house in Nebraska. And if you collect objects from a single place stretching over a span of thousands of years—say, pottery from Egypt—you'll find one set of objects morphing into new sets. But if you try to reach back millions of years, the evidence gets scarce.

The technological record of humans and their ancestors reaches back

3.3 million years. In 2015, researchers in Kenya discovered small chipped stones that could be used to chop, hammer, or cut. It's possible that hominins used these tools to scavenge meat from dead carcasses. A million years later, hominins in Tanzania were still making these tools. But the shape of these pieces of rock—known as Oldowan tools—reveals a change of thought. The hominins who made them understood how to knock off a sharp flake from a rock while leaving the rock intact enough to yield another flake. They could thus produce a number of tools of about the same size and shape, all made from a single rock. This skill probably had to be taught. Young hominins likely needed a teacher to demonstrate how to knock off flakes, to communicate its goal, and to help them improve their technique.

It wasn't until about 1.8 million years ago that new kinds of technology emerged. The most striking of the next generation of tools was a large, teardrop-shaped hand ax. The species *Homo erectus* carried these hand axes to Asia and Europe and kept making them for over a million years. They made the axes from flint, basalt, quartz, and even obsidian. But the overall shape stayed pretty much the same.

These hand axes hint that *Homo erectus* was approaching our level of cultural prowess. The tools may look crude to our modern eye, but a twenty-first-century human would be hard-pressed to make them from scratch. First you'd need to find the right supply of rocks (I'm guessing you don't know where the nearest vein of obsidian is to be found). And then you'd need to strike one rock with another many times over to craft the proper shape. Paleoanthropologists sometimes run experiments to see how much prowess is required for making hand axes. They give people a finished model and some rocks they can thwack together. Their subjects always fail. It takes years for aficionados of the Pleistocene to learn how to make a good hand ax.

That hard-earned skill has led Alex Mesoudi, a paleoanthropologist at the University of Exeter, and his colleagues to the conclusion that *Homo erectus* used extremely careful imitation to make their hand axes. A careless hand-ax maker might have been blinded by a flying shard or maimed by a badly aimed chop. Observing skilled toolmakers would give novices the knowledge necessary to avoid such injuries. Careful imitation could also explain why hand axes managed to stay so similar to each other for over a million years. If *Homo erectus* simply looked over old hand axes to guess how to make them, they would have accidentally introduced little variations to their craft. Over a few thousand years, those mismatches would have caused the hand ax to drift far away from its original shape.

Scientists cannot find clear evidence of full-blown cumulative culture until our own species emerged about 300,000 years ago in Africa. As *Homo sapiens* spread out across the continent, our ancestors developed new shapes and styles for their stone tools. The tools in Morocco became distinctive from those of Kenya and South Africa. Early humans then started using these stone tools to work other materials, like antlers and eggshells. They combined their creations into new inventions, like spear throwers or nets to hunt for game. Artwork—from geometric decorations on eggshells to human figurines—also emerged.

Neanderthals and Denisovans seem to have never crossed this threshold. Language might be one reason for the difference. Our own ancestors may have gained the power of full-blown speech, making it much easier to cooperate on a hunt for game or a search for tubers. It would have made teaching more effective, too, bringing a precision and depth to lessons.

The benefits of language could account for why our ancestors reached a higher population density than Neanderthals and Denisovans. (You can estimate this by measuring the genetic diversity in samples of ancient DNA.) And our sheer numbers may have then helped bring about cumulative culture. With more people around, our ancestors had more opportunities to meet people and encounter new ideas they could adopt.

To get a rough estimate of what this early social network would have looked like, Kim Hill, an anthropologist at Arizona State University, and his colleagues interviewed hundreds of hunter-gatherers from two tribes: the Ache of Paraguay and the Hadza of Tanzania. Both tribes are made up of small bands; Hill and his colleagues drew up a list of people in a number of bands and asked the Ache and Hadza men if they had ever met anyone on the list. They concluded that each man had a social universe of about a thousand people. That's far bigger than the social universe of any other primate; male chimpanzees interact with only twenty other males in their entire lifetime.

Once all these pieces fell into place, cumulative culture promptly exploded. Humans could inherit complicated cultural practices, tinker with them, and pass them accurately onward to future generations. It was the dawn of a new form of heredity, with some striking parallels to the dawn of the first heredity systems on the early Earth. Life's early genomes were so prone to errors as they replicated that they couldn't get large. Once life gained a faithful form of inheritance, it could leap to complex cells. Humans, likewise, may have crossed a threshold and shot into a new universe of complex culture.

———

An Aboriginal Australian girl born fifty thousand years ago received a tremendous inheritance from her ancestors. It was made up of genes from her parents, possibly along with some epigenetic marks. She inherited some of her microbiome from her mother, along with the mitochondria that were once independent bacteria billions of years ago. As an infant, she began to hear the language of her group and to learn it. She inherited customs for interacting with people, including rules that applied within her family, and others for strangers. Her mother and sisters probably taught her the vast compendium of knowledge previous generations had amassed about how to prepare meals, how to deliver babies, how to use plants to heal the sick. She inherited an Aboriginal cosmology, one that placed her and everyone she knew in a meaningful place in the world.

She also inherited something else: a human-altered environment.

The environment shapes every species that inhabits it. As species struggle to survive, they evolve adaptations to their surroundings, whether they're fish gaining antifreeze in the Arctic Ocean, or hummingbirds evolving oxygen-hungry blood for flying over the Andes. But some species can reverse the equation. Even as the environment shapes them, they shape their environment. Elephants, for example, tear down tree branches and split their trunks down the center. Lizards, insects, and other animals can then invade these trees, which were previously off-limits to them. The rampages of the elephants open light into dense forests, allowing small plants to sprout up, providing food to animals like gorillas and bush pigs. Elephants can convert open woodlands into savannas and keep them cleared and fertilized with their dung. The elephants thus live in a habitat of their own making.

At first, our ancestors had little effect on their environment. They were just apes that could walk on two legs in search of fruit, seeds, and tubers, along with the occasional carcass they scavenged along with hyenas and vultures. But then they started to alter their surroundings. Their first environment-changing tool was probably fire.

The oldest known evidence for the use of fire lies in a cave in South Africa. There, scientists have found bits of burned bone and plant matter dating back a million years. For hundreds of thousands of years, hominins probably made fires only in hearths, probably to cook food. But *Homo sapiens* discovered additional uses for it. By 164,000 years ago, people in South Africa were lighting fires in order to bake soil, turning it into a rocklike material that could be sculpted into tools. And by 75,000 years ago, there's evidence in South Africa that people were setting grasslands on fire, possibly to clear them for earlier hunting. The flames wiped out the aboveground plants, but also stimulated the underground tubers to grow back at a much greater density.

By the time people reached Australia, they were using fire to reshape entire landscapes. Aboriginals walked through grasslands with fire sticks perpetually lit. The continent itself still records the first fire sticks, with a thin new layer of charcoal buried several yards in the earth. Fire was so much a part of Aboriginal life that it enveloped their cosmology. In one creation story, people and animals started out looking very different. A sacred spirit used a fire stick to set the entire world ablaze, and only after they were scorched did living things take on the appearance they have today. When the first Europeans finally set eyes on Australia, it was still burning. "We saw either smoke by day, or fires by night, wherever we came," Captain James Cook wrote in his journal in 1770.

Aboriginals also used fire as a hunting weapon, torching grasslands to flush out kangaroos, lizards, snakes, and other game. These burns could last for days. Elsewhere they burned forests to foster the growth of plants they wanted to harvest and animals they wanted to eat. These fires were set carefully, so as not to destroy sacred trees. Aboriginals inherited the rules for burning along with the rest of their culture, and they also inherited the landscape that the fires of their ancestors produced.

Fire was just one of many tools that humans used to hunt. They also invented spears, snares, nets, and fishhooks. They taught their children how to make these new tools, and also taught them how to use them. Culture helped turn humans into umatched hunters, driving species like mammoths and ground sloths to extinction. These large mammals were themselves ecosystem engineers, and so their disappearance had profound impacts on their environments. Some trees grew giant fruits, to ensure their seeds would be spread in dung of the giant creature. Now their seeds fell to the ground close by them. Some ecologists have argued that once Siberia's grasslands lost their mammoths and other giant mammals, moss took over and established today's tundras.

But humans also influenced plants simply through the cultural practices they developed to eat them. People who lived in rain forests gathered fruits from wild trees and brought them to their camps to prepare for meals. After they moved on, the seeds stayed behind, growing into wild orchards that they could revisit in later years. In Iran, foragers brought wild beans from the hillsides to grow alongside rivers where they'd be easier to harvest. When the foragers collected seeds for the next growing season, they unconsciously favored the plants with the variants that let them grow faster in the new environment. The evolution of the plants was now guided by humanity.

Starting roughly ten thousand years ago, with the end of the last Ice Age, some of these tended plants evolved under human care into domesticated crops. In the Fertile Crescent, wheat, millet, beans, and other plants were transformed. In China it was rice; in Africa, sorghum. In Mexico a weed called teosinte became corn. And in some of these same places, wild animals were domesticated into livestock such as cows, goats, and sheep.

The same capacity for cumulative culture that had already spread humans to all the continents save Antarctica now allowed them to convert the wild lands around them into farm fields and grazing pastures. Children in these agricultural communities inherited traditions for farming, and they also inherited lands that had been converted from wilderness long before they were born. The Agricultural Revolution lofted our species to a far bigger population than before, leaving some farmers desperate for land. They moved into open territories still inhabited by hunter-gatherers, bringing with them the entire package of agriculture: not just the seeds for crops but also their livestock, their saddles, their hoes, and their inherited wisdom about how to use all of it to harvest food, brew beer, sew leather shoes, and all the rest of their cultural practices. And these farmers continued to accumulate new steps to their traditions. As they learned how to work metals, they could make sickle blades or horseshoes. The environment in which much of humanity was now born had become a domesticated landscape, covered by farms as well as by houses, roads, villages, and cities.

It was cultural heredity that led to an agricultural revolution, and it was that revolution that fostered the practice that gave heredity its name. Heirs began to inherit great wealth from their ancestors.

There was nothing new about parents bequeathing valuable things to their offspring. You could argue that our reptilian ancestors were doing it 300 million years ago. The females stocked their eggs with protein-rich yolk, sacrificing some of their own physical resources to pass down to their young. Feeding on those provisions inside their eggs, our ancestors hatched in a stronger condition, more likely to survive to adulthood. When our ancestors evolved into mammals some 200 million years ago, mothers could bequeath milk as well. And when our ancestors evolved into lemur-like primates, their young grew dependent on even more gifts from their parents, in the form of food and protection. Our ancestors grew even more reliant on their parents over the last few million years, because they were growing increasingly big brains.

Brains, ounce for ounce, demand twenty times more energy than muscle. A human infant needs to channel almost half of the calories it gets every day to fueling its neurons. The human brain doesn't reach full size until about age ten, but even then it's not done developing. The teenage brain furiously prunes connections between neurons while building long-range links between distant regions. The unique anatomy of the human brain is essential for our unique capacity for cumulative culture. But in the time it takes for the brain to develop, human children need help from their parents to get the necessary fuel.

Children fifty thousand years ago couldn't just raid a refrigerator whenever they got hungry. Someone had to kill an animal for them, or harvest some plants and cook them over a fire. The few societies that survive today on hunting and gathering can offer a few hints about what things were like when all humans lived that way. At an early age, hunter-gatherer children start helping to find food and to prepare it. But they still eat more calories than they bring in. This deficit shrinks as they get older and can work harder. But it only turns to a surplus when they reach their late teens. Until then, families have to work together to make up for the deficit of their children—not just the parents, but grandparents as well.

Some families fare better at this work than others. They may make better arrows, which allow them to take down more game, for example. Hunter-gatherer societies keep this inequality in check with a system of moral judgments. A successful hunter who doesn't share some of his meat with other families will suffer a blow to his reputation. But such rules only rein in inequality; they don't eliminate it. In hunter-gatherer societies, children from successful families still end up getting more food and enjoy better health than children from other families. Their families ally themselves to a bigger social network, one that can provide more help during a drought or some other disaster. Under the right conditions, this inequality in hunter-gatherer societies can grow from one generation to the next.

Anthropologists have documented one particularly striking case of such inequality on Vancouver Island. The Nootka people have lived there for at least four thousand years, catching salmon that swim up the island's rivers to spawn. They could smoke enough fish to feed themselves and still have plenty left over to trade with inland tribes. It was a prosperous way of life, giving the Nootka enough resources to build massive wooden houses and totem poles to honor their ancestors. But it did not leave the Nootka in an egalitarian utopia.

The Nootka chiefs held sway over extended families, controlling the best fishing sites on the rivers. Heredity justified and extended their power. Each chief inherited his authority from a distant spirit ancestor. He marked the inheritance of that power by his children with a series of extravagant feasts. Powerful Nootka families grew more powerful over the generations, while others fell into debt. Some Nootka became so destitute that they had to surrender themselves into slavery, moving their families into the houses of their masters.

Whenever this sort of inequality emerged in a Nootka village, it could endure for a few centuries before it collapsed. Droughts and other disasters could wipe out the advantages that some families had over others. It's likely that inequality was just as fragile in all hunter-gatherer societies across the world. But the rise of agriculture allowed inequality to explode and endure. In the Near East, for example, foragers gathered wild plants together and stored them in communal granaries. As they switched to farming, they claimed rights to land for growing crops and built private granaries to store their harvests. Farmers who planted superior crops ended up with extra food that they could barter for goods. They could then pass down their land and those goods to their children, who could start farming with huge advantages.

———

It's obvious that inheriting PKU from one's parents is not the same as inheriting the knowledge of how to make nardoo bread. The first inheritance involves the copying of genetic information that gets brought together in an embryo. The second comes about through years of teaching, expressed in actions and language. Yet they are not utterly different either. Both kinds of inheritance are part of the variation from person to person and from population to population. And both can sustain that variation across generations.

At the dawn of the twentieth century, scientists came to limit the word *heredity* to genes. Before long, this narrow definition spread its influence far beyond genetic laboratories. It hangs like a cloud over our most personal experiences of heredity, even if we can't stop trying to smuggle the old traditions of heredity into the new language of genes.

We call genetic disorders family curses. A family of rich real estate barons rationalizes its wealth with mysterious genes for success. Genealogy began as an ancient practice of legitimization through ancestry, but we still depend on it to explain how we came to be. When birth certificates and immigration records fail us, we take up DNA to draw branches further back and farther out. We exult in discovering a link to some famous figure in history, as if carrying their alleles makes us special—ignoring the fact that beyond a few generations, we share little or no DNA with our ancestors. We take DNA tests to find out if we're full-blooded Irish or Cherokee or Egyptian, using sixteenth-century terms that started out as ways to describe racehorses or to divide humans into arbitrary categories—despite the fact that the evidence from that DNA chews holes in those barriers like a termite invasion. There is certainly a history to each chunk of our genetic material, but each one hurtles back through human history in a different direction, leaping from continent to continent. A human genome is a shredded, shuffled sampling of DNA, both from humans and from our extinct near-human relatives.

We think of heredity as limited to what normally happens in human families. Two parents combine half their DNA to produce a new offspring, their genes obeying Mendel's Law. In truth, heredity does not stop at conception's door. Cells continue to divide, and their daughter cells inherit everything inside them—their mitochondria, their proteins, their chromosomes, and the epigenetic network that gives each cell its state. Our bodies are walking genealogies, the branches distinguished by locked-in networks of genes, mosaic mutations—even by chimeric origins in different people. Germ cells can extend this heredity into new lives, but August Weismann's barrier is far from absolute. An untold number of animals have gained a cancerous immortality by sending forth their tumors to burrow into new hosts for thousands of years on land and at sea.

None of this is to say that Weismann himself was wrong. He gave science a new way to think about heredity, one that prepared the ground for the recognition of Mendel's discoveries. By the early 1900s, Mendel's Law was allowing geneticists to begin making sense of human heredity, explaining the now-you-see-it, now-you-don't evasion of traits through the generations that left even Darwin baffled. The easiest mysteries it solved were diseases caused by a single dominant allele or a pair of recessive ones like PKU. But for other traits—including ones as seemingly simple as height—heredity has become an ocean of explanation. The effect that our biological past has on our height has been shattered into thousands of pieces of influence.

Mendel's Law also turned out to be exquisitely fragile. It is regularly broken. Gene drives regularly get the better of Mendel's Law, carrying genes throughout populations, even at risk of pushing species to extinction. Many species of animals, plants, and fungi themselves sometimes obey Mendel's Law and sometimes find a different way to pass genes down to future generations. Some become both father and mother. They use their own sperm to fertilize their own eggs. Or they simply recombine the chromosomes within a single cell, which then develops into an offspring. Others dispense with gametes altogether, cloning themselves. These species are not rare flukes; they live around us. They grow into forests and coral reefs. They make our bread and beer.

Our anthropocentrism also makes it easy to forget that Mendel's Law is young. The universe was born with the speed of light fixed at 186,000 miles a second. The mass of an electron was set. And Mendel's Law was nowhere in the cosmos to be found. It evolved much later, as far as we know on only one planet. But on that water-coated rock, heredity had already been unspooling for billions of years when Mendel's Law evolved. The microbes and viruses that until then were the sole residents of Earth for over half the history of life did not reproduce by combining germ cells. Nor do they do so today. They follow rules of their own. They can make near-identical copies of their DNA. But that DNA can also slip from one microbe to another, producing patchworks that might never have evolved if heredity could only travel vertically. Microbes evolved new kinds of heredity of their own. It was only in the early 2000s that scientists discovered that many species used CRISPR to gain a defense against viruses—a defense that their descendants could then inherit.

We cannot understand the natural world with a simplistic notion of genetic heredity. And some scientists have likewise argued that we must expand our definition of *heredity* again, to take into account other channels as well—be they culture, epigenetic marks, hitchhiking microbes, or channels we don't even know about yet. In the 1980s, Marcus Feldman and a number of other researchers began trying to build a theory of heredity that could include both culture and genes. Since then, others have tried to widen the theory even more. Russell Bonduriansky of the University of New South Wales in Australia and Troy Day of Queen's University in Ontario, Canada, believe it's time to build mathematical equations that can unite genetic and nongenetic forms of heredity in a single description. They lay out their vision in a 2018 book, *Not Only Genes.*

Bonduriansky and Day argue that the previous century of research on heredity was based on a fallacy. Geneticists didn't simply champion the gene by finding evidence for it. They rejected the possibility that anything other than genes could carry heredity. Yet the existence of one channel did not necessarily rule out the other. "This purely genetic concept of heredity was never firmly backed by evidence or logic," Bonduriansky and Day declare.

As the genetic concept took hold, research into nongenetic forms faded away. The field's reputation was stained by scandals, as some experiments that supposedly supported nongenetic heredity proved to be shoddy or even fraudulent. But even careful scientists were at a disadvantage. The effects of genes can have a downright arithmetic precision. Simply breeding peas or flies reveals profound facts about how genes are inherited. Nongenetic forms of heredity are harder to distinguish from influences of the environment, and they can be more prone to vanish. Researchers who have tried to document nongenetic heredity have thus had to play a long game of catch-up. Linnaeus laid his eyes on the monstrous *Peloria* in the 1740s. When I visited Robert Martienssen 170 years later, he was still struggling to discover what keeps them monstrous after all these generations.

The debate over heredity had reached a different stage. Both sides accepted the reality of the other. No champion of nongenetic heredity denies the power of genes. And of the geneticists I've spoken to, none has outright rejected the idea that heredity can be sustained by things other than genes. The fight now is over importance. Some geneticists don't see much to be gained by investigating nongenetic heredity. Other biologists believe that the only way to make sense of traits they find in organisms is to understand how they're transmitted—and sometimes that search will take them beyond genes.

One argument in favor of genes—and against nongenetic heredity—is that genes have sticking power. Our cells are packed with dedicated swarms of proteins and RNA molecules that work together to copy genes, ensuring that new copies are virtually identical to the original ones. The low rate of mutations in our DNA allows an allele to endure for generations. Its sticking power gives natural selection enough time to favor it over other alleles, and to drive evolutionary change.

Nongenetic heredity can be far more fleeting. Water fleas, tiny invertebrates that live in ponds and streams, use a form of nongenetic heredity to escape predators. If they pick up the odor of a predatory fish, they grow spikes on their heads and tails, making it more likely that a fish will spit them out rather than swallow them. Female water fleas will then produce offspring that grow spikes early in development, even in the absence of any odor of a predator. This shift will endure for several generations—but then it fades away. A new generation of water fleas emerges spike-free. This sort of heredity can't drive evolutionary change, because natural selection can't favor spiked water fleas over spikeless ones for very long.

Yet Bonduriansky and Day reject this argument against the importance of nongenetic heredity. In some cases, it can actually be quite durable. The fact that Linnaeus's monstrous toadflax are still monstrous demonstrates that heredity can flow down an epigenetic channel for centuries—at least in toadflax. If we accept culture as a form of heredity, then we can trace some cultural traditions for tens of thousands of years. And even when nongenetic heredity is fleeting, it can still have a big influence on a species. By programming their descendants with spikes and other protections, animals and plants can enjoy long-term evolutionary success even as their environment changes. This success can have effects that ripple outward across an entire food web. A water flea's spikes allow it to increase in numbers while keeping its predators hungry.

Nongenetic heredity matters for another reason: It can potentially steer evolution. Under some conditions, for example, natural selection may favor tall plants over short ones. Plants can reach great heights if they inherit the right alleles, but nongenetic factors can also influence how tall a plant gets. If nongenetic factors make tall plants even taller, they will be able to have even more offspring. In other cases, nongenetic heredity may work against genes, bringing evolution to a standstill.

In *Not Only Genes*, Bonduriansky and Day remain agnostic about just how important nongenetic heredity will turn out to be. They are not drumming up hype about epigenetic memories of past lives. They simply consider the question both important and unanswered. Bonduriansky and Day have developed mathematical equations and conceptual tools they hope will make it possible to study both forms of heredity at the same time, in the same organisms. They argue that a combination of both kinds of heredity could help address some of the biggest questions scientists still have about the history of life—why we get old, for example, how peacock tails and other extravagant courtship displays evolve, how new species arise. Even human history could benefit from an expanded view of heredity.

If you've recently enjoyed a cone of ice cream or a slice of Brie, for instance, you are experiencing one of the more bizarre results of the Agricultural Revolution. As a rule, mammals don't consume milk in any form once they stop nursing. After weaning, they stop producing lactase, the enzyme required to break down lactose sugar. The same is true for about two-thirds of people worldwide. For them, consuming milk can be an uncomfortable experience, leading to symptoms including bloating and diarrhea. But the remaining two billion or so people can continue to drink milk and eat dairy foods as adults. They have inherited mutations that lead them to persist in making lactase.

Scientists have found a number of these mutations in the same region of the genome. They alter a genetic switch that controls LCT, the gene for lactase, keeping it from shutting down after weaning. The mutations show signs of having been favored by natural selection within the past few thousand years. And they are found in people who can trace their ancestry back to places with a deep history of cattle herding, such as East Africa, the Near East, and northwestern Europe.

All this evidence aligns nicely. It suggests that after some people domesticated cattle, a mutation became common that allowed them to consume milk. At this point, it's tempting to declare the case closed. But Bonduriansky and Day point out a paradox embedded in this story. Before the domestication of cattle, very few people inherited LCT mutations. The early cattle herders, in other words, were mostly lactose intolerant. It's hard to see how they would have taken up the practice of consuming milk if it made most of them sick to their stomachs. And if they didn't, then there would be no opportunity for the lactose-tolerant people to thrive on milk and spread their alleles.

The way out of this paradox is to recognize that two kinds of heredity were playing out at the same time among early herders: the genetic heredity of LCT mutations, and the cultural heredity of milk-consuming traditions. The cultural practice of herding cattle, sheep, and goats started out mainly as a way to get meat. But later the herders used their creativity to discover other kinds of foods that their animals could provide. They began milking their animals. At first, the lactose-intolerant herders may have only drunk a little milk. But as the milk soured, they may also have discovered that it was easier to digest milk by first turning it into yogurt and cheese—two foods that have much less lactose than regular milk. In Poland, researchers have discovered a 7,200-year-old sieve with milk fat still embedded in it—a sign of early cheese-making.

Once this custom was in place, natural selection could favor people with mutations to LCT. In times of famine, they might be able to fall back on milk as a source of protein and carbohydrates, while others starved to death. Their descendants inherited their LCT mutations, and as lactose persistence became more common, it fostered the spread of the culture of consuming milk. In other words, genetic heredity and cultural heredity started off blocking each other, but in later generations they ended up pushing in the same direction.

Bonduriansky and Day see another advantage to a broader view of heredity. Putting on blinders and looking only at genes leaves scientists at risk of missing important new discoveries about biology.

In the nineteenth century, for example, physicians believed that parents who drank too much alcohol could pass down feeblemindedness and other disorders to their children. The physicians were fuzzy on the details of how this happened, falling back on the Bible's promise that God "punishes the children for the sin of the parents to the third and fourth generation."

To the early Mendelians, this talk was nothing but old-fashioned Lamarckism. It was impossible for a parent's alcoholism to extend its harm to a future generation. Instead of a cause, it was merely an effect. A faulty gene altered the brain, they assumed, causing hereditary feeblemindedness. The condition not only caused people to score poorly on intelligence tests; it also left them unable to resist dangerous pleasures such as drink.

"We may say that every feeble-minded person is a potential drunkard," Henry Goddard declared in 1914.

When Goddard assembled Emma Wolverton's pedigree to prove that feeblemindedness was hereditary, he eagerly noted every report of alcoholism in the family. He believed that they strengthened his case even more. In *The Kallikak Family*, Goddard popularized this explanation of alcoholism. But even as the Kallikaks appalled the world, other scientists were carrying out experiments that pointed in a different direction altogether.

At Cornell Medical School in New York, Charles Stockard and George Papanicolaou wafted alcohol fumes into the noses of guinea pigs shortly before they mated. The alcohol caused a host of troubles in their offspring. Some baby guinea pigs were deformed, and others had low birth weight; they tended to die in infancy, and the survivors had low fertility. Stockard and Papanicolaou even found the same troubles carried over for four generations of guinea pigs. They concluded that the alcohol had violated Weismann's barrier and affected the germ cells. "All future generations arising from this modified germ plasm will likewise be affected," Stockard and Papanicolaou concluded.

Goddard and other Mendelian-minded scientists weren't swayed by the new research. In 1920, Prohibition went into effect and brought medical research on alcohol to a standstill. By the time Prohibition was lifted in 1933, genetics had matured so far that scientists generally refused to take a fresh look at Stockard and Papanicolaou's work. "While alcohol does not make bad stock, many alcoholics come from bad stock," Howard Haggard and Elvin Jellinek explained in their 1942 book *Alcohol Explored*. "The fact is that no acceptable evidence has ever been offered to show that acute alcoholic intoxication has any effect whatsoever on the human germ, or has any influence in altering heredity."

It wasn't until the 1970s that doctors realized just how wrong Haggard and Jellinek had been. David Smith and Kenneth Jones, both pediatricians at the University of Washington, noticed a cluster of four children who all shared the same symptoms: They had small heads, short statures, and slow intellectual development. They had something else in common, too: Their mothers were all alcoholics. Smith and Jones discovered that other pediatricians were seeing similar symptoms in children. Together they put forward a new condition, known as fetal alcohol syndrome. Heavy drinking during pregnancy, doctors now recognize, can cause a spectrum of symptoms, from brain damage to hyperactivity to poor judgment. In the United States, the Centers for Disease Control and Prevention estimates, up to one in twenty schoolchildren has fetal alcohol syndrome or related disorders.

To understand the biology of the syndrome, scientists in recent years have studied what happens when pregnant rats consume alcohol. Those studies suggest that fetal alcohol syndrome is an epigenetic disease. The ethanol in the drinks alters the methyl groups and other molecules around the DNA in a fetus. As a result, some genes go quiet, while some become more active. It's also possible—although the evidence is thinner—that fathers who drink before conception can contribute to fetal alcohol syndrome, too. It's possible that ethanol can also change epigenetic patterns in their sperm, and that this change can carry over into an offspring. Even more tantalizing are experiments finding that male rats can pass down the same epigenetic alterations through two more generations.

The molecular details of fetal alcohol syndrome remain fairly mysterious. Yet the discovery of the syndrome itself did not have to wait as long as it did. We can only speculate how much earlier it might have been recognized if scientists were open to the possibility. If Henry Goddard could have learned about it when he trained as a psychologist in the early 1900s, history might have taken a different path. In 1995, a doctor named Robert Karp at the Children's Medical Center of Brooklyn looked over some of the materials Goddard collected to write *The Kallikak Family*. He examined the pictures of children whose parents had been alcoholics. Looking at some of them, Karp was struck by the faces: by their razor-thin upper lip, for example, and by the lack of a philtrum—the vertical groove below the nose. These are telltale signs of fetal alcohol syndrome. Goddard's records—such as the short stature of some of Emma Wolverton's relatives—only strengthened Karp's suspicions. But Goddard himself did not recognize the symptoms. Perhaps it was impossble for him to do so, thanks to his rigid view of heredity. Today, we are still making up for that lost time.

PART V

The Sun Chariot

CHAPTER 17

Yet Did He Greatly Dare

PHAETHON PAID a visit one day to his father, Phoebus, the sun god. He went to demand Greek mythology's equivalent of a paternity test. Rumors were swirling that Phoebus was not his father, and Phaethon wanted them put to rest. "Give me proof that all may know I am thy son indeed," he said.

Stepping down from his throne, Phoebus embraced Phaethon. He swore to do anything to prove his fatherhood. Phaethon asked him for the one thing that Phoebus wished he could deny: to ride the chariot of the sun across the sky.

Phoebus begged Phaethon to ask for something else—anything else. The horses were too strong for Phaethon to master, the course too hard. But Phaethon, supremely confident in his own skill and strength, refused to change his mind. Phoebus realized he was trapped by his own promise and led his son to his gold-wheeled chariot.

When Phaethon climbed aboard, the horses suddenly carried the chariot high above the Earth. Phaethon went blind with fear. Simply being the son of a god did not mean that Phaethon inherited his father's mastery. The horses galloped off course, dragging the sun down toward the Earth and far away again. Where they came too close to the land, it was scorched to desert. Where they rose too high, they left frozen wastelands behind.

Phaethon's wild ride did more than permanently alter the landscape. It also left its mark on humanity. When the lurching chariot passed over Africa, the sun dropped so close to the ground that it scorched the people living there. Their blood rose to their skin, turning it black. Their children would inherit their dark skin, as would all future generations.

Before long, the Earth cried out to Zeus for help, and he responded by hurling a bolt of lightning at the chariot. The sun god's son tumbled to Earth, blazing down like a shooting star. Nymphs buried his smoldering body and put a stone over his tomb. "Here Phaethon lies, his father's charioteer," it read. "Great was his fall, yet did he greatly dare."

The story of Phaethon, which survives today mostly through Ovid's telling in *Metamorphoses*, is many different stories bundled together. Among those stories is a tale about heredity. Phaethon's wild ride was an explanation for an inherited difference between people. Ancient philosophers and poets offered many such explanations for why children resembled their parents and why some diseases were inherited. Yet there's also a telling absence in their writings. As far as we know, Aristotle and other ancient scholars never offered instructions for how to alter heredity—how to extirpate inherited diseases or how to improve the animals and plants their lives depended on. Perhaps those ancient scholars thought humans could no more alter heredity than they could alter the course of the sun. And perhaps they thought that anyone who dared seize such power would be overwhelmed and die.

But there's yet another story hidden in Phaethon's tale. It's odd, when you think of it, that a god like Phoebus would have to use horses to pull his heavenly chariot. Certainly they must have been remarkable horses that could gallop across the sky, but they were horses nonetheless, complete with hooves, tails, and manes—the same animals that pulled the chariots of earthly Greeks in races and battles.

And yet the ancient Greeks and their fellow humans transformed their horses in a godlike way: They altered the DNA of the animals, steering them from the genes of their wild ancestors and toward new domesticated sequences. They reared the horses, raising foals to replace their parents in the traces. Each new generation of horses inherited traits from their parents that made them well adapted to this work: powerful hearts, strong bones in their legs, and a willingness to take commands from two-legged apes.

This particular combination of traits seems to have first come together about 5,500 years ago, when nomads in central Asia began to domesticate wild horses. They unknowingly picked out certain variants of certain genes for breeding. Domesticated horses then spread across much of Asia, Europe, and northern Africa in the millennia that followed. The horses of ancient Greece were thus the product of five thousand years of modification, and in later years, people continued transforming them into new breeds. Big workhorses like Clydesdales hauled heavy loads, while Thoroughbreds galloped swifly around racetracks. Every breed of horse inherited a particular combination of variants that altered everything about them—their size, their shape, and even their gait.

The Greeks and other ancient peoples had more control over heredity than Phaethon had had over his father's chariot, in other words. But they had little idea of what they were doing. They could not directly rewrite genes of horses to precisely meet their needs, creating permanent changes that would be passed down to future generations. They could only choose which horses to breed. The desirable genetic variants they blindly selected sat on stretches of DNA with harmful ones, too. Modern horses pay the price for this blind selection, inheriting genetic variants that make them worse at healing wounds than their ancestors, raise their risk of seizures, and create other vulnerabilities.

In the 1800s, a growing number of scientists tried to master heredity's chariot. They ran experiments to find its rules. And yet even in the early 1900s, controlling heredity still seemed like magic—in both the wondrous and dangerous senses of the word. It was no accident that Luther Burbank earned the nickname "the Wizard of Santa Rosa."

When the plant scientist George Shull spent time with Burbank, he realized that the wizard had no magic beyond a good eye for interesting flowers and fruits. And it was Shull, not Burbank, who would become the true pioneer of modern plant breeding. Back at Cold Spring Harbor, Shull ran an experiment on some Indian corn he had rescued from the lab's horse feed. He planted the kernels and then carefully pollinated each plant with its own pollen. In time, he created purebred lines of corn.

The two copies of every gene in each purebred plant were identical. Shull would pick out one line with a quality he liked, such as extra rows of kernels, and then breed it to another desirable line. Their hybrid offspring inherited a copy of each gene from each parent. Remarkably, the hybrid corn would show many of the traits that Shull selected in the inbred lines, while also growing bigger, healthier ears than their parents.

Shull painstakingly improved his inbred lines and found that when he crossed them, they produced even better hybrids. Scientists still argue about why his method worked. It may be that he could eliminate harmful recessive mutations without losing the traits he desired. It may also be that corn and other plants do better when they can use two versions of certain proteins rather than just one. What was immediately clear when Shull began publishing his experiments was that his method would allow farmers to get more food from their crops—what Burbank had originally claimed was his own life's mission.

By the 1920s, many plant scientists were following Shull's example, and before long farmers across the Midwest were filling their fields with hybrid corn. Not only did it produce more bushels per acre, but it also withstood the Dust Bowl droughts better than earlier strains. By the end of the twentieth century, plant breeders using Shull's methods had quintupled their yields. Yet enough genetic variation still remained in the corn plants to ensure that they could breed even better hybrid corn for many years to come.

———

U nderstanding Mendel's Law made Shull's hybrid corn possible. And yet Shull still worked for the most part in ignorance. He had no idea which genes he was selecting or how they made his corn better. He simply mixed the existing variations together. It was his combinations that were new.

Over the next century, scientists would gradually gain more control over heredity. Some dragged X-ray machines into cornfields and fired beams at the tassels. The radiation triggered new mutations that altered the descendants of the corn. Plant mutagenesis, as this method came to be known, threw out new varieties of pears, peppermint, sunflowers, rice, cotton, and wheat. Bombarding barley gave rise to new kinds of beer and whiskey. Scientists also hurled X-rays at mold, creating strains that could make superior penicillin.

Even these successes still depended on a lot of blind luck, though. Heredity remained a slot machine, and plant mutagenesis just gave scientists an extra bucket of coins to play it. More pulls of the arm raised their odds enough that, at some point, the reels would turn up three bars. It wasn't until the 1960s that microbiologists would discover molecular tools that gave them a more precise control over heredity.

Many species of bacteria make proteins called restriction enzymes that recognize a short sequence of DNA and cut the molecule wherever that sequence appears. These microbes use their restriction enzymes to defend themselves against attack—specifically, by destroying the DNA of invading viruses. Tinkering with these proteins, scientists found that they could also use them to cut other pieces of DNA, even genes inside human cells. Loading such a gene onto a plasmid—a ringlet of DNA—these researchers could then move the gene into a microbe.

By the end of the 1970s, researchers created strains of bacteria that carried the gene for human insulin. With the bacteria growing in fermentation tanks, scientists could now manufacture the insulin like living factories. Other researchers went on to use similar methods to do everything from giving crops resistance to viruses to giving mice humanlike hereditary diseases.

Behind these successes, however, were long stretches of effort and failure. It could take years for scientists to discover a gene worth moving from one species to another, and then years more to load it onto a vehicle that could carry it across the species boundary. And learning how to make that transfer to one species didn't help researchers with another. The tools that made it possible to import genes from jellyfish into rats were useless for moving daffodil genes into rice.

And even if scientists succeed in getting genes into a species, they might still fail. The scientists had little control over where a gene would get inserted in an organism's DNA. It might end up in a spot where it could operate smoothly, or it might drop into the middle of other genes, disrupting them and killing its new host. None of these challenges spelled doom for genetic engineering, but they did keep it expensive and limited to labs of scientists with hard-fought wisdom.

It wouldn't be until 2013, over a century after Shull discovered hybrid corn, that scientists would report their discovery of a versatile, cheap way to control the heredity of just about any species. They hadn't thought it up. Just like restriction enzymes before it, it was a system of molecules that bacteria had been using for billions of years to alter their own heredity.

In 2006, Jennifer Doudna was sitting in her office at the University of California, Berkeley, when she got a phone call out of the blue. A Berkeley microbiologist named Jill Banfield wanted to talk to her about something that sounded like *crisper.*

Doudna didn't understand what Banfield was talking about, or why she'd want to call her. But Banfield, who searched for new species of bacteria on mountaintops and ocean floors, seemed like a scientist worth talking to. At the time, Doudna studied the RNA molecules made by bacteria, humans, and other species. Most of her work took place in the quiet confines of a test tube. Banfield could enlighten her about the world beyond the tube.

The following week, Doudna and Banfield met at a café. Banfield introduced Doudna to CRISPR, at least as it was understood in 2006. She drew a diagram for Doudna in a notebook, showing the repeating sequences of DNA that some species of bacteria carried, with different bits of DNA wedged between them.

Banfield at the time was discovering CRISPR regions in the DNA of one species after another. And she could see that some of these bits of DNA had come from viruses. Other scientists had begun to explore the possibility that CRISPR was some sort of defense system that bacteria could use to fight viruses, a system they could pass down to their descendants. But nobody knew how it worked. One possibility was that bacteria made RNA molecules to seek out the viruses. Since Doudna was an expert on RNA in bacteria, Banfield wondered if she'd be willing to help find out.

Doudna took Banfield up on the offer. She hired a postdoctoral researcher named Blake Wiedenheft to work exclusively on CRISPR, and then gradually the rest of her lab switched over to studying it. A few other labs were also investigating CRISPR at the time. In 2011, Doudna joined forces with a French biologist named Emmanuelle Charpentier, and together they figured out that CRISPR, like restriction enzymes, destroys viral DNA.

But there was a profound difference between these two lines of defenses. Restriction enzymes had a shape that allowed them to recognize only a single short stretch of DNA, which could appear in many places in a genome. Microbes protected their own DNA from this attack by methylating their own sequences. Viruses, unable to methylate their genes, were left vulnerable to attack.

The Cas9 enzymes produced by the CRISPR system were far more sophisticated. Bacteria produced RNA guides that could lead the enzymes to one—and only one—stretch of DNA. By storing different RNA guides in their DNA, bacteria could precisely recognize several different strains of viruses.

Like any molecular biologist, Doudna was well aware of how restriction enzymes had helped create the biotechnology industry. She wondered if CRISPR might have a similar power. If it could recognize any stretch of DNA in a virus, perhaps Doudna and her colleagues could create RNA guides that would lead the enzymes to a particular spot in the DNA of a cucumber. Or a starfish. Or a human.

To test this idea, Doudna and her colleagues tried to cut out a piece of DNA from a jellyfish gene. (The gene is a common tool for molecular biologists, because it makes a glowing protein that can light up a cell like a microscopic jack-o'-lantern.) Doudna and her colleagues picked for their target a twenty-letter stretch. After synthesizing RNA molecules that matched the target, they mixed all the molecules together in a test tube. The RNA guides and Cas9 enzymes combined, and sought out the jellyfish genes. When

Doudna and her colleagues looked at that DNA afterward, they discovered it was now cut into precisely the fragments they had hoped to create. Four more trials, using RNA guides that sought different targets in the gene, worked just as well.

"We had built the means to rewrite the code of life," Doudna later recalled.

After Doudna and her colleagues published the details of the experiment in 2012, a CRISPR scramble began. Her team, as well as others, tried to get the CRISPR molecules into living cells. Researchers learned not only how to cut out pieces of DNA from those cells but how to repair it as well.

In one of these experiments, Feng Zhang and his colleagues at the Broad Institute in Cambridge, Massachusetts, delivered a pair of CRISPR systems into human cells. The molecules landed on two neighboring targets within a single gene and snipped the DNA at both sites, cutting out the short stretch in between. The cell's own repair enzymes then grabbed the two sliced ends and stitched them back together. The procedure, in other words, surgically removed a piece of DNA, leaving no scar behind. And when the cell divided, its descendants inherited that deletion.

Before long, scientists were starting to use CRISPR to replace stretches of genes with new sequences. Along with the Cas9 enzymes and RNA guides, the researchers would deliver small pieces of DNA to cells. After the enzymes cut out a section of DNA, the cells would patch the new pieces into the gap.

CRISPR was a drastic improvement on both X-ray mutagenesis and restriction enzymes. CRISPR did not introduce random mutations like mutagenesis. Nor was it limited to inserting an existing gene from one species into another. Since researchers were now able to synthesize short pieces of DNA from scratch, CRISPR could potentially let them make any sort of change they wanted to any species' own genes.

In the 1970s, Rudolf Jaenisch, a biologist at MIT, had used restriction enzymes to engineer mice for the first time. With the advent of CRISPR, he wondered if he could create new lines of mice with that tool as well. Collaborating with Feng Zhang, he and his graduate students and postdoctoral researchers played around with CRISPR until they found a chemical recipe they could use to slip the molecules into a fertilized mouse egg. They were able to alter as many as five different genes at once by delivering five different RNA guides. Jaenisch and his colleagues then implanted these altered eggs in female mice, where they developed into healthy pups. Eighty percent of the time, Jaenisch's team successfully engineered precisely the changes they desired.

A new generation of graduate students silently thanked Jaenisch every day for making their lives easier. Many PhD projects had to start with the creation of a mouse model to study a gene or a disease. It typically took eighteen months to create a line of mice, and often it took more than one try to get the mouse right. Now, with CRISPR, Jaenisch needed only five months to get the job done.

———

I was working as a reporter during those frenzied years, and I did my best to keep up with CRISPR's advances. But very soon the parade of CRISPR animals became a stampede. Scientists were altering the DNA of zebrafish and butterflies, of beagles and pigs. By 2014, it dawned on me that I was witnessing the beginning of something enormous. Biologists began speaking about their life before and after CRISPR. But I didn't truly appreciate what CRISPR meant to scientists until I returned to Cold Spring Harbor one early spring day to spend an afternoon in a giant greenhouse with cathedral-like ceilings made of glass.

A plant scientist named Zachary Lippman led me down narrow aisles past rows of pots, each with a plant climbing a tall stake. Although he was still young, Lippman had a pair of gray patches in the dark beard framing his chin. I wondered if the six children he and his wife were raising might have something to do with them. "They say I do genetics at the lab and then do genetics at home," Lippman said.

Lippman has a long history of showing off his plants. Growing up in Connecticut, he worked on a farm where he learned how to grow giant pumpkins. At the peak of the growing season, they put on ten or fifteen pounds a day. "To me the interest was how the hell does this thing get so big, and how can I get it bigger?" Lippman said.

Heading to Cornell for college, Lippman majored in plant breeding and genetics. There he discovered that scientists had long been asking his boy-hood question—not just about pumpkins but about other fruits and vegeta-bles. One of the key changes to crops during the Agricultural Revolution was making them bigger—to turn the stubby fruits of teosinte into long ears of corn, to swell the thin pale roots of carrots into stout orange tubers.

Using traditional methods for studying genes, the scientists had found some of the mutations that made these changes possible. They had done a lot of this work on tomatoes, Lippman discovered, because their biology lends them well to genetic experiments. Lippman followed in their scien-tific footsteps by studying tomatoes, too.

"Look at these tiny little berries," Lippman said to me. He had stopped at a tomato plant towering over us. Grabbing a stem, he cradled its fruit. "These plants here are the closest that we know to the first domesticated forms of tomato," he said.

The domestication of tomatoes by the indigenous farmers of Peru turned blueberry-sized fruits into the larger kind we find in supermarkets and at farm stands. Lippman's research has helped reveal how those earlier breeders made tomatoes big. It turns out that they had to change the shape of tomato flowers.

When a bud on a tomato plant begins developing into a flower, it first divides up into wedges, called locules. From those locules will develop the petals of the tomato flower. And at the center of the flower, those same loc-ules will give rise to the sections of a tomato. One gene controls how many locules form on a tomato plant. Mutate the gene, and the plant makes more locules. And more locules develop into a bigger tomato.

This locule-controlling gene was not the only one to mutate during the domestication of tomatoes. Lippman's research has also revealed that the crops adapted to the length of day as the crops were moved to different parts of the world.

Lippman and his colleagues found that wild tomatoes, which grow at the equator in South America, are adapted to getting twelve hours of sunlight every day through the year. When they brought wild tomato plants from the Galápagos Islands north to Cold Spring Harbor, they discovered that the plants fared poorly, thanks to the long New York summer days. The plants responded to the extra sunlight with flower-suppressing proteins, delaying the growth of their fruits until the end of the season. But domesticated tomatoes that grow in Europe and North America have acquired mutations that caused the plants to make fewer anti-flowering proteins in the summer.

In 2013, Lippman learned that scientists had figured out how to use CRISPR to edit genes in a plant for the first time. He got hold of the molecules and tested them out on tomatoes to see how well they worked. No genetic tool he had used before came close. "It was black and white," Lippman said. "We just sat down and had brainstorming sessions, just saying, 'What can we do?'"

One of the first items on their list was to get tomatoes to stop making flower-suppressing proteins altogether in response to long summer days. They used CRISPR to cut out the genetic switch for this activity in domesticated tomatoes. When Lippman and his colleagues planted the seeds of these altered plants, they grew their flowers—and their tomatoes—two weeks ahead of schedule. They might thrive in places with much shorter summers. "Now you can start to think about growing some of your best tomato varieties in even more northern latitudes, like in Canada," Lippman said.

Lippman had, in effect, created a new crop variety in one step. He did not need the eye of Luther Burbank, scanning thousands of plants for a single promising mutant each year. Nor did he need to transfer a gene from some other species to create a genetically modified crop. He directly altered the tomato's own genes, using the knowledge he had gained about how tomatoes work.

This success made Lippman's brainstorming more ambitious. He wanted to turn a wild plant into a domesticated crop. And for his new experiments, he chose ground-cherries.

I couldn't really appreciate what he was doing, Lippman assured me, unless I ate some ground-cherries first. He brought me a plastic box filled with golden fruits the size and shape of marbles. When I bit into a ground-cherry, I tasted a rich flavor that hovered somewhere between pineapple and orange. The fruits were so delicious and so distinctive that I wondered why I hadn't had one before. The reason, Lippman explained to me, is that they're wild.

Ground-cherries (known scientifically as *Physalis*) live across much of North and South America. They grow into bushes and develop their fruit inside a lantern-shaped husk. Native Americans gathered ground-cherries to make sauces, and European settlers followed suit. Some collected the seeds and planted them in their gardens. Today you can buy a packet of ground-cherry seeds, and sometimes you can find the fruits for sale at a farmers' market or a gourmet store. But because they're wild, ground-cherries remain an oddity rather than a crop. The fruits ripen one by one through a long season, and gardeners have to wait for them to drop to the ground before collecting them—hence their name.

Lippman has long had a scientific curiosity about ground-cherries, because they belong to the same family as tomatoes. Their close evolutionary relationship means that they have a lot of biology in common. Both ground-cherries and tomatoes form their flowers and fruits from locules, for example, and they use related versions of the same genes to build them. It's intriguing to Lippman that tomatoes were domesticated but their cousins, the ground-cherries, never were.

One possibility for this difference may be that ground-cherry DNA doesn't lend itself to easy domestication. Tomatoes, like humans, have two copies of each chromosome. But ground-cherries have four. To breed ground-cherries for some particular trait, farmers need to find plants that inherited the same mutation on all four copies of one of their genes. It occurred to Lippman that he could use CRISPR to edit mutations directly into ground-cherries instead.

Lippman scooted down an aisle, the leaves brushing against his shoulders. He found a ground-cherry bush that he had edited with CRISPR. It had flowered a few days before, and by now the petals had fallen off. The sepals—the leaflike petals that surround the flower—had expanded to form the papery lanterns inside which the fruit would now develop.

On an ordinary ground-cherry plant, these lanterns would have five sepals. Lippman peeled off the sepals on his edited plant, counting them as he went: "One, two, three, four, five, six, seven."

Once he had pulled away all the sepals, Lippman revealed a tiny young ground-cherry fruit inside. It had seven locules now instead of the normal five.

"We could never do this with traditional breeding," Lippman said. "And we got this"—he snapped his fingers—"in one generation. All four copies of the gene mutated."

Lippman was soon going to test other edits. He would edit genes that controlled when the fruits fell from the bushes, so that farmers wouldn't have to rummage on the ground for them. He would make a change to get the plants to ripen their fruits in batches rather than a few at a time. He would adjust the plants' response to sunlight so they would start producing fruit early in the growing season. They would grow to a fixed height so that farmers could use machines to gather them.

Lippman planned on starting by editing plants for one trait at a time. If he succeeded, he would then create RNA guides that could alter all the traits at one shot, in one plant. When that ground-cherry plant reproduced, its offspring could inherit all the genetic machinery required to be a domesticated plant instead of a wild one.

"I know this sounds a little ridiculous," Lippman confessed, "but I think this will be the next berry crop." Having listened to his plan, I didn't think it was ridiculous at all. I thought Lippman was being too modest. He was trying to replay the Agricultural Revolution on fast-forward. Instead of a thousand years, he might only need a single growing season.

———

To CRISPR, ground-cherries and humans proved pretty much the same: Their DNA was equally easy to cut.

Scientists quickly began using CRISPR to edit the genes of human cells, to answer questions about ourselves that once seemed unanswerable. We each carry about twenty thousand protein-coding genes, and thousands more genes that encode important RNA molecules. But how many of those do we really need? When mutations shut down some genes, it leads to lethal hereditary diseases. Yet many of us walk about in good health despite inheriting some broken genes. Scientists have long wondered just how many genes in the human genome are absolutely essential to our survival. But they knew it was impossible to actually compile that catalog.

CRISPR made it possible. In 2015, three separate teams of scientists used CRISPR to shut down all the protein-coding genes in human cells, one at a time, to see if the cells could survive without them. They ended up with lists that were pretty much identical. About two thousand genes, only about 10 percent of all the protein-coding genes in the human genome, proved to be essential. The experiments showed that many genes were expendable because they had backup. If they failed, other genes could take over their jobs.

Other scientists began experimenting with CRISPR on human cells with a different goal: to invent new forms of medicine. In December 2013, a team of Dutch researchers demonstrated how CRISPR medicine might work. They took samples of cells from people with cystic fibrosis and raised colonies of them in dishes. The cells all shared the same defective mutation in a gene called CFTR. The scientists fashioned CRISPR molecules that could chop out the mutation, and then they stitched a working version of that DNA in its place.

It soon became clear that CRISPR's power was not limited to altering somatic cells. It could change the DNA in germ-line cells as well. In December 2013, a group of scientists at the Shanghai Institutes for Biological Sciences in China reported the results of an experiment on mice that suffered from hereditary cataracts. The scientists injected CRISPR molecules into mouse zygotes and they repaired the mutant gene. The altered mice grew up to be fertile adults, and their descendants gazed through clear eyes.

Jennifer Doudna's delight now began getting undercut by worry. CRISPR was turning out to be far more powerful than she had expected. Xingxu Huang, a geneticist at the Model Animal Research Center of Nanjing University in China, and his colleagues used CRISPR to alter three genes in monkey embryos. They implanted the embryos in a female monkey, and she later gave birth to a pair of healthy twins. If the monkeys had offspring of their own, they'd inherit the CRISPR-altered genes, too.

A reporter sent Doudna an advance copy of the monkey paper in January 2014 for her comment. After she read it, she couldn't stop wondering when the first experiments on human embryos would take place. And it was about then when the nightmares started.

Sometimes Doudna dreamed she was back in Hawaii, where she had grown up, standing alone on a beach. She saw a low wave in the distance coming toward her, and after a while she realized that it was actually a tsunami. At first she was terrified, but then she found a surfboard and swam straight at the wave.

In another recurrent dream, a fellow scientist asked her to meet with someone very powerful. She went into a room. The powerful person turned out to be Hitler. In Doudna's dream, he had the face of a pig. He kept his back turned to Doudna as he jotted down notes.

"I want to understand the uses and implications of this amazing technology," the pig-faced Hitler told her.

Doudna woke up with her heart pattering. What, she asked herself, have we done?

———

Doudna was hardly the only person getting visits from Hitler. In 2015, a reporter asked the inventor Elon Musk if he was considering getting into the business of reprogramming DNA. Musk is the sort of entrepreneur who blithely sets out to replace the world's fleet of gas-powered cars with electric ones while simultaneously building the first recyclable rockets. But gene editing gave him pause.

"How do you avoid the Hitler Problem?" Musk asked in reply to the reporter's question. "I don't know."

We must never forget Hitler's genocidal ideology. But we need to remember it for what it was, rather than wrap it around scientific advances that took place seventy years after he died. Hitler wanted Germany's scientists to conquer the future—to build the world's first atomic weapons, to create the first computers. But when it came to biology, he wanted Nazi scientists to revive a mythical past. He had no need for new genes, because Aryans already had all the genetic superiority they could hope for.

The genetic nostalgia of the Nazis was so powerful that it even extended to other species. Hermann Göring, Hitler's most powerful deputy, became a patron of a project to restore the wild ancestors of cattle. Known as aurochs, these giant animals had become extinct in the Middle Ages. Under Göring's direction, zoologists searched Nazi-held countries for cows that seemed to retain a few vestigial features of aurochs. They bred the cattle, looking among the calves for the ones that appeared to step even further back in time.

Göring's goal was to release the restored aurochs in Poland, where they would roam one of the last primeval European forests. He pictured himself as a modern Siegfried from Wagner's *Ring of the Nibelung*, hunting the same noble beasts as his Aryan ancestors. To clear the path for his romantic vision, Göring emptied the Polish forests of Jews, Polish resistance fighters, and Soviet partisans.

The Nazi plans for humanity followed the same lines. The Aryan bloodline needed to be protected, revived, and purified. Systematic murder would protect future generations of Aryans from inferior heredity. And planned pregnancies would concentrate more Aryan blood in future generations, just as breeding would turn cows back into their auroch ancestors. The Nazis even forced blond, blue-eyed people to join an association known as Lebensborn, designed to produce children to restore the Aryan race.

After Hitler's defeat, Nazism and other forms of white supremacy did not disappear. As science advanced, latter-day Nazis kept distorting it to feed their genetic nostalgia. They took genetic ancestry tests in the hopes of demonstrating that they were indeed white. "Pretty pure damn blood," one member of a computer forum called Stormfront crowed when he got his results. The myth of white purity has endured even after the study of ancient DNA has proven that Europeans have inherited genes from wave after wave of migrants—people with ancestries separated by tens of thousands of years of history, mixing their genes together. A fair number of other Nazis discovered to their horror that they had some Jewish or African ancestry. They coped with the news either by dismissing the results as statistical noise or by arguing that all you need to do to know your past is look in the mirror—a kind of personal bald eagle test.

It's also a mistake to use Hitler as a label for all of eugenics. World War II and the horror of the Holocaust brought Hitler's particular version of eugenics to a halt. It also forced conservative forms of eugenics in places like the United States and Great Britain into retreat. But ever since Francis Galton coined the term, eugenics has taken on many different forms, each shaped by the politics and cultures of the people espousing it. And after World War II, a progressive form of eugenics survived. It even rose to prominence. The leading voice for this so-called reform eugenics was a protégé of Thomas Hunt Morgan, the American-born biologist Hermann Muller.

After Muller learned how to breed *Drosophila* flies in Morgan's lab at Columbia, he went to the University of Texas, where he used X-rays in the 1920s to create new mutations in the insects. He also came under the FBI's surveillance for advising a left-wing student newspaper that espoused suspicious goals such as social security for retired people, equal opportunity for women, and civil rights for African Americans. Muller grew disgusted by the American eugenics movement in the 1920s—its shoddy science, its push to sterilize the weak and ostracize immigrants—and became one of its most outspoken opponents.

In 1932, Muller was invited to speak at the Third International Eugenics Congress at the American Museum of Natural History in New York City. Charles Davenport and the other organizers of the meeting may have assumed he would limit his talk to his work on mutations in flies. To their horror, Muller planned instead to use his speech to burn the American eugenics movement to the ground. Davenport tried to get Muller to cut his talk, originally slated for an hour, down to fifteen minutes. Then he demanded it be cut to ten. Muller pushed back, accusing Davenport of trying to stifle dissent, and delivered his full address.

On August 23, Muller shocked his audience by condemning the idea that poverty and crime in the United States were due to heredity. Only in a society where people's needs were met—where children could grow up in the same environments—could eugenicists ever hope to improve humanity. In a country as rife with inequality as the United States, eugenics could do nothing of the sort. Instead, Muller said, it simply fostered "the naive doctrine that the economically dominant classes, races and individuals are genetically superior."

Muller grew so disenchanted with conditions in the United States that he accepted an invitation to do research in Germany. But once he got there, he realized he had made a very bad choice. When Hitler became chancellor, Nazis raided the institution where Muller worked, and he worried that his socialist tendencies and Jewish roots could put his life at risk. Another invitation seemed to offer him a new refuge: Muller left for the Soviet Union, where he was asked to establish a genetics lab in Leningrad.

At first, Muller was happy there, working with Soviet students on groundbreaking research. But his timing turned out once again to be disastrous. A plant scientist named Trofim Lysenko rose to prominence by arguing that genetics was a fraud, and that heredity was as malleable as clay. Muller took on Lysenko in a public debate, but the audience of three thousand scientists and farmers shouted him down. When Stalin began arresting and executing scientists, Muller fled the Soviet Union.

He went to Spain to serve as a doctor in the civil war and then traveled to Scotland to teach at the University of Edinburgh. Finally, in 1940, Muller returned to the United States. There he found stability at last, becoming a professor at Indiana University. His work on mutations had proven to be some of the most important research in modern biology, and in 1946, Muller was awarded the Nobel Prize for his work on mutations. Not long after, he was elected the first president of the newly created American Society of Human Genetics. Now one of the most prominent scientists in the postwar United States, Muller made the most of his fame by promoting his own vision for social progress.

With the fall of Nazi Germany, Muller believed, the fallacies of its eugenics were exposed. But, he warned his fellow scientists, "it is by no means all dead and buried yet, but represents a continuing peril, to be vigilantly guarded against by all serious students of human genetics." Muller urged a fight against American eugenicists—"racist propagandists," as Muller called them—who would try to smuggle their old ideology into postwar genetics.

But Muller also used his newfound megaphone to call for a different kind of eugenics. "Eugenics, in the better sense of the term, 'the social direction of human evolution,' is a most profound and important subject," he said.

In his research on mutations, Muller had made an important discovery: From one generation to the next, a species can become burdened with a growing load of mutations. Every new offspring may spontaneously gain new mutations, most of which will be fairly harmless. But added together, they could create diseases and lower fertility. In the wild, natural selection eliminated many of these new mutations. In our own species, Muller worried that our mutation load would become dangerously large. Thanks to medicine and other advances, natural selection had grown weak in humans, unable to strip out many harmful mutations from the gene pool.

It was naive to deny the existence of humanity's mutational load, Muller argued, but it was even more naive to try to blame it on some despised race, or on people with some form of intellectual disability. "None of us can cast stones," Muller said, "for we are all fellow mutants together."

Still, something had to be done—something beyond "making the best of human nature as it is, the while allowing it to slide genetically downhill, at an almost imperceptible pace in terms of our mortal time scale, hoping trustfully for some miracle in the future," Muller said.

Muller had a plan. He called it "Germinal Choice."

Traditionally, children could inherit genes only from people who had sex. But in the mid-1900s, sexless reproduction slowly began to emerge. Animal breeders had led the way, perfecting the art of artificial insemination. A prize bull could father countless calves without ever leaving his stanchion. Once breeders figured out how to safely freeze semen, bulls began fathering calves long after their deaths.

Doctors quietly started imitating veterinarians, using donated sperm to help couples when the husband was infertile. When the practice came to light, it was roundly condemned. The pope declared donor insemination adultery. In a 1954 divorce case, an Illinois judge ruled that a child produced by donated sperm was born out of wedlock. But the practice grew more common, and less controversial. By 1960, about fifty thousand children in the United States had been conceived with donated sperm.

Muller's Germinal Choice plan would turn artificial insemination into a national—perhaps even global—campaign against the mutation load. The sperm from the finest specimens of manhood would be collected and stored away in subterranean freezers, to protect their DNA from radiation and cosmic rays. The sperm from a single man could theoretically produce hundreds, perhaps thousands, of children. In the 1950s, eggs were proving harder for scientists to handle than sperm, but Muller was optimistic that, at some point, the underground germ-cell bunkers could store samples from superior women as well.

The public would then be educated about the coming mutational catastrophe and be invited to use the superior eggs and sperm to build their own families. Forward-minded couples would appreciate the scale of the threat and be the first to step forward. To avoid any awkward encounters with the biological parents, Muller would offer gametes only from people who had been dead for twenty years.

The volunteers would have to be ready to withstand the ignorant mockery of others. But as they raised their obviously glorious children—with the "innate quality of such men as Lenin, Newton, Beethoven, and Marx," Muller promised—other parents would follow in their path. "They will form a growing vanguard that will increasingly feel more than repaid by the day-by-day manifestations of their solid achievements, as well as by the profound realization of the value of the service they are rendering," Muller predicted.

Muller's Germinal Choice plan was met with nods and curiosity. Leading scientific journals asked him to write about it. Conferences invited him to speak. Newspapers and magazines ran interviews with him. Muller's fellow Nobelists considered Germinal Choice a step in the right direction.

For all its science-fiction luster, however, Muller's Germinal Choice was still a fairly traditional form of eugenics. It was what Francis Galton had called for in the nineteenth century, an imitation of what animal and plant breeders had been doing for centuries: combining existing genetic variations into better arrangements that could be inherited by descendants. Germinal Choice didn't require rewriting genes. Apparently, that was too much even for Hermann Muller to imagine.

———

While Muller's vision of a public Germinal Choice program never came to pass, a private version did. Sperm banks emerged from the shadows, taking on not just married heterosexual couples but single women and lesbian couples as well. By the early 2000s, over a million children had been born with donated sperm in the United States alone. While sperm banks tended to keep their donors anonymous, they allowed customers to pick men based on certain traits—traits, it must be said, that probably don't get transmitted in the chemistry packed inside a sperm cell.

The Fairfax Cryobank, located in Virginia, lets customers search by a donor's astrological sign, favorite subject in school (arts, history, languages, mathematics, natural science), religion, favorite pet (bird, cat, dog, fish, reptile, small animal), personal goals (community service, fame, financial security, further education, God/religion), and hobbies (musical, team sports, individual sports, culinary, craftsman). Looking at that list, I picture parents putting their child in a wood shop and waiting for freshly carved coffee tables to pile up.

In order to become a sperm donor, men have to get through a screening for sexually transmitted diseases, as required by the FDA. But there aren't any regulations about checking DNA, and so genetic screening varies from clinic to clinic. Many of them look at a prospective donor's family history for signs of a hereditary disease. Many also carry out a few genetic tests to see if donors are carriers of diseases such as cystic fibrosis. Sometimes a dangerous variant can slip through the screening. In 2009, for example, a Minnesota cardiologist discovered that a young patient with a hereditary heart condition had been the result of sperm donation. He tracked down the donor and discovered that he carried a dangerous variant. Out of twenty-two children he fathered with donated sperm, nine inherited his faulty gene. One of them died of a heart attack at age two.

The falling price of DNA sequencing may be able to eliminate most of these disasters. Rather than look for a handful of common disease variants, it is now possible to scan every protein-coding gene in a potential sperm donor. Men with a dominant disease-causing mutation could be barred altogether from donating sperm. To avoid recessive diseases, doctors could sort through sperm and eggs to make sure they couldn't combine two dangerous mutations.

Muller was right to expect eggs to be harder to use for Germinal Choice. In the 1930s, scientists were able to fertilize rabbit eggs in a dish with rabbit sperm and coax the embryos to start dividing. But it wasn't until the 1960s, however, that two scientists—a University of Cambridge physiologist named Robert Edwards and a gynecologist named Patrick Steptoe—figured out how to get viable eggs from women. Their next challenge was to find a chemical cocktail—the "magic fluid," as Edwards called it—that could keep the eggs alive in a dish long enough to be fertilized. In 1970, Edwards and Steptoe announced that they had succeeded at last. After they fertilized human eggs, they managed to keep them alive for two days, during which they divided into as many as sixteen cells.

Edwards spoke about this milestone at a meeting in Washington, DC, in 1971. One of his fellow panelists was a professor of religion named Paul Ramsey. After Edwards finished speaking, Ramsey declared the procedure an abomination that must be banned. He believed it would bring the world closer to "the introduction of unlimited genetic changes into human germinal material." Heredity, in other words, was a sanctuary that humans dare not enter.

Edwards and Steptoe weren't scared away by Ramsey's warnings. Instead, they invited couples having trouble getting pregnant to come to them for help. In 1978, they had their first success: a healthy girl named Louise Joy Brown. Louise's birth led people to see in vitro fertilization not as the first step to human genetic engineering but as a treatment for infertility. In the 1980s, in vitro fertilization clinics opened up across the world, capitalizing on the pent-up demand of struggling couples. The procedure remained hit-or-miss, though, with many embryos failing to implant. To improve their odds, fertility doctors would produce batches of embryos and then pick out the healthiest among them.

In time, it became possible to inspect an embryo's DNA, too. Scientists would analyze a single cell removed from an embryo in its first few days of life and analyze its genes. (If a cell is removed at that stage, the remaining ones can still proliferate into a healthy fetus.) Fertility doctors could use this method to reduce the odds that parents would pass down a genetic disorder to their children.

In one early experiment, a team of British doctors treated women who carried disease-causing mutations on one of their X chromosomes. The women themselves were healthy, thanks to their second X chromosome, which lacked the mutation, but any sons they might have would run a 50 percent chance of developing the disorder.

There was a straightforward way to avoid this suffering: make sure the women had only daughters. In 1990, the British doctors screened the embryos of two women who carried X-linked mutations. One had a variant causing mental retardation, the other a potentially devastating nerve disorder. The women both underwent in vitro fertilization, and then the doctors plucked a cell from each embryo to examine. They used molecular probes that could detect a distinctive segment of DNA repeated many times over on the Y chromosome, and only the Y chromosome. The doctors set aside any embryo that tested positive and used the rest for implantation. Nine months later, each mother gave birth to twin girls. Because the babies carried a normal X chromosome from their fathers, all of them turned out healthy.

By the early 2000s, it became possible to test embryos for mutations on other chromosomes, too. Karen Mulchinock, a woman living in the English city of Derby, had grown up knowing that Huntington's disease ran in her family. Her grandmother had died of it, and she had watched her father decline through his fifties, dying at age sixty-six. Mulchinock got a test at age twenty-two and found she carried a faulty copy of the HTT gene as well. She and her husband decided to use in vitro fertilization to prevent the next generation of her family from inheriting it. In 2006, she had her eggs harvested, and her doctors tested the embryos for the Huntington's mutation. Over the course of five rounds of IVF, she gave birth to two children, neither of whom had to worry about the disease. "The curse is finally broken," she said.

Couples who worried about other inherited disorders began coming to fertility doctors to ask for the same tests, tailored instead for their own mutations. When an English couple had their first child, they were surprised that she had PKU. Like Pearl and Lossing Buck, neither of the parents knew before that they carried a faulty copy of the PAH gene. The parents decided to have more children, using preimplantation genetic diagnosis to prevent their other children from inheriting it.

After fertility doctors produced a set of embryos from the parents, they tested the PAH in each one. They successfully detected the mutations in some of the embryos, and implanted the mutation-free ones in the mother. In 2013, they reported that she gave birth to a healthy boy. He was free not only of PKU but of the worry of passing down a faulty PAH gene to his own children.

Preimplantation genetic diagnosis has grown more popular as the years have passed, not just in Europe and the United States but also in developing economies such as China. But the procedure is still rare. Despite the gripping stories of parents like Karen Mulchinock, only a tiny fraction of people affected by Huntington's disease (roughly 200,000 people worldwide) use the procedure. The cost puts it out of reach of many. Even in Europe, where the procedures are covered by government-provided health insurance, few people with Huntington's disease follow Mulchinock's example. Between 2002 and 2012, only an estimated one in one thousand cases of Huntington's disease was prevented.

Many children with Huntington's disease don't get tested themselves, since there's no treatment they could get if they turned out positive. Because

Huntington's disease doesn't affect people until after fifty, they usually start their families long before they find out if they inherited the allele. If they have a 50 percent chance of having Huntington's disease, the odds for their children are one in four. They may be so busy caring for their sick parent that they don't want to bother with the time, money, and frustration that in vitro fertilization demands.

In other words, we are not living in Muller's eugenic utopia. Nor are we living in a nightmare of the sort Aldous Huxley imagined in his 1932 novel *Brave New World*. Of the small fraction of people who are using in vitro fertilization, an even smaller fraction is using it to control the inheritance of their children. We have the technology right now to effectively eradicate Huntington's disease from the planet, along with many other genetic disorders. But the messy realities of human existence—of economics, emotions, politics, and the rest—override the technological possibilities.

———

I n April 1963, a microbiologist named Rollin Hotchkiss traveled from New York City to the town of Delaware, Ohio. He had been invited to take part in a meeting that must have felt at the time like a feverish dream. As Hotchkiss later put it, he and nine other biologists spent a day together at Ohio Wesleyan University "to consider whether man can and will change his own inheritance."

One of Hotchkiss's fellow speakers was Hermann Muller. Muller described to the audience his plan for sperm banks and Germinal Choice. To Hotchkiss and the other scientists, there probably wasn't much surprising in what Muller had to say, given that he had been talking about his reform eugenics for more than thirty years. But when it was Hotchkiss's turn to talk, he described something that was fundamentally different from Muller's Germinal Choice—or from any of the eugenic breeding schemes bruited over the previous century.

Hotchkiss raised the possibility of directly altering human DNA. He used a new term to describe what he and some of the other scientists at the meeting envisioned: *genetic engineering*.

It might seem odd for a microbiologist to talk about changing humanity's inheritance. But in 1963, Hotchkiss had come as close as anyone on Earth to carrying out genetic engineering. In the 1950s, he had begun working with Oswald Avery, following up on the original experiments on the "transforming principle" that turned harmless bacteria into killers. Hotchkiss and his colleagues performed more sophisticated versions of Avery's original experiments, proving beyond a shadow of a doubt that the transforming principle was DNA. By moving this DNA into bacteria, Hotchkiss was effectively engineering their genes. In later years, Hotchkiss discovered how to transform bacteria in other ways—moving genes into microbes to make them resistant to penicillin, for example.

At the Delaware meeting, Hotchkiss predicted that the same procedure would be used on humans. "I believe it surely will be done, or attempted," he said.

After all, Hotchkiss pointed out, our species was always searching for improvements. We started by finding better food and shelter, and now our search had evolved into modern medicine. When Hotchkiss spoke in 1963, doctors were celebrating their recent victory over PKU, thanks to their ability to identify babies with the hereditary condition and treat them with a brain-protecting diet. "We cannot resist interfering with the heritable traits of the phenylketonuric infant by feeding him tyrosine at the right time to form a normal nervous system," Hotchkiss said. If scientists learned how to rewrite the faulty genes that caused PKU, Hotchkiss predicted, it would be hard to resist the temptation to do it in people. "We are going to yield when the opportunity presents itself," he said.

Hotchkiss left the Delaware meeting convinced that the world had to get ready for that opportunity. Humanity had to think ahead to all the benefits and risks that the opportunity might bring. Hotchkiss gave lectures and wrote scientific prophecies. Genetic engineering would not follow the traditional eugenic game plans, he said, driven by government decree. It would instead be driven by consumers. People would be persuaded by seductive ads for the latest "gene replacements" to alter their own DNA.

At first, Hotchkiss predicted, doctors would use genetic engineering to cure hereditary diseases like PKU in both adults and children, changing their genes as Hotchkiss changed genes inside bacteria. "One would presumably want to act at the earliest possible time in the development of the organism," Hotchkiss said. "Even *in utero*."

Hotchkiss could see the appeal of genetically engineering embryos. Doctors might be able to fix a genetic defect across much of the body if they could manipulate just a tiny clump of cells. But these doctors might accidentally alter germ cells, too. If those unborn patients grew up and had children, they might very well inherit the gene replacement. And they would pass the gene replacement down to their own children.

"Now one is meddling with the gene pool of the entire race," Hotchkiss warned.

If a gene replacement turned out to be harmful, it would be bad enough to make one patient suffer. But if that gene was replaced in the germ line, future generations could inherit the suffering as well. To Hotchkiss, the decision to alter the genes of people yet to be conceived was an assault on liberty. No man should ever be given total discretion to determine his brother's fate. The same held true for his great-grandchild.

Hotchkiss turned out to be a pretty good prophet. To look into the future of genetic engineering from 1964, he could rely only on what little hard evidence existed at the time—much of it coming from his own crude experiments on bacteria. Within a decade, some of the pieces of his vision would already fall into place. Rudolf Jaenisch was breeding mice with DNA he had pasted into their genomes. Robert Edwards and Patrick Steptoe were growing human embryos in petri dishes. And by the mid-1970s, some scientists were even trying to cure hereditary diseases in people, with Hotchkiss's gene replacements. They called it gene therapy.

One of the pioneers of gene therapy was a hematologist at UCLA named Martin Cline. He developed a method for getting genes into mouse cells, jolting the cells with a pulse of electricity to open temporary pores in their membranes. As a hematologist, he turned his mind immediately to blood disorders. Blood cells came from lineages of stem cells nestled in bone marrow. If Cline could slip a working version of a broken gene into a patient's stem cells, their daughter cells would inherit it. He could thus create a lineage of healthy blood.

Cline tested this idea first on mice. After injecting the engineered stem cells back into the bones of the animals, he waited two months to let them multiply. Cline then drew blood circulating through the mice and inspected the cells. Half of the cells had inherited the gene he had added.

That was good enough for Cline. He immediately set out to use gene therapy to cure people. He chose for his disease beta-thalassemia. This hereditary disorder disables a gene called HBB, leaving blood cells unable to build hemoglobin. People with beta-thalassemia can die because their blood can't deliver enough oxygen around their bodies. To Cline, the need for a cure justified trying out his new gene therapy techniques on people. But when he submitted his proposal to UCLA, the university judged it reckless and turned him down. That didn't stop Cline. He went abroad, recruiting a patient in Israel and another in Italy.

In 1980, Cline performed gene therapy on both patients. He extracted cells from their bone marrow and added working HBB genes to them. Then Cline injected the altered cells back into the patients' bones, where they could proliferate. The new bone marrow cells would inherit the HBB cells and produce working blood cells.

At least, that was the plan. After the procedure, Cline's patients didn't experience any improvement in their symptoms. And when word got out of the experiment, the scientific community roundly condemned Cline. Not only had he leaped far beyond his studies on mice, but he had changed his protocol midway through the human experiment without telling anyone— not his colleagues, not the committees overseeing his research, not even his two patients.

In the wake of the scandal, the National Institutes of Health took away Cline's grants, and UCLA forced him to resign as chairman of his department. In an editorial headlined "The Crime of Scientific Zeal," the *New*

York Times even singled him out for condemnation. "He was rightly punished," they declared.

With all the news about Cline's reckless experiments, test-tube babies, and human-bacteria chimeras, a general alarm about genetic engineering began to blare. In 1980, President Jimmy Carter dispatched a commission to explore the ethical landscape of human genetic engineering, and soon after, Congress asked the Office of Technology Assessment to also look into the matter. They followed Hotchkiss's lead fifteen years earlier. To dissect the ethics of genetic engineering, they split it into August Weismann's two fundamental categories: somatic and germ line.

Somatic genetic engineering—otherwise known as gene therapy—got a green light. Politicians and scientists alike agreed that it might eventually cure thousands of hereditary diseases. As long as the research went carefully, as long as the treatments were safe, no one saw any serious ethical concerns.

The green light in the 1980s prompted a number of scientists to begin work on gene therapy. They needed first of all to find a new way to deliver genes to cells. Cline's method worked only for types of cells that could be removed from a patient, altered in a petri dish, and then put back in. If someone needed gene therapy for a brain disease, no one would try pulling chunks of gray matter out of their head.

Viruses offered a promising solution. Scientists figured out how to paste human genes into viruses, which could then infect cells to deliver their payload. By the 1990s, scientists were getting promising results from viruses in experiments on mice, and they had even been encouraged by a few human trials. But it turned out the viruses were not as safe as once thought. An ill-fated trial for a metabolic disorder in 1999 brought gene therapy research to a halt for several years. One of the volunteers, a nineteen-year-old man named Jesse Gelsinger, had an intense immune response to the viruses. He developed so much inflammation that he died in a matter of days.

After Gelsinger's death, the clinical trials of gene therapy stopped. The few researchers who remained in the field took a step back, searching for safer viruses. Within a few years, new clinical trials started, and after a few more years they began to deliver promising results. Philippe Leboulch of the Paris Descartes University and his colleagues tackled beta-thalassemia, the disease that had bested Martin Cline thirty years earlier. The French researchers extracted bone marrow cells from a boy, infected them with a virus carrying HBB, and then injected the cells back into his bones. In 2010, they reported that his cells started to make normal hemoglobin again, and he stopped needing a monthly transfusion of blood to stay alive.

People who suffer from other diseases, such as muscular dystrophy and hemophilia, are waiting in the hope that gene therapy will help them as well. Instead of struggling with difficult diets, many people with PKU look to gene therapy as a true cure.

In the debates over genetic engineering in the 1980s, almost everyone agreed that somatic cells were a promising target, but germ cells had to stay out of bounds. "Deciding whether to engineer a profound change in an expected or newborn child is difficult enough," President Carter's commission concluded in its 1982 report. "If the change is inheritable, the burden of responsibility could be truly awesome."

When the Office of Technology Assessment looked into the matter, they agreed. There were too many uncertainties—both medical and ethical—to even investigate a technology that might someday alter heredity. "The question of whether and when to begin germ line gene therapy must therefore be decided in public debate," they concluded. In 1986, the US Recombinant DNA Advisory Committee, which determined what sort of research in genetic engineering could be funded, cut off the money. They flatly declared that they "will not at present entertain proposals for germ line alteration."

And that, for the next three decades, was pretty much that. From time to time, a few scientists would rattle the regulatory cage, arguing that germ line engineering would be a boon to humanity, not a threat. In 1997, the American Association for the Advancement of Science revisited the matter with a forum about germ line intervention. The assembled scientists and philosophers granted that tinkering with the germ line might bring good changes. But they weren't willing to give a full-throated endorsement for the idea. They warned that genetic engineering "might some day offer us the power to shape our children and generations beyond in ways not now possible, giving us extraordinary control over biological and behavioral features that contribute to our humanness."

And yet, even as the forum tried to peer into the future from 1997, that day had already come. A few doctors had gone ahead and tampered with human heredity in a way no one had ever imagined, without asking anyone's permission.

———

I n 1996, a woman named Maureen Ott went to Saint Barnabas Medical Center in Livingston, New Jersey, in the hopes of having a baby. Seven years of in vitro fertilization had failed. Her eggs seemed healthy, but whenever her doctors implanted embryos in her uterus, they stopped dividing. Now, at age thirty-nine, Ott was watching her biological clock wind down. She came to Livingston because she had heard that doctors there had found a way to rejuvenate eggs.

The Saint Barnabas team, led by a French-born doctor named Jacques Cohen, had run some encouraging trials on mice. In the experiments, they drew off a little of an egg's jellylike filling—its ooplasm—and injected it into a second, defective egg. This microscopic injection raised the odds that the defective egg would develop into a healthy mouse embryo. The researchers speculated that the procedure worked because molecules from the donor egg undid some unknown damage.

If the procedure worked on mice, it might also work on humans. Cohen and his colleagues recruited healthy young women to donate eggs for a study. The doctors would draw off ooplasm from the donated eggs and inject it into the eggs of women struggling to have children—women like Ott.

Ott's doctors warned her that the procedure was far from a sure thing. Ooplasm contains many different kinds of molecules. Some of those might rejuvenate an embryo, but others might cause harm. It was even possible that the doctors might inject some mitochondria from the donor eggs into the eggs of the patients. If that happened, any children born through the procedure would inherit the mitochondrial DNA from the donor. Its genetic inheritance would come from three people instead of two.

To Ott, the prospect of her child inheriting someone else's genes wasn't a reason to stop. Mitochondria were merely responsible for making fuel and for other basic tasks in the cell. They didn't create the traits that defined a person. "If I was doing something like, say, I only wanted a blond-haired girl, I would feel that was unethical," Ott later explained to a reporter.

Cohen's team injected ooplasm into fourteen of Ott's eggs. They fertilized the treated eggs with her husband's sperm, and nine of the embryos began to divide. Nine months later, in May 1997, Ott gave birth to a healthy baby girl, Emma. A cursory look at Emma's cells revealed no sign of the donor's mitochondria.

Two months after Emma's birth, Cohen's team published an account of Ott's unprecedented experience in the *Lancet*. Newspapers reported in awe about the revival of Ott's flagging eggs. Other couples struggling to get pregnant besieged Cohen's team. Fertility specialists in the United States and abroad used the *Lancet* paper as a cookbook recipe for performing ooplasm transfers of their own.

But it didn't take long for the enthusiasm to curdle into suspicion. In June 1998, a *Sunday Times* journalist named Lois Rogers reported how doctors in California were offering ooplasm transfer to their own patients. Rogers didn't portray the effort as a way to help would-be parents. In her articles, it turned into a dangerous experiment in heredity.

Rogers said that embryologists and politicians were fretting "that the treatment is being given without a full debate over the biological and ethical implications of a child inheriting genes from two mothers." What the doctors were actually doing, Rogers wrote, was creating "the three parent child."

That turn of phrase locked onto the public consciousness and proved impossible to shake off. A Canadian columnist named Naomi Lakritz railed against the quest for "three-parent children," attacking doctors for concerning themselves only about the science involved. "Never mind science,"

Lakritz declared. "What about the ethical implications of cooking up a human omelet which results in the hapless child carrying the genetic material of two mothers?"

In 2001, the fear of three-parent babies flared even higher when Cohen and his colleagues published a new paper on their work. They closely examined the DNA of some of the babies that had developed from their rejuvenated eggs. Cohen's team found two children with a mix of mitochondria from their mother and the ooplasm donor.

"This report is the first case of human germline genetic modification resulting in normal healthy children," the researchers announced.

By then, dozens of children had been born through ooplasm transfers. Some of them were probably genetically modified as well. Despite two decades of government-sanctioned hand-wringing over genetic engineering, fertility specialists had waltzed right over Weismann's barrier. The rules that had been put in place had applied only to research funded with government grants. Cohen and his colleagues, working at a clinic, had done their work in private.

That freedom didn't last much longer. A month after Cohen and his colleagues published their report, they received a letter from the FDA. So did all the other American fertility clinics that were performing ooplasm transfers. The FDA pointed out that it had jurisdiction over the procedure. From now on, it would give ooplasm transfer an official status as an Investigational New Drug. That designation meant that anyone who wanted to try to carry it out would first have to fill out a mountain of paperwork and follow a lengthy set of procedures to make sure it was safe. Big pharmaceutical companies could handle those demands, but not small fertility clinics. Cohen and the other ooplasm transfer practitioners in the United States gave up.

But two American doctors, Jamie Grifo and John Zhang of New York University School of Medicine, refused to stop. When the FDA sent out their letter in 2001, Grifo and Zhang had been working on an advanced version of ooplasm transfer. Instead of using a bit of ooplasm from a donor egg, they wanted to try using all of it. They would put a woman's nucleus into a donor's nucleus-free egg.

Now unable to carry out their work in the United States, Grifo and Zhang went to China. There they collaborated with doctors at Sun Yat-sen University, looking for struggling parents who would volunteer for a study.

The doctors carefully removed the entire nucleus from each donated egg. They replaced the original nucleus with a nucleus from one of the patient's unfertilized eggs. After they fertilized the egg with the partner's sperm, the new zygote began to divide in a normal fashion.

Chinese doctors at a local hospital implanted embryos from this procedure into a thirty-year-old woman. Three of the embryos went on to develop regular heartbeats. Grifo, Zhang, and their colleagues urged the woman and her partner to come to the United States to get better medical care, but the couple decided to stay at the local hospital. About a month later, their doctors decided to remove one of the embryos to improve the odds of survival for the other two, despite protests from the Sun Yat-sen team. The remaining twins developed for another four months. Then one fetus's amniotic sac burst and it died soon after an early delivery. The woman developed an infection—possibly due to the delivery of the first baby—and the second baby died as well.

Although the experiment ended in heartbreak for the would-be parents, Grifo and Zhang saw it as a step forward for the procedure. With a fresh supply of ooplasm, the woman's embryos had developed normally, and they might have even survived if she had gotten better prenatal care. In 2003, the scientists decided to present their results at a conference and share the news with reporters from the *Wall Street Journal*.

Interviewed thirteen years later by the *Independent*, Zhang said he regretted that decision. "I think some of my team members were so eager to be famous," he said. "They wanted to let the whole world know."

The world greeted the news not with celebration they hoped for but with dismay. Critics said that Zhang and his colleagues were veering recklessly toward the manufacture of human clones. The Chinese government responded by banning the procedure, effectively ending all research on ooplasm. The experience was such a catastrophe that Zhang and his colleagues didn't even publish details of the case until 2016. "There was too much heat," Zhang said.

This line of research might have ground to a complete halt if not for a few scientists in England and the United States who quietly continued running experiments on mice. They did not want to invent a way to help infertile parents have children, however. They wanted to stop mitochondrial diseases.

The discovery of mutations that caused mitochondrial diseases made it possible to diagnose mysterious illnesses and make sense of their strange heredity. But it didn't immediately point to any straightforward cure. In 1997, a British biologist named Leslie Orgel proposed a different line of attack. Rather than curing the disease, doctors could block its inheritance. In the journal *Chemistry & Biology*, Orgel published a diagram showing how to move the nucleus of a fertilized egg into another egg that had its nucleus removed. No longer burdened with defective mitochondria, the egg could give rise to a healthy child. Orgel gave his speculative procedure a name: mitochondrial replacement.

By the mid-2000s, scientists had learned enough about mitochondria and manipulating cells to try to turn Orgel's idea into real medicine. In the United States, Shoukhrat Mitalipov launched a series of experiments at Oregon Health & Science University. He found that mitochondrial replacement therapy could cure sick mice. Then he successfully carried out the same procedure on monkeys. The monkeys grew to adulthood without any sign of harm. Meanwhile, at Newcastle University in England, Doug Turnbull led an effort on a different method that also yielded promising results. Both teams then experimented on human embryonic cells, and found that mitochondrial replacement therapy might work in our species, too. With these results in hand, Mitalipov and Turnbull went to their respective governments to ask for permission to push mitochondrial replacement therapy toward clinical trials.

Their requests opened up a fresh debate about the wisdom of human genetic engineering. Much of the debate revolved around strictly medical questions: Would mitochondrial replacement therapy be safe and effective?

Thanks to the ooplasm transfers of the 1990s, it was possible to get a few clues. By the early 2010s, the children born through the procedures had become genetically modified teenagers. Cohen and his colleagues tracked down fourteen of the children and found that they were going to school, trying out for cheerleading squads, getting braces, taking piano lessons, and doing all the things that teenagers typically do. Some of them had medical conditions, ranging from obesity to allergies. But these were nothing beyond what you'd expect from a group of ordinary teenagers. Still, promising as the survey might be, it was too small to put concerns about safety to rest. And there was no telling whether the teenagers might develop medical problems later in life.

Some critics also raised doubts about how effective mitochondrial replacement therapy could ever be. When researchers hoisted a nucleus out of a woman's egg, it didn't slip out cleanly. Sometimes mitochondria remained stuck to its sides. After the researchers plunged the nucleus into a donated egg, the embryo that resulted would sometimes end up with a mixture of old and new mitochondria. Even if 99 percent of the mitochondria in the embryo came from the donor egg, the 1 percent from the mother might still put a child at risk. Making matters worse, that dangerous 1 percent might increase as the cells in the embryo divided.

Even if doctors could scrub all the old mitochondria from a nucleus, there might still be risks from mitochondrial replacement therapy. Many of the proteins that generate fuel inside mitochondria are encoded in genes in the nucleus. Once the cell makes those proteins, it has to ferry them into the mitochondria, where they have to cooperate with the mitochondria's own proteins. Some researchers wondered if mitochondrial replacement therapy would create a mismatch between the proteins, causing the mitochondria to malfunction.

To test for this mismatch, scientists carried out mitochondrial replacement therapy on mice. They gave some of their animals mitochondria from a genetically identical donor, while others got genetically distinct ones. In some trials, genetic mismatches led to troubling problems. Some mice became obese. Others had trouble learning. Some differences in the mice—such as the amount of fat in their heart and liver—appeared only late in life. Since mice have such short lives, the researchers had to wait only a matter of months to see the symptoms emerge. In humans, it might take decades to discover such side effects. Some researchers urged that only genetically similar donors provide their mitochondria.

But a lot of the debates over mitochondrial replacement therapy were driven not by concerns over safety but by deeper passions. After all, people were talking about making three-parent babies. To tamper with heredity was deeply frightening to many people. In a 2014 congressional hearing, Jeff Fortenberry, a representative from Nebraska, condemned mitochondrial replacement therapy as "the development and promotion of genetically modified human beings with the potential for unknown, unintended, and permanent consequences for future generations of Americans." If it sounded like Fortenberry was describing a living nightmare, he didn't mind. "These scenarios scare people," he said, "and I would be very worried if it didn't scare people."

Fortenberry cast genetic engineering as a kind of hereditary plague. Once a tampered gene was introduced into one child's DNA, it would sweep through a country like a vicious strain of influenza. But that's not how heredity works. It's been estimated that only 12,423 women in the United States are at risk of passing down mitochondrial diseases to their children. If every last one of those women underwent mitochondrial replacement therapy before having children, the result would be barely detectable. The mitochondrial DNA from the donors would become slightly more common among people in the United States. As the next generation had children of their own, that donated DNA would not become any more common. The gene pool of the United States (and the rest of the world, for that matter) would barely ripple.

To opponents like Fortenberry, mitochondrial replacement therapy not only threatened the future but desecrated the past. Every child born through the procedure had inherited their genes in a manner unlike anyone born before 1997. And any person who was a source of genes became, by definition, a parent. The "three-parent" label, first stamped on ooplasm transfer, was now attached to mitochondrial replacement therapy. "The creation of three-parent embryos is not an innocuous medical treatment," Fortenberry warned. "It is a macabre form of eugenic human cloning."

There's nothing macabre about the teenagers who inherited mitochondria through ooplasm transfer, and there's nothing about the egg donors that merits the word *parent*, no matter how often congressmen and head-line writers use it. We don't hand out such an important title so carelessly. When women get eggs from donors—complete with nuclear DNA from another woman—they still get to be called their children's mothers.

While opponents of mitochondrial replacement therapy have resorted to fearmongering at times, its supporters have sometimes lapsed into faulty logic of their own. One common strategy is to play down the importance of mitochondrial DNA. In 1997, Maureen Ott convinced herself that she wasn't crossing a moral line because she wasn't picking out meaningful traits, like the color of her daughter's hair. Seventeen years later, the United Kingdom's Department of Health made much the same argument in a favorable report on mitochondrial replacement therapy they published in 2014.

"Mitochondrial donation techniques do not alter personal characteris-tics and traits of the person," the report declared. "Genetically, the child will, indeed, have DNA from three individuals, but all available scientific evidence indicates that the genes contributing to personal characteristics and traits come solely from the nuclear DNA, which will only come from the proposed child's mother and father."

When I read the report, I scoured it for a definition of *personal charac-teristics and traits*. I found nothing. As best I could tell, the authors dismissed mitochondria as doing nothing more than producing fuel. Meanwhile, I could only conclude, the genes inside the nucleus handled the really impor-tant stuff, like coloring our hair.

Ranking genes this way is absurd. The entire point of mitochondrial replacement therapy is to change something profound about a person: to take away a mitochondrial disease. People who inherit mitochondrial mu-tations may end up with traits that profoundly influence their lives and their identity, from a short stature to weakened muscles to blindness. In-deed, the huge range of symptoms that mitochondrial diseases can cause reveals just how many parts of our lives are influenced by the way we pro-duce fuel for our cells.

And there's no organ where fuel production matters more than in the brain, where neurons have to burn a lot of it to fire signals. Some mitochon-drial mutations affect the way certain parts of the brain function. Others slow down the migration of neurons through the brain during its development, so that they fail to reach their destinations. If altering the brain can't affect "personal characteristics and traits," I can't imagine what could.

Mitochondria have also turned out to be important for other functions besides generating fuel. Some proteins that mitochondria produce make their way into the nucleus, where they relay signals to thousands of genes there. Genetic variations in mitochondria can thus do much more than cause rare hereditary diseases. They can influence how long we live, how fast we run, how well we handle breathing at high altitudes. Genetic variants in mitochondrial DNA can influence our ability to remember things. Some mutations have been implicated in psychiatric disorders such as schizophrenia.

The UK report helped persuade Parliament to take up the matter of mitochondrial replacement therapy. The health minister, Jane Ellison, reassured MPs that the procedure merely replaced a cell's battery packs. In 2015, Parliament voted to approve mitochondrial replacement therapy, and in March 2017, a fertility clinic in Newcastle won the first license to perform the procedure.

The deliberations in the United States veered off onto a different path. In a survey of people with mitochondrial mutations, they overwhelmingly said mitochondrial replacement therapy research was worthwhile. The National Academy of Sciences examined the evidence and came out in 2016 with a cautious endorsement. It might be wise to start using the procedure only on sons, they said, since they wouldn't be able to pass on their altered mitochondria to their own children. Scientists would also need to keep careful tabs on the children born to women who underwent the mitochondrial replacement therapy, to make sure they didn't suffer unexpected harm years later.

But all these discussions ended up being moot. Somebody in Congress— no one has ever figured out who exactly—slipped a ten-line provision into an enormous 2016 spending bill that blocked the FDA from evaluating mitochondrial replacement therapy. Without any debate in Congress, the ban went into effect.

In that same year, however, John Zhang—who had gone to China back in 2003 to continue his research—announced that he and his colleagues had taken another trip outside of the United States to perform the first human mitochondrial replacement therapy.

Now working at the New Hope Fertility Center in New York City, Zhang had been contacted by a couple in Jordan for help. Their two children had developed Leigh syndrome, a rare mitochondrial disease that weakens the muscles, damages the brain, and usually leads to death in childhood. The couple's first child died at age six, their second at just eight months.

Before having children, the mother had no idea that she carried Leigh syndrome. In her own cells, only about a quarter of her mitochondria carried the mutation, while the rest functioned normally. In both her children, her mutant mitochondria became more common and crossed the deadly threshold.

The couple came to Zhang in the hopes of having another child, one without Leigh syndrome. He knew he couldn't use mitochondrial replacement therapy in the United States. But he also knew that Mexico had no regulations in place that would block him. He traveled there with the couple, carrying out the procedure at a branch of his clinic. Zhang's team transferred five nuclei from the woman's eggs to donor eggs, which they then fertilized. When they implanted one of the embryos in the woman's uterus, it developed normally, and in April 2016 she gave birth to a boy.

When Zhang and his colleagues examined the boy, his health seemed good. But the doctors found that his mother's faulty mitochondria had not been entirely replaced. Two percent of the mitochondria in cells they sampled from his urine came from his mother. In cells from his foreskin, that fraction jumped to 9 percent. No one could say for sure what the levels were in his heart or his brain. And it was doubtful that anyone would ever find out. Barring some unforeseen medical emergency, his parents refused any further testing. The boy slipped away from science's gaze, another genetically engineered child brought into this world.

CHAPTER 18

Orphaned at Conception

T<span></span>O JENNIFER DOUDNA, mitochondrial replacement therapy looked like a faint foreshadowing of what CRISPR might deliver. Doctors like Zhang were only replacing faulty genes in embyros with healthy ones. CRISPR might allow them to rewrite any of the twenty thousand or so protein-coding genes sitting on an embryo's chromosomes. And that change could be inherited by their descendants.

The last thing Doudna wanted was for CRISPR to replay the botched history of mitochondrial replacement therapy. That treatment had crept into practice without any public discussion of its ethics, and when the conversation finally got off to a late start, it was distorted by lurid visions of Frankenstein and charged language about three-parent babies. In 2014, Doudna decided that she would try to avoid such a debacle by starting a public conversation.

It was not a role Doudna relished. She felt comfortable talking about the inner workings of bacteria, not about the potential dangers of retooling human embryos. Her new role "felt foreign," she later said. "Almost transgressive."

Doudna started small. In January 2015, she hosted a meeting at a cozy inn in the Napa Valley, about an hour from Berkeley. Among the eighteen people who assembled in wine country to talk about CRISPR were David Baltimore and Paul Berg, two eminent biologists who had led similar meetings in the 1970s to discuss recombinant DNA. Now as then, Weismann's barrier split the conversation in two. The Napa group divided their time talking about tinkering with somatic cells and germ cells.

CRISPR might prove superior to viruses as gene therapy, the researchers speculated, because doctors could use it to fix somatic cells with more precision. As for the germ line, some people at the Napa meeting weren't bothered by the prospect of using CRISPR to alter it. Others considered Weismann's barrier a line never to be crossed.

Despite their differences, everyone at the meeting knew that they couldn't just let their differences lay idle. There wasn't time. There were rumors that scientists in China had already used CRISPR on a human embryo. A paper from the scientists was supposedly circulating among journals for publication. Any day now, the news might break. Most of the participants at the Napa meeting agreed to coauthor a commentary, which they'd submit to a journal. On March 19, Doudna and seventeen fellow scientists published a piece in *Science* called "A Prudent Path Forward for Genomic Engineering and Germline Gene Modification."

The scientists didn't call for an outright ban on human germ line engineering, but they did strongly discourage it for now. They also proposed that a public meeting take place, bringing experts together from around the world to drill deeper into the risks and benefits of the new technology. Even such a big gathering as that would not be enough to settle matters, Doudna and her coauthors warned. "At present, the potential safety and efficacy issues arising from the use of this technology must be thoroughly investigated and understood before any attempts at human engineering are sanctioned, if ever, for clinical testing," they declared.

The *Science* piece drew so much attention that the National Academy of Sciences agreed to host an international meeting just a few months later, and the Royal Society and the Chinese Academy of Sciences signed on to participate. Things were unfolding as Doudna had hoped, at least for a few weeks.

In April, she discovered that the rumors she'd been hearing were true. Junjiu Huang, a biologist at Sun Yat-sen University, and his colleagues reported that they had crossed the line. They had used CRISPR to alter human embryos.

Depending on how you looked at it, Huang's experiment was a historic achievement or a botched nonstory. As the Chinese scientists explained in the journal *Protein & Cell*, they set out to alter the HBB gene, the gene in which mutations can cause beta-thalassemia. They designed the experiment to sidestep ethical concerns about tinkering with viable embryos. When fertility clinics fertilize eggs, they sometimes make mistakes, allowing two sperm to fuse to a single egg. These embryos end up with three sets of chromosomes—hence their name, triploid—and they fail to develop for more than a few cell divisions. Huang and his colleagues got hold of dozens of triploid embryos to study, confident in the knowledge that these embryos could never be used to start a pregnancy even if someone wanted to.

Huang and his colleagues built CRISPR molecules that could cut part of HBB genes, allowing them to be replaced with a new stretch of DNA. They injected the mix into the triploid embryos and waited for them to divide into eight cells. Huang and his colleagues analyzed fifty-four of the embryos to see how well the CRISPR molecules had worked. They had only managed to cut the HBB genes in twenty-eight. And of those embryos, a smaller fraction had replaced the old DNA with the new material. In other embryos, the cells had accidentally copied similar genes elsewhere in their DNA.

A number of the embryos, Huang and his colleagues found, ended up as mosaics. Some cells in the mosaic embryos had an altered version of the HBB gene, and some didn't. It turned out that the CRISPR molecules needed a lot of time to find their targets in the human DNA. By the time they found the HBB gene, the single-cell zygote had divided into several new cells, some of which ended up without any of the CRISPR molecules inside.

When news of Huang's paper broke, I asked Doudna what she thought of it. She told me that it "simply underscores the point that the technology is not ready for clinical application in the human germline."

Doudna was choosing her words carefully. From the Napa meeting onward, she wanted to avoid turning public opinion against CRISPR in general. Her fellow scientists would have to restrain their curiosity and not carry out experiments that might seem grotesque or reckless. On the other hand, Doudna didn't want to rule out germ line engineering altogether. Perhaps someday in the future, it would be worth considering.

Her careful parsing couldn't stop the story from spiraling out of control. It caused such a worldwide commotion that Francis Collins, the director of the National Institutes of Health, released a blunt statement a few days later invoking the policies put in place in the 1980s. "The concept of altering the human germline in embryos for clinical purposes has been debated over many years from many different perspectives, and has been viewed almost universally as a line that should not be crossed," he declared. The NIH was pouring over $250 million a year into gene therapy research in the hopes of curing diseases by changing genes in somatic cells. But they would not pay anyone to leap to the germ line, full stop.

When scientists like Collins talked about a line that should not be crossed, they made it sound blindingly clear. Yet scientific research was only making it harder to find. Collins spoke of altering embryos "for clinical purposes," for example. One could argue that an experiment like Huang's had no clinical purpose whatsoever, because the triploid embryos he used could never develop into children. Was his experiment beyond the line anyway, because someone might use the knowledge he gained to alter an embryo's HBB gene and eliminate beta-thalassemia from a line of descendants?

In September 2015, the line got even harder to discern. A scientist named Kathy Niakan at the Francis Crick Institute in London applied to the British government for permission to use CRISPR on human embryos. Niakan planned to use CRISPR to shut down genes believed to be crucial for the early development of embryos. By seeing how the embryos then developed could give Niakan clues about the jobs the genes carry out. She would study the embryos only up till they were about a week old and then destroy them. But, unlike Huang's experiment, Niakan would be carrying out CRISPR on viable embryos. Did their viability make her research an affront to all that is decent, even if she destroyed them when they were still a microscopic clump of altered cells?

The news about Niakan and Huang brought an intense urgency to the

International Summit on Human Gene Editing when it opened on December 1, 2015. At the National Academy of Sciences in Washington, David Baltimore welcomed the five hundred attendees with a provocative kickoff, echoing Hotchkiss's words fifty-one years beforehand.

"Today, we sense that we are close to being able to alter human heredity. Now we must face the questions that arise," Baltimore said. "The overriding question is when, if ever, will we want to use gene editing to change human inheritance?"

———

The conference never managed to live up to Baltimore's provocation. "It was an extremely low-key meeting," the reporter Sharon Begley observed, "with more and more empty seats as it went on."

For some scientists, the low-key tone seemed like an intentional strategy. They hewed to jargon-rich prepared remarks, not wanting to stumble over any ethical trip wires. At the end of the meeting, Baltimore, Doudna, and the rest of the organizing committee came on stage to deliver a consensus statement. It didn't go much beyond what the Napa group had agreed to eleven months earlier. The committee endorsed CRISPR for somatic cells—gene therapy, in other words—as well as CRISPR-based research on early human embryos. They came out against germ line modification in embryos used to establish a pregnancy. But they didn't close the door. "The clinical use of germline editing should be revisited on a regular basis," they declared.

For some of the scientists at the meeting, the real question about germ line editing was whether it was even worth trying. "If we really care about helping parents avoid cases of genetic disease," said Eric Lander, the director of the Broad Institute in Cambridge, Massachusetts, "germline editing is not the first, second, third, or fourth thing that we should be thinking about."

Lander argued that parents would be better helped in almost every case by preimplantation genetic diagnosis. If they risked giving their child a hereditary disease, they could get the help of doctors to sort through embryos—or perhaps even eggs and sperm—to make sure their children were not born with the disease. Only after parents tried these measures without success would it make sense to edit the germ line.

A few of the speakers came out strongly against even this restrained approach to CRISPR. Marcy Darnovsky, the executive director of the Center for Genetics and Society, painted a dark future very different from Lander's. She envisioned an unregulated marketplace where germ line editing would run rampant. "It's a radical rupture with past human practices," she warned.

The risks—of CRISPR missing its target and rewriting a different gene, for example—were simply too great. And not only was it dangerous to change a child's DNA, Darnovsky argued, but it was simply wrong. An embryo could not give consent for the procedure. And altering the child's genes was, when you came down to it, an affront to the child's individuality. Even if CRISPR worked splendidly, its very success might create unprecedented social woes. Darnovsky could picture a world in which the rich engineered their children's genes to escape the burden of disease, while poorer children could not. And once parents started trying to get rid of certain traits in their children, where would they stop?

Deafness is not lethal, for example, but when Harry Laughlin of the Eugenics Record Office published his Model Eugenical Sterilization Law in 1922 he put it on the list. Would fertility doctors offer to knock out deafness mutations? Would other conditions, such as dwarfism, be judged intolerable burdens, too? Disabled communities might feel besieged before long. And once parents got used to altering more and more aspects of their children's biology, Darnovsky predicted, they might well start changing genes to enhance their children.

"The temptation to enhance future generations is profoundly dangerous," Darnovsky warned. "I ask you to think about how business competition might kick in with fertility clinics offering the latest upgrades for your offspring. Think about how the market works."

Most of the scientists at the meeting shied away from even mentioning enhancement. One exception was a towering, long-bearded geneticist from Harvard named George Church. Enhancement was coming, Church said, and it would begin not with embryos but with old people.

Here's one way that might happen. Nine percent of older people suffer from brittle bones due to osteoporosis. Cells in their skeleton start to break down the surrounding bone, releasing the minerals into the bloodstream. Some drugs can slow down the decline by sticking to cells, making it harder for them to make contact with bone. But there may be a better way to treat the disease.

One reason that cells break down bone is that in old age they make less of a protein called TERT. In 2012, researchers at the Spanish National Cancer Research Centre coaxed old mice to make extra TERT by giving them an extra copy of the gene. Their osteoporosis reversed, and their bones strengthened. It's conceivable that doctors could use CRISPR gene therapy to treat people as well. CRISPR molecules would home in on the TERT gene in bone cells and edit it. The gene would behave as it did when the patients were younger, strengthening their bones.

But gene therapy for TERT could do a lot of other kinds of good, too. The Spanish researchers who tried it out on mice found that it also reversed aging in their muscles, their brains, and their blood. It extended the life span of old mice by 13 percent. When the scientists treated younger mice with TERT gene therapy, the animals lived 24 percent longer.

If CRISPR-based gene therapy got approved for osteoporosis in humans, people might soon clamor to get the treatment for a longer, healthier life. And if it proved to be more effective the earlier it was administered, then some parents might want their children to have that benefit from the very start. Why end life with good TERT genes when you can start with them?

TERT was just one candidate among many that Church saw for enhancement. Editing genes to treat wasting muscles might lead to people enhancing their strength. Researchers are discovering ways to fight the decline in memory and learning that comes with Alzheimer's. Gene therapy could conceivably deliver the same results. And editing those genes in healthy people might enhance their cognition. Parents might choose to edit those same genes in their children to give them an edge in school and at work.

"I think enhancement will creep in the door," Church told the audience.

I n one sense, Doudna's campaign was a success. She had succeeded in getting a conversation started. By 2016, CRISPR had cracked open its cocoon and was a full-blown media butterfly, the subject of regular coverage on television and in the newspapers. Doudna hopped from city to city to explain the awesome tools she had helped invent and to provoke discussions about its use.

That conversation's center of gravity soon shifted in a profound way. The National Academy of Sciences brought together a committee of twenty-two experts—including biologists, bioethicists, and social scientists—to grapple with the science, ethics, and governance of human genome editing. In February 2017, they released a 260-page report on their deliberations. Instead of kicking the germ line editing can even farther down the road, they came to a startling agreement. They endorsed clinical trials for treating serious diseases for which there were no other alternative treatments.

In their report, the committee asked readers to consider the case of Huntington's disease. Karen Mulchinock was able to prevent her children from inheriting the disease by using preimplantation genetic diagnosis. Because she carried only one defective copy of the HTT gene, she could pick out embryos that inherited her good copy. If two people with Huntington's have children, however, their children run a 25 percent risk of inheriting the disease-causing copy of the gene from both parents. In these rare cases, preimplantation genetic diagnosis offers no help. Genome editing might. "The number of people in situations like those outlined above might be small, but the concerns of people facing these difficult choices are real," the committee observed.

Shortly after the genome editing committee released their report, Shoukhrat Mitalipov published an experiment that suggested that these clinical trials could very well succeed. Mitalipov followed Junjiu Huang's example and tried to erase a genetic disorder by editing human embryos. Unlike Huang, however, Mitalipov and his colleagues manipulated viable embryos rather than doomed ones. They did everything they could to keep the embryos viable. And while Huang's team used relatively crude CRISPR tools, Mitalipov took advantage of newer, more precise ones.

For their experiment, Mitalipov chose a genetic disease of the heart known as hypertrophic cardiomyopathy. Mutations to a gene called MYBPC3 cause the heart's walls to thicken and falter. Without warning, people with the disease may die of a sudden heart attack. The disorder is dominant, meaning that a child need inherit only one copy of the faulty gene to develop a faulty heart.

Mitalipov's team got sperm from a man suffering from hypertrophic cardiomyopathy and used it to fertilize healthy eggs. They also delivered CRISPR molecules tailored to seek out the mutation on the MYBPC3 gene. Out of fifty-four embryos treated this way, thirty-six ended up with two healthy copies of the gene. Another thirteen ended up as mosaics.

By the time Mitalipov's CRISPR molecules made their way inside the fertilized eggs, they had already made a new copy of their DNA. The CRISPR molecules apparently edited only one copy, so that when the embryo divided into two cells, one had the defective MYBPC3 gene and the other didn't. As the embryo continued to grow, new cells inherited one version of the gene or the other.

To avoid making mosaics, Mitalipov's team tried a new method. They edited the mutation-carrying sperm, and then used them to fertilize eggs. In these trials, 72 percent of the embryos lost the mutation. Mitalipov and his colleagues found that their edited embryos developed normally for eight days. If they had implanted those embryos instead in a woman's uterus, they might well have developed into babies with healthy hearts.

I was on vacation with my family in a little English village in August 2017 when the news broke of Mitalipov's experiment. I had hoped to take a break from CRISPR, filling my days with footpath walks and visits to castles. But one day I walked into the local grocery store and spotted a newspaper with four-cell human embryos scattered across its front page.

"The Cells That Could End Genetic Disease," it blared. That evening, I turned on the television at our cottage, only to encounter Mitalipov talking about his experiment. CRISPR was becoming inescapable.

F or all the attention the world gave Mitalipov's research, he didn't promise much from it. Parents who carry variants for hypertrophic cardiomyopathy can already use preimplantation genetic diagnosis to identify embryos that won't develop the disease. CRISPR might help them deal with simple Mendelian inheritance, which leaves them with only 50 percent of their embryos to implant, lowering the odds of a successful birth.

"Gene correction would rescue mutant embryos, increase the number of embryos available for transfer and ultimately improve pregnancy rates," Mitalipov and his colleagues wrote in their paper.

By narrowing his focus, Mitalipov made the ethical challenges of CRISPR seem manageable. He just wanted to improve the odds that parents could have healthy families. Likewise, other scientists who contemplated germ line engineering with CRISPR simply wanted to cure hereditary diseases in the womb, rather than wait to treat somatic cells later in life.

Within this narrow focus, a lot of the ethical concerns that have been raised about CRISPR seem less like dystopian nightmares than the everyday challenges that conventional medicine already poses. If CRISPR turns out to work reliably, we might well face a world where hereditary diseases are a bigger burden on those who can't afford it. But the cost of medicine has been a grave problem for generations, and many recent advances have made this inequality more dire. As gene therapy has inched its way to the clinic, companies have begun floating astonishing price tags for the treatment. A single shot of gene-carrying viruses might cost a million dollars or more. Yet no one has responded to this figure by demanding that gene therapy be banned. There's nothing wrong with gene therapy in itself, only with the ability of some people to get it while many others can't. That's a problem of politics, of economics, of regulations. If we are worried some people can't get CRISPR, then the solution is obvious: CRISPR for everyone.

The question of consent isn't new to germ line engineering either. We don't require that children give their consent in order to get vaccines or antibiotics. That's what parents are for. If conventional medicine fails to help sick children, their parents may give their consent to experimental treatments, knowing full well their children may not be cured and may even suffer as a result. Early in the history of gene therapy research, parents started enrolling their children in studies. It was a profound decision for the parents to make, weighing the grave diseases their children suffered against the possibility that some side effect would emerge. It was especially serious for gene therapy studies, since the children would be carrying genes in their cells possibly for the rest of their lives. Yet no one has responded to these difficult ethical choices by calling for all gene therapy to be banned.

As for the ethics of enhancements, we already live in a world in which parents try to enhance their own children's prospects. And many of those enhancements are already spread out unfairly. In 2010, American parents whose income was in the top 10 percent spent more than $7,000 on young children, including books, computers, and musical instruments. Parents in the bottom 10 percent spent less than $1,000.

In some cases, societies have managed to spread enhancements beyond the wealthy few. Vaccinations enhance our immune systems by priming them to fight measles and other diseases. The world has committed itself to getting as many children vaccinated as possible. That's a noble achievement. But, as a society, we still fall shamefully short in other respects.

While some enhancements should be spread more fairly, other enhancements are simply misguided. Some parents insist on having their short children treated with human growth hormone—not because they're sick, but because their parents want to give them social advantages that come with being tall. This push for enhancement can lead instead to insecurity, as children feel inadequate in their parents' eyes, despite being perfectly healthy.

If we treated embryo editing in this way—as very early gene therapy— we would probably muddle our way to a new normal. We'd allow it much as we allow in vitro fertilization. We'd debate whether this or that use of CRISPR is acceptable. We'd ban some, approve others. Altering some genes might turn out to have dangerous side effects, and regulations would need to be put in place to keep children safe. Somebody would certainly sneak off and try a reckless treatment, and we'd try to make sure no one ever did so again. And, in time, CRISPR would become a responsible form of medicine.

But editing embryos is not merely another form of medicine. As David Baltimore made plain at the outset of the international meeting in 2015, what made it so unsettling was the possibility that we could use CRISPR to alter the future of human heredity. We were climbing into the chariot of the sun without knowing if we were wise enough to control its course.

———————

I t's heredity that matters most, and yet we have given surprisingly little reflection about why it matters so much, or how our actions would actually alter it.

One of the few people to think through the ethics of altering heredity was a theologian named Emmanuel Agius. In 1990, years before CRISPR even had a name, Agius argued that germ line editing would rob future generations of their inheritance.

"The collective human gene pool knows no national or temporal boundary, but is the biological heritage of the entire human species," Agius said. "No generation has therefore an exclusive right of using germ line therapy to alter the genetic constitution of the human species."

But what would it really mean to alter the genetic constitution of a new species? People have sketched out many scenarios: a utopia without disease; a dystopia where the rich enjoy genetically enhanced intelligence and health while the poor endure nature's miseries. Some have even claimed that we will turn *Homo sapiens* into a new species altogether.

These are dreams. Sometimes dreams prove prophetic. Sometimes they prove to be fantasies. Hermann Muller's dream of Germinal Choice got some parts of the future right, and some parts wrong. He assumed a socialist government would protect the future of the human gene pool. He may well have been shocked if he had lived long enough to see sperm banks, in vitro fertilization, and preimplantation genetic diagnosis take hold thanks to capitalism instead. Parents did not volunteer for duty as he envisioned. They became consumers.

Whichever future we end up in, the path will have to start from where we are today. Preimplantation genetic diagnosis might be the beginning of a major shift in how children are born. In 2014—thirty-six years after the first test-tube baby was born—only 65,175 babies in the United States were born from in vitro fertilization. That's about 1.6 percent of the American babies born that year. Of those, only a small fraction went through preimplantation genetic diagnosis (it's hard to get precise numbers). Worldwide, there might be tens of thousands of children who went through it. Together, they make up only a hair-thin fraction of the 130 million babies born each year world-wide. But with each year, more parents are choosing the procedure—in some cases, encouraged by their governments. In a 2010 study, researchers inves-tigated how much money would be saved if a couple used preimplantation genetic diagnosis to avoid having a child with cystic fibrosis. A $57,500 pro-cedure could avoid $2.3 million in medical costs over a lifetime.

Today, parents typically use preimplantation genetic diagnosis if they already know their children are at risk of particular diseases. As DNA se-quencing speeds up, it will become possible for doctors to scan every gene in an embryo, detecting hereditary diseases that parents may not know they carry. It would be a seductive offer. I can imagine how it would work by looking at my own genome. It turns out, for example, that one of my cop-ies of a gene called PIGU has a mutation that puts people at greater risk of skin cancer. If I had my druthers, I'd prefer my children to inherit my good copy and not my bad one. Either way, they'd still be inheriting my DNA.

And it would be hard to stop there. I have also discovered I have a vari-ant of a gene called IL23R that dramatically *lowers* my risk of certain disor-ders. Those disorders include Crohn's disease, a chronic inflammation of the gut, and ankylosing spondylitis, which fuses the vertebrae in the spine, causing chronic pain and forcing people to hunch forward. What they have in common is the immune system going haywire and attacking the body's own tissue. No one knows exactly what triggers this attack, but it appears my variant of IL23R—found in only 8 percent of people with European ancestry—tamps down the immune system's communication network. My variant is so potent, it turns out, that drugmakers used its biology as the basis for drugs for autoimmune disorders. As a parent, I would do anything I could to lower the risk that my children get diseases that put their back into howling pain or that give them a lifetime of intestinal distress. To give my children my protective copy of IL23R, rather than the ordinary one, would be the least I could do.

If governments allowed it, some parents might ask if they could pick out their variants that could influence other traits in their children. Preimplantation genetic diagnosis on its own would usually fail to produce big results. But there might be exceptions. Scientists have found a mutation on a gene called STC2 that alters a hormone our bodies make, called stanniocalcin. One copy of the mutant version can make a person three-quarters of an inch taller. But only one in a thousand people carry it.

It's hard to predict how far parents would go as they picked out natural variations. A physicist named Stephen Hsu at Michigan State University has claimed that parents could raise their children's intelligence by selecting from embryos. Their doctors could check for which versions of a hundred genes influencing intelligence each embryo had. The embryo with the highest score could be implanted. Hsu estimated that this selection might, on average, raise a child's IQ score by five to ten points.

Geneticists generally scoff at Hsu's claims. We still know precious little about the genes that influence intelligence. While scientists have zeroed in on some of the genes that likely play a role, it's entirely possible that the true players are nearby genes or gene switches. And since we know so little about how genes for intelligence interact with the environment, picking out certain alleles to give to embryos could wind up having no effect at all.

That skepticism didn't stop Hsu. In 2011, he joined researchers at a Chinese DNA-sequencing center called BGI to found their Cognitive Genomics Lab. They set out to get DNA from two thousand of the world's smartest people and find variants they shared. In 2013, reporters got wind of the project and described it in breathless tones. "Why Are Some People So Smart? The

Answer Could Spawn a Generation of Superbabies" was the headline of a *Wired* article. "China Is Engineering Genius Babies," *Vice* announced.

Vice claimed that the BGI team was close to finding intelligence alleles and that China had "developed a state-endorsed genetic-engineering project." *Wired*'s John Bohannon suggested that a generation of superbabies might be spawned if a government like Singapore's encouraged parents to use preimplantation genetic diagnosis to pick embryos with high genetic scores for intelligence. Hsu himself found the coverage outrageous. In an interview with the journalist Ed Yong, he simply said, "That's nuts."

But Hsu had Muller-size dreams of his own. Imagine that preimplantation genetic diagnosis for intelligence genes became widespread in a country. Now imagine that the children produced from that selection used the procedure on their own children. In a 2014 essay for *Nautilus*, Hsu argued that this would be no different from what happens when cattle breedings select animals for size or milk yield. With so many variants influencing intelligence, it would be possible to raise intelligence test scores for generations until today's tests would no longer be able to measure it. "Ability of this kind would far exceed the maximum ability among the approximately 100 billion total individuals who have ever lived," Hsu promised.

In 2017, I e-mailed Hsu to see how his dream was faring. Six years had passed since the Chinese intelligence project had launched. And in that time I had heard of no concrete results. When I contacted Hsu, he told me that BGI had sequenced about half of the two thousand people in the study. Then they got in a business dispute with the company supplying them with their DNA-sequencing equipment.

"As a peripheral consequence," Hsu said, BGI "cut our project off a few years ago. So, we still to this day have not sequenced all of our samples."

The first generation of superbabies would have to wait.

———

Preimplantation genetic diagnosis already allows parents to choose which of their own variants their children can inherit. It may open the door to CRISPR, just as Mitalipov has proposed. If that does happen, it will not alter heredity any more than we're already altering it by blocking some disease-causing mutations from getting into future generations. At least not at first.

If CRISPR became a standard tool in fertility clinics, people might lose their suspicions of it—just as people lost their suspicions of in vitro fertilization in the 1980s. Before long, people might be willing to entertain a new use for CRISPR. Doctors might edit beneficial changes into an embryo's genes. They might protect children from Crohn's disease by rewriting the IL23R gene. There are other rare variants that show signs of protecting people from Alzheimer's disease, various kinds of cancer, and infectious diseases like tuberculosis.

None of these variants are artificial, since they were discovered in people's DNA. Parents could give their children all the advantages that scientists have found in our species' genetic variations. But more variants keep coming to light with more research. If this practice became popular enough, the Australian philosopher Robert Sparrow has speculated, parents might hold off having children, in the same way people wait to buy a phone until a new model is released. Sparrow wonders if future generations might find themselves stuck in an "enhanced rat race."

The choices that parents make about editing embryos would not just affect their children. The alterations could be inherited by their grandchildren. For parents with Huntington's disease, it would probably be a great relief to know their descendants wouldn't be tormented by a faulty HTT gene—unless, of course, they inherited it from another ancestor.

But when we try to look far forward, over the course of many generations, heredity doesn't necessarily work the way we imagine it to. Introducing a single edited gene into the human gene pool does not guarantee that it will take over the human species as Agius promised. In fact, the science of population genetics has found that it's far more likely for a new variant to eventually disappear. Bequeathing an Alzheimer's-fighting variant to your children may seem like a wonderful gift, but it's not a gift that can be reliably passed down through the generations. Imagine your daughter, equipped with two copies of the variant, marries a man who lacks them. Their children will inherit only one copy apiece. When they have children with people who don't have a copy, many of your great-grandchildren will probably not have any protection left at all.

Natural selection won't raise the allele's odds of surviving, either. While we may all want to avoid Alzheimer's disease, evolution doesn't care about our desires. Alleles get spread over the generations if they help people reproduce more. An allele that lowers the odds of dementia at age seventy doesn't help at all. Within a few generations, the variant you paid so dearly to give your children might disappear entirely.

When people wring their hands about what genetic engineering might do to the human gene pool, they often forget that it's actually more like a human gene ocean. If I strain my science-fiction faculties to their limit, I can imagine a worldwide dictatorship that forces every parent on Earth to submit to CRISPR and introduce the same variants into every child. But just because I can imagine the movie doesn't mean I think it's likely.

Gene-pool arguments are flawed for another reason. They treat the collective DNA of our species as if it were inscribed in stone tablets long ago and passed down unchanged ever since. In fact, the human gene pool has always been changing, and will continue to change, regardless of what we do to it. Each one of the 130 million babies born each year gains dozens of new mutations. Some will gain mutations so toxic that they will never get the chance to have children of their own, while others will choose not to. The rest will pass down some of those new mutations to future generations. Some mutations will lead to slightly larger families, on average, and those will become more common in the human gene pool. Over time, other mutations will fade back. The variants that succeed in one part of the world will sometimes be different from those in other places. Some variants are beneficial at high altitudes but not low; some tend to spread in places with malaria and not in places free of the parasite.

Amidst all this churning change, another transformation has also been steadily occurring in our species. As Muller feared, the human gene pool is indeed gaining a burden of harmful mutations. Muller first proposed the concept of a mutation load at a time when scientists knew next to nothing about the biological details of mutations. For the most part, Muller just re-lied on math. It wasn't until long after his death in 1967 that biologists be-gan making precise measurements of the mutation load by surveying people's DNA. It turns out that our species does carry a substantial burden of harmful genetic variants. While extreme mutations are rare, mildly harmful ones are abundant. They are accumulating in our DNA as we find more ways to shield ourselves from suffering and death.

In 2017, Alexey Kondrashov, a geneticist at the University of Michigan, got so worried about the emerging research on our mutation load that he published a book-length warning, called *Crumbling Genome*. It's possible, Kondrashov said, that each generation will inherit a more burdened gene pool than the previous one. Depending on how quickly the mutation load grows, it might someday drag down our collective health.

Muller's Germinal Choice plan might sound absurd, but Kondrashov believes that the mutation load is a threat we cannot ignore. He suggests there are some ethically uncomplicated things we might do today to defend against it. As men get older, their sperm accumulate more mutations. If they freeze sperm as young men, they can pass on less of a burden to future generations. If the mutation load gets worse despite such measures, our spe-cies might have to use CRISPR or some other gene editing tool to plug the rising tide.

"I hope that 'War on Mutation' is declared soon," Kondrashov wrote.

———

The future probably won't match the most extreme visions we can dream up. But it will disorient us. It will take what we thought were iron laws of heredity and stretch them in strange figures. In fact, the disori-entation has already begun.

In the early 2000s, for example, fertility doctors began producing so-called savior siblings. When children developed leukemia or some other disease re-quiring a bone marrow transplant, some families would go through rounds of in vitro fertilization until they produced a baby with just the right combi-nations of HLA alleles to be a donor.

In 2011, a seventeen-year-old Israeli girl named Chen Aida Ayash was killed in a car accident. After her death, her parents asked for doctors to collect some eggs from her cadaver. They had to go to court to get permission, explaining to a judge that they wanted to fertilize Chen's eggs, after which Chen's aunt would bear them to term. After her own death, Chen would give her parents grandchildren.

These cases are carrying us into realms where old customs and rules start to sputter and fail. The words that we used to use to talk about heredity lose their old meanings, or take on new ones. And when people fight about those words, judges struggle to figure out who is right. In 2012, the US Supreme Court found itself in such a bind when they heard a case brought by a Florida woman named Karen Capato. Her husband, Robert, was diagnosed in 1999 with esophageal cancer. He immediately began depositing his sperm in a sperm bank, so that if his chemotherapy left him infertile, Karen could still become pregnant through in vitro fertilization. The treatment failed, and Robert died in 2002.

Karen did not have his frozen sperm destroyed after his death. Nine months later, she used some of it to fertilize her eggs and gave birth to twins. Karen filled out paperwork so that the twins could get Social Security benefits for their father's death. But the Florida state government rejected her application. They pointed to their state laws, and held that children conceived after the death of their father couldn't inherit his personal property.

After hearing Karen Capato's appeal, the Supreme Court ended up ruling against her. But their 9–0 decision came only after hours of maddening oral arguments. The judges and the lawyers got bogged down in debates over the definition of a child. It was clear that the congressmen who laid down the rules about inheritance in the Social Security Act in 1939 couldn't have imagined children being conceived months after their father's death. "They never had any inkling about the situation that has arisen in this case," grumbled Associate Justice Samuel Alito.

If genetic engineering ever becomes commonplace, the Supreme Court will probably find itself in even harder quandaries, where old laws provide even worse guidance to the new ways of tinkering with heredity.

A few cases have been brought by children against their parents for allowing them to be born with congenital diseases. According to these "wrongful birth" lawsuits, the parents were negligent for ignoring tests on the fetus before birth and going ahead with it anyway. Some ethicists now wonder if children in the future may sue their parents for not using mitochondrial replacement therapy to cure Leigh syndrome or some other devastating mitochondrial disease. If parents have the genome of an embryo sequenced and choose not to edit out a variant that puts people at a high risk of dementia, their children might hold them accountable.

It's hard to say if such children would win. In some forms of mitochondrial replacement therapy, the nucleus from an unfertilized egg is moved to a new egg. Only then do doctors fuse a sperm to it. For a child in this case to claim they were harmed by coming into existence, they have to show they're worse off as a result of the procedure. But if not for the therapy, somebody else would have been born—in other words, an embryo that inherited a different combination of genetic variants from its parents.

As a society, we are probably not prepared to handle these ethical dilemmas. But there are even more profound challenges to our concepts of heredity coming fast over the horizon. Fertilizing eggs months after a father's death seems strange because it stretches the timing by which one generation produces the next. But the process of heredity that takes place is utterly conventional. Karen and Robert Capato, for instance, produced lineages of cells in their bodies that gave rise to germ cells. They shuffled their chromosomes through meiosis as cells have done for billions of years. The germ cells combined, joining their genes together to produce embryos. And then the embryos developed into children with germ cells of their own.

Not even gene editing with CRISPR changes this series of events. Once scientists began using CRISPR on ground-cherries and mice, their descendants still inherited DNA. The only difference was that some of the variants they inherited were put in place by people, rather than through spontaneous mutations. It's as if scientists were rerouting a river: Even in its new configuration, the river still flows.

But some recent advances in research may alter heredity itself in a far more profound, puzzling way. In one of them, scientists accidentally broke through Weismann's barrier.

———————

I n 1999, a Japanese biologist named Shinya Yamanaka opened a new lab at the Nara Institute of Science and Technology, hoping to find a way to make a mark for himself in a crowded field. Before coming to Nara, Yamanaka had discovered some genes that were active in the early embryos of mice. Many other scientists were also studying mouse embryos, figuring out how embryonic cells take on different identities. They pinpointed proteins that could push a lineage of stem cells to become muscles or neurons or other types of tissue. In the 1990s, research on embryonic cells raised hopes for a new way to treat diseases. Scientists could pluck a single cell from an embryo made in a fertility clinic and use it to make a colony of embryonic cells in a laboratory dish. With the right chemical signals, the embryonic cells could keep dividing into new embryonic cells for six months. A number of scientists began predicting that this method would make it possible to grow tissue on demand. People with Parkinson's disease could get transplants of healthy neurons. After a heart attack, doctors could repair a patient's damaged cardiac muscle with new cells.

Yamanaka thought that if he joined the chase, he'd get trampled into obscurity. So he decided to turn around and head in the opposite direction. Instead of figuring out how to turn embryonic cells into adult cells, Yamanaka would try to turn adult cells back into embryonic cells.

No one else was trying to pull off this trick, and with good reason. Turning back developmental time seemed impossible. If you trace a branch of the human body's pedigree from the fertilized egg to any cell in the adult body, you travel a long, twisted route. There may be hundreds or thousands of branching points along the way where one cell divided in two. And within each generation of cells, there was a flurry of biochemistry that made it possible for a different flurry to take over in daughter cells. To turn an adult skin cell back into an embryonic cell would seem to require traveling through all that history to the beginning, running all that biochemistry backward.

But Yamanaka suspected that our inner heredity might not be so hard to override after all. A few experiments carried out over the years gave him some hope. In 1960, for example, a British biologist named James Gurdon destroyed the nucleus in a frog's egg and replaced it with the nucleus from the animal's intestines. The egg began dividing, and eventually it grew into another frog. With this experiment, Gurdon had cloned the first animal. And in the process, he also showed that the genes in an adult cell could be reprogrammed to build an embryo all over again. In 1996, the Scottish biologist Ian Wilmut and his colleagues achieved much the same thing, this time in a sheep, creating a clone they dubbed Dolly.

Yamanaka wondered if there might be a simpler way to reprogram an adult cell to make it embryonic. To understand what makes embryonic cells embryonic, he looked for genes that were active only early in life and became silent in adulthood. Yamanka discovered some genes for proteins that acted like master switches, grabbing onto many genes in the cells and either shutting them down or turning them on. Yamanaka contemplated the idea of flooding adult somatic cells with proteins like these. They might seize control, forcing the cells back to an embryonic state once more.

It was, Yamanaka knew, a long shot. While he was aware of a few proteins that were active in embryonic cells, he had no idea how many others he would have to manipulate. There might be dozens, even hundreds. "We thought at that time that the project would take 10, 20, 30 years or even longer to complete," Yamanaka said.

Yamanaka organized his lab to start hunting for the proteins in mouse embryos. Five years of searching brought them two dozen. The scientists then tested each of the genes to see if it could reprogram an adult cell. They would add extra copies of a given gene to a skin cell from an adult mouse. The extra genes would flood the cell with extra copies of their protein. But the adult cell always stubbornly remained adult.

As the disappointments piled up, a graduate student named Kazutoshi Takahashi suggested that they stop testing the proteins one at a time. Instead, they should flood cells with all twenty-four of their proteins at once. Perhaps the combination of all the proteins might be able to deliver a little nudge to the cells. Even such a tiny sign of hope would tell them their work wasn't in vain.

Yamanaka gave his blessing to the experiment, although he was sure Takahashi would fail. Takahashi inserted all twenty-four genes into the skin cells and waited to see what happened. Four weeks later, Takahashi came to Yamanaka with news. The adult skin cells had turned themselves into what looked like full-blown embryonic cells.

"I thought this might be some kind of mistake," Yamanaka said. He had Takahashi rerun the experiment many times over. Time and again, the cells turned embryonic.

It was impressive enough that the cells looked like embryonic cells and made the key embryonic cell proteins. But Yamanaka wondered if they could behave like embryonic cells, too. His team injected a few of the reprogrammed cells into early mouse embryos to find out. The embryos developed into healthy adults, and the scientists found that the reprogrammed cells had given rise to normal adult cells scattered throughout the body.

This success led Yamanaka to wonder if flooding cells with all twenty-four proteins was overkill. He launched a new experiment, creating cocktails containing only some of the proteins and leaving others out. His lab found they needed only four proteins. Working with James Thomson at the University of Wisconsin–Madison, Yamanaka demonstrated that human cells became embryonic with the same simple recipe.

In his reports on the experiments, Yamanaka referred to his reprogrammed cells as induced pluripotent stem cells. Other scientists began testing out these cells, hoping they would prove even better than embryonic cells for medical treatments. It was easy to imagine doctors taking skin cells from patients, reprogramming them, and then coaxing the induced pluripotent stem cells into any type of adult cell they needed. Because the cells belonged to patients themselves, there wouldn't be any worry about rejection of foreign tissue.

In 2012, Yamanaka won the Nobel Prize. The prize not only recognized the practical promise of induced pluripotent stem cells; it also honored his discovery of a new way to think about time. August Weismann had pictured the body as a branching tree of cells, the branches splitting through time. We could split our development into milestones: day 1, fertilization; day 2, two totipotent cells; and so on through the calendar of life. Each milestone had to come after the previous ones, because it depended on them. The heart could not appear before the three germ layers, because the heart had to form from one of those layers. Time gradually stiffened our inner heredity, committing each lineage to a single fate till death.

Yamanaka showed that time is not actually essential to the difference between an embryonic cell and a cell in the gall bladder or a hair cell in the ear. Our ancestors evolved a way to develop over time, to build one biochemical reaction on another in lineages of cells. But we can just push cells from one state to another.

Yamanaka didn't just undermine the power of time with his research; he also undermined some long-held beliefs about the germ line. The germ line has come to be seen as an all-important thread of heredity that is the sole link from one generation to the next. But this is a convenient fiction. When sperm and egg combine, they produce an embryo that has no distinct germ cells at all. Any cell in the embryo can, at that stage, give rise to new germ cells (or any other kind of cell). The germ line breaks, in other words, and only later in an embryo's life is it rebuilt. By turning somatic cells into germ cells, Yamanaka could sneak around Weismann's barrier.

Induced pluripotent stem cells behave much like the earliest cells in the embryo before the germ line reappears. With the right signals, they can develop into germ cells, just as they can become other types of tissue. In 2007, Yamanaka and his colleagues injected induced pluripotent stem cells into male mouse embryos, and found that some of the injected cells developed into sperm. The chimeric mice could even father mouse pups of their own with these sperm, which had come from a different mouse.

In order for the induced pluripotent stem cells to become sperm, a mouse's body sent it a particular series of chemical signals, guiding their development. Yamanaka's experiment led other researchers to wonder if they could deliver the same signals to cells sitting in a dish rather than in a mouse. In 2012, the Japanese biologist Katsuhiko Hayashi managed to coax induced pluripotent stem cells to develop into the progenitors of eggs. If he implanted them in the ovaries of female mice, they could finish maturing. Over the next few years, Hayashi perfected the procedure, transforming mouse skin cells into eggs entirely in a dish. When he fertilized the eggs, some of them developed into healthy mouse pups. Other researchers have figured out how to make sperm from skin cells taken from adult mice.

Translating those results into experiments on human cells has proven hard. Some researchers have managed to turn a man's skin cells into a precursor of sperm called spermatids. But these transformed cells don't easily undergo meiosis, shuffling their DNA and pulling it into two sets.

Nevertheless, the success that Yamanaka and other researchers have had with animals is grounds for optimism—or worry, depending on what you think about how we might make use of this technology. It's entirely possible that, before long, scientists will learn how to swab the inside of people's cheeks and transform their cells into sperm or eggs, ready for in vitro fertilization.

If scientists can perfect this process—called in vitro gametogenesis—it will probably be snapped up by fertility doctors. Harvesting mature eggs from women remains a difficult, painful undertaking. It would be far easier for women to reprogram one of their skin cells into an egg. It would also mean that both women and men who can't make any sex cells at all wouldn't need a donor to have a child. A man left infertile by chemotherapy, for example, could use a skin cell to make sperm instead.

Some researchers think that in vitro gametogenesis will trigger an explosion in the test-tube-baby business. Henry Greely, a bioethicist at Stanford University Law School, explored this possibility in his 2016 book *The End of Sex and the Future of Human Reproduction*. Greely speculated about a future world "where most pregnancies, among people with good health coverage, will be started not in bed but in vitro and where most children have been selected by their parents from several embryonic possibilities."

Today, parents who use in vitro fertilization can choose from about half a dozen embryos. In vitro gametogenesis might offer them a hundred or more. Shuffling combinations of genes together so many times could produce a much bigger range of possibilities.

Even after ruling out the embryos with disease-causing mutations, parents would still have many embryos left to choose from. They might pick embryos with variants that could affect the color of their children's eyes. Or they might follow Stephen Hsu's call, and pick out embryos that have a combination of variants that have been linked to higher intelligence scores.

But the implications of in vitro gametogenesis go far beyond these familiar scenarios—to ones that Hermann Muller never would have thought of. Induced pluripotent stem cells have depths of possibilities that scientists have just started to investigate. Men, for instance, might be able to produce eggs. A homosexual couple might someday be able to combine gametes, producing children who inherited DNA from both of them. One man might produce both eggs and sperm, combining them to produce a family—not a family of clones, but one in which each child draws a different combination of alleles. It would give the term *single-parent family* a whole new meaning.

The possibilities go on. Instead of three-parent babies, one can envision a four-parent child. It might be possible someday for four people to swab their cheeks and have induced pluripotent stem cells produced. Scientists could then turn the cells into sperm or eggs, which could then be used to make embryos. Two people would produce one set of embryos, and the other two would produce a second.

At the earliest stages of development, when the embryos were just balls of cells, scientists could remove cells from each one and coax them to develop in a dish into more eggs and sperm. And those could be used to produce a new embryo. If that embryo was then implanted in a surrogate mother and allowed to develop, the child would inherit a quarter of its DNA from each of the donors.

We haven't reached the age of multiplex parenting just yet. But it's close enough that philosophers have been thinking seriously about what it would signify. It would make mitochondrial replacement therapy look like ethical child's play. It would also leave children struggling to make sense of their own heredity. With some help from humans, any somatic cell can now gain the germ line's immortality and give rise to a new organism.

Hayashi's experiments push our language of kinship to the breaking point. The mouse pups he produced have a mother of sorts, although they descended from her skin rather than from one of her original eggs. The same might be said of a human child born through in vitro gametogenesis. But once scientists start carrying out rounds of fertilization in their labs, it will be hard to say exactly what their pedigree is. Can your parents be eight-cell embryos? Since those original embryos won't be implanted, they will never become human beings. Robert Sparrow has argued that the embryos produced this way would be orphaned at conception.

Strange possibilities have a way these days of becoming real. To make sense of them and to make ethical judgments, we need a deep sense of heredity, of the full scope of what the word means. We have to recognize Mendel's Law as one of many ways that genes move naturally from ancestors to their offspring, something that we are learning to manipulate. We have to recognize that the cells in our bodies have ancestors and descendants of their own—ones that can become mosaics or mingle together in chimeras. We have to loosen the boundaries of what we call heredity, to consider other ways in which today correlates with yesterday—be they molecules other than DNA that slip into future generations, or microbes that hitchhike along as well, or the tools and traditions of human culture, or even the environment into which our children are born. Only then will we have the language to talk about the ways in which we can control heredity for our benefit, and the dangers that we leave to the future.

CHAPTER 19

The Planet's Heirs

W HEN VALENTINO GANTZ first heard about CRISPR, it sounded like a godsend. At the time, he was getting his PhD at the University of California, San Diego, studying genes in *Drosophila* and related flies. He tinkered with their genes and observed whether he could change how their embryos developed. But the best tools he could use were clumsy and crude. In 2013, Gantz heard that researchers had figured out how to use CRISPR to alter a gene in *Drosophila* with easy precision.

"It was one of the things I was waiting for," Gantz told me when I visited him at his laboratory on a eucalyptus-covered hillside by the Pacific. After hearing the news, he had immediately ordered CRISPR molecules of his own and started trying them out. He had no idea he was about to discover a way to use CRISPR to alter heredity on a species-wide scale.

Gantz decided to try out CRISPR by altering a *Drosophila* gene called *yellow*. It was, in a sense, an inherited choice. The *yellow* gene was discovered just over a century earlier in the lab of none other than Thomas Hunt Morgan. One day in 1911, Morgan's team of students were inspecting their grayish flies when they spotted a single golden insect. They bred the fly and determined that it had a recessive mutation in a gene they dubbed *yellow*.

The *yellow* gene proved useful to Morgan's team, because they could see with the naked eye which allele of the gene any given fly carried. Morgan's students bred a line of *yellow* flies, and when they became professors years later, they taught their own students how to breed the flies for experiments. Those students started their own careers, and took the *yellow* flies with them. Over the twentieth century, each new generation inherited this knowledge, just as earlier generations had learned how to make stone tools and how to plant barley. There's even a website, called FlyTree—The Academic Genealogy of Drosophila Genetics, that chronicles this particular line of cultural inheritance.

The entire tree begins, of course, with Thomas Hunt Morgan. It branches down to his many graduate students. Their ranks included Max Delbrück, a physicist who turned to the mysteries of life, traveling from Germany to Caltech in 1937 to study under Morgan. Delbrück went on to become a professor at Caltech himself, and among his own graduate students he trained another physicist turned biologist, Lily Y. Jan. Jan got a job at the University of California, San Diego. In the 1980s, Ethan Bier joined her lab as a postdoctoral researcher. There he bred countless yellow flies, and became a professor at San Diego as well. Two decades later, Valentino Gantz arrived in Bier's lab and learned the *yellow* craft. He became Morgan's scientific great-great-grandchild.

To try out CRISPR, Gantz fashioned an RNA guide to alter the *yellow* gene in *Drosophila* embryos, introducing a mutation to make them golden. He added the molecules to the fly cells, let them develop into adults, and bred them together, hoping to produce flies with two CRISPR-altered copies of the *yellow* gene. To his delight, among the grayish flies, there were some golden ones. The technology had worked just as advertised. "I was sold," Gantz said.

Gantz then began playing around with the CRISPR molecules to see if he could use them on another species of fly he was studying for his PhD, called *Megaselia scalaris*. Unlike *Drosophila melanogaster*, it has a distinctively hunched thorax, which has earned it the common name of the humpbacked fly. It behaves differently, too, for which it has earned other names. It's known as the scuttle fly for the way it runs along the ground in bursts. And it's sometimes called the coffin fly for the way its maggots dig deep into the ground in search of food, sometimes traveling all the way down into buried caskets.

Gantz tailored the *Drosophila* CRISPR molecules so that they would target the *yellow* gene in humpbacked flies. But when he used them on the flies, the experiment failed. "It was very frustrating that we weren't recovering any mutants," said Gantz.

Gantz wasn't sure what went wrong. It was still possible that he had managed to edit one copy of the *yellow* gene in a few flies. But the CRISPR molecules were succeeding so rarely that Gantz could never bring together two flies that both carried an altered copy of the *yellow* gene.

When Gantz told Bier the news, his advisor was disappointed but not surprised. Failure is common in science. Bier headed off on a much-needed vacation in Italy and didn't give the disappointment any more thought.

But as soon as Bier stepped back into his office in San Diego, Gantz bounded in. He immediately started describing an idea for how to get CRISPR to work. The idea was simple. Gantz would get the flies to edit their own DNA.

First, Gantz would use CRISPR molecules to chop out a gene, which he would replace with a segment of DNA. That segment would include not only an altered gene but genes for CRISPR molecules, too. The fly's own cells would then make CRISPR molecules that would seek out the matching chromosome and edit the second copy of the gene. Gantz would then have humpbacked flies with two mutated copies of the same gene. He could then use them to start a line of mutants. Gantz dubbed the process a "mutagenic chain reaction."

The idea seemed far-fetched to Bier. He doubted it would work reliably enough to deliver the genetic changes Gantz hoped for. But if it did work, Bier could tell, it might become a powerful tool for studying genetics not just in humpbacked flies but in many other species as well. And as Bier and Gantz talked about the mutagenic chain reaction more, it occurred to them that it might be more powerful than Gantz had initially thought.

"We realized, 'Whoa, this could go through the germ line,'" Bier told me.

If Gantz mated a CRISPR-carrying fly to an ordinary fly, it could break

Mendel's Law. It would pass down one of its chromosomes that carried its genes for CRISPR, along with the gene that the CRISPR molecules were designed to copy. In the embryos of the second-generation flies, the CRISPR molecules would rewrite its partner chromosome. The result would be that the second generation would not be hybrids, each with a single copy of the CRISPR genes. They would all carry two copies. And the result would be the same when they bred with ordinary flies, too.

"Imagine that a blond person married a dark-haired person, and all their kids were blond," Bier told me. "All their grandkids were blond. All their great-grandkids were blond, and that went on forever. Imagine something like that."

Bier told Gantz to hold off on testing the mutagenic chain reaction for a while. He wanted Gantz to reflect on the risks and benefits first. In his mandatory reflection, it occurred to Gantz that if a single altered animal got into the wild—either intentionally released or accidentally allowed to escape—it could mate with other members of its species. Its CRISPR genes could drive themselves further into a population with every new generation.

Under the right conditions, that might be a good thing. Instead of targeting the *yellow* gene in *Drosophila*, scientists could target genes in insects that destroy crops or spread disease. Other researchers had been searching for years for gene drives that might fight pests, and Gantz may have found one at last. But if an animal slipped out of a lab while Gantz was still doing basic experiments on the mutagenic chain reaction, he had no idea what sort of unplanned changes he might unleash.

Gantz devised a way to safely test the mutagenic chain reaction. He would attempt to edit the *yellow* gene in *Drosophila*. But he would prevent the flies from sneaking out of his lab by working in a secure room, housing the insects in shatterproof plastic vials sealed with tight plugs, which would go inside larger tubes, which would be sealed in turn inside plastic boxes.

Gantz and Bier invited a group of senior geneticists at the university to hear them out. They described their concept of the mutagenic chain reaction, and their plan for a safe experiment. They wanted to know if it sounded crazy to someone else.

"'Yeah, do it'" was the consensus, Bier recalled. "'Be careful, but do it.'"

In October 2014, Gantz used CRISPR to modify some *Drosophila* larvae. If the molecules worked as he hoped, they would replace one copy of the *yellow* gene with a different stretch of DNA. That new stretch carried an altered version of the gene, along with genes for CRISPR molecules. Gantz hoped that the fly's own cells would then use its genes to alter its other copy of *yellow*. But the only way to know for sure if the procedure worked would be to breed the flies.

Gantz let the flies mature and then bred them with ordinary mates. The female flies laid their eggs, which grew into larvae and then built cases called pupae around themselves. Inside the pupae, they matured into adults. And when the mature flies broke out at last, some of them—both males and females—were golden.

"This was the first indication that things were going the right way," Gantz recalled.

But now came the real test of the mutagenic chain reaction: Could it carry over to the next generation? Gantz picked out golden females, which carried the altered *yellow* gene and its CRISPR package on both of their X chromosomes. He put them in tubes with ordinary male flies and waited for them to mate. In a conventional experiment with golden females and gray males, the results would reliably follow Mendel's Law. Their sons would all be golden, because they always inherited their one X chromosome from their mother. Their daughters, on the other hand, would inherit an X chromosome from each parent. And since *yellow* is recessive, the daughters would all be gray.

If the mutagenic chain reaction worked, Gantz would see a very different result. The next generation of flies would also make CRISPR molecules. The molecules would alter the second X chromosome in the daughters, and they would turn out golden instead.

Once the flies laid their eggs, Gantz and Bier could only wait for them to mature into adults. "For the whole two weeks I drove my wife crazy, saying 'yellow-yellow-yellow' and knocking on wood everywhere," said Bier. His colleagues prepared him for failure. "Don't get too excited," one of

Bier's colleagues told him. "I'll bet you almost everything that in the next generation it'll just be Mendelian."

Gantz knew that when the eggs of *yellow* mutants hatch, the larvae sometimes take on a faint golden cast. He squinted through his boxes and tubes, hoping to glimpse it on some of the maggots. But as best as he could tell, none of them looked yellow at all.

"I told everyone, 'Okay, this is not working,'" said Gantz. "I'm a very pessimistic person."

Just to be thorough, though, Gantz let the larvae pupate and unfurl their wings. He gassed his newly adult flies with carbon dioxide to put them to sleep. Then he sat down at a lab bench and dumped them out onto a pad. He would inspect their bodies before declaring the experiment an official failure.

But when Gantz looked down at the pad, all he saw was gold. The females had converted their own genes as he had hoped. When he stepped into Bier's office to deliver the news, Bier screamed and jumped in the air. Bier had bred thousands of *yellow* mutants, watching Mendel's Law work every time. Now suddenly the rules had changed.

"You walk into a room, you're normally used to walking on the floor, and all of a sudden you're walking on the ceiling," said Bier. "I mean, that's how weird it was to me."

O ne of the first people Bier called about the experiment was a biologist named Anthony James. "Holy mackerel," James replied.

He had been searching for twenty years for what Bier and Gantz had just found. In the 1980s, James had set out to fight diseases carried by mosquitoes, such as malaria and dengue fever. He would wage his personal war by studying mosquito genes. James set up an insectarium not far from Bier and Gantz, at the University of California campus in Irvine. There he raised mosquitoes on warm blood and experimented on their DNA.

James started by mapping their genes, since the mosquito genome at the time was terra incognita. As James and his colleagues pinned down the location of individual genes, he would be able to experiment with them. Perhaps there were genes that controlled exactly which pathogens can survive inside mosquitoes and use the insects to get to a new victim. Over 200 million people each year develop malaria because they're bitten by mosquitoes that carry the single-celled parasite *Plasmodium*. But no one has ever gotten the flu from a mosquito bite. James wondered if there might be genetic variants in certain mosquitoes that made them resistant to malaria.

"If we could just figure out how to get those genes out there in the populations at high enough frequencies, then—you know, game over," James told me when I visited him in Irvine.

James and his colleagues succeeded in mapping some mosquito genes. But the work was so slow—thanks in part to the challenge of raising blood-sucking mosquitoes—that he began to despair of ever finding a way to fight mosquito-borne diseases. He thought about how he could use what he had learned, to fight them in a different way.

"I thought, 'Well, we'll just *make* genes,'" James said.

He had an idea of the right gene to make. In the late 1960s, Ruth Nussenzweig, a biologist at New York University, discovered that mice don't get sick with human malaria. She found that the immune cells in the mice produce an antibody that clamps onto the parasite, essentially suffocating it. In later experiments, scientists fed Nussenzweig's antibody to mosquitoes, mixed in their meals of blood. Somehow, the antibody escaped being digested inside the mosquitoes and attacked the parasites inside the insects. After this treatment, the mosquitoes couldn't transmit malaria.

James and his colleagues set out to genetically engineer mosquitoes to make the mouse antibody for themselves. The scientists created a gene that encoded the antibody, which they could insert into a mosquito's DNA. James wanted to make the gene safe for the mosquitoes. He worried that if the gene stayed on all the time, the insects would swell up with mouse antibodies and get sick from them. So James and his colleagues connected the antibody gene to a switch. Now the gene would turn on only in response to blood coming into a female mosquito's body.

James and his colleagues inserted the gene and its switch into the DNA of mosquitoes. When they fed the insects *Plasmodium*-laced blood, they started making antibodies and wiped out the parasites.

Impressive as this feat might be, it wouldn't put a dent in the worldwide burden of malaria. If James simply released some of his malaria-proof mosquitoes in Africa or India, Mendel's Law would work against him. The engineered mosquitoes would almost always mate with ordinary mosquitoes, and soon their defenses would get diluted in the vast mosquito gene pool.

To spread his malaria-fighting genes, James would need a way to break Mendel's Law. Since the 1960s, scientists had wondered if they could harness gene drives for this purpose. The idea was simple: Link your gene to a gene drive, and with each generation, it would become more common in a population. But no one had yet figured out how to make it work.

James decided to take a crack at the problem. For his gene drive, he picked a piece of DNA called the P element. Carried by *Drosophila* flies, it spreads by occasionally getting its host cell to make a new copy of itself, which gets inserted at another spot in the fly's DNA. The P element first came to the attention of scientists in the mid-1900s, and over the next few decades it spread like genetic wildfire in North American flies. Margaret Kidwell, a biologist at Brown University, put flies with the P element into tubes with flies that lacked it. Within ten or twenty generations, all her flies carried it.

James and his colleagues thought that if they link their antibody gene to a P element, it could drive resistance to *Plasmodium* into an entire population of mosquitoes, and they would keep passing it down to later generations. That was the idea, at least, but no matter how the scientists adjusted the experiment, it never worked.

In hindsight, James thinks that evolution worked against him. Mosquitoes, like other animals, have probably faced many attacks by gene drives in their long history. And they escaped extinction only because they evolved defenses—defenses that James had no idea how to overcome. "We were probably doomed at the outset on that one," he told me with a laugh.

When James first heard about CRISPR, he was intrigued. He wondered if he might be able to adapt it to block malaria. But James didn't get much further than these idle thoughts, when Bier called him. James immediately recognized that the mutagenic chain reaction might be able to spread genes in mosquitoes.

Bier and Gantz drove north up the California coast an hour and a half to visit James and plan out a new experiment. Plastic boxes and shatter-proof tubes might be good enough to keep *Drosophila* in check, but for mosquitoes, they'd need much better security. James would need to run the experiment in the safety of his insectarium.

As excited as James was, he knew the odds of success were slim. He and Bier and Gantz would have to design a long piece of DNA containing a number of genes. The DNA would have to carry a gene for the mouse anti-body, as well as its switch to turn on during blood meals. It would also have to carry the CRISPR genes to copy all that DNA to other chromosomes. In order to see if they succeeded, the scientists would also need to add a gene that turned the mosquitoes' eyes red.

That was an awful lot to load onto one piece of DNA, especially when the scientists were trying to engineer mosquitoes in a manner no one had ever tried before. James proposed to his colleagues that they break the DNA into smaller chunks. They could then test one chunk at a time, and only later combine them for a final test. Gantz pushed for them to test the whole thing at once. He didn't want to crawl forward when he might leap. When James and Bier agreed to the plan, Gantz worked as fast as he could to assemble all the parts into a single piece of DNA. He then delivered it to James to see if it worked.

During my visit, James took me to the basement of his building, and we walked up to a door marked A. JAMES TRANSGENIC MOSQUITO FACILITY. He pulled the sleeves of a blue paper smock over his flannel shirt. James had short gray hair, bright big teeth, and broad shoulders that made it hard for him to get the smock on. After a little struggle, he gave up mid-humerus. The smock looked like a newspaper that had slapped against his chest on a windy day.

James waited for me to slide my smock on, and then opened the door. We stepped into a windowless vestibule. A silvery mesh—fine enough to block a mosquito—covered the ceiling vents. Once we had both stepped into the vestibule, James closed the outer door and waited a moment until he could open the inner one. We entered the insectarium—a small cluster of rooms in which James and his staff raise tens of thousands of mosquitoes.

The first thing I saw when we stepped inside was a row of yellow movie theater popcorn buckets. They sat on a table, each with a gray cylinder attached to the top, trailing a power cord. When I leaned over the buckets, I could see adult female mosquitoes clinging to the inner walls. Each had a swollen belly. The gray cylinders contained warm calf's blood. James unscrewed an empty cylinder to show me how he lined it with a membrane. The mosquitoes could pierce the membrane with their needlelike mouthparts and drink deeply, filling their tear-shaped abdomens with blood.

Once a mosquito is sated on blood, she will grow hundreds of eggs inside her body. James and his staff of technicians have found that the mosquitoes prefer to lay their eggs in the dark, and so they transfer the insects from the popcorn buckets into a lightless room. After the mosquitoes are done laying, James's technicians gather the eggs and attach them to strips of paper.

Now they can alter the genes of the new generation of mosquitoes. They pierce the soft eggs with fine glass needles, injecting DNA. They have only a few hours to work on the mosquitoes before the eggs tan and harden.

"It's like an assembly line," James said as he showed me the microscope where his team manipulates the DNA inside the eggs. "You can do three or four hundred, maybe five hundred a day, depending on how good you are."

In the wild, a mosquito lays her eggs in water, where they float together like a raft. After a few days, the eggs hatch, and the larvae swim away to spend their first stage of life underwater. James has to re-create that stage in his insectarium, too. He led me through a transparent vinyl-strip door into his larva room. Metal shelves reached from the floor to the ceiling, and on many of them were plastic tubs half-filled with water. Each was like a pond filled with hundreds of larvae. They swam about like hairy miniature snakes.

We stopped to look closely at one tub of larvae. There was a tag taped to the rim, with the number 29 written in thick marker ink. The larvae twirled and twitched. Each one had a pair of pinprick-size eyes. All the eyes were red.

"So these are the famous gene drive ones, here," James said.

The number 29 referred to the twenty-nine generations of mosquitoes James had reared since beginning his experiment with Gantz and Bier. After Gantz created a new piece of DNA with all the gene drive pieces, James and his team injected them into the soft eggs of mosquitoes. To his delight, the larvae hatched with red eyes, meaning that they carried two copies of the malaria-resistant gene. James and his colleagues mated the male larvae with ordinary females, and the following generation was red-eyed as well. I was now looking at generation twenty-nine, at the end of an unbroken chain of heredity.

In November 2015, James and his colleagues announced that they had successfully used gene drive in mosquitoes, just seven months after Gantz and Bier had published the original mutagenic chain reaction paper. "People said, 'That went fast!'" James told me as I looked at his red-eyed mosquitoes. "Well, it didn't really go fast, because we'd been laboring for years." James had all the tools he needed for the experiment by the time Bier called him. "It was just another piece of DNA for us to inject," he said.

The mutagenic chain reaction hit the news amidst jolting stories about experiments with CRISPR on human embryos. Human genetic engineering had been the stuff of speculation for more than fifty years, since Rollin Hotchkiss had worried over it. But the idea of overriding heredity to cure diseases with gene drive came as a bigger surprise. Even most scientists who worked on CRISPR hadn't seen it coming.

There were exceptions: George Church and one of his colleagues at Harvard, Kevin Esvelt, had been musing about the idea. In 2014, they and some of their colleagues published a couple of speculative pieces. But they called CRISPR-based gene drive only a "theoretical technology."

Once Bier and Gantz revealed the mutagenic chain reaction, the technology was no longer theoretical. Esvelt and his colleagues reported that they could use CRISPR in yeast to override Mendel's Law as well. The technology might conceivably work in just about any sexually reproducing species scientists might want to alter.

As Jennifer Doudna and her colleagues grappled with CRISPR's use on people, Bier, Gantz, Esvelt, and other scientists began working through the implications of gene drives. At conferences and in scientific reviews, they laid out some of the ways the technology could make life better. Making mosquitoes malaria-proof could save thousands, or even hundreds of thousands, of lives every year. Esvelt traveled to Nantucket to propose to its residents that he use CRISPR on the island's mice. It might be possible to render them resistant to Lyme disease, breaking the disease's cycle. Plant scientists speculated about fighting the evolution of herbicide-resistant weeds. They could use gene drive to eradicate the genes that made them resistant, replacing them with genes that make the plants vulnerable once more.

It might even be possible to use CRISPR to drive a population, or even an entire species, extinct. Scientists could give genes to an undesired animal that made it less fertile. The animals inheritng these genes would have fewer offspring, but thanks to CRISPR, they would end up in a growing fraction of the population. Eventually, the population would cross a tipping point and collapse.

Conservation biologists had long dreamed of this kind of power to fight invasive species. When snakes and rats are introduced to remote islands, for example, they can wipe out local bird species by eating their eggs. A team of Australian scientists calculated that the introduction of a hundred CRISPR-altered rodents to an island could wipe out a population of fifty thousand. It would take only five years.

But gene drive might also wreak havoc. If scientists unleashed a gene drive in the wild, it might not work as they had planned. If it caused harm, it might be impossible to undo the damage. A committee organized by the National Academy of Sciences issued a report in 2016 in which it warned that gene drives had the potential to cause "irreversible effects on organisms and ecosystems."

Artificial gene drives represent a profound ethical quandary—arguably a bigger one than using CRISPR to genetically engineer human embryos.

They may be able to alter heredity not just in the genetic sense of the word but in other senses as well. We might drastically alter the genes that animals or plants inherit far into the future. We would also leave an ecological inheritance to our descendants that they might curse us for. To judge the wisdom of this tool, we would do well to look back at how the tools we've already invented have altered our ecological inheritance over the past ten thousand years.

————

Human cumulative culture allowed hunter-gatherers to learn over generations how to harvest plants and control animals. Mostly without knowing it, some of them engineered new environments where agriculture could arise. Their descendants became farmers, sowing crops and raising livestock. Each new generation inherited more than just the knowledge required to farm. Humans now left an ecological inheritance to their descendants.

Before about ten thousand years ago, children were born into a world sculpted by fire, hunting, and foraging. Farmers began reworking the land on a greater scale, and at an accelerating pace. By clearing fields to plant, they could grow enough food to feed their families, with surplus they could sell to others. Farmers stopped moving, settling into villages with sturdy houses and granaries where they could store their extra food. A farmer could now pass down this accumulated wealth to his children, along with the land that he used to generate it.

But this new form of inheritance created an inescapable tension: A family's land could be divided among the children in small portions, or passed on to just one of them, leaving the rest to find other kinds of work. That tension drove some members of the family to find more land to clear. It also drove them to discover and adopt new cultural practices that let them get bigger harvests out of a given parcel, such as plows drawn by horses or oxen. By the Bronze Age, kilns were invented. Their fires reached temperatures humans had never managed before. Miners could smelt ores, and smiths could work metals. They discovered that coal was a better fuel than wood.

Out of these intense fires came new metal tools, including axes that farmers could use to clear more forests, and plows to plant more crops.

Yet these advances did not free farmers from the feedback loop between their culture and the environment. The short-term benefits they got from new farming equipment came at the long-term expense of the land's fertility. As fields eroded and became less productive, people cleared forests to work soils that would have previously been considered too poor to bother with. This feedback continued to raise the world's population, fostering more cultural innovations. And those innovations allowed people to convert even more wildlands to farms and cities.

The Industrial Revolution, which came about ten thousand years after the Agricultural Revolution, was an acceleration of this feedback. Instead of using animal-drawn plows, farmers could run tractors powered with new fuels like gasoline. Instead of spreading manure from their own livestock, they could spread fertilizers extracted from mines or produced from petroleum. The cotton gathered by New World slaves no longer had to be woven by people; it was now turned over to coal-powered looms. As railroads cut across continents, cattle could be grazed on lands thousands of miles from the people who would ultimately eat them. The influence of human culture now produced a worldwide ecological inheritance.

By some measures, this cultural feedback loop has been a great success. Before the Agricultural Revolution, a square kilometer of land could typically feed fewer than ten hunter-gatherers. Today, if it's intensively farmed, it can feed thousands. In the early 1800s, more than 90 percent of the world lived in extreme poverty, scraping by on the equivalent of about two dollars a day. Today, less than 10 percent are. A child born in the United States in 1900 had an average life expectancy just short of fifty years. Children born in 2016 will live, on average, to age seventy-nine.

I feel fortunate that my children get to inherit this world created by cumulative culture. But I can also see that they inherit an environment that is suffering in many respects. Since the dawn of agriculture, three-quarters of the terrestrial biosphere has been converted from wilderness. Somewhere between a quarter and a third of the planet's biological productivity—its ability to convert sunlight into biomass—is now appropriated for human use. If the same cultural practices that have reworked the planet so dramatically over the past ten thousand years are inherited by future generations, we may push many species to extinction and threaten our own well-being.

Our cumulative culture has even altered the atmosphere. We're not the first organisms to change the chemistry of the air—photosynthetic bacteria began pumping oxygen into the sky two billion years ago, and every generation of living things has had to adapt to an oxygen-rich planet ever since. But it's unheard-of for just one species of animal, using tools of its own making, to manage such a feat.

When hunter-gatherers set fire to meadows or forests, they could loft carbon dioxide and other molecules into the air. Because their numbers were so low, early humans barely nudged the atmosphere's makeup. But once farmers started clearing land for planting, the soils released carbon dioxide at a greater rate. By three thousand years ago, mining operations were belching particles of lead and other pollutants into the air. The traces of this pollution are trapped in Bronze Age layers of ice in Greenland.

The same forces that led to the destruction of most wild land on Earth also polluted the air. By the Industrial Revolution, the pollution had become so thick in cities that it cut millions of lives short. As people began burning coal, oil, and gas, they also flooded the atmosphere with so much carbon dioxide that it began trapping extra heat—enough to raise the average temperature of the entire planet. By the early twenty-first century, humans had raised the level of carbon dioxide in the atmosphere to its highest level in millions of years. In response, the planet had warmed about 2 degrees Fahrenheit since 1880.

Much of the pollution that humans put into the atmosphere washes out swiftly. The lead-laced fumes from gasoline, for instance, disappeared soon after it was barred. Carbon dioxide is different. It hangs in the atmosphere for centuries, still trapping heat and warming the planet. If tomorrow we cut our emissions of carbon dioxide to zero, the planet would keep warming another degree or more. Future generations would inherit a planet with endangered coastlines, increased wildfires, and farmland threatened with drought.

Three million years ago, the bipedal apes who were teaching each other how to break apart rocks were a small part of a vast ecosystem. But the open-ended power of cultural heredity, transferring knowledge across generations, gave humans the power to rework the land into its own ecological inheritance, which has now led to a climatic inheritance.

———

Today, we sense that we are close to being able to alter human heredity," David Baltimore declared at the international gene-editing meeting in 2015. He was speaking shorthand, one that an audience at a meeting about CRISPR intuitively understood. To them, *human heredity* was the transmission of genes from human parents to human children. And to them, the looming ability to alter it was a new chapter in human history to be entered with awe and fear.

We certainly need to come to a collective decision about using CRISPR on human embryos, to use it only in ways that help people without creating serious dangers of their own. But this shorthand about heredity poses dangers, too. We risk coming to see ourselves as merely the product of the genes we inherited from our parents, and the future as nothing more than carrying those genes forward. The prospect of altering genetic heredity becomes wildly thrilling or terrifying. Soon no one will suffer from a genetic disease again, we're promised. Soon China will breed an army of supergeniuses, we're assured. This shorthand makes it hard to think clearly about genetic heredity. It leads us to overvalue our ambiguous knowledge of how genes work and dismiss the other factors that shape our lives—and could be reshaped to improve the world.

None of this is to say we should dismiss the power of heredity, or shy away from altering it. Instead, we can switch from a shorthand to a longhand. Thinking about heredity more broadly could lead plant scientists to better crops, for example. The first plant breeders manipulated the genetic makeup of crops, picking out good plants to cross to make better varieties.

In recent decades, plant breeders have become more aware of the genes that their crops inherit. The epigenetic side of plant biology has only started to emerge, and some plant breeders are starting to investigate how they can tinker with it to improve crops even more.

Plants sometimes naturally change their epigenetic profile. The methylation that decorates its DNA may get altered, a methyl group falling away from a gene, for example. That change may awaken the gene and improve a plant's growth. Scientists are searching for such changes and trying to propagate the plants so that the new generations inherit the same epigenetic profile.

Transgenerational epigenetic inheritance is a real phenomenon in plants, but many scientists are skeptical that it matters very much in nature. It's wrong to call it Lamarckian, because Lamarck had a very different vision for the inheritance of acquired characters. He thought that inheritance could produce intricate adaptations. Skeptics like Robert Martienssen see little evidence for such adaptations in wild plants.

Yet that doesn't mean such adaptations are impossible. In fact, Martienssen told me, he thinks we know enough now about epigenetics to try to build one.

Martienssen can imagine a plant that could respond to disease outbreaks by turning on immune defenses, and then pass down RNA molecules to their offspring to keep those defenses turned on. If, over the generations, the disease faded away, the plants could shut the resistance genes down so that they didn't have to use their energy to make proteins they no longer needed.

"We could easily engineer a plant to be epigenetically adaptive—to be Lamarckian," Martienssen said.

———

Thinking broadly about heredity might help us outside of laboratories as well. In the United States, it has proven all too tempting over the centuries to blame poverty and inequality on biology. A woman like Emma Wolverton could be institutionalized for life because she was judged a genetically doomed moron. The relative poverty of African Americans could be written off, even by some psychologists, as the result of their inheriting the wrong genes.

Others have argued that the gulfs in the United States are the product of the environment into which people are born and grow up. But the word *environment* is too bland to help us understand much about this problem. The stubborn inequalities in the United States are not the result of some people living in a physical environment. Their environment is built by social forces, and those forces last for centuries because they are regenerated across the generations.

After blacks were emancipated from slavery, they still had to contend with structural racism as well as the racist attitudes of individuals. This racism did not keep springing out of the void year after year. Children learned it, either implicitly or outright, from parents and other adults, and then passed it on to their own children. The social environment then shaped the physical environment into which later generations of blacks were born. Housing discrimination and segregation created neighborhoods where children ended up in poorly performing schools, had to contend with far greater odds of getting shot, and had fewer opportunities for work.

Cumulative culture allowed our species to make giant leaps in technological progress, but it also made us prone to inequality. Hunter-gatherers tend to keep these differences in check, although in a society like the Nootka of Vancouver Island, some people ended up impoverished slaves serving wealthy masters. Once farmers began building up surpluses of food, the gulf could begin to open. They could grow not just over the course of one farmer's lifetime but across generations, because now there were goods to inherit. At first, children might inherit farms and stores of grain from their parents; later, gold and houses and other goods enriched them. The Industrial Revolution made the entire world richer, but some people became vastly richer than others. Francis Galton's ancestors built an empire on guns and banking, which left him able to hire all the math tutors he cared to.

In 1931, the historian James Truslow Adams contrasted the United

States with countries like Great Britain by what he called "the American dream." He defined the dream as being "that life should be made richer and fuller for everyone and opportunity remain open to all." For much of the 1900s, the United States lived up to that dream fairly well. Immigrants fared better there than they had in their home countries. As the United States grew wealthier, much of that wealth flowed to the poorest half of American citizens and allowed them to climb the economic ladder. Raj Chetty, a Stanford economist, has estimated that Americans born in 1940 had a 90 percent chance of making more money than their parents at age thirty.

But Chetty and his colleagues have found that those odds then steadily dropped. Americans born in 1984 had only a 50 percent chance of making more than their parents. The shift was not the result of the United States suddenly running out of money. It's just that wealthy Americans have been taking much of the extra money the economy has generated in recent decades. Chetty's research suggests that if the recent economic growth in the United States was distributed more broadly, most of the fading he has found would disappear. "The rise in inequality and the decline in absolute mobility are closely linked," he and his colleagues reported in 2017.

Inheritance has helped push open that gulf. About two-thirds of parental income differences among Americans persist into the next generation. Economists have found that American children who are born to parents in the ninetieth percentile of earners will grow up to make three times more than children of the tenth percentile.

This inheritance is not simply what parents leave in their wills but the things that they can buy for their children as they grow up. In the United States, affluent parents can afford a house in a good public school district, or even private school tuition. They can pay for college test prep classes to increase the odds their children will get into good colleges. And if they do get in, their parents can cover more of their college tuitions.

Poor parents have fewer means to prepare their children to get into college. Even if their children do get accepted, they have fewer funds, and they're more vulnerable to layoffs or medical bankruptcy. Their children may graduate saddled with steep college debts or drop out before getting a degree.

The gifts that children inherit can keep coming well into adulthood. Parents may help cover the cost of law school, or write a check to help out with a septic tank that failed just after their children bought their first house. Protected from catastrophes that can wipe out bank accounts, young adults from affluent families can get started sooner on building their own wealth.

Inheritance also goes a long way to explain the gap in wealth between races in the United States. In 2013, the median white American household had thirteen times the wealth of the median black household, and ten times that of the median Latino household. In 2017, a team of researchers from Brandeis University and the public policy group Demos sifted through a number of hypotheses that might account for the differences. Going to college didn't close the gap. In fact, the researchers found, the median wealth of white people who didn't finish high school was greater than that of blacks who went to college. Black families actually save money at a greater rate than their white counterparts. Nevertheless, the median white single parent has 2.2 times more wealth than the median black household where there are two parents.

The one big difference the researchers did find was inheritance. Whites are five times as likely to receive major gifts from relatives, and when they do, their value is much greater. These gifts can, among other things, allow white college students to graduate with much less debt than blacks or Latinos. And the effects of these inheritances have compounded through the generations as blacks and Latinos were left outside the wealth feedback loop that benefited white families.

Left unmanaged, these cultural inheritances will roll on, and future generations will be born into systems of economic inequality. The same is true for the environmental inheritance we leave. One of the most important things each new generation learns from the previous one is how to get enough energy to survive. That usually means liberating the Earth's supply of organic carbon and putting some of it in the air. Some people learn how to cut down forests to make charcoal. Others pilot cargo ships across the ocean, leaving a diesel plume trailing behind. If we carry on this way, we may manage to burn through the remaining 12 billion tons of fossil fuels tucked away inside our planet by 2250.

In the process, we would be raising the concentration of carbon dioxide in the atmosphere to levels not seen in the last 200 million years, raising the temperature of the planet to levels far beyond what we humans—a species of ape that evolved in the modest swings of Ice Ages—could handle. And the day that the final gas tank ran dry, the last lightbulb winked out, the planet would not immediately set itself back to the way it was before cultural heredity became such a titanic force. It would take thousands of years for the planet to naturally draw down the carbon dioxide to levels close to what they were before the Agricultural Revolution.

To solve a problem like global warming, we cannot come up with a clever technological fix. We are not being threatened by a giant volcano belching out carbon dioxide from the depths of the Earth, which we can simply cover with a titantic plug. Global warming is a problem of cultural inheritance. To fix it, we need a social form of CRISPR—a means to alter the practices and the values that make their way from one generation to the next.

A cynic may say that there are no systems that can possibly put the brakes on the problems we have made for ourselves. But the environmental scientist Erle Ellis has observed that history records many examples of cultures that transmitted customs through the generations that allowed them to thrive while not destroying their environment in the process. The Maasai of East Africa, for example, have herded cattle for centuries on a landscape that also supported elephants, zebra, lions, and many other wild animals. The long-term health of their ecosystem was the direct result of the culture that the Maasai inherited from their ancestors. Much of their cultural identity is wrapped up in herding cattle—which means they have no need to hunt wild game. To lose a herd of cows and have to hunt marks a huge fall in status. The result was that East Africa could support the great diversity of large mammals on Earth.

"This is a gift to every one of us on Earth now and in the future," Ellis wrote in a 2017 essay. "The megafauna and landscapes they helped to sustain might yet outlast the Great Pyramids or New York City."

When the rest of us look at a culture like the Maasai, we should ask what sort of world we want to leave as a legacy, and then figure out how to do so. It may be that CRISPR can be one tool that we can use toward that end. But we must be confident it reworks the world as we truly need it to.

———

B y the time I paid a visit to Anthony James's insectarium in 2017, gene drive was already becoming something of a Manhattan Project. James and other researchers were getting massive grants from the United States Department of Defense, as well as from major foundations around the world. Yet neither James nor any other gene drive researcher had yet released a CRISPR-bearing creature into the wild. And they were in no rush to do so. They were all too aware of past attempts to fix environmental problems that had turned into ecological disasters. And since those introduced species could keep reproducing, every new generation inherited a warped ecosystem.

Starting in the late 1800s, for example, Australian farmers established sugarcane farms, but they fell into a constant fight with cane beetles. In the early 1930s, an Australian entomologist named Reginald Mungomery got an idea for how to win the battle. He heard stories of the giant marine toad. Native to South and Central America, it had a tremendous appetite for insects, and some people had brought the toads to Hawaii to control sugarcane pests there. He got hold of the toads and raised 2,400 of them. And then, in 1935, he set them loose.

Mungomery didn't understand that the toads were catholic in their tastes. Soon the enormous amphibians—known in Australia as cane toads—were hopping out of the plantations and feeding on small mammals. Australian snakes and other predators sometimes tried to eat the cane toads, but a poisonous secretion in their skin made that impossible. At best, the predators spat out the toads and never tried eating them again. At worst, they died. The cane toad spread relentlessly across Australia, pushing a number of native species toward extinction. Australian researchers have tried all sorts of ways to stop them—poisoning the frogs, training native species not to try to eat them—but so far, nothing has worked.

No one wants to be the Reginald Mungomery of the CRISPR age. It's possible that gene drives could go wrong by hopping from a species we want to eradicate to a related one we want to save. It's possible that changing how mosquitoes and other animals respond to one disease could lead them to carry others. Perhaps getting rid of mosquitoes might disrupt ecosystems in ways we can't yet imagine.

Jennifer Kuzma and Lindsey Rawls, two legal scholars at North Carolina State University, have started to examine the ethics of gene drive as a kind of inheritance. Altering the heredity of disease-carrying insects in the short-term could be tremendously valuable in the number of lives it saves and in the suffering it eliminates. But we also owe future generations a careful, forward-looking consideration of the world they will inherit.

Kuzma and Rawls suggest that, by this standard, some gene drives will prove to be justified and others not. They suggest that saving endangered birds should get ranked over altering weeds. The birds deserve a higher priority because they may very well go extinct if we do nothing. Their disappearance will itself be a permanent legacy we leave to future generations.

When I visited James and his colleagues, I asked them about these ethical issues. They didn't have a lot to say. It's not that they didn't care. They just had more pressing problems at hand. They weren't sure if CRISPR gene drives would work at all.

After all, the natural world was littered with the remains of dead gene drives. They had evolved, raced through populations, and then stopped. In some cases, mutations destroyed them. In other cases, animals evolved defenses that kept them in check. Some biologists have argued that it would be easy for mosquitoes to evolve resistance to a CRISPR gene drive. Some of the insects might gain mutations that changed the sequence of DNA that the CRISPR guide molecules searched for. Their descendants would inherit those mutations and might outbreed the ones carrying the gene drive.

"It's probably easier to break because it isn't an evolved system," Bier told me. "The system we're making is all completely synthetic. It's frail."

Meanwhile, James was toiling away in his insectarium, trying to figure out how to make CRISPR work better. When he put Gantz's malaria-fighting gene drive into a mosquito, all of its offspring inherited it. In the second generation, though, it faltered. Almost all the males inherited it but only some of the females did.

James could still carry forward the gene drive to a new generation by mating the male mosquitoes with normal females. The hairy larvae that I inspected in James's insectarium were all males from his twenty-ninth generation, ready to produce the thirtieth. But James still puzzled over why the female mosquitoes were proving a weak link in the hereditary chain.

The answer may have to do with how mosquitoes develop from a single egg. As a female mosquito develops, it requires many divisions before some of its cells develop into a new batch of eggs. Along the way, a chromosome inside a cell may break. Cells repair this sort of damage by copying DNA from the undamaged copy of the chromosome. James suspected that, during these bouts of repair, the female mosquitoes were editing out their own CRISPR genes. Male mosquitoes, on the other hand, may not lose their gene drives because they set aside sperm cells earlier in their development. If James and his colleagues were right in this hunch, it was hard to see how to overcome it. The inner heredity of mosquitoes isn't easily altered.

After James had shown me all his mosquitoes and answered all my questions, it was time for us to leave the insectarium. We stepped back out into the vestibule, and he closed the inner door loudly behind us. On the other side were thousands of mosquitoes drinking blood, and thousands of larvae writhing in tubs. Here, in the quiet of the vestibule, it was just us two humans, as far as I could tell.

James turned to the pale door to the insectarium and stared at its blankness. The blue smock still hung from his arms.

"The protocol is for us to stand here for a little while," he said. "See if anybody followed us out."

The mosquitoes that James raises come from India. They're adapted to the wet, tropical climate there. If a CRISPR-infused mosquito managed to escape James's insectarium, buzz down the hallways, slip up an elevator shaft, and dart through the doorways into the dry hills around Irvine, it would almost certainly die. And yet, even with such safeguards in place, James still stared at the door, to be sure all his mosquitoes were still penned in his insectarium. As time passed, we both grew quiet. On the other side, a potential new chapter of heredity was crawling, swimming, flying.

Once James felt satisfied that no mosquitoes had escaped, he turned away from the pale inner door. He opened the outer door and we stepped out into the basement hallway. We cast off our smocks into a bin and took the elevator up to the mosquito-killing California sun. We left the next chapter penned in its underground cell, at least for now.

GLOSSARY

Allele: A variant form of a gene. In some cases, different alleles will produce variations in an inherited trait.

Amino acids: The building blocks of proteins.

Bases: The four building blocks of DNA (A, C, G, and T).

Cell lineage: The cells that share common descent with a common ancestral cell in the body.

Chromosome: A threadlike structure of DNA and proteins. Humans have twenty-three pairs of chromosomes.

CRISPR (clustered regularly interspaced short palindromic repeats): A naturally occurring mechanism that gives bacteria immunity to viruses, allowing them to identify and destroy specific sequences of foreign DNA. Adapted to edit DNA.

DNA: The double-stranded molecule that encodes genes.

Dominant: A kind of allele that has an effect when either one or two copies are inherited.

Endosymbiont: A microbe that can exist only within a host and has to be transmitted from mother to offspring.

Enzyme: A protein that catalyzes a chemical reaction in the cell, such as breaking down nutrients.

Epigenetic: Related to molecules such as transcription factors and methylation that affect the expression of genes by altering their DNA sequence.

Epigenome: The physical factors that affect the expression of genes without affecting the actual DNA sequences of the genome.

Eukaryotes: A lineage of species that evolved about 1.8 billion years ago, characterized by features such as the nucleus. Includes animals, plants, fungi, and protozoans.

Gametes: Sperm or eggs.

Gemmule: A hypothetical hereditary particle that Charles Darwin proposed streamed from somatic cells to the gametes.

Gene: A segment of DNA that encodes a protein or a functional RNA molecule.

Gene drive: A system of biased inheritance that allows a genetic element to pass from parents to offspring more than Mendel's Law would allow.

Gene expression: The production of proteins or functional RNA molecules from a gene.

Gene flow: The transfer of DNA from one population into another population.

Gene therapy: A method to treat genetic disorders by delivering correct versions of genes to somatic cells.

Genetic engineering: Introduction of DNA, RNA, or proteins that are manipulated by humans to effect a change in an organism's genome or epigenome.

Genome: The complete sequence of DNA in an organism.

Genome-wide association study: An analysis of a group of genomes that can reveal unusually common genetic variants in people who have the same condition.

Germ cell: A cell in a lineage that produces gametes. Distinguished from cells in the rest of the body (somatic cells).

Germ line: A cellular lineage in sexually reproducing organisms that produces gametes, which transmit genetic material to the next generation.

Germ line engineering: Altering DNA in the germ line (in gametes or embryos) to create changes that can be inherited by descendants.

Haplogroup: A group of people who can trace their ancestry to a single person, sharing a set of genetic variants.

Heritability: The proportion of variance in a trait in a population due to genetic variance, measured from 0 to 100 percent.

Hybrid: The offspring of two plants or animals of different species or varieties.

Meiosis: A type of cell division leading to the development of gametes. Meiosis reduces the number of chromosomes in the mother cell by half and produces four gamete cells. During meiosis, chromosomes can undergo recombination.

Mendelian: A trait that follows Mendel's Law, with a three-to-one ratio of dominant and recessive alleles.

Methylation: An epigenetic mechanism to silence a gene through the addition of a methyl group ($-CH_3$) to a site on a DNA molecule.

Microbiome: The set of microbes that resides in a host.

Mitochondria: Fuel-generating organelles inside the cell containing a small amount of DNA. Mitochondria are inherited only through the maternal line.

Mitochondrial replacement therapy: A treatment for mitochondrial disorders in which the nucleus of a healthy egg or zygote is inserted into a donor egg whose nucleus has been removed.

Mosaicism: Genetic variation among somatic and germ cells in a single multicellular organism.

Mutation: A new genetic variation that arises in a cell and can be inherited by its offspring.

Nucleus: A sac containing chromosomes found in cells of humans and other eukaryotes.

PKU: Phenylketonuria, a recessive hereditary disorder caused by a faulty enzyme.

Pluripotent: An embryonic cell that can develop into a wide range of (but not all) types of cells.

Protein: A long chain of amino acids encoded in a gene.

Recessive: A kind of allele that has an effect only when two copies are inherited.

Recombination: An exchange of DNA between pairs of chromosomes during meiosis.

RNA: A single strand of bases. The production of RNA is a step in the production of proteins, but RNA molecules can also act on their own to catalyze chemical reactions in the cell.

Single-nucleotide polymorphism: A site in DNA where a single base varies in a population.

Somatic cell: A cell that is not in the germ line, typically unable to carry genes to the next generation.

Stem cell: A cell that can generate other types of cells, either in an embryo or in an adult.

STRUCTURE: A computer program first developed by Jonathan Pritchard and his colleagues to trace the ancestry of individuals to unknown populations.

Totipotent: The earliest cells in an embryo, which can develop into any type of cell in the embryo or placenta.

Transcription factors: Proteins that bind to DNA to alter the expression of genes.

X and Y chromosomes: The sex chromosomes in mammals. Females have two X chromosomes; males have an X and a Y.

Zygote: A fertilized egg.

NOTES

Epigraph
ix. Darwin 1868.

Prologue
2. "most wondrous map": National Human Genome Research Institute 2000.
2. genetic material pieced together from a mix of people: Wade 2002.
3. a gene called HEXA: US National Library of Medicine 2017.

Part I: A STROKE ON THE CHEEK
1. The Light Trifle of His Substance
11. The emperor, clad in black: Curtis 2013; Parker 2014; Prescott 1858.
11. sat before the assembly: Belozerskaya 2005.
12. "May the Almighty bless you with a son": Quoted in Prescott 1858, p. 15.
12. The Romans did not use their word: Du Plessis, Ando, and Tuori 2016.
12. "that person's assets pass to us": Quoted in Du Plessis 2016.
13. The Apinayé of Brazil had it both ways: Maybury-Lewis 1960.
13. the sons split their father's land: Müller-Wille and Hans-Jörg Rheinberger 2007.
13. Venice's Great Council created the Golden Book: Johnson 2013.
14. a pageant that was put on in 1432: Osberg 1986.
14. a pair of trees: Kingsford 1905; Klapisch-Zuber 1991.
14. the teachings of ancient Greeks and Romans: Cobb 2006.
15. Aeschylus: from *Eumenides*.
15. a tribe known as the Longheads: Zirkle 1946, p. 94.
15–16. "The people of cold countries . . .": Quoted in Eliav-Feldon, Isaac, and Ziegler 2010, p. 40.
16. In the 1200s, the philosopher Albertus Magnus: Ibid., p. 197.
16. the Malaysian island of Langkawi: Carsten 1995.
17. falcons had the noblest blood: Oggins 2004.
17. animals that shared the same blood: a *race*: Eliav-Feldon, Isaac, and Ziegler 2010.
17. tips for providing a "good race" of horse: Quoted in Eliav-Feldon, Isaac, and Ziegler 2010, p. 249.
17. "The good man of good race always returns to his origins . . .": Ibid., p. 250.
18. carried in their blood and embedded in their seed: Johnson 2013, p. 131.
18. "From the days of Alexander . . ." Quoted in Eliav-Feldon, Isaac, and Ziegler 2010, p. 248.
18. Jews and *conversos* alike as the Jewish "race": Martínez 2011.
18. blue blood: Pratt 2007.
18. a clothes merchant or a moneylender: Martínez 2011.
19. "I have found no monsters": Columbus, "Santangel Letter."
19. They declared Native Americans to be natural slaves: Pagden 1982.
19. "For them there is no tomorrow . . .": Quoted in Pagden 1982, p. 42.
19. "Nature proportioned their bodies . . .": Ibid.
20. African slaves: Sweet 1997.
20. "possess attributes that are quite similar to those of dumb animals": Quoted in Kendi 2016, p. 20.
20. straight to Brazil, Peru, and Mexico: Smedley 2007.

20. the descendants of Ham, one of Noah's sons: Haynes 2007 and Robinson 2016.
20. "... And from this race these blacks ...": Quoted in Haynes 2007, p. 34.
21. a small upper jaw that failed to develop: Peacock et al. 2014.
21. Don Carlos: See Hodge 1977; Parker 2014.
22. a higher rate of infant mortality: Álvarez, Ceballos, and Quinteiro 2009.
23. "... how could the light trifle of his substance ...": Montaigne 1999.
23. People did not reproduce; they were engendered: Jacob 1993.
24. Luis Mercado: See Mercado and Musto 1961; Müller-Wille and Rheinberger 2012.
24. "modest in dress ...": Mercado and Musto 1961, p. 350.
25. A hereditary disease was like a stamp: Müller-Wille and Rheinberger 2012.
25. "... teach the deaf and dumb to speak ...": Quoted in Mercado and Musto 1961, p. 371.
26. "most beautiful in features ...": Quoted in Langdon-Davies 1963, p. 15.
26. "He seems extremely weak ...": Ibid., p. 62.
27. "He has a ravenous stomach ...": Quoted in Cowans 2003, p. 189.
27. "without allowing the least dismemberment ...": Quoted in Langdon-Davies 1963, p. 256.

2. Traveling Across the Face of Time

29. his clothes were foul: Schwartz 2008.
29. a blue-and-white sign informing visitors: Dare 1905.
30. Luther Burbank: For biographical details on Burbank, see Beeson 1927; Burbank and Hall 1939; Dare 1905; Dreyer and Howard 1993; Janick 2015; Pandora 2001; Smith 2009; Stansfield 2006; Sweet 1905; Thurtle 2007.
30. "His results are so stupendous ...": de Vries 1905, p. 340.
30. Burbank's postman brought him thirty thousand letters: Dare 1905.
30. "the wizard of horticulture": Ibid.
30. "Such a knowledge of Nature ...": Quoted in Palladino 1994.
30. "... a scale so extensive as to suggest magic ...": Quoted in Eames 1896.
30. "... there was no stone in the plum ...": de Vries 1905, p. 334.
31. "the inherent constitutional life force ...": Burbank 1904, p. 35.
32. Merino: Müller-Wille and Rheinberger 2012; Wood and Orel 2001.
33. Robert Bakewell: Pawson 1957.
33. "the Mr. Bakewell who invented sheep": Ibid., p. 7.
33. Mr. Bakewell was born in 1725: Wykes 2004.
33. "the smaller the bones ...": Quoted in Young 1771, p. 111.
34. He chalked his data on slates: Wood 1973.
34. Bakewell's new breed: Wood and Orel 2001.
34. "only fit to glide down the throat of a Newcastle coal-heaver": Ibid., p. 109.
34. "My people want fat mutton and I gave it to them": Quoted in Wood 1973, p. 235.
35. "this prince of breeders": Quoted in Wood and Orel 2001, p. 232.
35. "had been making observations": Ibid., p. 106.
35. "He has convinced the unbelievers of the truth ...": Quoted in Wykes 2004, p. 55.
35. Frederick Augustus: Wood and Orel 2001.
36. "The Association of Friends, Experts and Supporters ...": Poczai, Bell, and Hyvönen 2014.
36. Thomas Andrew Knight: Kingsbury 2011.
37. "None appeared so well calculated to answer my purpose": Knight 1799, p. 196.
37. "... any number of new varieties may be obtained": Ibid., p. 196.
37. "A single bushel ...": Quoted in Kingsbury 2011, p. 81.
37. "genetic rules of nature": Quoted in Poczai et al. 2014.
38. Napp and his friars got into the breeding business: Allen 2003.
38. to pay off the priory's massive debts: Endersby 2009.
38. "a lengthy, troublesome and random affair": Quoted in Orel 1973, p. 315.
38. The trouble would not go away: Gliboff 2013.
38. "What we should have been dealing with . . .": Quoted in Müller-Wille, Staffan, and Rheinberger 2007, p. 241.
38. semaphore flags or telegraph messages: Gliboff 2013.
39. a pair of "antagonistic elements": See Van Dijk and Ellis 2016.
40. "... dealing only with individual phenomena ...": Ibid.
41. Abbot Mendel got so ensnared in tax battles: Schwartz 2008.

41. By 1837, there were a million Merinos in Vermont alone: Vermont Historical Society.
41. climbed beyond a thousand dollars: Smith 2009.
41. hen fever: Burnham 1855.
41. Jesse Hiatt: See Friese 2010; Kingsbury 2011; Pollan 2001.
42. while she gathered strawberries: Beeson 1927.
42. "the wood to bring, weeds to pull . . .": Ibid., p. 58.
42. ". . . not second-hand, but first-hand . . .": Quoted in Dreyer and Howard 1993, p. 49.
42. "Nature was calling me to the land . . .": Quoted in Burbank and Hall 1927, p. 9.
43. The textbooks Burbank read in school: Dreyer and Howard 1993, p. 270.
43. if a woman "has a small, taper waist . . .": Cutter 1850, p. 242.
43. ". . . it is impossible for most people to realize the thrills . . .": Quoted in Beeson 1927, p. 74.
44. "The laws governing inheritance": Darwin 1859, p. 14.
44. jotting down notes and questions: Geison 1969; Bartley 1992.
44. Bakewell's famous rules: Wood 1973.
44. a short pamphlet entitled *Questions . . .* : Darwin 1839.
45. "the greatest treat, in my opinion . . .": Quoted in Secord 1981, p. 166.
45. alienists: López-Beltrán 2004; López-Beltrán 1995; Noguera-Solano and Ruiz-Gutiérrez 2009.
45. "mental alienation is the most eminently hereditary": Quoted in Porter 2018.
45. *Treatise on Natural Inheritance*: Discussed in Churchill 1987.
46. a constellation of disorders: See Álvarez, Ceballos, and Berra 2015; Hayman et al. 2017.
46. "wretched contemptible invalid": Quoted in Berra, Álvarez, and Ceballos 2010, p. 376.
46. "It is the great drawback to my happiness . . .": Ibid., p. 377.
46. he saved that profound matter for a book of its own: Geison 1969.
46. the strange ways in which animals and plants reproduced: Müller-Wille 2010.
46. "minute granules or atoms": Darwin 1868.
47. something that combined *cells* with *genesis*: Browne 2002.
47. "merely a provisional hypothesis . . .": Quoted in Deichmann 2010, p. 92.
47. "It has thrown a flood of light on my mind . . .": Quoted in Browne 2002, p. 286.
48. improved breeds of cattle grew small lungs: Darwin 1868, p. 299.
48. ". . . said to have been trying an experiment . . .": Ibid., p. 3.
48. "While I had been struggling along . . .": Burbank and Hall 1927, p. 74.
49. His cabbage seeds and sorghum won prizes: Smith 2009.
49. "Stored in every cherished seed . . .": Burbank and Hall 1927, p. 12.
50. "In short I was a product of all my heredity": Ibid., p. 20.
50. "an inherited sensitiveness about money": Quoted in Dreyer and Howard 1993, p. 78.
50. "These were indeed dark days": Ibid., p. 77.
50. "this already famous Potato": Ibid., p. 78.
51. "Something must happen to 'stir up their heredities' . . .": Burbank and Hall 1939, p. 121.
51. ". . . it is like stirring up an ant-hill . . .": Ibid., p. 95.
52. "In his laboratory garden he has done for Nature . . .": Dare 1905.
52. "one California town—Vacaville . . .": Ibid.
52. "he stands unique in the world": Jordan and Kellogg 1909, p. 79.
52. "In the present state of science . . .": Quoted in James 1868, p. 367.
53. "made a marked epoch in my own mental development": Quoted in Galton 1909.
53. "It seems hardly credible now . . .": Ibid.
53. "I find that talent is transmitted by inheritance . . .": Quoted in Galton 1865, p. 157.
54. "Men and women of the present day . . .": Ibid., p. 166.
54. Galton became convinced that pangenesis "is the only theory . . .": Galton 1870.
54. "Good rabbit news!": Quoted in Bulmer 2003, p. 118.
54. The experiments proved "a dreadful disappointment": Ibid.
54. "The conclusion from this large series of experiments . . .": Galton 1870, p. 404.
55. "I have not said one word about the blood": Darwin 1871.
55. August Weismann: Churchill 2015.
56. "This substance transfers its hereditary tendencies . . .": Weismann 1889, p. 74.
56. "Ever since I began to doubt the transmission . . .": Ibid., p. 319.
57. "All such 'proofs' collapse": Ibid., p. 434.

58. "We talked for a short time about all kinds of things . . .": Quoted in Van der Pas 1970.
59. a thirty-five-year-old paper by "a certain Mendel": Quoted in Schwartz 2008, p. 84.
60. A British doctor named Archibald Garrod: Comfort 2012.
60. "whole problem of heredity . . .": Quoted in Schwartz 2008, p. 114.
61. "perhaps as original as Darwin's": Quoted in Pandora 2001, p. 504.
61. "The sole aim of all his labors . . .": Quoted in de Vries 1905, p. 333.
62. retiring to the village of Lunteren: Schwartz 2008.
62. ". . . your experimental investigations . . .": Quoted in Dreyer and Howard 1993, p. 132.
63. "Environment is the architect of heredity": Burbank 1906.
63. the young botanist seemed to be preparing to explode his legend: Glass 1980.
64. "great contributions to good taste": Quoted in Pandora 2001, p. 496.
64. The painter Frida Kahlo: Giese 2001.
66. "All things—plants, animals, and men . . .": Quoted in Clampett 1970.

3. This Race Should End with Them

69. Emma Wolverton: See Allen 1983; Doll 2012; Smith 1985; Smith and Wehmeyer 2012a, 2012b; Zenderland 1998.
70. "no peculiarity . . .": Quoted in Goddard 1912, p.2.
70. ". . . the connection between nature and their being": The Vineland Training School 1899, p. 28.
71. ". . . cause these songs of savagery to become the songs of civilization": Ibid.
71. "off for camp": Smith and Wehmeyer 2012b.
71. "We are doing God's work": The Vineland Training School 1898.
71–72. ". . . study of the deficient and delinquent classes . . .": Ibid.
72. "disreputable lives": The Vineland Training School 1899, p. 12.
72. "She is an almost perfect worker": Quoted in Goddard 1908.
73. "How many cents have I?": See Goddard 1908, 1910, 1911.
74. "Nobody knew me or cared a whit . . .": Quoted in Zenderland 1998, p. 342.
74. "Quaker jail": Ibid., p. 20.
74. "In all my adult life": Ibid., p. 23.
75. "a law of child nature . . .": Ibid., p. 52.
75. "I never dreaded anything more": Goddard 1931, p. 56.
75. "a great family . . .": Quoted in Goddard 1910b, p. 275.
75. ". . . a great human laboratory": Ibid., p. 275.
76. "Degeneracy is increasing . . .": The Vineland Training School 1906, p. 28.
76. ". . . some way of exercising these brains . . .": The Vineland Training School 1907, p. 39.
76. "After two years my work was so poor . . .": Quoted in Zenderland 1998, p. 91.
77. a new exam called the Simon-Binet test: Goldstein, Princiotta, and Naglieri 2015.
77. "otherwise called good sense . . .": Ibid., p. 158.
78. ". . . a mathematical proof . . .": The Vineland Training School 1911, p. 311.
78. "It cannot be cured": The Vineland Training School 1909, p. 41.
78. "after-admission blank": See Goddard 1910a.
79. "collect data on heredity": Quoted in Zenderland 1998, p. 154.
79. a geneticist by the name of Charles Davenport: Porter 2018.
79. Davenport had leaped to fame only a few years before: Witkowski 2015.
79. ". . . from the field of speculative sciences . . .": Davenport 1899, p. 39.
79. ". . . wide fields of unexpected facts . . .": de Vries 1904, p. 41.
80. Inspecting his canaries, he concluded: Davenport 1908.
80. "I can hardly express my enthusiasm . . .": Quoted in Porter 2018.
81. "As to the nature of the 'genes'": Quoted in Falk 2014.
81. "a pleasing manner and address . . .": Goddard 1914, p. 24.
81. "seem to conform perfectly to the Mendelian law": The Vineland Training School 1909, p. 42.
81. ". . . famous the world over and for all time": Ibid., p. 43.
82. ". . . the color of your hair . . .": Goddard 1916, p. 269.
82. "She would lead a life that would be vicious . . .": Goddard 1912, p. 12.
82. "Her philosophy of life is the philosophy of the animal": Ibid.
83. "The biologist could hardly plan . . .": Ibid., p. 69.

83. ". . . the most valuable that have ever been contributed . . .": The Vineland Training School 1910, p. 35.
83. ". . . our best people must replenish the Earth": Goddard 1916, p. 270.
84. "a brief word . . .": Galton 1883, p. 24.
84. "a gentle painless death": McKim 1899, p. 188.
85. "the salvation of the race through heredity": Davenport 1911, p. 260.
85. "The Elimination of Feeble-Mindedness": Goddard 1911a.
85. "but all these causes combined . . .": Ibid., p. 510.
85. "boy crazy": Hill and Goddard, 1911.
85. sterilization: Reilly 1991, 2015.
86. ". . . a carefully worded sterilization law . . .": Goddard 1911a, p. 270.
86. ". . . bad stock": Goddard 1912, p. 12.
87. "No amount of education or good environment . . .": Ibid., p. 53.
87. "I doubt if there is in all literature . . .": See "How One Sin Perpetuates Itself" 1916, p. 6.
87–88. "imbeciles, feeble-minded and persons with physical or mental defects . . .": Quoted in Zenderland 1998, p. 266.
88. "We were in fact most inadequately prepared for the task": Goddard 1917, p. 271.
88. "The same arguments which induce us to segregate . . .": Ibid., p. 264.
89. "They can hardly stand by themselves as valid": Ibid., p. 274.
89. ". . . we are getting now the poorest of each race": Ibid., p. 266.
89. "Morons beget morons": Ibid., p. 270.
89. "If the latter, as seems likely . . .": Ibid., p. 280.
89. ". . . testing of the 1,700,000 men . . .": Goddard 1931, p. 59.
90. "moron majority": White 1922.
90. "And then will come perfect government": Goddard 1920, p. 99.
91. "My home": Quoted in Smith 1985.
91. "dignified courtesy": Quoted in Smith and Wehmeyer 2012a, p. 205.
92. "You could put more pep in it": Quoted in Doll 2012, p. 32.
92. "kindly dismissed by a lenient justice-of-the-peace": Quoted in Smith 1985, p. 31.
92. ". . . she would return pregnant": Quoted in Allen 1983, p. 79.
92. "It isn't as if I'd done anything really wrong": Quoted in Smith 1985, p. 33.
92. "fatal error": Quoted in Smith and Wehmeyer 2012a, p. 127.
93. "Such children should never be born": Gosney and Popenoe 1929, p. viii.
93. Carrie Buck: Cohen 2016.
93. "a social and racial crime of the first magnitude": Quoted in Moses and Stone 2010.
93. ". . . lest we admit more degenerate 'blood'": Laughlin 1920.
94. a government-run program to breed the best parents: Weiss 2010.
94. "The head of the German ethno-empire . . .": Quoted in Poliakov 1974, p. 298.
94. "Questions which were only cautiously touched upon . . .": Quoted in Kühl 2002, p. 41.
95. ". . . can not perpetuate his suffering . . ." Stephen Spielberg Film and Video Archive.
95. "There is only one answer: heredity": Quoted in Kühl 2002, p. 42.
95. "racial hygiene" laws: Reilly 2015.
95. a suitcase, books, bottles, and other objects: Proctor 1988.
95. a program to kill children judged to be idiots: Lifton 2000.
96. "Can you name the four seasons?": Quoted in Burleigh 2001, p. 370.
96. Walter Lippmann: Lippmann 1922.
97. Abraham Myerson: Myerson 1925.
97. "used his germplasm in orthodox fashion . . .": Ibid., p. 78.
97. " . . any definite information about my great-great-grandfather . . .": Ibid., p. 79.
97. *Drosophila melanogaster*: See Endersby 2009; Schwartz 2008.
98. many genes could influence a single trait: Morgan 1915.
98. "It is of the utmost importance . . .": See "Mendelism Up to Date" 1916, p. 20.
99. "It is extravagant to pretend . . .": Morgan 1925, p. 201
99. "In reality, our ideas are very vague": Ibid., p. 208.
99. extra legs if they were born in the winter: Ibid., p. 41.
99–100. ". . . demoralizing social conditions that might swamp a family . . .": Ibid., p. 201.
100. "The student of human heredity . . .": Ibid., p. 205.
100. "futile system": Quoted in Allen 2011, p. 317.

100. "a worthless endeavor from top to bottom": Quoted in Yudell 2014, p. 195.
100. "her chances of going insane were no better than my own": Dunlap 1940, p. 225.
101. Amram Scheinfeld: Scheinfeld 1944.
101. "As for myself": Quoted in Zenderland 1998, p. 326.
102. "Half of the world must take care of the other half": Goddard 1931, p. 59.
102. ". . . all I could stand for one day!": Quoted in Zenderland 1998, p. 323.
102. "Much in the way of polish is lacking . . .": Goddard 1942.
102. "The author's conclusion . . .": Quoted in Associated Press 1957.
103. ". . . a feeble-minded tavern girl": Quoted in Garrett 1955.
103. "the vote of the feeble-minded person . . .": Tucker 1994.
103. curious investigators: Smith and Wehmeyer 2012a; Straney 1994.
103. A pair of genealogists, David Macdonald and Nancy McAdams: Macdonald and McAdams 2001.
103. Public records show he was a landowner: Smith and Wehmeyer 2012a.
105. "Emma was tall and reticent": Quoted in Allen 1983, p. 52.
105. *I'm a gypsy*: Doll 2012.
105. "I guess after all I'm where I belong": Quoted in Smith and Wehmeyer 2012a.
106. "for a dear, wonderful friend . . .": Quoted in Smith 1985, p. 30.
106. "She was devoted to the people . . .": Quoted in Allen 1983, p. 52.
106. "The nicest thing about it": Quoted in Zenderland 1998, p. 339.

4. Attagirl

107. Pearl Buck: See Buck 1950; Conn 1996; Finger and Christ 2004; Harris 1969; Paul and Brosco 2013; Spurling 2011.
107. "Doesn't she look very wise . . .": Buck 1950, p. 32.
108. "He has never seen or understood anything": Quoted in Conn 1996, p. 182.
110. "I realized I must leave her in some place": Quoted in Spurling 2011, p. 181.
110. "I had found out enough to know . . .": Buck 1950, p. 59.
111. "I saw children playing around the yards behind the cottages . . .": Ibid., p. 45.
111. "Only the thought of a future with the child grown old . . .": Quoted in Spurling 2011, p. 182.
112. "a creature hopelessly mongrel": Quoted in Conn 1996, p. 230.
112. "It is not a shame at all but something private . . .": Quoted in Finger and Christ 2010, p. 45.
112. "I would gladly have written nothing . . .": Quoted in Conn 1996, p. 132.
113. "I feel toward her as tenderly as ever . . .": Quoted in Harris 1969, p. 279.
114. "I have been a long time making up my mind . . .": Buck 1950, p. 106.
114. "It was my child who taught me to understand so clearly . . .": Ibid., p. 52.
114. "Though the mind has gone away . . .": Ibid., p. 43.
115. Borgny Egeland: On the discovery of PKU, see Centerwall and Centerwall 2000; Harper 2008; Kaufman 2004; Messner 2012; Paul and Brosco 2013.
117. Lionel Penrose: See Comfort 2012; Harper 1992; Harris 1974; Kevles 1995; Laxova 1998; Valles 2012; Wellcome Library.
118. "pretentious and absurd": Penrose 1949, p. 22.
118. lurid tales like *The Kallikak Family*: Penrose 1933.
118. "That mental deficiency . . .": Quoted in Penrose 1933, p. 146.
118. "The first consideration in the prevention of mental deficiency . . .": Ibid., p. 164.
119. a single sample turned green: Penrose 1935.
119. "preferable to the original . . .": Quoted in Penrose 1946, p. 949.
119. "an abominable abbreviation.": Quoted in Paul and Brosco 2013, p. 15.
120. "I was informed that this patient . . .": Ibid.
121. "Phenylketonuria: A Problem in Eugenics": See Penrose 1946.
124. Rosalind Franklin: Maddox 2002.
126. Later generations of scientists: Robson et al. 1982; Woo et al. 1983.
126. "Her mother was not at all impressed . . .": Quoted in Bickel 1996, p. S2.
128. filming a silent movie: New England Consortium of Metabolic Programs 2010.
129. 25 percent of people with PKU lived to the age of thirty: Paul and Brosco 2013.
130. "Attagirl," the president said: Hunter 1961.
130. memorialized by an official White House photograph: White House Photographs 1961.
130. Sheila and Kammy appeared in *Life*: "New Way to Detect a Dread Disease" 1962.

130. "... a child can live a normal life": "U.S. Panel Urges Testing at Birth" 1961.
131. the first person with PKU to gain a PhD: Beck 1998.
131. "In Carol's case nothing matters ...": Quoted in Paul and Brosco 2013, p. 226.
131. "... the same unusual odor": Centerwall and Centerwall 2000, p. 89.
132. Pearl "had trouble accepting ...": Buck 1992, p. 97.
132. Crayons and coloring books: Conn 1996.
133. "The possession of a genetic map ...": Quoted in Paul and Brosco 2013.
133. "If you simply remove foods ...": Quoted in Collins, Weiss, and Kathy 2001.
133. "PKU is the example where the paradigm was proven": Quoted in Paul and Brosco 2013.
133. deep flaws: See Panofsky 2014; Yudell 2014.
134. "... both disingenuous and misleading": Rose 1972.
134. "It turns out," he cheerfully wrote: See Wright 1995.
134. children growing up on a low-phenylalanine diet: Paul and Brosco 2013.

PART II: WAYWARD DNA
5. An Evening's Revelry

137. one of the loftiest titles in science: "Mendel's Law": Bateson and Saunders 1902.
138. Life likely emerged: See Adami 2015; Baross and Martin 2015; Joyce 2012; Kun et al. 2015; Pressman, Blanco, and Chen 2015; Sojo et al. 2016; Szostak, Wasik, and Blazewicz 2016.
140. may belong to a trillion different species: Locey and Lennon 2016.
140. horizontal inheritance: Daubin and Szöllősi 2016.
141. *Enterococcus faecium*: Lester et al. 2006.
143. a system of molecules called CRISPR-Cas: Zimmer 2015a.
144. a genuine case of Lamarckian heredity: Koonin and Wolf 2009.
144. These microbial monsters were eukaryotes: Dacks et al. 2016.
145. meiosis: See Baudat, Imai, and de Massy 2013; Coop et al. 2008; Hunter 2015; Lenormand et al. 2016; Mézard et al. 2015; Sung and Klein 2006; Zickler and Kleckner 2015, 2016.
145. meiosis is the same inside a woman's body: Evans and Robinson 2011.
146. polar bodies: Schmerler and Wessel 2011.
146. When Mendel crossed tall and short pea plants: Reid and Ross 2011.
146. "an evening's revelry": Hurst 1993.
146. Frans Alfons Janssens: See Koszul et al. 2012; Centre of Microbial and Plant Genetics n.d.
147. "Are we being presumptuous": Janssens 2012, p. 329.
148. why meiosis evolved: See Lenormand et al. 2016; Mirzaghaderi and Hörandl 2016; Niklas, Cobb, and Kutschera 2014; Wilkins and Holliday 2009.
149. Sometimes, yeast have sex instead: See Casselton 2002; Hodge 2010.
150. The differences ... were clear: McDonald, Rice, and Desai 2016.
151. Peter Visscher: Visscher et al. 2006.
152. meiosis has crumbled away: Mirzaghaderi and Hörandl 2016; van Dijk et al. 2016.
152. hawkweed: Koltunow et al. 2011; Bicknell et al. 2016.
153. Sergey Gershenson: See Gershenson 1928; Wasser 1999.
153. "... it seemed impossible to explain them ...": Gershenson 1928, p. 490.
154. fixing Mendel's dice: Lindholm et al. 2016. On plants: Fishman and Willis 2005. Fungi: Grognet et al. 2014. Mammals: Didion et al. 2016.
154. still unclear in humans: Huang, Labbe, and Infante-Rivard 2013.
155. Some of those databases include genetic markers: Meyer et al. 2012.
155. A few promising genes have turned up: Liu et al. 2013.
155. push a population to extinction: Unckless and Clark 2015.

6. The Sleeping Branches

159. *History of the Goodspeed Family*: Goodspeed 1907.
163. The American obsession with genealogy: Zerubavel 2012.
163. steeped in traditional European customs: See Weil 2013; Zerubavel 2012.
164. "... what I have been told were the family arms ...": Quoted in Weil 2013, p. 27.
164. "I am the youngest Son ...": Quoted in Jordan et al. 1899, p. 6.
164. "One of the strongest natural proofs ..." Paine 1995.
164. their high status in the new republic: Morgan 2010a, 2010b.

164. Electa Fidelia Jones: Morgan 2010a.
165. John Randolph, an early US senator: Weil 2013, p. 82.
165. ". . . the traditions of chivalry and knighthood": See Order of the Crown of Charlemagne.
166. "When I talk to a genealogist": Quoted in Weil 2013, p. 47.
166. "some cotton and tobacco, and negroes, etc.": Quoted in Warren 2016, p. 7.
166. ". . . the son of an African Chief . . .": Quoted in Armistead 1848, p. 510.
167. an interest in genealogy as keen as that of their white counterparts: Gatewood 1990.
167. ". . . the more intellectual and high-class branch . . .": Hughes 1940, p. 208.
167. "a negro boy . . . known by the name of 'You-Boy'": Quoted in Hatfield 2015, p. 79.
167. When the journalist Alex Haley was growing up: See Haley 1972; Norrell 2015.
168. "'Rooting' a Negro family . . .": Quoted in Norrell 2015, p. 98.
169. historian Willie Lee Rose: Rose 1976.
169. sometimes he dodged the questions by calling *Roots* "faction": Page 1993.
169. paying out $650,000: Nobile 1993.
169. An expert on African oral history: Wright 1981.
169. a catalog of errors: See Mills and Mills 1984; Mills and Mills 1981.
169. "Suddenly, white Americans were tuning in . . .": Page 1993.
170. "I believe that Haley sold out": Mills 1984, p. 41.
170. the second-most-popular search topic: Falconer 2012.
170. The photos are of heaps of bodies: US Holocaust Memorial Museum Photo Archives 1944.
171. *pater est quem nuptiae demonstrant*: Browne-Barbour 2015.
171. "the privity between a man and his wife . . .": Quoted in Baker 2004, p. 24.
171. "bald eagle evidence": See Shapiro, Reifler, and Psome 1992.
171. Charlie Chaplin: See Ackroyd 2014; Friedrich 2014; Louvish 2010.
172. their blood type: See Lederer 2013; Zimmer 2014a.
172. Ludwik Hirszfeld: See Geserick and Wirth 2012; Lederer 2013; Mikanowski 2012; Okroi and McCarthy 2010; Pierce 2014; Starr 1998.
173. "to find the real father . . .": Hirschfeld and Hirschfeld 1919, p. 676.
173. "Showing none of the temperament . . .": See "The Case of Carol Ann" 1945.
174. "with charts, diagrams, and elaborate explanations": Associated Press 1944.
174. "In accordance with the well accepted laws of heredity": *Berry v. Chaplin* 1946.
174. "You'll sleep well . . .": Quoted in Ackroyd 2014, p. 211.
174. "Unless the verdict is upset": Quoted in Benson 1981.
174. "This is magic": Quoted in Ackroyd 2014, p. 211.
175. bits of our genetic material: Roewer 2013.
175. Tsar Nicholas II: Coble et al. 2009.
176. enzymes that shred its own mitochondrial DNA: Zhou et al. 2016.
177. Philip's mitochondrial DNA: Gill et al. 1994.
177. some skeptics questioned the identity of the bones: Zhivotovsky 1999.
178. Once again, Gill examined the remains: Coble et al. 2009.
178. cheek swabs from 188 Jewish men: Hammer et al. 1997.
179. Hammer and his colleagues published their new research: Hammer et al. 2009.
180. ". . . a piece of me missing": See "Roots Revisited" 2016.

7. Individual Z

186. DNA.Land could confidently recognize 112 identical segments: For more on identity by descent, see Browning and Browning 2012; Donnelly 1983.
187. Elias Gottesman: Mandel 2014.
188. Graham Coop: See Coop 2013c.
188. 628 chunks: Coop, personal communication. See also Coop 2013a and 2013b.
189. Joseph Chang: Chang 1998.
189. Coop and his colleague Peter Ralph: Ralph and Coop 2013.
190. Everyone who was alive five thousand years ago: Thomas 2013.
190. E1b1b1c1: Lucotte and Diéterlen 2014.
191. That woman lived in Africa: Fu et al. 2013
192. he lived in Africa 190,000 years ago: Poznik et al. 2016.
194. Noble families struggled: Martinez 2001.
194. fine distinctions: Jordan 2014.

195. gradually singled out the people from Africa: Smedley 2007.
195. as surely as they inherited their color: Haynes 2007.
195. Ham's curse: Robinson 2016.
195. "The Blacks born here . . .": Long 1774.
195. Carl Linnaeus defined four races: Smedley 2007.
196. Johann Friedrich Blumenbach: Frederickson 2003.
196. "abominable mixture and spurious issue": Quoted in Jordan 2014.
197. ". . . a question for the solution of a jury": Quoted in Morris 2004, p. 28.
197. "My father was a white man . . .": Douglass 1855.
197. Douglass's mother was a Maryland: Barnes 2013.
198. Racial Integrity Act: Zimmer 2014a.
199. "The mistakes the negroes made . . .": Galton 1869, p. 349.
199. the question of race was far more urgent and intimate: Yudell 2014.
200. A scientist named Harvey Jordan: Dorr 2008.
200. "as a leaven in lifting . . .": Quoted in Yudell 2014, p. 38.
200. "I have been wondering if I could be of service . . .": Quoted in Dorr 2008, p. 55.
201. Milton Bradley Company: Keevak 2011.
201. "There isn't the least doubt, I think . . .": Quoted in Davenport and Davenport 1910.
201. "Skin color in negro x white crosses . . .": Ibid.
202. "melodic endowment": Quoted in Jordan 1913, p. 579.
202. "The Effects of Race Intermingling": Davenport 1917.
203. "The human species so shade and mingle . . .": Dubois 1906, p. 16.
203. Ludwik Hirszfeld: See Geserick and Wirth 2012; Lederer 2013; Mikanowski 2012; Okroi and McCarthy 2010; Starr 1998.
203. "the most crowded and cosmopolitan spot . . .": Quoted in Owen 1919.
203. samples from 8,400 people: Hirschfeld and Hirschfeld 1919.
204. "Our biochemical index . . .": Hirszfeld and Hirszfeldowa 1918.
204. Theodosius Dobzhansky: See Adams 2014; Dobzhansky 1941; Ford 1977; Gannett 2013; Mather and Dobzhansky 1939; Sturtevant and Dobzhansky 1936.
206. "not only a line but a wide gulf . . .": Quoted in Yudell 2014, p. 82.
206. "had no basis in biology": Quoted in Yudell 2014.
206. "The idea of a pure race . . .": Dobzhansky 1941, p. 162.
206. "The laws of heredity are the most universally valid ones . . .": Ibid.
206. humans certainly varied: Gannett 2013.
206. Dobzhansky's new allies pushed the attack further: Yudell 2014.
207. Lewontin, working with John Lee Hubby: Hubby and Lewontin 1966.
208. "to which human perceptions are most finely tuned . . .": Quoted in Lewontin 1972.
209. In 2015, for example, three scientists: Hunley, Cabana, and Long 2016.
210. Senegal, Nigeria, Angola, even Madagascar: Patin et al. 2017.
210. "to keep Negros quiet": Quoted in Warren 2016, p. 70.
211. embarrassing blunders: Duster 2015.
211. "There is no race which is so subject . . .": Thomas 1904.
211. "The 'Jewish nose' . . .": Dillingham 1911, p. 74.
211. Jews, doctors came to agree, had diabetes: Tuchman 2011.
211. "some hereditary defect": Quoted in Tuchman 2011.
211. William Osler, the most important clinical doctor of the early 1900s: Ibid.
212. Hispanics are 60 percent more likely: US Department of Health and Human Services 2017.
212. more sensitive to the blood-thinning drug warfarin: Lam and Cheung 2012.
212. Richard Cooper: Cooper 2013.
213. some variants in African Americans and Nigerians: Zhu et al. 2011.
213. "a sideshow and a distraction": Quoted from Rosenberg and Edge (in press).
213. What matters is ancestry: Nielsen et al. 2017.

8. Mongrels

214. The Taita thrush: Pritchard, Stephens, and Donnelly 2000.
215. They had named the program STRUCTURE: November 2016.
216. chopstick effect: See Lander and Schork 1994.
217. William Knowler, a researcher: Knowler et al. 1988.

217. families with people outside the tribe: Williams et al. 1992.
218. people are not Taita thrushes: Pritchard et al. 2000.
218. genetic variations in 1,056 people from around the planet: Rosenberg et al. 2002.
219. Aristotle also threw together species: Rosenberg and Edge (in press).
222. software, known as RFMix: Maples et al. 2013.
223. came together in Poland to seek refuge: Behar 2013.
224. Shai Carmi of Hebrew University: Xue et al. 2016.
224. as dubious as terms like *black* and *Hispanic*: Mathias et al. 2016.
225. 2,400-year-old mummy: Pääbo 1985.
225. assemble it into an entire genome: Der Sarkissian et al. 2015.
225. one kind of bone was far better than the rest: Pinhasi et al. 2015.
226. a kind of genetic transect: Fu et al. 2016.
227. the first farmers: Lazaridis et al. 2016, 2014.
227. southern edges of Europe: Brandt et al. 2015.
228. skin colors: Jablonski and Chaplin 2017.
229. Sarah Tishkoff: Crawford et al. 2017.
231. a hunter-gatherer who lived in Spain: Olalde et al. 2014.
231. This long lag is puzzling: See Beleza et al. 2013; Martiniano et al. 2017; Mathieson et al. 2015a, 2015b, 2017; Olalde et al. 2017.
231. mergings of deeply different peoples: See Hellenthal et al. 2014; Pickrell and Reich 2014; Slatkin and Racimo 2016.
232. ancient DNA from African skeletons: Skoglund et al. 2017. See also Beltrame, Rubel, and Tishkoff 2016; Kwiatkowski et al. 2016; Nielsen et al. 2017.
233. a single village in Borneo: Brucato et al. 2016.
235. Henry Fairfield Osborn: Regal 2002.
235. "confined to native Americans": Quoted in Barkan 1992, p. 68.
235. "Heredity and racial predisposition are stronger . . .": Grant 1916, p. xi.
235. "an enormous head . . .": Osborn 1915, p. 243.
236. "the superior organization of the brain . . .": Ibid., p. 236.
236. "The Neanderthals represent a side branch . . .": Ibid., p. 257.
236. "They were armed with weapons . . .": Ibid., p. 258.
236. "probably belonged to the Caucasian stock": Ibid., p. 492.
236. "three absolutely distinct stocks": Osborn 1926, p. 4.
237. ". . . intelligence of the average adult Negro . . .": Ibid., p. 5.
237. ". . . a state of arrested development": Ibid., p. 6.
237. ". . . a 1,250,000-word treatise . . .": "Dr. Henry F. Osborn Dies in His Study" 1935.
237. clear that humans originated in Africa: Gibbons 2006.
238. Research on Neanderthals since Osborn's death: See Roebroeks and Soressi 2016; Villa and Roebroeks 2014.
238. switched on a sterile electric saw: Pääbo 2014.
241. "I am very proud . . .": "Find Your Inner Neanderthal" 2011.
242. "statistical noise": Sankararaman et al. 2014.
244. The earliest hints of interbreeding came to light in 2017: Posth et al. 2017.
245. Joshua Akey and his colleagues: Vernot et al. 2016.
246. Neanderthal genes that play a role in reproduction: See Harris et al. 2016; Juric et al. 2016.
246. Neanderthal immune genes are more common: Dannemann, Andrés, and Kelso 2016.
247. the Denisovans: Reich et al. 2010.
247. molars dug up in the same Siberian cave: Slon et al. 2017.
248. Denisovan DNA was as ill-suited to modern humans: Sankararaman 2016.
248. One . . . gene is called EPAS1: Huerta-Sánchez and Casey 2015.
248. Emilia Huerta-Sanchez: Huerta-Sánchez et al. 2014.

9. Nine Foot High Complete

252. Og, king of Bashan: Wood 1868.
253. "nine foot high complete": Quoted in ibid., p.108.
253. an Irishman named Charles Byrne: See Bergland 1965; Muinzer 2014; Wood 1868.
253. "This truly amazing phenomenon . . .": Quoted in Wood 1868, p. 157.
253. people with dwarfism were also singled out: Adelson 2005.

254. entire races of miniature people: Leroi 2003.
255. ". . . there must be some peculiarity in Ireland . . .": Prichard 1826.
255. "Such as dwell in places which are low-lying . . .": Hippocrates, *On Airs, Waters, and Places.*
256. "In about five years . . .": Aristotle, *On the Generation of Animals.*
256. the Italian physician Pavisi: Tanner 2010.
256. ". . . kept some Persons from being turn'd out of the Service": Wasse 1724.
256. The military's cherishing of height: Hall 2006.
256. In England, the gap was staggering: Blum 2016.
257. ". . . conscience and humanity": Quoted in Tanner 2010, p. 548.
258. "Numbers rule the world": Ibid., p. 548.
258. Quetelet turned his attention to people: Rose 2015.
259. "represent all which is grand, beautiful, and excellent": Quoted in Hall 2006, p. 229.
259. "It reigns with serenity . . .": Galton 1889, p. 66.
260. "The large do not always beget the large": Ibid., p. 2.
260. a new way of studying heredity: Bulmer 2003.
260. "I had to collect all my data for myself": Quoted in Galton 1889, p. 71.
261. ". . . those who desire to be accurately measured . . .": Quoted in Johnson et al. 1985.
262. Karl Pearson: Pearson 1895, 1904.
262. opposite sides of the same coin: Visscher, McEvoy, and Yang 2010.
263. heritability feeds the world: See Khush 1995; Okuno et al. 2014; Peiffer et al. 2014; Teich 1984.
264. Scientists who study human heritability: Vinkhuyzen et al. 2013.
265. "The one element that varies in different individuals . . .": Quoted in Galton 1883, p. 173.
265. Hermann Werner Siemens: See Boomsma, Busjahn, and Peltonen 2002; Rende, Plomin, and Vandenberg 1990; Siemens 1924.
266. Karri Silventoinen: Silventoinen et al. 2003.
267. 11,214 pairs of regular siblings: Visscher et al. 2007.
267. "a decrease, however slight, in misery": Quoted in Tanner 1979, p. 163.
268. Gravettian men stood on average six feet tall: Grasgruber et al. 2014.
268. When English people emigrated to the American colonies: Steckel 2009, 2013.
268. starting around 1870: Hatton 2014.
269. Latvian women: NCD Risk Factor Collaboration 2017.
269. an international network of reseachers: NCD Risk Factor Collaboration 2016.
269. *Homo erectus* grew as tall as five foot seven: Gallagher 2013.
270. Gert Stulp: Stulp and Barrett 2016.
270. A child's growing body demands fuel: Steckel 2016.
270. Diseases can also stunt a child's growth: Danaei et al. 2016.
270. the height of children at age three: Hübler 2016.
270. When the Industrial Revolution came: Craig 2016.
271. affordable milk and meat: Hatton 2014.
271. the size of families shrank: Dalgaard and Strulik 2016.
271. 36 percent of all two-year-olds: Danaei et al. 2016.
272. the country's economic inequality is partly to blame: Hadhazy 2015.
272. When Jaime Guevara-Aguirre was growing up: See Berg et al. 1992; Guevara-Aguirre et al. 2011; Taubes 2013; Rosenbloom et al. 1990.
273. not what Byrne would have wished for: Bergland 1965.
273. "resurrectionists," as they were known: Greenfieldboyce 2017.
274. Harvey Cushing and Arthur Keith: Keith 1911.
274. other scientists took X-rays: Landolt and Zachmann, 1980.
275. Márta Korbonits: Leontiou et al. 2008.
275. this mutation arose in Ireland roughly 2,500 years ago: Chahal et al. 2011; Radian et al. 2016.
276. 2,327 people from 483 families: Hirschhorn et al. 2001.
276. But very often, the links would melt away: See Egeland et al. 1987; Kelsoe et al. 1989; Robertson 1989.
276. "Has the genetic study of complex disorders . . .": Risch and Merikangras 1996.
277. a genome-wide association study: See Price, Spencer, and Donnelly 2015; Visscher et al. 2012, 2016.
277. age-related macular degeneration: See Dennis 2003; Klein et al. 2005.
277. Hoh's findings were later confirmed: Van Lookeren Campagne, Strauss, and Yaspan 2016.

278. diseases such as diabetes and arthritis: Wellcome Trust Case Control Consortium 2007.
278. a gene called HMGA2: Weedon et al. 2007.
278. an eight-year-old boy: Ligon et al. 2005.
279. examined the height of tens of thousands of people: Wood et al. 2014.
280. a study on more than 700,000 people: Marouli et al. 2017.
280. Missing heritability dogged many studies: Nolte et al. 2017.
280. the geneticist Joseph Nadeau: Maher 2008.
280. "Garbage-In Garbage-Out Syndrome": See Génin and Clerget-Darpoux 2015.
280–81. may be far more than two: Edwards et al. 2014.
281. missing heritability is hiding beyond genes: Ibid.
281. Animal breeders study the heritability of cows: See Madrigal 2012; Van Eenennaam 2014.
282. "negligible": Yang et al. 2015.
284. *omnigenic*: Pritchard 2017.

10. Ed and Fred

286. a wealthy British childhood: See Fancher 1987; Gillham 2001.
286. Galton inherited that wealth: See Pearson 1930; Smith 1967.
287. Maryland's ironworks: See Lewis 1974; Maryland State Archives 2007.
287. Guns and slavery grew even more intertwined: Richards 1980.
287. ". . . the fruit of God's blessing . . .": Quoted in Pearson 1930, p. 41.
287. "The Trade devolved upon me . . .": Quoted in Smith 1967.
288. "Why, university honors, to be sure": Quoted in Fancher 1987, p. 21.
289. "A mill seemed to be working inside my head": Galton 1909, p. 79.
289. "the highest intellects of their age": Ibid., p. 80.
289. Galton's obsession with heredity: Browne 2016.
289. *Hereditary Genius*: Galton 1869.
290. "The average mental grasp . . .": Ibid., p. 21.
291. "even as we inherit their stature, forearm and span": Pearson 1904, p. 156.
291. Goddard's collaborator Lewis Terman: Fancher 1987.
291. "The immigrants who have recently come to us . . .": Terman 1922.
292. Porteus found: Goldstein et al. 2015; Porteus 1937.
292. a deep-seated feature: Cooper 2015.
292. "a catch-all word . . .": Haier 2017, p. 11.
293. whether they can recognize a word like *defenestrate*: Ritchie 2015, p. 26.
293. best-replicated findings: Ritchie 2015.
293. the government of Scotland: Deary 2009.
293. "This will change our lives": Ibid., p. x.
295. how efficiently the brain processes information: Haier 2017; Zimmer 2008b.
295. a shape flashes on a computer screen: Grudnik and Kranzler 2001; Osmon and Jackson 2002.
295. some people needed just 0.02 seconds: Stough et al. 1996.
296. half as likely to have died from heart disease: Calvin 2017.
296. "system integrity": Deary 2012.
296. "A person can no more be trained to have it . . .": Quoted in Goldstein, Princiotta, and Naglieri 2015.
297. Frank Freeman: Newman, Freeman, and Holzinger 1937.
297. referred to only as Ed and Fred: Fancher 1987.
297. "We shall be satisfied if we have succeeded . . .": Newnan et al. 1937, p. 363.
298. Cyril Burt: See Hearnshaw 1979; Mackintosh 1995.
298. "a thing inborn and not acquired": Quoted in Tucker 2007.
298. "Within ten minutes of starting to read Burt": Quoted in Fancher 1987, p. 207.
299. William Tucker offered an explanation: Tucker 2007.
299. "The superior proficiency at Intelligence tests . . .": Burt 1909.
299. twin studies have come to the same conclusion: McGue and Gottesman 2015.
299. they drew criticisms of their own: Kamin and Goldberger 2002; Panofsky 2014.
299. In a 2015 study: Fosse, Joseph, and Richardson 2015.
299. these effects are weak or nonexistent: Polderman et al. 2015.
300. Dalton Conley: Conley et al. 2013.
300. from smoking to divorce rates to watching television: Panofsky 2014.

300. studies of intellectual disorders such as PKU: Ellison, Rosenfeld, and Shaffer 2013.
301. "... many areas of psychology will be awash ...": Quoted in Plomin and Crabbe 2000.
301. Michael Egan: Egan et al. 2001.
302. failed to find any effect: Chabris et al. 2012; Plomin, Kennedy, and Craig 2006.
302. other candidate genes: Chabris et al. 2012.
302. Our brains use 84 percent: Lein and Hawrylycz 2014.
303. they failed to find even one gene: Davies et al. 2011.
303. a staggering 94 percent: Trzaskowski et al. 2014a.
303. educational attainment is a modestly heritable trait: Cesarini and Visscher 2017.
303. correlated with their educational attainment: Rietveld et al. 2013.
304. they reported three variants: Rietveld et al. 2015.
304. nearly 80,000 people: Sniekers 2017.
305. venturing into a daunting wilderness: Turkheimer 2012.
305. They ramify each other: Turkheimer 2012, 2015.
305. fetal alcohol syndrome: Mattson, Corcker, and Nguyen 2011.
305–306. vulnerable to toxins such as lead paint: Skerfving et al. 2015.
306. Brenda Eskenazi: Bouchard et al. 2011.
306. give people iodine: Feyrer, Politi, and Weil 2013; Zimmermann 2008.
306. It can also lead to cretinism: Syed 2015.
306. neurons crawl to their proper location: See de Escobar, Obregón, and del Rey 2004.
306. A third of the world's population: Li and Eastman 2012.
306. Sarah Bath: See Bath et al. 2013; Rayman and Bath 2015.
307. James Feyrer: Feyrer et al. 2013.
307. recruits for World War II: Pendergrast, Milmore, and Marcus 1961.
307. Robert DeLong: See DeLong 2010; O'Donnell et al. 2002.
308. "We are driven to the absurd conclusion ...": Quoted in Flynn 2009.
309. Alan Kaufman: Kaufman et al. 2014.
309. Our behaviors are shaped by our experiences: Nisbett 2013.
309. Christian Brinch and Taryn Ann Galloway: Brinch and Galloway 2012.
309. the enrollment rate in the early 1900s was 50 percent: Baker et al. 2015.
310. Eric Turkheimer: Turkheimer et al. 2003.
311. although others have not: Tucker-Drob and Bates 2016.
311. measuring their heritability at different ages: Haworth et al. 2010.
312. "For the last forty years": Pearson 1904, p. 160.
312. German psychiatrists: Burleigh 2001, p. 366.
312. "It is quite possible that he has never had the chance ...": Ibid., p. 356.
313. Helen Barrett and Helen Koch: Cravens 1993.
313. "One of the few fixed stars in the creed ...": "I.Q. Control" 1938.
313. World War II ... focused the country's attention abroad: Vinovskis 2008.
314. white person after a lobotomy: Tucker 1994.
314. Garrett was an ardent supporter of segregation: Winston 1998.
314. reporting on the "Communistic theories": Jackson 2005.
314. "the scientific hoax of the century": Quoted in Garrett 1961.
315. "The fundamental theoretical basis ...": Quoted in Castles 2012, p. 114.
315. many benefits to the Head Start program: Bauer 2016.
315. only to drift back down: Montialoux 2016.
315. an educational psychologist named Arthur Jensen: Jensen 1967.
315. have rejected them: Colman 2016; deBoer 2017; Lewontin 1970; Nisbett 2013; Nisbett et al. 2012.
316. The Flynn effect did not leave behind American blacks: Nisbett 2013; Rindermann and Pichelmann 2015.
316. Dorothy Roberts: Roberts 2015.
316. When education researchers test out new programs: Cesarini and Visscher 2017.
317. tailor school programs to each child: See Asbury 2015; Asbury and Plomin 2013.
317. "If a simple blood test at birth could spot children ...": Quoted in Asbury 2015.
317. "Precision education," as this approach has been called: See Hart 2016.
317. it's only a placeholder of an idea: Panofsky 2015.
317. dubbed this kind of thinking "genetic essentialism": Dar-Nimrod and Heine 2011.

317. our minds instinctively sort things into categories: Gelman 2003.
318. nothing we do in our lives: Cheung, Dar-Nimrod, and Gonsalkorale 2014.
318. manipulate people's genetic essentialism: Dar-Nimrod et al. 2014.

PART III: THE PEDIGREE WITHIN
11. Ex Ovo Omnia

324. how many cell types there are: See Regev et al. 2017; Yong 2016a.
324. Aristotle asked himself this question: Leroi 2014.
324. "This point beats and moves . . .": Aristotle, *The History of Animals*.
325. much as the mind produces thoughts: Cobb 2012.
326. He drew the head of a sperm: Lawrence 2008.
326. Caspar Friedrich Wolff: Aulie 1961.
326. they were all variations on a theme: See Harris 1999; Mazzarello 1999.
326. "And this principle . . .": Schwann 1847, p. 166.
327. the overgrowth of their two parents: See Amundson 2007; Churchill 2015.
328. "How is it," he asked, "that such a single cell can reproduce . . .": Quoted in Churchill 2015, p. 303.
328. a mysterious thing Weismann called "hereditary tendencies": Dröscher 2014.
329. Weismann also recognized another kind of heredity: Churchill 1987; Griesemer 2005.
329. a "theoretical illustration": Weismann 1893, p.103.
329. draw trees of their own: Maienschein 1978.
329. Edwin Grant Conklin: Clement 1979.
330. "They called it cellular bookkeeping": Bonner and Bell, 1952, p. 81.
330. "came behind me while I was anxiously studying . . .": Conklin 1968, p. 115.
331. an essential part of embryology: See Buckingham and Meilhac 2011; Kretzschmar and Watt 2012; Stern and Fraser 2001.
331. "The 'Wanderlust' of geneticists . . .": Quoted in Harrison 1937.
331. Conrad Waddington: See Baedke 2013; Slack 2002; Stern 2003.
332. It used its many genes to produce many proteins: Henikoff and Greally 2016.
332. ". . . a rough and ready picture of the developing embryo . . .": Waddington 1957.
332. like visions from the future: See Allis and Jenuwein 2016; Felsenfeld 2014.
333. Mary Lyon: See Cooper 2011; Fisher and Peters 2015; Gartler 2015; Gitschier 2010; Harper 2011; Kalantry and Mueller 2015; Nightingale 2015; Opitz 2015; Rastan 2015a, 2015b; Vines 1997.
334. She carried out elegant experiments: Silvers 1979.
334. "They wanted me to get married at one point": Quoted in Gitschier 2010.
335. "always tried to stick to the mouse work": Quoted in Genetics and Medicine Historical Network 2004.
336. seven paragraphs: Lyon 1961.
337. "He may not have realized I wasn't a PhD student . . .": Quoted in Vines 1997, p. 269.
337. "It is concluded": Grüneberg 1967, p. 255.
338. Ronald Davidson: Davidson, Nitowsky, and Childs 1963.
338. molecules that shut down X chromosomes: See Jegu and Lee 2017; Payer 2016; Vacca et al. 2016; Vallot, Ouimette, and Rougeulle 2016.
339. rolls the genetic dice: See Galupa and Heard 2015; Xu, Tsai, and Lee 2006.
339. Jeremy Nathans: Wu et al. 2014.
341. more steps on the journey: Henikoff and Greally 2016.
341. master genes also sustain each other: Moris, Pina, and Arias 2016; Semrau and Van Oudenaarden 2015.
342. right back to controlling the DNA in the two new cells: Teves et al. 2016.
342. These fluctuations can throw the cell's feedback loops out of whack: Goolam 2016.
343. Leila Boubakar: Boubakar et al. 2017.
345. survive for only four months: Milo and Phillips 2015.
345. hidden refuges of stem cells: Goodell, Nguyen, and Shroyer 2015.
345. So-called satellite cells: Yablonka-Reuveni 2011.
345. Stem cells need to hide in their refuges: Adam and Fuchs 2016.

345. manipulating the way their daughter cells inherit their molecules: Knoblich 2008.
345. one of the last places where they were discovered: in the brain: See Bergmann and Frisén 2013; Bergmann et al. 2012; Bergmann, Spalding, and Frisén 2015; Bhardwaj et al. 2006; Spalding et al. 2013, 2005.
346. "Everything may die; nothing may be regenerated": Quoted in Rubin 2009, p. 410.
347. this tiny infusion may make an important difference: See Anacker and Hen 2017; Bergmann and Frisén 2013.

12. Witches'-Broom

348. translated into English as witches'-broom: Fordham 1967.
348. a pair of Boston horticulturalists: "Dwarf Alberta Spruce."
349. they dubbed it a bud sport: Bossinger and Spokevicius 2011; Marcotrigiano 1997.
349. all pink grapefruits descend: da Graca, Louzada, and Sauls 2004.
349. "the spark which ignites a mass of combustible matter.": Quoted in Darwin 1868.
350. T. D. A. Cockrell wrote in 1917: Cockerell 1917.
350. animals can be mosaics, too: Spinner and Conlin 2014.
350. German dermatologist Alfred Blaschko: Kouzak, Mendes, and Costa 2013.
351. "Brace yourselves up to witness . . .": Quoted in Howell and Ford 2010, p. 74.
351. "the most disgusting specimen of humanity . . .": Treves 1923.
351. "an animal in a cattle market": Quoted in Howell and Ford 2010, p. 77.
352. Theodor Boveri: See Balmain 2001; Boveri 2008; Dietel 2014; Gull 2010; Heim 2014; McKusick 1985; Meijer 2005; Ried 2009; Wright 2014.
353. "The skepticism with which my ideas were met . . .": Boveri 2008.
354. The altered chromosomes drove cells to become cancerous: Nowell 1960.
354. sequencing entire genomes from tumor cells: Griffith et al. 2015.
354. They steal mitochondria genes from healthy cells: Tan et al. 2015.
355. simple arithmetic: See Campbell et al. 2014; Forsberg, Gisselsson, and Dumanski 2016.
355. On August 5, 1959: Hirschhorn, Decker, and Cooper 1960.
356. a team of Israeli geneticists: Chemke, Rappaport, and Etrog 1983.
356. Proteus syndrome: See Biesecker 2005, 2006; de Souza 2012; Tibbles and Cohen 1986; Wiedeman 1983.
356. Leslie Biesecker: Lindhurst et al. 2011.
357. it may one day become curable: Lindhurst 2015.
357. the genetic causes of more mosaic diseases: See Campbell et al. 2015; Lupski 2013.
357. A mutation may arise at any stage: Frank 2014.
358. Epidermal cells stream in rivers: See Happle 2002; Kouzak et al. 2013.
358. Jonathan Pevsner: See Freed, Stevens, and Pevsner 2014; Shirley et al. 2013.
359. hemimegalencephaly: Flores-Sarnat et al. 2003.
359. brain tissue taken from eight people: Poduri et al. 2013, 2012.
359. a long stretch of chromosome I was duplicated: D'Gama et al. 2015.
360. her third child, a daughter named Astrea: Dusheck 2016; Priest et al. 2016.
362. genetic purgatory: Ackerman 2015.
366. mosaicism can heal: See Gajecka 2016; Lai-Cheong, McGrath, and Uitto 2011; Pasmooij, Jonkman, and Uitto 2012.
367. over half of its cells end up with the wrong number of chromosomes: See Freed et al. 2014; Oetting et al. 2015; Spinner and Conlin 2014; Vanneste et al. 2009.
367. reject the embryo altogether: Freed et al. 2014.
367. can survive with some variety in their chromosomes: Rutledge and Cimini 2016.
367. Markus Grompe: Duncan et al. 2012.
368. all down to their descendants as a mosaic legacy: See Frank 2014; Ju 2017.
368. reconstruct the cell lineages of the brain: See Evrony 2016; Linnarsson 2015; Lodato et al. 2015.

13. Chimeras

370. "the bull-calf becomes a very proper bull": Hunter 1779, p. 279.
370. "The flesh of a fatted free martin . . .": Mills 1776, p. 262.
371. Frank Lillie started dissecting cow fetuses: Capel and Coveney 2004.
372. Cows were Owen's life: Owen 1983.

373. "It was a kind of bio-business venture.": Ibid., p. 11.
373. a Maryland cattle farmer got in touch: Owen 1959.
374. carried proteins matching both bulls: Martin 2015.
375. Peter Medawar: Martin 2007b; Martin 2015.
376. punched out bits of skin: Anderson et al. 1951.
377. Mrs. McK: See Dunsford et al. 1953; Martin 2007a, 2007b.
379. "The Uniqueness of the Individual": Medawar 1957.
379. "to lift the phenomenon out of the 'freak' category.": Quoted in Martin 2007a.
379. Patricia Tippett: Tippett 1983.
380. 8 percent of the twins were chimeras: Van Dijk, Boomsma, and de Man 1996.
380. a thirty-year-old woman in Germany: Sudik 2001.
381. a girl was born in a Seattle hospital: Gartler, Waxman, and Giblett 1962; Waxman, Gartler, and Kelley 1962.
382. a tetragametic chimera: Yunis et al. 2007.
382. to produce a girl or a boy with ordinary genitals: Malan et al. 2007.
382. Lydia Fairchild: See Arcabascio 2007; Martin 2007a; *ABC News* 2016; Wolinsky 2007.
383. Karen Keegan: Yu et al. 2002.
384. ". . . part of me hadn't passed on to them": Quoted in Baron 2003.
385. a doctor first took note of this traffic: Jeanty, Derderian, and Mackenzie 2014.
385. Christian Georg Schmorl: Lapaire et al. 2007.
385. Rajendra Desai and William Creger: Desai and Creger 1963.
385. Desai and colleagues in Boston ran the reverse experiment: Desai et al. 1966.
387. new method to sort the cells of a fetus: Herzenberg et al. 1979.
387. Diana Bianchi: See Bianchi 2007; Bianchi et al. 1996; Martin 2010.
388. up to half of mothers: Forsberg et al. 2016.
388. a legacy in reverse: Khosrotehrani and Bianchi 2005.
388. 42 percent of children: Jeanty, Derderian, and Mackenzie 2014.
388. blood samples from 154 girls: Müller et al. 2016.
388. a group of researchers at Leiden University: Rijnink et al. 2015.
389. Lee Nelson: Chan et al. 2012.
389. Gerald Udolph: Zeng et al. 2010.
389. what effects their divided inheritance had on them: Martin 2010.
389. scleroderma: Nelson et al. 1998.
390. good for your health: Bianchi 2007; Falick Michaeli, Bergman, and Gielchinsky 2015; Martin 2010.
390. Peter Geck: Dhimolea et al. 2013.
391. Ruth Fischbach and John Loike: Fischbach and Loike 2014; Loike and Fischbach 2013.
391. The Tasmanian Devil couldn't have been dead long: Murchison 2016.
392. A singular epidemic: See Ostrander, Davis, and Ostrander 2016; Ujvari, Gatenby, and Thomas 2016b; Ujvari, Papenfuss, and Belov 2016; Ujvari et al. 2014.
393. The genome in the tumors: Murchison et al. 2012.
393. a type of nerve known as a Schwann cell: Murchison et al. 2010.
395. Delabere Blaine: Blaine 1810.
395. Mstislav Novinski: Shabad and Ponomarkov 1976; Shimkin 1955.
395. In 1934, a veterinarian and a pathologist reported: Stubbs and Furth 1934.
396. Robin Weiss: Murgia et al. 2006.
397. humans have spread CTVT: Strakova et al. 2016.
397. in the sea: Metzger et al. 2015.
398. a second devil facial tumor: Pye et al. 2015.
399. journey from one host to another: See Tissot et al. 2016; Ujvari, Gatenby, and Thomas 2016a.
399. genes linked to immune surveillance: Ostrander et al. 2016.
399. CTVT cells have stolen mitochondria: Strakova et al. 2016.
400. the eight cases identified so far: Riquet 2017.
400. cases of cancer cells moving from one person to another: Lazebnik and Parris 2015.
400. passed some of their cancerous immune cells to their fetuses: Isoda 2009.
401. got infected with a tapeworm: Muehlenbachs et al. 2015.

Part IV: OTHER CHANNELS
14. *You, My Friend, Are a Wonderland*

405. the one-fin flashlight fish: See Haneda and Tsuji 1971; Haygood, Tebo, and Nealson 1984; Hendry et al. 2016; Hendry, de Wet, and Dunlap 2014; Meyer-Rochow 1976.
407. about 37 trillion human cells and about the same number of bacteria: Sender, Fuchs, and Milo 2016.
407. we are no different from any other animal: Yong 2016b.
407. this cycle of renewal looks a lot like heredity: Hurst 2017.
408. The genes of bacteria can take strikingly similar routes: Bright and Bulgheresi 2010.
408. thick beds of vesicomyid clams thrive: Funkhouser and Bordenstein 2013.
409. S. Craig Cary and Stephen Giovannoni: Cary and Giovannoni 1993.
409. cockroaches: See Funkhouser and Bordenstein 2013; Sabree, Kambhampati, and Moran 2009.
410. a gradual slide: Bennett and Moran 2015.
411. 80 percent of their genome: Hendry et al. 2016.
411. some of heredity's essential features: See Bordenstein 2015; Gilbert 2014; Theis et al. 2016.
412. "You, my friend, are a wonderland": Quoted in Zimmer 2011.
413. this seeding: Blaser and Dominguez-Bello 2016; Rosenberg and Zilber-Rosenberg 2016.
413. Breasts foster microbes: McGuire and McGuire 2017.
413. the strains that move most successfully through nursing into the babies: Asnicar et al. 2017.
413. Mothers may thus transmit microbes to future generations: Foster et al. 2017.
413. Howard Ochman: Moeller et al. 2016.
414. oligosaccharides in human milk: Browne et al. 2017; Urashima et al. 2012.
414. *Helicobacter pylori*: Cellini 2014.
415. The history recorded in its branches: Moodley et al. 2012.
415. intertwined with our own heredity: See Goodrich et al. 2016; Van Opstal and Bordenstein 2015.
416. was only discovered in 2012: Morotomi 2012.
416. weight is heritable: Silventoinen et al. 2016.
417. mitochondria, the tiny pouches that produce fuel inside our cells: See Archibald 2015; Ball, Bhattacharya, and Weber 2016; Gray 2012; Martin et al. 2016; McCutcheon 2016; Rogers et al. 2017.
417. Edmund Cowdry: Cowdry 1953.
418. alphaproteobacteria: Wang and Wu 2015.
419. how this merger happened: Martin et al. 2017.
420. mitochondria's distinctive patterns of heredity: Stewart and Chinnery 2015.
420. It wasn't until the late 1980s: Holt, Harding, and Morgan-Hughes 1988.
421. Ink cap mushrooms: Breton and Stewart 2015.
421. Douglas Wallace: Sharpley et al. 2012a.
421. Limiting mitochondrial heredity: Christie, Schaerf, and Beekman 2015.
421. inspect their eggs: Haig 2016.

15. *Flowering Monsters*

422. Magnus Zioberg: See Coen 1999; Gustafsson 1979.
424. Enrico Coen: Cubas, Vincent, and Coen 1999.
426. "If acquired characters cannot be transmitted . . .": Weismann 1889, p. 403.
426. ". . . convincing disproof of the loose and vague arguments . . .": Morgan 1925, p. 177.
427. that's unfair both to him and to history: See Burkhardt 2013; Deichmann 2016.
427. "The law of nature . . .": Quoted in Burkhardt 2013, p. 796.
427. some scientists continued to fight for conceptual room: See Bonduriansky and Day 2009; Day and Bonduriansky 2011; Uller and Helanterä 2013.
427. a few cases came to light: Schaefer and Nadeau 2015.
427. Överkalix: See Bygren et al. 2014; Epstein 2013; Lim and Brunet 2013.
428. Michael Skinner: See Nilsson and Skinner 2015; Skinner 2015; Tollefsbol 2014.
428. Brian Dias: See Dias and Ressler 2014; Hughes 2014.

430. behaviors could be acquired: See Bale 2014, 2015; Bohacek and Mansuy 2015; Rodgers and Bale 2015.
430. in vitro fertilization alone: Lim and Brunet 2013.
431. helpless until the morning's army of molecules wakes them: Papazyan et al. 2016.
431. When we develop an infection: See Allis and Jenuwein 2016; Busslinger and Tarakhovsky 2014.
431. cellular memory: Henikoff and Greally 2016.
431. The memories we store in our brains: Kim and Kaang 2017.
431. it will respond to drought more strongly: Lämke and Bäurle 2017.
432. Michael Meaney: See Moore 2015; Provençal and Binder 2015; *Science* News Staff 1997; Zimmer 2010.
433. epigenetic clock: See Chen et al. 2016; Gibbs 2014; Horvath 2013; Horvath and Levine 2015; Walker et al. 2015.
434. the result of epigenetic changes: Bourrat 2017.
434–35. "If the 20th century belonged to Charles Darwin": Kaufman 2014.
435. If you let your imagination run wild: Juengst et al. 2014.
435. including some of the chemicals in plastics: Schaefer and Nadeau 2015.
435. "The New Theory That Could Explain Crime . . .": Johnson 2014.
436. Wolynn declared on his website: Wolynn, *The Family Constellation Institute.*
436. how much skepticism there is in scientific circles: See Heard and Martienssen 2014; Whitelaw 2015.
436. the causes may have nothing to do with inherited epigenetic marks: Greally 2015.
436. the most potent attacks: Francis 2014.
437. primordial germ cells: See Guo et al. 2015; Tang et al. 2016.
437. methylation suffers radical amnesia: Chen, Yan, and Duan 2016.
437. Kevin Mitchell, a neurogeneticist: Mitchell 2016.
438. "Transgenerational Epigenetic Inheritance: Myths and Mechanisms": Heard and Martienssen 2014.
440. Azim Surani: Tang et al. 2015.
441. Tracy Bale: Rodgers et al. 2015.
441. the effect of stress that male mice experienced early in life: Gapp et al. 2014a.
441. Antony Jose: See Devanapally, Ravikumar, and Jose 2015; Marré, Traver, and Jose 2016.
442. spurring young worms to make more copies of themselves: Rechavi and Lev 2017.
442. tiny bubbles, called exosomes: See Bohacek and Mansuy 2015; Eaton et al. 2015; Smythies, Edelstein, and Ramachandran 2014.
442. embryos may use exosomes to send signals: McGough and Vincent 2016.
442. Heart cells may release them: Sahoo and Losordo 2014.
442. Cristina Cossetti: Cossetti et al. 2014.
443. researchers at Cornell University: Rasmann et al. 2011.
443. Dandelions: Wilschut et al. 2015.
444. "is only the sum of all past environment": Burbank 1906.

16. The Teachable Ape

445. my older daughter, Charlotte: I first wrote about this experience in Zimmer 2005. For the results of the experiment, see Lyons, Young, and Keil 2007.
447. adapted for inheriting culture: Dean et al. 2014.
448. the Yandruwandha: Henrich 2016. See also Boyd 2017; Cathcart 2013.
448. The ancestors of the Yandruwandha: Clarkson 2017.
451. "I am weaker than ever . . .": Quoted in Boyd 2017, p.9.
452. ". . . a new kind of replicator . . .": Dawkins 2016, p. 248.
452. lodged itself in the heads of many readers: Aunger 2006.
452. "The Net has effectively become a meme factory": Quoted in Burman 2012.
454. "The word meme . . .": Dawkins 2016, p. 423.
454. ". . . appear to be a dead end": Ehrlich and Feldman 2003.
454. biological building blocks: Haidle and Conard 2016.
455. Lars Chittka: Alem et al. 2016.
456. Swaythling: Aplin, Sheldon, and Morand-Ferron 2013; Fisher and Hinde 1949; Laland 2017.
457. birds known as great tits: Aplin 2015.

458. more animal traditions: Whiten, Caldwell, and Mesoudi 2016.
458. James Hain: Hain et al. 1982.
458. Luke Rendell: Allen et al. 2013.
459. Lewis Dean: Dean et al. 2012.
460. cumulative culture: Dean et al. 2014.
461. a peculiar feature of Western societies: Hewlett and Roulette 2016.
462. Meerkats are one: Thornton and McAuliffe 2006.
462. chimpanzee teachers . . . in the Republic of Congo: Musgrave et al. 2016.
462. four years to master the skill: Byrne and Rapaport 2011.
462. all their efforts are wasted: Laland 2017.
463. extreme imitation: Mesoudi 2016; Nielsen et al. 2014.
463. trustworthy experts: Chudek, Muthukrishna, and Henrich 2015; Ross, Richerson, and Rogers 2014.
463. "The theory of natural pedagogy": Heyes 2016.
464. known as Oldowan tools: Gärdenfors and Högberg 2017.
464. Their subjects always fail: Chudek, Muthukrishna, and Henrich 2015.
464. Alex Mesoudi: Lycett et al. 2015.
465. full-blown cumulative culture: Mesoudi and Aoki 2015.
465. our sheer numbers: Hill et al. 2014; Muthukrishna et al. 2014.
466. the dawn of a new form of heredity: Andersson 2013.
466. a human-altered environment: Ellis 2015.
467. to bake soil: Brown et al. 2009.
467. using fire to reshape entire landscapes: Flannery and Marcus 2012.
467. walked through grasslands with fire sticks: Pyne and Cronon 1998.
468. influenced plants simply through the cultural practices: See Laland 2017; Rowley-Conwy and Layton 2011; Smith 2011.
470. when they reach their late teens: Hooper et al. 2015.
470. such rules only rein in inequality: See Bowles, Smith, and Borgerhoff Mulder 2010; Flannery and Marcus 2012; Shennan 2011; Smith et al. 2010a, 2010b.
474. *Not Only Genes*: Bonduriansky and Day 2018.
477. a sign of early cheese-making: Marciniak and Perry 2017.
478. ". . . every feeble-minded person is a potential drunkard": Goddard 1914, p. 11.
479. ". . . no acceptable evidence . . .": Haggard and Jellinek 1942.
479. fathers who drink before conception: Sarkar 2016.
480. The molecular details: Armstrong and Abel 2000; Pauly 1996; Warner and Rosett 1975.
480. Robert Karp: Karp et al. 1995.
480. telltale signs of fetal alcohol syndrome: Paul 2010.

PART V: THE SUN CHARIOT
17. Yet Did He Greatly Dare

483. Phaethon paid a visit one day: Zirkle 1946.
483. "Give me proof that all may know I am thy son indeed": Ovid 2008, p. 26.
484. toward new domesticated sequences: Librado et al. 2016, 2017.
485. "the Wizard of Santa Rosa": Quoted in Smith 2009, p. 184.
485. an experiment on some Indian corn: See Glass 1980; Shull 1909.
487. pears, peppermint, sunflowers, rice, cotton, and wheat: Broad 2007.
487. creating strains that could make superior penicillin: Adrio and Demain 2006; Raper 1946.
487. It wasn't until the 1960s that microbiologists would discover: Zimmer 2008a.
488. Jennifer Doudna: Doudna and Sternberg 2017; Doudna and Charpentier 2014.
490. "We had built the means": Doudna and Sternberg 2017, p. 84.
490. Rudolf Jaenisch: Wang et al. 2013.
491. the parade of CRISPR animals became a stampede: Ledford 2016.
492. "To me the interest was how . . .": Cold Spring Harbor Lab 2013.
493. Lippman and his colleagues planted the seeds of these altered plants: Soyk et al. 2016.
496. About two thousand genes: Boone and Andrews 2015.
496. cells from people with cystic fibrosis: Schwank et al. 2013.
496. mice that suffered from hereditary cataracts: Wu et al. 2013.

497. wondering when the first experiments on human embryos would take place: Doudna 2015.
497. the nightmares started: Doudna and Sternberg 2017.
497. "... the Hitler Problem?": Quoted in Urban 2015.
498. vestigial features of aurochs: Wang 2012.
498. "Pretty pure damn blood": Quoted in Boodman 2017.
499. many different forms: Bashford and Levine 2010.
499. Hermann Muller: Carlson 1983, 2009.
500. "the naive doctrine ...": Muller 1933, p. 46.
500–501. "it is by no means all dead and buried ...": Muller 1949, p. 2.
501. "for we are all fellow mutants together": Muller 1950, p. 169.
501. "making the best of human nature ...": Muller 1961a.
501. perfecting the art of artificial insemination: Richards 2008.
502. "They will form a growing vanguard ...": Muller 1961b.
503. The Fairfax Cryobank: See Fairfax Cryobank "Donor Search."
503. But there aren't any regulations about checking DNA: Silver et al. 2016.
503. Sometimes a dangerous variant: Mroz 2012.
503–504. a young patient with a hereditary heart condition: Maron et al. 2009.
504. scan every protein-coding gene in a potential sperm donor: Silver et al. 2016.
504. "magic fluid": Quoted in Franklin 2013, p. 106.
504. they had succeeded at last: Edwards 1970.
504. "the introduction of unlimited genetic changes ...": Henig 2004, p. 72.
505. screened the embryos of two women: Handyside et al. 1990.
506. "The curse is finally broken": See "Derbyshire Sisters Rose and Daisy Picked as Embryos to Beat Killer Disease" 2014.
506. successfully detected the mutations in some of the embryos: Lavery 2013.
506. one in one thousand cases of Huntington's disease: Schulman and Stern 2015.
507. "... change his own inheritance": Quoted in Sonneborn 1965, p. 38.
507. *genetic engineering*: Ibid., p. 40.
508. "... the heritable traits of the phenylketonuric infant ...": Ibid., p. 38.
509. "Even *in utero*": Hotchkiss 1965, p. 201.
509. Martin Cline: See Beutler 2001; Wade 1980, 1981a, 1981b.
511. "He was rightly punished": "The Crime of Scientific Zeal" 1981.
512. Philippe Leboulch: Cavazzana-Calvo et al. 2010.
512. gene therapy as a true cure: Harding 2017.
512. "Deciding whether to engineer a profound change . . .": See President's Commission 1982, p. 65.
512. "entertain proposals for germ line alteration": National Institutes of Health 1990.
512. a forum about germ line intervention: American Association for the Advancement of the Sciences 1997.
513. Maureen Ott: See Tingley 2014; Weintraub 2013.
513. encouraging trials on mice: Pratt and Muggleton-Harris 1988.
514. "... something like, say, I only wanted a blond-haired girl ...": Quoted in Tingley 2014.
514. Ott's unprecedented experience in the *Lancet*: Cohen et al. 1997.
514. Newspapers reported in awe: Galant 1998.
514. Lois Rogers: Rogers 1998.
514. Naomi Lakritz: Lakritz 1998.
515. "... the first case of human germline genetic modification ...": Barritt et al. 2001.
515. they wanted to try using all of it: Cohen and Malter 2016; Johnston 2016; Regalado and Legget 2003; Zhang et al. 2016.
516. "... so eager to be famous": Quoted in Johnston 2016.
516. "There was too much heat": Ibid.
517. Leslie Orgel: Orgel 1997.
517. Shoukhrat Mitalipov launched a series of experiments: Tavernise 2014.
518. genetically modified teenagers: Chen et al. 2016; Marchione 2016; Tingley 2014; Weintraub 2013.
518. it didn't slip out cleanly: See Neimark 2016; Sharpley et al. 2012.
518. To test for this mismatch: See Dunham-Snary and Ballinger 2015; Latorre-Pellicer et al. 2016.

519. ". . . promotion of genetically modified human beings . . .": Budget Hearing, Food and Drug Administration 2014.
519. only 12,423 women in the United States: Adashi and Cohen 2017.
520. United Kingdom's Department of Health: See Department of Health, UK. 2014.
521. do much more than cause rare hereditary diseases: See Latorre-Pellicer et al. 2016; Picard, Wallace, and Burelle 2016.
521. influence our ability to remember things: Hamilton 2015.
521. implicated in psychiatric disorders such as schizophrenia: Hjelm et al. 2015.
521. merely replaced a cell's battery packs: Connor 2015.
521. The deliberations in the United States: Adashi and Cohen 2017.
521. mitochondrial replacement therapy research was worthwhile: Engelstad et al. 2016.
521. an enormous 2016 spending bill that blocked the FDA: Reardon 2015.
522. the first human mitochondrial replacement therapy: Hamzelou 2016.
522. Zhang and his colleagues examined the boy: Reardon 2016, 2017; Zhang et al. 2017.

18. Orphaned at Conception

524. "A Prudent Path Forward . . .": Baltimore et al. 2015.
524. Junjiu Huang: Liang et al. 2015.
526. "The concept of altering the human germline . . .": Collins 2015.
526. $250 million: National Institutes of Health 2017.
526. A scientist named Kathy Niakan: Ball 2016.
527. "Today, we sense that we are close . . .": Quoted in Olson 2015.
527. "It was an extremely low-key meeting": Begley 2015.
527. "The clinical use of germline editing should be revisited . . .": Quoted in Olson 2015, p. 7.
527. Lander argued: Quoted passages from Lander, Darnovsky, and Church from meeting video recordings.
530. a committee of twenty-two experts: National Academy of Sciences 2017.
530. these clinical trials could very well succeed: Ma et al. 2017.
531. "The Cells That Could End Genetic Disease": Front page of the *Independent*, August 3, 2017.
533. American parents: Kornrich 2016.
534. a theologian named Emmanuel Agius: Agius 1990.
535. could avoid $2.3 million in medical costs: Cyranowski 2017.
536. "Why Are Some People So Smart? . . ." Bohannon 2013.
537. "China Is Engineering Genius Babies": Eror 2013.
537. "That's nuts": Quoted in Yong 2013.
537. In a 2014 essay for *Nautilus*: Hsu 2014.
538. "enhanced rat race": Sparrow 2015.
539. the concept of a mutation load: Henn et al. 2015, Kondrashov 2017, Lynch 2016.
540. "I hope that 'War on Mutation' . . .": Kondrashov 2017.
540. savior siblings: Kakourou et al. 2017.
541. a seventeen-year-old Israeli girl named Chen Aida Ayash: Even 2011.
541. a Florida woman named Karen Capato: *Astrue v. Capato* 2012.
543. Shinya Yamanaka: Scudellari 2016.
544. "We thought at that time that the project would take 10, 20, 30 years . . .": Yamanaka 2012.
546. Katsuhiko Hayashi: Hayashi et al. 2012.
547. transforming mouse skin cells into eggs entirely in a dish: Hikabe et al. 2016.
547. in vitro gametogenesis: Hendriks et al. 2015; Imamura et al. 2014; Moreno et al. 2015; Segers et al. 2017.
547. *The End of Sex . . .* : Greely 2016.
547. ". . . most pregnancies . . .": Ibid., p. 191.
549. orphaned at conception: Sparrow 2012.

19. The Planet's Heirs

551. FlyTree: Academic Family Tree.
551. *Megaselia scalaris*: See Disney 2008; Varney and Noor 2010.
557. wiped out the parasites: Isaacs et al. 2011.
557. the P element: Kelleher 2016.
560. they had successfully used gene drive in mosquitoes: Grantz et al. 2015.

560. The mutagenic chain reaction hit the news: Gantz and Bier 2015.
560. "theoretical technology": Esvelt et al. 2014.
561. wipe out a population of fifty thousand: Prowse 2017.
561. "irreversible effects on organisms and ecosystems": National Academy of Sciences 2016.
562. our ecological inheritance: Ellis 2015.
562. plows drawn by horses or oxen: Williams 2003.
564. It hangs in the atmosphere for centuries: Lewis and Maslin 2015.
564. the planet would keep warming: Mauritsen and Pincus 2017.
566. The epigenetic side of plant biology: Springer and Schmitz 2017.
568. Raj Chetty: Chetty et al. 2016, 2017.
568. two-thirds of parental income differences: Russell Sage Foundation 2016.
568. ninetieth percentile of earners: Pinsker 2015.
569. the gap in wealth: Coy 2017; Traub et al. 2017.
570. the remaining 12 billion tons: Foster, Royer, and Lunt 2017.
570. Erle Ellis: Ellis 2017.
572. Jennifer Kuzma and Lindsey Rawls: Kuzma and Rawls 2016.
572. might outbreed: Bull and Malick 2017.

Glossary

575. Many definitions are adapted from National Academy of Sciences 2016.

BIBLIOGRAPHY

ABC News. 2016. "She's Her Own Twin." August 15. http://abcnews.go.com/Primetime /shes-twin/story?id=2315693 (accessed October 28, 2016).

Abyzov, Alexej, Jessica Mariani, Dean Palejev, Ying Zhang, Michael Seamus Haney, Livia Tomasini, Anthony F. Ferrandino, and others. 2012. "Somatic Copy Number Mosaicism in Human Skin Revealed by Induced Pluripotent Stem Cells." *Nature* 492:438–42.

Academic Family Tree. FlyTree—The Academic Genealogy of Drosophila Genetics. https:// academictree.org/flytree/ (accessed May 10, 2017).

Ackerman, Michael J. 2015. "Genetic Purgatory and the Cardiac Channelopathies: Exposing the Variants of Uncertain/unknown Significance Issue." *Heart Rhythm: The Official Journal of the Heart Rhythm Society* 12:2325–31.

Ackroyd, Peter. 2014. *Charlie Chaplin: A Brief Life*. New York: Doubleday.

Adam, Rene C., and Elaine Fuchs. "The Yin and Yang of Chromatin Dynamics in Stem Cell Fate Selection." *Trends in Genetics* 32:89–100.

Adami, Christoph. 2015. "Information-Theoretic Considerations Concerning the Origin of Life." *Origins of Life and Evolution of Biospheres* 45:309–17.

Adams, Mark B., ed. 2014. *The Evolution of Theodosius Dobzhansky: Essays on His Life and Thought in Russia and America*. Princeton: Princeton University Press.

Adashi, Eli Y., and I. Glenn Cohen. 2017. "Mitochondrial Replacement Therapy: Unmade in the USA." *JAMA* 317:574–75.

Adelson, Betty M. 2005. *The Lives of Dwarfs: Their Journey from Public Curiosity Toward Social Liberation*. New Brunswick, NJ: Rutgers University Press.

Adrio, Jose L., and Arnold L..Demain. 2006. "Genetic Improvement of Processes Yielding Microbial Products." *FEMS Microbiology Reviews* 30:187–214.

Agius, E. 1990. "Germ-line Cells—Our Responsibilities for Future Generations." In *Our Responsibilities towards Future Generations*. Edited by S. Busuttill and others. Valletta, Malta: Foundation for International Studies.

Akbari, Omar S., Hugo J. Bellen, Ethan Bier, Simon L. Bullock, Austin Burt, George M. Church, Kevin R. Cook, and others. 2015. "Safeguarding Gene Drive Experiments in the Laboratory." *Science* 349:972–79.

Alem, Sylvain, Clint J. Perry, Xingfu Zhu, Olli J. Loukola, Thomas Ingraham, Eirik Søvik, and Lars Chittka. 2016. "Associative Mechanisms Allow for Social Learning and Cultural Transmission of String Pulling in an Insect." *PLOS Biology* 14:e100256.

Allen, Elizabeth Cooper. 1983. *Mother, Can You Hear Me?: The Extraordinary True Story of an Adopted Daughter's Reunion with Her Birth Mother After a Separation of Fifty Years*. New York: Dodd, Mead.

Allen, Garland E. 2003. "Mendel and Modern Genetics: The Legacy for Today." *Endeavour* 27:63–68.

———. 2011. "Eugenics and Modern Biology: Critiques of Eugenics, 1910–1945." *Annals of Human Genetics* 75:314–25.

Allen, Jenny, Mason Weinrich, Will Hoppitt, and Luke Rendell. 2013. "Network-based Diffusion Analysis Reveals Cultural Transmission of Lobtail Feeding in Humpback Whales." *Science* 340:485–88.

Allis, C. David, and Thomas Jenuwein. 2016. "The Molecular Hallmarks of Epigenetic Control." *Nature Reviews Genetics* 17:487.

———, Marie-Laure Caparros, Thomas Jenuwein, and Danny Reinberg. 2015. *Epigenetics*. Cold Spring Harbor, NY: Cold Spring Harbor Laboratory Press.

Álvarez, Gonzalo, Francisco C. Ceballos, and Celsa Quinteiro. 2009. "The Role of Inbreeding in the Extinction of a European Royal Dynasty." *PLOS One* 4:e5174.

———, Francisco C. Ceballos, and Tim M. Berra. 2015. "Darwin Was Right: Inbreeding Depression on Male Fertility in the Darwin Family." *Biological Journal of the Linnean Society* 114:474–83.

American Association for the Advancement of the Sciences. 1997. "Guidelines for Human Germ-Line Interventions Topic of AAAS Forum." September 12. https://www.eurekalert .org/pub_releases/1997-09/AAft-GFHG-120997.php (accessed March 28, 2017).

Amundson, Ronald. 2007. *The Changing Role of the Embryo in Evolutionary Thought: Roots of Evo-devo.* New York: Cambridge University Press.

Anacker, Christoph, and René Hen. 2017. "Adult Hippocampal Neurogenesis and Cognitive Flexibility—Linking Memory and Mood." *Nature Reviews Neuroscience.* doi:10.1038 /nrn.2017.45.

Anderson, D., Rupert E. Billingham, G. H. Lampkin, and Peter Brian Medawar. 1951. "The Use of Skin Grafting to Distinguish Between Monozygotic and Dizygotic Twins in Cattle." *Heredity* 5:379–97.

Andersson, Claes. 2013. "Fidelity and the Emergence of Stable and Cumulative Sociotechnical Systems." *PaleoAnthropology* 2013:88–103.

Aplin, Lucy M., Ben C. Sheldon, and Julie Morand-Ferron. 2013. "Milk Bottles Revisited: Social Learning and Individual Variation in the Blue Tit, Cyanistes Caeruleus." *Animal Behaviour* 85:1225–32.

Arcabascio, Catherine. 2007. "Chimeras: Double the DNA—Double the Fun for Crime Scene Investigators, Prosecutors, and Defense Attorneys." *Akron Law Review* 40:435.

Archibald, John M. 2015. "Endosymbiosis and Eukaryotic Cell Evolution." *Current Biology* 25:R911–21.

Aristotle. *The History of Animals.* Translated by D'Arcy Wentworth Thompson. MIT Internet Classics Archive. http://classics.mit.edu/Aristotle/history_anim.6.vi.html (accessed October 2, 2017).

———. *On the Generation of Animals,* Book I. Translated by Arthur Platt. https://en.wikisource .org/wiki/On_the_Generation_of_Animals/Book_I (accessed August 2, 2017).

Armistead, Wilson. 1848. *A Tribute for the Negro: Being a Vindication of the Moral, Intellectual, and Religious Capabilities of the Coloured Portion of Mankind; with Particular Reference to the African Race.* Manchester, UK: W. Irwin.

Armstrong, Elizabeth M., and Ernest L. Abel. 2000. "Fetal Alcohol Syndrome: The Origins of a Moral Panic." *Alcohol and Alcoholism* 35:276–82.

Asbury, Kathryn. 2015. "Can Genetics Research Benefit Educational Interventions for All?" *Hastings Center Report* 45, Suppl 1, S39–S42.

———, and Robert Plomin. 2013. *G Is for Genes: The Impact of Genetics on Education and Achievement.* New York: Wiley-Blackwell.

Asnicar, Francesco, Serena Manara, Moreno Zolfo, Duy Tin Truong, Matthias Scholz, Federica Armanini, Pamela Ferretti, and others. 2017. "Studying Vertical Microbiome Transmission from Mothers to Infants by Strain-Level Metagenomic Profiling." *mSystems* 2:00164–16.

Associated Press. 1944. "Doctor Backs Chaplin." *New York Times,* December 28, p. 24.

———. 1957. "Dr. Henry Goddard, Psychologist, Dies; Author of 'The Kallikak Family' Was 90." *New York Times,* June 22, p. 15.

Astrue v. Capato, Ed. 2d 887 (2012).

Aulie, Richard P. 1961. "Caspar Friedrich Wolff and His 'Theoria Generationis,' 1759." *Journal of the History of Medicine and Allied Sciences* 16:124–44.

Aunger, Robert. 2006. "What's the Matter with Memes?" In *Richard Dawkins: How a Scientist Changed the Way We Think.* Edited by Alan Grafen and Mark Ridley. Oxford: Oxford University Press.

Baedke, Jan. 2013. "The Epigenetic Landscape in the Course of Time: Conrad Hal Waddington's Methodological Impact on the Life Sciences." *Studies in History and Philosophy of Biological and Biomedical Sciences* 44 Pt B, 756–73.

Baker, David P., Paul J. Eslinger, Martin Benavides, Ellen Peters, Nathan F. Dieckmann, and Juan Leon. 2015. "The Cognitive Impact of the Education Revolution: A Possible Cause of the Flynn Effect on Population IQ." *Intelligence* 49:144–58.

Baker, Katharine K. 2004. "Bargaining or Biology—The History and Future of Paternity Law and Parental Status." *Cornell Journal of Law and Public Policy* 14:1. http://scholarship.law .cornell.edu/cjlpp/vol14/iss1/1.

Bale, Tracy L. 2014. "Lifetime Stress Experience: Transgenerational Epigenetics and Germ Cell Programming." *Dialogues in Clinical Neuroscience* 16:297–305.

———. 2015. "Epigenetic and Transgenerational Reprogramming of Brain Development." *Nature Reviews Neuroscience* 16:332–44.

Ball, Philip. 2016. "Kathy Niakan: At the Forefront of Gene Editing in Embryos." *Lancet* 387:935.

Ball, Steven G., Debashish Bhattacharya, and Andreas P. M. Weber. 2016. "Pathogen to Powerhouse." *Science* 351:659–60.

Balmain, Allan. 2001. "Cancer Genetics: From Boveri and Mendel to Microarrays." *Nature Reviews Cancer* 1:77–82.

Balmer, Jennifer. 2014. "Smoking Mothers May Alter the DNA of Their Children." *Science,* July 28. http://www.sciencemag.org/news/2014/07/smoking-mothers-may-alter-dna-their -children (accessed August 4, 2017).

Baltimore, David, Paul Berg, Michael Botchan, Dana Carroll, R. Alta Charo, George Church, Jacob E. Corn, and others. 2015. "Biotechnology: A Prudent Path Forward for Genomic Engineering and Germline Gene Modification." *Science* 348:36–38.

Barkan, Elazar. 1992. *The Retreat of Scientific Racism: Changing Concepts of Race in Britain and the United States Between the World Wars.* Cambridge: Cambridge University Press.

Barnes, L. Diane. 2013. *Frederick Douglass: Reformer and Statesman.* New York: Routledge.

Baron, David. 2003. "DNA Tests Shed Light on 'Hybrid Humans.'" National Public Radio, August 11. http://www.npr.org/templates/story/story.php?storyId=1392149 (accessed February 21, 2017).

Baross, John A., and William F. Martin. 2015. "The Ribofilm as a Concept for Life's Origins." *Cell* 162:13–15.

Barritt, J. A., C. A. Brenner, H. E. Malter, and J. Cohen. 2001. "Mitochondria in Human Offspring Derived from Ooplasmic Transplantation." *Human Reproduction* 16:513–16.

Bartley, Mary M. 1992. "Darwin and Domestication: Studies on Inheritance." *Journal of the History of Biology* 25:307–33.

Bashford, Alison, and Philippa Levine, eds. 2010. *The Oxford Handbook of the History of Eugenics.* Oxford: Oxford University Press.

Bateson, William, and Edith Rebecca Saunders. 1902. "The Facts of Heredity in the Light of Mendel's Discovery." *Reports to the Evolution Committee of the Royal Society* 1:125–60.

Bath, Sarah C., Colin D. Steer, Jean Golding, Pauline Emmett, and Margaret P. Rayman. 2013. "Effect of Inadequate Iodine Status in UK Pregnant Women on Cognitive Outcomes in Their Children: Results from the Avon Longitudinal Study of Parents and Children (ALSPAC)." *Lancet* 382:331–37.

Baudat, Frédéric, Yukiko Imai, and Bernard de Massy. 2013. "Meiotic Recombination in Mammals: Localization and Regulation." *Nature Reviews Genetics* 14:794–806.

Bauer, Lauren, and Diane Whitmore Schanzenbach. 2016. "The Long-term Impact of the Head Start Program." The Hamilton Project of the Brookings Institution. http://www .hamiltonproject.org/papers/the_long_term_impacts_of_head_start (accessed September 11, 2017).

Bauer, Tobias, Saskia Trump, Naveed Ishaque, Loreen Thürmann, Lei Gu, Mario Bauer, Matthias Bieg, and others. 2016. "Environment-induced Epigenetic Reprogramming in Genomic Regulatory Elements in Smoking Mothers and Their Children." *Molecular Systems Biology* 12:861.

Beck, Tracey L. 1998. "My Life with PKU." Article written for *National PKU News,* Spring. http://www.stsci.edu/~tbeck/mystory.html (accessed August 24, 2017).

Beeson, Emma Burbank. 1927. *The Early Life & Letters of Luther Burbank, by His Sister Emma (Burbank) Beeson.* San Francisco: Harr Wagner Publishing Company.

Begley, Sharon. 2015. "Dare We Edit the Human Race? Star Geneticists Wrestle with Their Power." *STAT,* December 2.

Behar, Doron M., Mait Metspalu, Yael Baran, Naama M. Kopelman, Bayazit Yunusbayev, Ariella Gladstein, Shay Tzur, Hovhannes Sahakyan, Ardeshir Bahmanimehr, and Levon Yepiskoposyan. 2013. "No Evidence from Genome-Wide Data of a Khazar Origin for the Ashkenazi Jews." *Human Biology* 85:859–900.

Beleza, Sandra, António M. Santos, Brian McEvoy, Isabel Alves, Cláudia Martinho, Emily Cameron, Mark D. Shriver, Esteban J. Parra, and Jorge Rocha. 2013. "The Timing of Pigmentation Lightening in Europeans." *Molecular Biology and Evolution* 30:24–35.

Belozerskaya, Marina. 2005. *Luxury Arts of the Renaissance.* London: Thames & Hudson.

Beltrame, Marcia Holsbach, Meagan A. Rubel, and Sarah A. Tishkoff. 2016. "Inferences of African Evolutionary History from Genomic Data." *Current Opinion in Genetics & Development* 41:159–66.

Bennett, Gordon M., and Nancy A. Moran. "Heritable Symbiosis: The Advantages and Perils of an Evolutionary Rabbit Hole." *Proceedings of the National Academy of Sciences of the United States of America* 112:10169–76.

Benson, Fred. 1981. "Blood Tests Showing Nonpaternity-Conclusive or Rebuttable Evidence?: The Chaplin Case Revisited." *American Journal of Forensic Medicine and Pathology* 2:221–24.

Berg, Mary Anne, Jaime Guevara-Aguirre, Arlan L. Rosenbloom, Ron G. Rosenfeld, and Uta Francke. 1992. "Mutation Creating a New Splice Site in the Growth Hormone Receptor Genes of 37 Ecuadorean Patients with Laron Syndrome." *Human Mutation* 1:24–34.

Bergland, Richard M. 1965. "New Information Concerning the Irish Giant." *Journal of Neurosurgery* 23:265–69.

Bergman, Yehudit, and Howard Cedar. 2013. "DNA Methylation Dynamics in Health and Disease." *Nature Structural & Molecular Biology* 20:274–81.

Bergmann, Olaf, and Jonas Frisén. 2013. "Why Adults Need New Brain Cells." *Science* 340:695–96.

——, Jakob Liebl, Samuel Bernard, Kanar Alkass, Maggie S. Yeung, Peter Steier, Walter Kutschera, and others. 2012. "The Age of Olfactory Bulb Neurons in Humans." *Neuron* 74:634–39.

——, Kirsty L. Spalding, and Jonas Frisén. 2015. "Adult Neurogenesis in Humans." *Cold Spring Harbor Perspectives in Biology* 7:a018994.

Bernstein, Joseph. 2016. "This Man Helped Build the Trump Meme Army—Now He Wants to Reform It." BuzzFeed News, December 30.

Berra, Tim M., Gonzalo Álvarez, and Francisco C. Ceballos. 2010. "Was the Darwin/Wedgwood Dynasty Adversely Affected by Consanguinity?" *BioScience* 60: 376–83.

Berry v. Chaplin, 74 Cal. App 2d 652 (1946).

Beutler, Ernest. 2001. "The Cline Affair." *Molecular Therapy* 4:396.

Bhardwaj, Ratan D., Maurice A. Curtis, Kirsty L. Spalding, Bruce A. Buchholz, David Fink, Thomas Björk-Eriksson, Claes Nordborg, and others. 2006. "Neocortical Neurogenesis in Humans Is Restricted to Development." *Proceedings of the National Academy of Sciences of the United States of America* 103:12564–68.

Bianchi, D. W., G. K. Zickwolf, G. J. Weil, S. Sylvester, and M. A. DeMaria. 1996. "Male Fetal Progenitor Cells Persist in Maternal Blood for as Long as 27 Years Postpartum." *Proceedings of the National Academy of Sciences of the United States of America* 93:705–08.

Bianchi, Diana W. 2007. "Robert E. Gross Lecture. Fetomaternal Cell Trafficking: A Story That Begins with Prenatal Diagnosis and May End with Stem Cell Therapy." *Journal of Pediatric Surgery* 42:12–18.

Bickel, H. 1996. "The First Treatment of Phenylketonuria." *European Journal of Pediatrics* 155:S2–S3.

Bicknell, Ross, Andrew Catanach, Melanie Hand, and Anna Koltunow. 2016. "Seeds of Doubt: Mendel's Choice of *Hieracium* to Study Inheritance, a Case of Right Plant, Wrong Trait." *Theoretical and Applied Genetics.* 129:2253–66.

Biesecker, Leslie. 2005. "Proteus Syndrome." In *Management of Genetic Syndromes.* Edited by Suzanne B. Cassidy. New York: Wiley-Liss.

——. 2006. "The Challenges of Proteus Syndrome: Diagnosis and Management." *European Journal of Human Genetics* 14:1151–57.

Black, Sandra E., Paul J. Devereux, Petter Lundborg, and Kaveh Majlesi. 2015. "Poor Little Rich Kids? The Determinants of the Intergenerational Transmission of Wealth." *National Bureau of Economic Research Working Paper Series,* working paper 21409. http://www.nber.org /papers/w21409 (accessed September 11, 2017).

Blaine, Delabere. 1810. *A Domestic Treatise on the Diseases of Horses and Dogs: So Conducted as to Enable Persons to Practise with Ease and Success on Their Own Animals.* London: Printed for T. Boosey.

Blaser, Martin J., and Maria G. Dominguez-Bello. 2016. "The Human Microbiome before Birth." *Cell Host and Microbe* 20:558–60.

Blum, Matthias. 2016. "Inequality and Heights." In *The Oxford Handbook of Economics and Human Biology.* Edited by John Komlos and Inas R. Kelly. Oxford: Oxford University Press.

Boddy, Amy M., Angelo Fortunato, Melissa Wilson Sayres, and Athena Aktipis. 2015. "Fetal Microchimerism and Maternal Health: A Review and Evolutionary Analysis of Cooperation and Conflict beyond the Womb." *BioEssays* 37:1106–18.

Bohacek, Johannes, and Isabelle M. Mansuy. 2015. "Molecular Insights into Transgenerational Non-Genetic Inheritance of Acquired Behaviours." *Nature Reviews Genetics* 16:641–52.

Bohannon, John. 2013. "Why Are Some People So Smart? The Answer Could Spawn a Generation of Superbabies." *Wired*, July 16.

Bonduriansky, Russell, and Troy Day. 2009. "Nongenetic Inheritance and Its Evolutionary Implications." *Annual Review of Ecology, Evolultion, and Systematics* 40:103–25.

———. 2018. *Not Only Genes: Nongenetic Inheritance in Evolution and Human Life*. Princeton: Princeton University Press.

Bonner, J. T., and W. J. Bell Jr. 1984. "'What Is Money For?': An Interview with Edwin Grant Conklin." *Proceedings of the American Philosophical Society* 128:79–84.

Boodman, Eric. 2017. "White Nationalists Are Flocking to Genetic Ancestry Tests. Some Don't Like What They Find." *STAT*, August 16. https://www.statnews.com/2017/08/16/white -nationalists-genetic-ancestry-test/ (accessed September 10, 2017).

Boomsma, Dorret, Andreas Busjahn, and Leena Peltonen. 2002. "Classical Twin Studies and Beyond." *Nature Reviews Genetics* 3:872–82.

Boone, Charles, and Brenda J. Andrews. 2015. "The Indispensable Genome." *Science* 350:1028–29.

Bordenstein, Seth R. 2015. "Rethinking Heritability of the Microbiome." *Science* 349:1172–73.

Bossinger, Gerd, and Antanas Spokevicius. 2011. "Plant Chimaeras and Mosaics." In *Encyclopedia of Life Sciences*. Chichester, UK: Wiley.

Boubakar, Leila, Julien Falk, Hugo Ducuing, Karine Thoinet, Florie Reynaud, Edmund Derrington, and Valérie Castellani. 2017. "Molecular Memory of Morphologies by Septins during Neuron Generation Allows Early Polarity Inheritance." *Neuron* 95:834–51.

Bouchard, Maryse F., Jonathan Chevrier, Kim G. Harley, Katherine Kogut, Michelle Vedar, Norma Calderon, Celina Trujillo, and others. 2011. "Prenatal Exposure to Organophosphate Pesticides and IQ in 7-Year-Old Children." *Environmental Health Perspectives* 119:1189–95.

Bourrat, Pierrick, Qiaoying Lu, and Eva Jablonka. 2017. "Why the Missing Heritability Might Not Be in the DNA." *BioEssays*. doi:10.1002/bies.201700067.

Boveri, Theodor. 2008. "Concerning the Origin of Malignant Tumours by Theodor Boveri." Translated and annotated by Henry Harris. *Journal of Cell Science* 121, Suppl 1, 1–84.

Bowles, Samuel, Eric Alden Smith, and Monique Borgerhoff Mulder. 2010. "The Emergence and Persistence of Inequality in Premodern Societies." *Current Anthropology* 51:7–17.

Boyd, Robert. 2017. *A Different Kind of Animal: How Culture Transformed Our Species*. Princeton: Princeton University Press.

Brandt, Guido, Anna Szécsényi-Nagy, Christina Roth, Kurt Werner Alt, and Wolfgang Haak. 2015. "Human Paleogenetics of Europe—The Known Knowns and the Known Unknowns." *Journal of Human Evolution* 79:73–92.

Breton, Sophie, and Donald T. Stewart. 2015. "Atypical Mitochondrial Inheritance Patterns in Eukaryotes." *Genome* 58:423–31.

Bright, Monika, and Silvia Bulgheresi. 2010. "A Complex Journey: Transmission of Microbial Symbionts." *Nature Reviews Microbiology* 8:218–30.

Brinch, Christian N., and Taryn Ann Galloway. 2012. "Schooling in Adolescence Raises IQ Scores." *Proceedings of the National Academy of Sciences of the United States of America* 109:425–30.

Broad, William J. 2007. "Useful Mutants, Bred with Radiation." *New York Times*, August 28, F1.

Brown, Kyle S., Curtis W. Marean, Andy I. R. Herries, Zenobia Jacobs, Chantal Tribolo, David Braun, David L. Roberts, Michael C. Meyer, and Jocelyn Bernatchez. 2009. "Fire as an Engineering Tool of Early Modern Humans." *Science* 325:859–62.

Browne, E. J. 2002. *Charles Darwin: The Power of Place*. New York: Alfred A. Knopf.

Browne, Hilary P., B. Anne Neville, Samuel C. Forster, and Trevor D. Lawley. 2017. "Transmission of the Gut Microbiota: Spreading of Health." *Nature Reviews Microbiology* 15:531–43.

Browne, Janet. 2016. "Inspiration to Perspiration: Francis Galton's Hereditary Genius in Victorian Context." In *Genealogies of Genius*. Edited by Joyce E. Chaplin and Darrin M. McMahon. New York: Springer.

Browne-Barbour, Vanessa S. 2015. "Mama's Baby, Papa's Maybe: Disestablishment of Paternity." *Akron Law Review* 48:263.

Browning, Sharon R., and Brian L. Browning. 2012. "Identity by Descent between Distant Relatives: Detection and Applications." *Annual Review of Genetics* 46:617–33.

Brucato, Nicolas, Pradiptajati Kusuma, Murray P. Cox, Denis Pierron, Gludhug A. Purnomo, Alexander Adelaar, Toomas Kivisild, Thierry Letellier, Herawati Sudoyo, and François-Xavier Ricaut. 2016. "Malagasy Genetic Ancestry Comes from an Historical Malay Trading Post in Southeast Borneo." *Molecular Biology and Evolution* 33:2396–2400.

Bryc, Katarzyna, Eric Y. Durand, J. Michael Macpherson, David Reich, and Joanna L. Mountain. 2015. "The Genetic Ancestry of African Americans, Latinos, and European Americans across the United States." *American Journal of Human Genetics* 96:37–53.

Buck, Carol. 1992. Introduction to *The Child Who Never Grew*. 2nd ed. Rockville, MD: Woodbine House.

Buck, Pearl. 1950. *The Child Who Never Grew*. New York: J. Day.

Buckingham, Margaret E., and Sigolène M. Meilhac. "Tracing Cells for Tracking Cell Lineage and Clonal Behavior." *Developmental Cell* 21:394–409.

Budget Hearing—Food and Drug Administration | Committee on Appropriations, U.S. House of Representatives. March 27, 2014. http://appropriations.house.gov/calendar/eventsingle .aspx?EventID=373227 (accessed March 29, 2017).

Bull, James J., and Harmit S. Malik. 2017. "The Gene Drive Bubble: New Realities." *PLOS Genetics* 13:e1006850.

Bulmer, M. G. 2003. *Francis Galton: Pioneer of Heredity and Biometry*. Baltimore: Johns Hopkins University Press.

Burbank, Luther. 1904. "Some Fundamental Principles of Plant Breeding." *Proceedings of the International Conference on Plant Breeding and Hybridization*. New York: Horticultural Society of New York.

———. 1906. *The Training of the Human Plant*. New York: Century Co.

———. 1939. *Partner of Nature*. New York: Appleton-Century Co.

———, and Wilbur Hall. 1927. *The Harvest of the Years*. Boston: Houghton Mifflin.

Burkhardt, Richard W. 2013. "Lamarck, Evolution, and the Inheritance of Acquired Characters." *Genetics* 194:793–805.

Burleigh, Michael. 2001.*The Third Reich: A New History*. New York: Hill and Wang.

Burman, Jeremy Trevelyan. 2012. "The Misunderstanding of Memes: Biography of an Unscientific Object, 1976–1999." *Perspectives on Science* 20:75–104.

Burnham, George Pickering. 1855. *The History of the Hen Fever: A Humorous Record*. New York: James French.

Burt, Austin, and Robert Trivers. 2006. *Genes in Conflict: The Biology of Selfish Genetic Elements*. Cambridge: Belknap Press of Harvard University Press.

Burt, Cyril. 1909. "Experimental Tests of General Intelligence." *British Journal of Psychology* 3:94–177.

Bushman, Diane M., and Jerold Chun. 2013. "The Genomically Mosaic Brain: Aneuploidy and More in Neural Diversity and Disease." *Seminars in Cell & Developmental Biology* 24:357–69.

Busslinger, Meinrad, and Alexander Tarakhovsky. 2014. "Epigenetic Control of Immunity." *Cold Spring Harbor Perspectives in Biology* 6:a019307.

Bygren, Lars O., Petter Tinghög, John Carstensen, Sören Edvinsson, Gunnar Kaati, Marcus E. Pembrey, and Michael Sjöström. 2014. "Change in Paternal Grandmothers' Early Food Supply Influenced Cardiovascular Mortality of the Female Grandchildren." *BMC Genetics* 15:12.

Byrne, Richard W., and Lisa G. Rapaport. 2011. "What Are We Learning from Teaching?" *Animal Behaviour* 82:1207–11.

Calvin, Catherine M., G. David Batty, Geoff Der, Caroline E. Brett, Adele Taylor, Alison Pattie, Iva Čukić, and Ian J. Deary. 2017. "Childhood Intelligence in Relation to Major Causes of Death in 68 Year Follow-up: Prospective Population Study." *BMJ* 357:j2708.

Campbell, Ian M., Bo Yuan, Caroline Robberecht, Rolph Pfundt, Przemyslaw Szafranski, Meriel E. McEntagart, Sandesh CS Nagamani, Ayelet Erez, Magdalena Bartnik, Barbara Wiśniowiecka-Kowalnik, and others. 2014. "Parental Somatic Mosaicism Is Underrecognized and Influences Recurrence Risk of Genomic Disorders." *American Journal of Human Genetics* 95:173–82.

Campbell, Ian M., Chad A. Shaw, Pawel Stankiewsicz, and James R. Lupski. 2015. "Somatic Mosaicism: Implications for Disease and Transmission Genetics." *Trends in Genetics* 31:382–92.

Capel, Blanche, and Doug Coveney. 2004. "Frank Lillie's Freemartin: Illuminating the Pathway to 21st Century Reproductive Endocrinology." *Journal of Experimental Zoology. Part A, Comparative Experimental Biology* 301:853–56.

Carlson, Elof Axel. 1981. *Genes, Radiation, and Society: The Life and Work of H. J. Muller*. Ithaca: Cornell University Press.

———. 2009. *Hermann Joseph Muller, 1890–1967*. Washington, DC: National Academy of Sciences.

Carsten, Janet. 1995. "The Substance of Kinship and the Heat of the Hearth: Feeding, Personhood, and Relatedness Among Malays in Pulau Langkawi."*American Ethnologist* 22:223–41.

Cary, S. C., and S. J. Giovannoni. 1993. "Transovarial Inheritance of Endosymbiotic Bacteria in Clams Inhabiting Deep-sea Hydrothermal Vents and Cold Seeps." *Proceedings of the National Academy of Sciences of the United States of America* 90:5695–99.

"Case of Carol Ann, The." 1945. *Life*, January 8, p. 30.

Casselton, L. A. 2002. "Mate Recognition in Fungi." *Heredity* 88:142–47.

Castles, Elaine E. 2012. *Inventing Intelligence: How America Came to Worship IQ*. Santa Barbara: Praeger.

Cathcart, Michael. 2013. *Starvation in a Land of Plenty: Will's Diary of the Fateful Burke and Wills Expedition*. Canberra: National Library of Australia.

Cavazzana-Calvo, Marina, Emmanuel Payen, Olivier Negre, Gary Wang, Kathleen Hehir, Floriane Fusil, Julian Down, and others. 2010. "Transfusion Independence and HMGA2 Activation after Gene Therapy of Human B-Thalassaemia." *Nature* 467:318–22.

Cellini, Luigina. 2014. "*Helicobacter pylori*: A Chameleon-like Approach to Life." *World Journal of Gastroenterology* 20:5575–82.

Centerwall, Siegried A., and Willard R. Centerwall. 2000. "The Discovery of Phenylketonuria: The Story of a Young Couple, Two Retarded Children, and a Scientist." *Pediatrics* 105:89–103.

Centre of Microbial and Plant Genetics. "Short Biography of F. A. Janssens." KU Leuven, Belgium. https://www.biw.kuleuven.be/m2s/cmpg/About/fajanssens (accessed August 4, 2017).

Cesarini, David, and Peter M. Visscher. 2017. "Genetics and Educational Attainment." *Science of Learning* 2:4.

Chabris, Christopher F., Benjamin M. Hebert, Daniel J. Benjamin, Jonathan Beauchamp, David Cesarini, Matthijs van der Loos, Magnus Johannesson, and others. 2012. "Most Reported Genetic Associations with General Intelligence Are Probably False Positives." *Psychological Science* 23:1314–23.

Chahal, Harvinder S., Karen Stals, Martina Unterländer, David J. Balding, Mark G. Thomas, Ajith V. Kumar, G. Michael Besser, A. Brew Atkinson, Patrick J. Morrison, and Trevor A. Howlett. 2011. "AIP Mutation in Pituitary Adenomas in the 18th Century and Today." *New England Journal of Medicine* 364:43–50.

Chan, William F. N., Cécile Gurnot, Thomas J. Montine, Joshua A. Sonnen, Katherine A. Guthrie, and J. Lee Nelson. 2012. "Male Microchimerism in the Human Female Brain." *PLOS One* 7:e45592.

Chang, Joseph. 1999. "Recent Common Ancestors of All Present-Day Individuals." *Advances in Applied Probability* 31:1002–26.

Chemke, J., S. Rappaport, and R. Etrog. 1983. "Aberrant Melanoblast Migration Associated with Trisomy 18 Mosaicism." *Journal of Medical Genetics* 20:135–37.

Chen, Brian H., Riccardo E. Marioni, Elena Colicino, Marjolein J. Peters, Cavin K. Ward-Caviness, Pei-Chien Tsai, Nicholas S. Roetker, Allan C. Just, Ellen W. Demerath, and Weihua Guan. 2016. "DNA Methylation-Based Measures of Biological Age: Meta-Analysis Predicting Time to Death." *DNA* 8:9.

Chen, Qi, Wei Yan, and Enkui Duan. 2016. "Epigenetic Inheritance of Acquired Traits Through Sperm RNAs and Sperm RNA Modifications." *Nature Reviews Genetics* 17:733–43.

Chen, Serena H., Claudia Pascale, Maria Jackson, Mary Ann Szvetecz, and Jacques Cohen. 2016. "A Limited Survey-based Uncontrolled Follow-up Study of Children Born After Ooplasmic Transplantation in a Single Centre." *Reproductive Biomedicine Online* 33:737–44.

Chetty, Raj, David Grusky, Maximilian Hell, Nathaniel Hendren, Robert Manduca, and Jimmy Narang. 2017. "The Fading American Dream: Trends in Absolute Income Mobility Since 1940." *Science* 356:398–406.

Cheung, Benjamin Y., Ilan Dar-Nimrod, and Karen Gonsalkorale. 2014. "Am I My Genes? Perceived Genetic Etiology, Intrapersonal Processes, and Health." *Social and Personality Psychology Compass* 8:626–37.

Christie, Joshua R., Timothy M. Schaerf, and Madeleine Beekman. 2015. "Selection Against Heteroplasmy Explains the Evolution of Uniparental Inheritance of Mitochondria." *PLOS Genetics* 11:e1005112.

Chudek, Maciej, Michael Muthukrishna, and Joseph Henrich. 2015. "Cultural Evolution." In *The Handbook of Evolutionary Psychology*. 2nd ed. Edited by David M. Buss. Hoboken, NJ: Wiley.

Churchill, Frederick B. 1987. "From Heredity Theory to Vererbung: The Transmission Problem, 1850–1915." *Isis* 78:337–64.

———. 2015. *August Weismann: Development, Heredity, and Evolution*. Cambridge: Harvard University Press.

Clampett, Frederick W. 1970. *Luther Burbank "Our Beloved Infidel": His Religion of Humanity*. Westport, CT: Greenwood.

Clarkson, Chris, Zenobia Jacobs, Ben Marwick, Richard Fullagar, Lynley Wallis, Mike Smith, Richard G. Roberts, and others. 2017. "Human Occupation of Northern Australia by 65,000 Years Ago." *Nature* 547:306–10.

Claussnitzer, Melina, Simon N. Dankel, Kyoung-Han Kim, Gerald Quon, Wouter Meuleman, Christine Haugen, Viktoria Glunk, and others. 2015. "FTO Obesity Variant Circuitry and Adipocyte Browning in Humans." *New England Journal of Medicine* 373:895–907.

Clement, Anthony C. 1979. "Edwin Grant Conklin." *American Zoologist* 19:1255–59.

Cobb, Matthew. 2006. "Heredity Before Genetics: A History." *Nature Reviews Genetics* 7:953–58.

———. 2012. "An Amazing 10 Years: The Discovery of Egg and Sperm in the 17th Century." *Reproduction in Domestic Animals* 47, Suppl 4, 2–6.

Coble, Michael D., Odile M. Loreille, Mark J. Wadhams, Suni M. Edson, Kerry Maynard, Carna E. Meyer, Harald Niederstätter, and others. 2009. "Mystery Solved: The Identification of the Two Missing Romanov Children Using DNA Analysis." *PLOS One* 4:e4838.

Cockerell, Theodore Dru Alison. 1917. "Somatic Mutations in Sunflowers." *Journal of Heredity* 8:467–70.

Coen, Enrico. 1999. *The Art of Genes: How Organisms Make Themselves*. Oxford: Oxford University Press.

Cohen, Adam. 2016. *Imbeciles: The Supreme Court, American Eugenics, and the Sterilization of Carrie Buck*. New York: Penguin Press.

Cohen, Jacques, and Henry Malter. 2016. "The First Clinical Nuclear Transplantation in China: New Information about a Case Reported to ASRM in 2003." *Reproductive Biomedicine Online* 33:433–35.

———, Richard Scott, Tim Schimmel, Jacob Levron, and Steen Willadsen. 1997. "Birth of Infant After Transfer of Anucleate Donor Oocyte Cytoplasm into Recipient Eggs." *Lancet* 350:186–87.

Cold Spring Harbor Lab. 2013. "CSHL Associate Professor Zach Lippman at the Secret Science Club, Brooklyn, NY, July 16, 2013." YouTube video, August 2. https://www.youtube.com /watch?v=gY6IrR2FUH4 (accessed May 13, 2017).

Collins, Francis S. 2015. "Statement on NIH Funding of Research Using Gene-Editing Technologies in Human Embryos." *NIH Director*, April 28. https://www.nih.gov/about-nih /who-we-are/nih-director/statements/statement-nih-funding-research-using-gene-editing -technologies-human-embryos (accessed August 24, 2017).

———, Lowell Weiss, and Hudson Kathy. 2001. "Heredity and Humanity." *New Republic*, June 25. https://newrepublic.com/article/61291/heredity-and-humanity (accessed September 11, 2017).

Colman, Andrew M. 2016. "Race Differences in IQ: Hans Eysenck's Contribution to the Debate in the Light of Subsequent Research." *Personality and Individual Differences* 103:182–89.

Columbus, Christopher. "Santangel Letter." Early Modern Spain website of King's College London. http://www.ems.kcl.ac.uk/content/etext/e022.html (accessed July 23, 2017).

Comfort, Nathaniel C. 2012. *The Science of Human Perfection: How Genes Became the Heart of American Medicine*. New Haven: Yale University Press.

Conklin, Edwin Grant. 1968. "Early Days at Woods Hole." *American Scientist* 56:112–20.

Conley, Dalton, Emily Rauscher, Christopher Dawes, Patrik K. E. Magnusson, and Mark L. Siegal. 2013. "Heritability and the Equal Environments Assumption: Evidence from Multiple Samples of Misclassified Twins." *Behavior Genetics* 43:415–26.

Conn, Peter. 1996. *Pearl S. Buck: A Cultural Biography.* Cambridge: Cambridge University Press.

Connor, Steve. 2015. "'Three-Parent Babies': Britain Votes in Favour of Law Change." *Independent,* February 3.

Coop, Graham. 2013a. "How Much of Your Genome Do You Inherit from a Particular Grandparent?" Gcbias blog, October 20. https://gcbias.org/2013/10/20/how-much-of-your -genome-do-you-inherit-from-a-particular-grandparent/ (accessed July 27, 2017).

———. 2013b. "How Much of Your Genome Do You Inherit from a Particular Ancestor?" Gcbias blog, November 4. https://gcbias.org/2013/11/04/how-much-of-your-genome-do-you -inherit-from-a-particular-ancestor/ (accessed July 27, 2017).

———. 2013c. "How Many Genomic Blocks Do You Share with a Cousin?" Gcbias blog, December 2. https://gcbias.org/2013/12/02/how-many-genomic-blocks-do-you-share -with-a-cousin/ (accessed July 27, 2017).

———, Xiaoquan Wen, Carole Ober, Jonathan K. Pritchard, and Molly Przeworski. 2008. "High-Resolution Mapping of Crossovers Reveals Extensive Variation in Fine-Scale Recombination Patterns Among Humans." *Science* 319:1395–98.

Cooper, Colin. 2015. *Intelligence and Human Abilities: Structure, Origins and Applications.* New York: Routledge.

Cooper, David N. 2011. "Lionizing Lyonization 50 Years On." *Human Genetics* 130:167–68.

Cooper, Richard S. 2013. "Race in Biological and Biomedical Research." *Cold Spring Harbor Perspectives in Medicine* 3:a008573.

Cossetti, Cristina, Luana Lugini, Letizia Astrologo, Isabella Saggio, Stefano Fais, and Corrado Spadafora. 2014. "Soma-to-Germline Transmission of RNA in Mice Xenografted with Human Tumour Cells: Possible Transport by Exosomes." *PLOS One* 9:e101629.

Cowans, Jon. 2003. *Early Modern Spain: A Documentary History.* Philadelphia: University of Pennsylvania Press.

Cowdry, E. V. 1953. "Historical Background of Research on Mitochondria." *Journal of Histochemistry & Cytochemistry* 1:183–87.

Coy, Peter. 2017. "The Big Reason Whites Are Richer than Blacks in America." *Bloomberg BusinessWeek,* February 8.

Craig, Lee. 2016. "Antebellum Puzzle: The Decline in Heights at the Onset of Modern Economic Growth." In *The Oxford Handbook of Economics and Human Biology.* Edited by John Komlos and Inas R. Kelly. New York: Oxford University Press.

Cravens, Hamilton. 1993. *Before Head Start: The Iowa Station & America's Children.* Chapel Hill: University of North Carolina Press.

Crawford, Nicholas G., Derek E. Kelly, Matthew E. B. Hansen, Marcia H. Beltrame, Shaohua Fan, Shanna L. Bowman, Ethan Jewett, and others. 2017. "Loci Associated with Skin Pigmentation Identified in African Populations." *Science* doi:10.1126/science.aan8433.

"Crime of Scientific Zeal, The." 1981. *New York Times,* June 5, editorial.

Cubas, P., C. Vincent, and E. Coen. 1999. "An Epigenetic Mutation Responsible for Natural Variation in Floral Symmetry." *Nature* 401:157–61.

Curtis, Benjamin. 2013.*The Habsburgs: The History of a Dynasty.* London: Bloomsbury.

Cutter, Calvin. 1850. *A Treatise on Anatomy, Physiology, and Medicine: Designed for Colleges, Academies, and Families; with One Hundred and Fifty Engravings.* Boston: Benjamin B. Mussey.

Cyranoski, David. 2017. "China's Embrace of Embryo Selection Raises Thorny Questions." *Nature* 548:272.

Dacks, Joel B., Mark C. Field, Roger Buick, Laura Eme, Simonetta Gribaldo, Andrew J. Roger, Céline Brochier-Armanet, and Damien P. Devos. 2016. "The Changing View of Eukaryogenesis—Fossils, Cells, Lineages and How They All Come Together." *Journal of Cell Science* 129:3695–3703.

da Graca, J. V., E. S. Louzada, and J. W. Sauls. 2004. "The Origins of Red Pigmented Grapefruits and the Development of New Varieties." *Proceedings of the International Society of Citriculture* 1:369–74.

Dalgaard, Carl-Johan, and Holger Strulik. 2016. "Physiology and Development: Why the West Is Taller than the Rest." *Economic Journal* 126, no. 598, 1–32.

Danaei, Goodarz, Kathryn G. Andrews, Christopher R. Sudfeld, Günther Fink, Dana Charles McCoy, Evan Peet, Ayesha Sania, Mary C. Smith Fawzi, Majid Ezzati, and Wafaie W. Fawzi. 2016. "Risk Factors for Childhood Stunting in 137 Developing Countries: A Comparative Risk Assessment Analysis at Global, Regional, and Country Levels." *PLOS Medicine* 13:e1002164.

Dannemann, Michael, Aida M. Andrés, and Janet Kelso. 2016. "Introgression of Neandertal- and Denisovan-like Haplotypes Contributes to Adaptive Variation in Human Toll-like Receptors." *American Journal of Human Genetics* 98:22–33.

Dare, Helen. 1905. "Luther Burbank: The Wizard of Horticulture." *San Francisco Sunday Call*, June 25.

Dar-Nimrod, Ilan, and Steven J. Heine. 2011. "Genetic Essentialism: On the Deceptive Determinism of DNA." *Psychological Bulletin* 137:800–18.

———, Ilan, Benjamin Y. Cheung, Matthew B. Ruby, and Steven J. Heine. 2014. "Can Merely Learning about Obesity Genes Affect Eating Behavior?" *Appetite* 81:269–76.

Darwin, Charles. 1839. *Questions About the Breeding of Animals*. London: Stewart & Murray. http://darwin-online.org.uk/content/frameset?itemID=F262&viewtype=text&pageseq=1 (accessed July 23, 2017).

———. 1859. *On the Origin of Species by Means of Natural Selection*. London: John Murray.

———. 1868. *The Variation of Animals and Plants Under Domestication*. London: John Murray.

———. 1871. "Pangenesis." *Nature* 3:502–03.

Daubin, Vincent, and Gergely J. Szöllősi. 2016. "Horizontal Gene Transfer and the History of Life." *Cold Spring Harbor Perspectives in Biology*. doi:10.1101/cshperspect.a018036.

Davenport, Charles B. 1899. *Statistical Methods, with Special Reference to Biological Variation*. New York: Wiley.

———. 1908. *Inheritance in Canaries*. Carnegie Institution of Washington.

———. 1911. *Heredity in Relation to Eugenics*. New York: H. Holt.

———. 1917. "The Effects of Race Intermingling." *Proceedings of the American Philosophical Society* 56:364–68.

Davenport, Gertrude C., and Charles B. Davenport. 1910. "Heredity of Skin Pigmentation in Man." *American Naturalist* 44:641–72.

Davidson, R. G., H. M. Nitowsky, and B. Childs. 1963. "Demonstration of Two Populations of Cells in the Human Female Heterozygous for Glucose-6-Phosphate Dehydrogenase Variants." *Proceedings of the National Academy of Sciences of the United States of America* 50:481–85.

Davies, G., A. Tenesa, A. Payton, J. Yang, S. E. Harris, D. Liewald, X. Ke, and others. 2011. "Genome-Wide Association Studies Establish That Human Intelligence Is Highly Heritable and Polygenic." *Molecular Psychiatry* 16:996–1005.

Dawkins, Richard. 1976. *The Selfish Gene*. Oxford: Oxford University Press, 1976.

———. 2016. *The Selfish Gene*. 40th Anniversary Edition. Oxford: Oxford University Press.

Day, Troy, and Russell Bonduriansky. 2011. "A Unified Approach to the Evolutionary Consequences of Genetic and Nongenetic Inheritance." *American Naturalist* 178:E18–36.

Dean, Lewis G., Gill L. Vale, Kevin N. Laland, Emma Flynn, and Rachel L. Kendal. 2014. "Human Cumulative Culture: A Comparative Perspective." *Biological Reviews of the Cambridge Philosophical Society* 89:284–301.

———, Rachel L. Kendal, Steven J. Schapiro, Bernard Thierry, and Kevin N. Laland. 2012. "Identification of the Social and Cognitive Processes Underlying Human Cumulative Culture." *Science* 335:1114–18.

Deary, Ian J. 2012. "Looking for 'System Integrity' in Cognitive Epidemiology." *Gerontology* 58:545–53.

———, and Lawrence J. Whalley. 2009. *A Lifetime of Intelligence: Follow-up Studies of the Scottish Mental Surveys of 1932 and 1947*. Washington, DC: American Psychological Association.

deBoer, Fredrik. 2017. "Disentangling Race from Intelligence and Genetics." Blog, April 10. https://fredrikdeboer.com/2017/04/10/disentangling-race-from-intelligence-and-genetics / (accessed September 11, 2017).

Deichmann, Ute. 2010. "Gemmules and Elements: On Darwin's and Mendel's Concepts and Methods in Heredity." *Journal for General Philosophy of Science* 41:85–112.

———. 2016. "Why Epigenetics Is Not a Vindication of Lamarckism—and Why That Matters." *Studies in History and Philosophy of Biological and Biomedical Sciences* 57:80–82.

DeLong, G. Robert. "A Career in Child Neurology: Explorations at the Frontiers." *Journal of Child Neurology* 25:1051–62.

Dennis, Carina. 2003. "Special Section on Human Genetics: The Rough Guide to the Genome." *Nature* 425:758–59.

Department of Health, UK. 2014. "Mitochondrial Donation: Government Response to the Consultation on Draft Regulations to Permit the Use of New Treatment Techniques to Prevent the Transmission of a Serious Mitochondrial Disease from Mother to Child." London: Public Health Directorate/Health Science and Bioethics Division.

"Derbyshire Sisters Rose and Daisy Picked as Embryos to Beat Killer Disease." 2014. *Derby Telegraph*, May 15.

Der Sarkissian, Clio, Morten E. Allentoft, María C. Ávila-Arcos, Ross Barnett, Paula F. Campos, Enrico Cappellini, Luca Ermini, and others. 2015. "Ancient Genomics." *Philosophical Transactions of the Royal Society B* 370:1660, 2013038. doi:10.1098/rstb.2013.0387.

Desai, R. G., and W. P. Creger. 1963. "Maternofetal Passage of Leukocytes and Platelets in Man." *Blood* 21:665–73.

———, E. McCutcheon, B. Little, and S. G. Driscoll. 1966. "Fetomaternal Passage of Leukocytes and Platelets in Erythroblastosis Fetalis." *Blood* 27:858–62.

Devanapally, Sindhuja, Snusha Ravikumar, and Antony M. Jose. 2015. "Double-Stranded RNA Made in C. Elegans Neurons Can Enter the Germline and Cause Transgenerational Gene Silencing." *Proceedings of the National Academy of Sciences of the United States of America* 112:2133–38.

D'Gama, Alissa M., Ying Geng, Javier A. Couto, Beth Martin, Evan A. Boyle, Christopher M. LaCoursiere, Amer Hossain, and others. 2015. "Mammalian Target of Rapamycin Pathway Mutations Cause Hemimegalencephaly and Focal Cortical Dysplasia." *Annals of Neurology* 77:720–25.

Dhimolea, Eugen, Viktoria Denes, Monika Lakk, Sana Al-Bazzaz, Sonya Aziz-Zaman, Monika Pilichowska, and Peter Geck. 2013. "High Male Chimerism in the Female Breast Shows Quantitative Links with Cancer." *International Journal of Cancer* 133:835–42.

Dias, Brian G., and Kerry J. Ressler. 2014. "Parental Olfactory Experience Influences Behavior and Neural Structure in Subsequent Generations." *Nature Neuroscience* 17:89–96.

Didion, John P., Andrew P. Morgan, Liran Yadgary, Timothy A. Bell, Rachel C. McMullan, Lydia Ortiz de Solorzano, Janice Britton-Davidian, and others. 2016. "R2d2 Drives Selfish Sweeps in the House Mouse." *Molecular Biology and Evolution.* doi:10.1093/molbev /msw036.

Dietel, Manfred. 2014. "Boveri at 100: The Life and Times of Theodor Boveri." *Journal of Pathology* 234:135–37.

Dillingham, William P. 1911. *Dictionary of Races or Peoples.* Washington, DC: US Government Printing Office.

Disney, R. H. 2008. "Natural History of the Scuttle Fly, *Megaselia scalaris*." *Annual Review of Entomology* 53:39–60.

Dobzhansky, Theodosius. 1941. "The Race Concept in Biology." *Scientific Monthly* 52:161–65.

Doll, Edgar. 2012. "Deborah Kallikak, 1889–1978: A Memorial." *Intellectual and Developmental Disabilities* 50:30–32.

Donnelly, Kevin P. 1983. "The Probability That Related Individuals Share Some Section of Genome Identical by Descent." *Theoretical Population Biology* 23:34–63.

Dorr, Gregory Michael. 2008. *Segregation's Science: Eugenics and Society in Virginia.* Charlottesville: University of Virginia Press.

Doudna, Jennifer A. 2015. "Genome-Editing Revolution: My Whirlwind Year with CRISPR." *Nature* 528:469.

———, and Emmanuelle Charpentier. 2014. "The New Frontier of Genome Engineering with CRISPR-Cas9." *Science* 346:1258096.

———, and Samuel H. Sternberg. 2017. *A Crack in Creation: Gene Editing and the Unthinkable Power to Control Evolution.* New York: Houghton Mifflin Harcourt.

Douglass, Frederick. 1848. *The North Star*, September 15.

———. 1855. *My Bondage and My Freedom.* New York: Auburn.

"Dr. Henry F. Osborn Dies in His Study." 1935. *New York Times*, November 7, p.23.

Dreyer, Peter, and W. L. Howard. 1993. *A Gardener Touched with Genius: The Life of Luther Burbank.* Santa Rosa: L. Burbank Home & Gardens.

Dröscher, Ariane. 2014. "Images of Cell Trees, Cell Lines, and Cell Fates: The Legacy of Ernst Haeckel and August Weismann in Stem Cell Research." *History and Philosophy of the Life Sciences* 36:157–86.

Du Bois, W. E. B. 1906. *The Health and Physique of the Negro American.* Atlanta: Atlanta University Press.

Duncan, Andrew W., Amy E. Hanlon Newell, Leslie Smith, Elizabeth M. Wilson, Susan B. Olson, Matthew J. Thayer, Stephen C. Strom, and Markus Grompe. 2012. "Frequent Aneuploidy Among Normal Human Hepatocytes." *Gastroenterology* 142:25–28.

Dunham-Snary, Kimberly J., and Scott W. Ballinger. 2015. "Mitochondrial-Nuclear DNA Mismatch Matters." *Science* 349:1449–50.

Dunlap, Knight. 1940. "Antidotes for Superstitions Concerning Human Heredity." *Scientific Monthly* 51:221–25.

Dunsford, I., C. C. Bowley, A. M. Hutchison, J. S. Thompson, R. Sanger, and R. R. Race. 1953. "A Human Blood-Group Chimera." *British Medical Journal* 2:81.

Du Plessis, Paul J., Clifford Ando, and Kaius Tuori, eds. 2016. *The Oxford Handbook of Roman Law and Society.* Oxford: Oxford University Press.

Dusheck, Jenny. 2016. "Girl's Deadly Arrhythmia Linked to Mosaic of Mutant Cells." *Stanford Medicine News Center,* September 26.

Duster, Troy. 2015. "A Post-Genomic Surprise. The Molecular Reinscription of Race in Science, Law and Medicine." *British Journal of Sociology* 66:1–27.

"Dwarf Alberta Spruce." Boston: The Arnold Arboretum of Harvard University. http:// arboretum.harvard.edu/wp-content/uploads/Picea_glauca.pdf.

Eames, Ninetta. 1896. "California's Great Plant Specialist: Luther Burbank, the Wizard of Horticulture." *San Francisco Call,* March 8.

Eaton, Sally A., Navind Jayasooriah, Michael E. Buckland, David Ik Martin, Jennifer E. Cropley, and Catherine M. Suter. 2015. "Roll Over Weismann: Extracellular Vesicles in the Transgenerational Transmission of Environmental Effects." *Epigenomics.* doi:10.2217 /epi.15.58.

Edwards, Matthew D., Anna Symbor-Nagrabska, Lindsey Dollard, David K. Gifford, and Gerald R. Fink. 2014. "Interactions between Chromosomal and Nonchromosomal Elements Reveal Missing Heritability." *Proceedings of the National Academy of Sciences of the United States of America* 111:7719–22.

Edwards, R. G., P. C. Steptoe, and J. M. Purdy. 1970. "Fertilization and Cleavage in Vitro of Preovulator Human Oocytes." *Nature* 227:1307–09.

Egan, Michael F., Terry E. Goldberg, Bhaskar S. Kolachana, Joseph H. Callicott, Chiara M. Mazzanti, Richard E. Straub, David Goldman, and Daniel R. Weinberger. 2001. "Effect of COMT Val108/158 Met Genotype on Frontal Lobe Function and Risk for Schizophrenia." *Proceedings of the National Academy of Sciences of the United States of America* 98:6917–22.

Egeland, Janice A., Daniela S. Gerhard, David L. Pauls, James N. Sussex, Kenneth K. Kidd, Cleona R. Alien, Abram M. Hostetter, and David E. Housman. 1987. "Bipolar Affective Disorders Linked to DNA Markers on Chromosome 11." *Nature* 325:783–89.

Ehrlich, Paul, and Marcus Feldman. 2003. "Genes and Cultures: What Creates Our Behavioral Phenome?" *Current Anthropology* 44:87–101.

Eliav-Feldon, Miriam, Benjamin Isaac, and Joseph Ziegler, eds. 2010. *The Origins of Racism in the West.* Cambridge: Cambridge University Press.

Ellis, Erle C. 2015. "Ecology in an Anthropogenic Biosphere." *Ecological Monographs* 85:287–331.

———. 2017. "Nature for the People." *Breakthrough Journal* no. 7, Summer. https:// thebreakthrough.org/index.php/journal/issue-7/nature-for-the-people (accessed September 16, 2017).

Ellison, Jay W., Jill A. Rosenfeld, and Lisa G. Shaffer. 2013. "Genetic Basis of Intellectual Disability." *Annual Review of Medicine* 64:441–50.

Endersby, Jim. 2009. *A Guinea Pig's History of Biology.* Cambridge: Harvard University Press.

Engelstad, Kristin, Miriam Sklerov, Joshua Kriger, Alexandra Sanford, Johnston Grier, Daniel Ash, Dieter Egli, and others. 2016. "Attitudes toward Prevention of MtDNA-Related Diseases through Oocyte Mitochondrial Replacement Therapy." *Human Reproduction* 31:1058–65.

Epstein, David. 2013. "How an 1836 Famine Altered the Genes of Children Born Decades Later." *io9,* August 26. http://io9.gizmodo.com/how-an-1836-famine-altered-the-genes-of-children-born-d-1200001177 (accessed July 24, 2017).

Eror, Aleks. 2013. "China Is Engineering Genius Babies." *Vice,* March 15.

Esvelt, Kevin M., Andrea L. Smidler, Flaminia Catteruccia, and George M. Church. 2014. "Concerning RNA-Guided Gene Drives for the Alteration of Wild Populations." *eLife* 3:e1601964.

Evans, Janice P., and Douglas N. Robinson. 2011. "The Spatial and Mechanical Challenges of Female Meiosis." *Molecular Reproduction and Development* 78:769–77.

Even, Dan. 2011. "Dead Woman's Ova Harvested After Court Okays Family Request." *Haaretz,* August 8.

Evrony, Gilad D. 2016. "One Brain, Many Genomes." *Science* 354:557–58.

Fairfax Cryobank. "Donor Search." https://www.fairfaxcryobank.com/search/ (accessed August 24, 2017).

Falconer, Bruce. 2012. "We Are Family—Ancestry.com Can Prove It." *SFGate,* September 21. http://www.sfgate.com/business/article/we-are-family-ancestry-com-can-prove-it-3884980.php (accessed August 6, 2017).

Falick Michaeli, Tal, Yehudit Bergman, and Yuval Gielchinsky. 2015. "Rejuvenating Effect of Pregnancy on the Mother." *Fertility and Sterility* 103:1125–28.

Falk, Raphael. 2014. "A Century of Mendelism: On Johannsen's Genotype Conception." *International Journal of Epidemiology* 43:1002–07.

Fancher, Raymond E. 1987. *The Intelligence Men: Makers of the IQ Controversy.* New York: W. W. Norton.

Felsenfeld, Gary. 2014. "The Evolution of Epigenetics." *Perspectives in Biology and Medicine* 57:132–48.

Feyrer, James, Dimitra Politi, and David N. Weil. 2013. "The Cognitive Effects of Micronutrient Deficiency." *National Bureau of Economic Research Working Paper Series,* working paper 19233. http://www.nber.org/papers/w19233.

"Find Your Inner Neanderthal." 2011. 23andMe blog, December 15. https://blog.23andme.com/ancestry/find-your-inner-neanderthal/ (accessed July 25, 2017).

Finger, Stanley, and Shawn E. Christ. 2004. "Pearl S. Buck and Phenylketonuria (PKU)." *Journal of the History of the Neurosciences* 13:44–57.

Fischbach, Ruth L., and John D. Loike. 2014. "Maternal-Fetal Cell Transfer in Surrogacy: Ties That Bind." *American Journal of Bioethics* 14:35-36.

Fisher, Elizabeth M. C., and Jo Peters. 2015. "Mary Frances Lyon (1925–2014)." *Cell* 160:577–78.

Fisher, James, and Robert A. Hinde. 1949. "The Opening of Milk Bottles by Birds." *British Birds* 42:57.

Fishman, Lila, and John H. Willis. 2005. "A Novel Meiotic Drive Locus Almost Completely Distorts Segregation in *Mimulus* (Monkeyflower) Hybrids." *Genetics* 169:347–53.

Flannery, Kent V., and Joyce Marcus. 2012. *The Creation of Inequality: How Our Prehistoric Ancestors Set the Stage for Monarchy, Slavery, and Empire.* Cambridge: Harvard University Press.

Flores-Sarnat, Laura, Harvey B. Sarnat, Guillermo Dávila-Gutiérrez, and Antonio Álvarez. 2003. "Hemimegalencephaly: Part 2. Neuropathology Suggests a Disorder of Cellular Lineage." *Journal of Child Neurology* 18:776–85.

Flynn, James Robert. 2009. *What Is Intelligence?: Beyond the Flynn Effect.* Cambridge: Cambridge University Press.

Ford, Edmund Brisco. 1977. "Theodosius Grigorievich Dobzhansky, 25 January 1900–18 December 1975." *Biographical Memoirs of Fellows of the Royal Society* 23:59–89.

Fordham, Alfred J. 1967. "Dwarf Conifers from Witches'-Brooms." *Arnoldia* 27:29–50.

Forsberg, Lars A., David Gisselsson, and Jan P. Dumanski. 2016. "Mosaicism in Health and Disease—Clones Picking Up Speed." *Nature Reviews Genetics* 18:128–42.

Fosse, Roar, Jay Joseph, and Ken Richardson. 2015. "A Critical Assessment of the Equal-Environment Assumption of the Twin Method for Schizophrenia." *Frontiers in Psychiatry* 6:62.

Foster, Gavin L., Dana L. Royer, and Daniel J. Lunt. 2017. "Future Climate Forcing Potentially without Precedent in the Last 420 Million Years." *Nature Communications* 8. doi:10.1038 /ncomms14845.

Foster, Kevin R., Jonas Schluter, Katharine Z. Coyte, and Seth Rakoff-Nahoum. 2017. "The Evolution of the Host Microbiome as an Ecosystem on a Leash." *Nature* 548:43–51.

Francis, Gregory. 2014. "Too Much Success for Recent Groundbreaking Epigenetic Experiments." *Genetics* 198:449–51.

Frank, Steven A. 2014. "Somatic Mosaicism and Disease." *Current Biology* 24:R577–81.

Franklin, Sarah. 2013. *Biological Relatives IVF, Stem Cells, and the Future of Kinship.* Durham and London: Duke University Press.

Freed, Donald, Eric L. Stevens, and Jonathan Pevsner. 2014. "Somatic Mosaicism in the Human Genome." *Genes* 5:1064–94.

Frederickson, George. 2002. *Racism: A Short History.* Princeton, NJ: Princeton University Press.

Friedrich, Otto. 2014. *City of Nets: A Portrait of Hollywood in the 1940s.* New York: Harper Perennial.

Friese, Kurt Michael. 2010. *A Cook's Journey: Slow Food in the Heartland.* Ice Cube Books.

Fu, Qiaomei, Alissa Mittnik, Philip L. F. Johnson, Kirsten Bos, Martina Lari, Ruth Bollongino, Chengkai Sun, and others. 2013. "A Revised Timescale for Human Evolution Based on Ancient Mitochondrial Genomes." *Current Biology* 23:553–59.

Fu, Qiaomei, Cosimo Posth, Mateja Hajdinjak, Martin Petr, Swapan Mallick, Daniel Fernandes, Anja Furtwängler, and others. 2016. "The Genetic History of Ice Age Europe." *Nature* 534:200–05.

Funkhouser, Lisa J., and Seth R. Bordenstein. 2013. "Mom Knows Best: The Universality of Maternal Microbial Transmission." *PLOS Biology* 11:e1001631.

Gajecka, Marzena. 2016. "Unrevealed Mosaicism in the Next-Generation Sequencing Era." *Molecular Genetics and Genomics* 291:513–30.

Galant, Debra. 1998. "The Egg Men." *New York Times*, March 1.

Gallagher, Andrew. 2013. "Stature, Body Mass, and Brain Size: A Two-Million-Year Odyssey." *Economics and Human Biology* 11:551–62.

Galton, Francis. 1865. "Hereditary Talent and Character." *Macmillan's Magazine* 12:157–66.

———. 1869. *Hereditary Genius: An Inquiry into Its Laws and Consequences.* London: Macmillan.

———. 1870. "Experiments in Pangenesis, by Breeding from Rabbits of a Pure Variety, into Whose Circulation Blood Taken from Other Varieties Had Previously Been Largely Transfused." *Proceedings of the Royal Society of London* 19:393–410.

———. 1883. *Inquiries into Human Faculty and Its Development.* London: Macmillan.

———. 1889. *Natural Inheritance.* London: Macmillan.

———. 1909. *Memories of My Life.* London: Methuen.

Galupa, Rafael, and Edith Heard. 2015. "X-Chromosome Inactivation: New Insights into Cis and Trans Regulation." *Current Opinion in Genetics & Development* 31:57–66.

Gannett, Lisa. 2013. "Theodosius Dobzhansky and the Genetic Race Concept." *Studies in History and Philosophy of Biological and Biomedical Sciences* 44:250–61.

Gantz, Valentino M., and Ethan Bier. 2015. "The Mutagenic Chain Reaction: A Method for Converting Heterozygous to Homozygous Mutations." *Science* 348:442–44.

Gantz, Valentino M., Nijole Jasinskiene, Olga Tatarenkova, Aniko Fazekas, Vanessa M. Macias, Ethan Bier, and Anthony A. James. 2015. "Highly Efficient Cas9-mediated Gene Drive for Population Modification of the Malaria Vector Mosquito Anopheles stephensi." *Proceedings of the National Academy of Sciences of the United States of America* 112:E6736–E6743.

Gapp, Katharina, Ali Jawaid, Peter Sarkies, Johannes Bohacek, Pawel Pelczar, Julien Prados, Laurent Farinelli, Eric Miska, and Isabelle M. Mansuy. 2014. "Implication of Sperm RNAs in Transgenerational Inheritance of the Effects of Early Trauma in Mice." *Nature Reviews Neuroscience* 17:667–69.

Gapp, Katharina, Saray Soldado-Magraner, María Alvarez-Sánchez, Johannes Bohacek, Gregoire Vernaz, Huan Shu, Tamara B. Franklin, David Wolfer, and Isabelle M. Mansuy. 2014. "Early Life Stress in Fathers Improves Behavioural Flexibility in Their Offspring." *Nature Communications* 5:5466.

Gärdenfors, Peter, and Anders Högberg. 2017. "The Archaeology of Teaching and the Evolution of *Homo docens.*" *Current Anthropology* 58:188–208.

Garrett, Henry E. 1955. *General Psychology.* New York: American Book Co.

———. 1961. "The Equalitarian Dogma." *Perspectives in Biology and Medicine* 4:480–84.

Gartler, Stanley M. 2015. "Mary Lyon's X-Inactivation Studies in the Mouse Laid the Foundation for the Field of Mammalian Dosage Compensation." *Journal of Genetics* 94:563–65.

Gartler, Stanley M., Sorrell H. Waxman, and Eloise Giblett. 1962. "An XX/XY Human Hermaphrodite Resulting from Double Fertilization." *Proceedings of the National Academy of Sciences of the United States of America* 48:332–35.

Gatewood, Willard B. 1990. *Aristocrats of Color: The Black Elite, 1880–1920*. Bloomington: Indiana University Press.

Geison, G. L. 1969. "Darwin and Heredity: The Evolution of His Hypothesis of Pangenesis." *Journal of the History of Medicine and Allied Sciences* 24:375–411.

Gelman, Susan A. 2003. *The Essential Child: Origins of Essentialism in Everyday Thought*. Oxford: Oxford University Press.

Genetics and Medicine Historical Network. "Interview with Dr. Mary Lyon." Interviewed by Peter Harper. Recorded October 11, 2004. https://genmedhist.eshg.org/fileadmin/content /website-layout/interviewees-attachments/Lyon%2C%20Mary.pdf (accessed August 24, 2017).

Génin, Emmanuelle, and Françoise Clerget-Darpoux. 2015. "The Missing Heritability Paradigm: A Dramatic Resurgence of the GIGO Syndrome in Genetics." *Human Heredity* 79:1–4.

Gershenson, S. 1928. "A New Sex-Ratio Abnormality in *Drosophila obscura*." *Genetics* 13:488–507.

Geserick, Gunther, and Ingo Wirth. 2012. "Genetic Kinship Investigation from Blood Groups to DNA Markers." *Transfusion Medicine and Hemotherapy* 39:163–75.

Gibbons, Ann. 2006. *The First Human: The Race to Discover Our Earliest Ancestors*. New York: Doubleday.

Gibbs, W. Wayt. 2014. "Biomarkers and Ageing: The Clock-Watcher." *Nature* 508:168.

Giese, Lucretia Hoover. 2001. "A Rare Crossing: Frida Kahlo and Luther Burbank." *American Art* 15:52–73.

Gilbert, Scott F. 2014. "A Holobiont Birth Narrative: The Epigenetic Transmission of the Human Microbiome." *Frontiers in Genetics* 5:282.

Gill, Peter, Pavel L. Ivanov, Colin Kimpton, Romelle Piercy, Nicola Benson, Gillian Tully, Ian Evett, Erika Hagelberg, and Kevin Sullivan. 1994. "Identification of the Remains of the Romanov Family by DNA Analysis." *Nature Genetics* 6:130–35.

Gillham, Nicholas W. 2001. *A Life of Sir Francis Galton: From African Exploration to the Birth of Eugenics*. New York: Oxford University Press.

Gitschier, Jane. 2010. "The Gift of Observation: An Interview with Mary Lyon." *PLOS Genetics* 6:e1000813.

Glass, Bentley. 1980. "The Strange Encounter of Luther Burbank and George Harrison Shull." *Proceedings of the American Philosophical Society* 124:133–53.

Gliboff, Sander. 2013. "The Many Sides of Gregor Mendel." In *Outsider Scientists: Routes to Innovation in Biology*. Edited by Oren Harman and Michael R. Dietrich. Chicago: University of Chicago Press.

Goddard, Henry H. 1908. "A Group of Feeble-Minded Children with Special Regard to Their Number Concepts." *Supplement to the Training School* 2:1–16.

———. 1910a. "Heredity of Feeble-Mindedness." *American Breeders Magazine* 1:165–78.

———. 1910b. "The Institution for Mentally Defective Children: An Unusual Opportunity for Scientific Research." *Training School* 7:275–78.

———. 1910c. "A Measuring Scale for Intelligence." *Training School* 6:146–55.

———. 1911a. "The Elimination of Feeble-Mindedness." *American Academy of Political and Social Science* 37:261–72.

———. 1911b. "A Revision of the Binet Scale." *Training School* 8:56–62.

———. 1911c. "Two Thousand Normal Children Tested by the Binet Scale." *Training School* 7: 310–12.

———. 1912. *The Kallikak Family. A Study in the Heredity of Feeble-Mindedness*. New York: Macmillan.

———. 1914. *Feeble-Mindedness: Its Causes and Consequences*. New York: Macmillan.

———. 1916. "The Menace of Mental Deficiency from the Standpoint of Heredity." *Boston Medical and Surgical Journal* 175:269–71.

———. 1917. "Mental Tests and the Immigrant." *Journal of Delinquency* 2:243–77.

———. 1920. *Human Efficiency and Levels of Intelligence: Lectures Delivered at Princeton University April 7, 8, 10, 11, 1919.* Princeton: Princeton University Press.

———. 1931. "Anniversary Address." In *Twenty-Five Years: The Vineland Laboratory 1906-1931.* Edited by Edgard A. Doll. Vineland: Smith Printing House.

———. 1942. "In Defense of the Kallikak Study." *Science* 95:574–76.

Goldstein, Sam, Dana Princiotta, and Jack A. Naglieri, eds. 2015. *Handbook of Intelligence: Evolutionary Theory, Historical Perspective, and Current Concepts.* New York: Springer.

Goodell, Margaret A., Hoang Nguyen, and Noah Shroyer. 2015. "Somatic Stem Cell Heterogeneity: Diversity in the Blood, Skin and Intestinal Stem Cell Compartments." *Nature Reviews Molecular Cell Biology* 16:5299–5330.

Goodrich, Julia K., Emily R. Davenport, Michelle Beaumont, Matthew A. Jackson, Rob Knight, Carole Ober, Tim D. Spector, Jordana T. Bell, Andrew G. Clark, and Ruth E. Ley. 2016. "Genetic Determinants of the Gut Microbiome in UK Twins." *Cell Host & Microbe* 19:731–43.

———, Jillian L. Waters, Angela C. Poole, Jessica L. Sutter, Omry Koren, Ran Blekhman, Michelle Beaumont, and others. 2014. "Human Genetics Shape the Gut Microbiome." *Cell* 159:789–99.

Goodspeed, Weston Arthur. 1907. *History of the Goodspeed Family, Profusely Illustrated: Being a Genealogical and Narrative Record Extending from 1380 to 1906, and Embracing Material Concerning the Family Collected during Eighteen Years of Research, Together with Maps, Plats, Charts, Etc.* Chicago: W. A. Goodspeed.

Goolam, Mubeen. 2016. "Heterogeneity in Oct4 and Sox2 Targets Biases Cell Fate in 4-Cell Mouse Embryos." *Cell* 165:61–74.

Gosney, E. S., and Paul Popenoe. 1929. *Sterilization for Human Betterment: A Summary of Results of 6,000 Operations in California, 1909-1929.* New York: Macmillan.

Grant, Madison. 1916. *The Passing of the Great Race: Or, the Racial Basis of European History.* New York: Charles Scribner's Sons.

Grasgruber, P., J. Cacek, T. Kalina, and M. Sebera. 2014. "The Role of Nutrition and Genetics as Key Determinants of the Positive Height Trend." *Economics and Human Biology* 15:81–100.

Gray, Michael W. 2012. "Mitochondrial Evolution." *Cold Spring Harbor Perspectives in Biology* 4:a011403.

Greally, John. 2015. "Over-Interpreted Epigenetics Study of the Week." "Epgntxeinstein" blog, August 23. http://epgntxeinstein.tumblr.com/post/127416455028/over-interpreted -epigenetics-study-of-the-week (accessed July 26, 2017).

Greely, Henry T. 2016. *The End of Sex and the Future of Human Reproduction.* Cambridge: Harvard University Press.

Greenfieldboyce, Nell. 2017. "Fate of Irish Giant's Bones Rekindles Debate over Rights After Death." NPR *All Things Considered*, March 13.

Griesemer, James R. 2005. "The Informational Gene and the Substantial Body: On the Generalization of Evolutionary Theory by Abstraction." In *Idealization XII: Correcting the Model. Idealization and Abstraction in the Sciences.* Edited by Martin R. Jones and Nancy Cartwright. Amsterdam: Rodopi.

Griffith, Malachi, Christopher A. Miller, Obi L. Griffith, Kilannin Krysiak, Zachary L. Skidmore, Avinash Ramu, Jason R. Walker, and others. 2015. "Optimizing Cancer Genome Sequencing and Analysis." *Cell Systems* 1:210–23.

Grognet, Pierre, Hervé Lalucque, Fabienne Malagnac, and Philippe Silar. 2014. "Genes That Bias Mendelian Segregation." *PLOS Genetics* 10:e1004387.

Grudnik, Jennifer L., and John H. Kranzler. 2001. "Meta-Analysis of the Relationship Between Intelligence and Inspection Time." *Intelligence* 29:523–35.

Grüneberg, Hans. 1967. "Sex-linked Genes in Man and the Lyon Hypothesis." *Annals of Human Genetics* 30:239–57.

Guevara-Aguirre, Jaime, Priya Balasubramanian, Marco Guevara-Aguirre, Min Wei, Federica Madia, Chia-Wei Cheng, David Hwang, and others. 2011. "Growth Hormone Receptor Deficiency Is Associated with a Major Reduction in Pro-Aging Signaling, Cancer, and Diabetes in Humans." *Science Translational Medicine* 3:70ra13.

Gull, Keith. 2010. "Boveri and Cancer: Prescient Views of Molecular Mechanisms." *Notes and Records of the Royal Society* 64:185–87.

Guo, Fan, Liying Yan, Hongshan Guo, Lin Li, Boqiang Hu, Yangyu Zhao, Jun Yong, and others. 2015. "The Transcriptome and DNA Methylome Landscapes of Human Primordial Germ Cells." *Cell* 161:1437–52.

Gustafsson, Å. 1979. "Linnaeus' *Peloria*: The History of a Monster." *Theoretical and Applied Genetics* 54:241–48.

Guyot, A., and C. C. Felton. 1852. *The Earth and Man: Or, Physical Geography in Its Relation to the History of Mankind*. London: J. W. Parker and Son.

Hadhazy, Adam. 2015. "Will Humans Keep Getting Taller?" BBC *Future*, May 14.

Haggard, Howard Wilcox, and E. M. Jellinek. 1942. *Alcohol Explored*. Garden City, NY: Doubleday, Doran & Company.

Haidle, Miriam N., Nicholas J. Conard, and Michael Bolus, eds. 2016. *The Nature of Culture: Based on an Interdisciplinary Symposium "The Nature of Culture," Tübingen, Germany*. Dordecht: Springer.

Haier, Richard J. 2017. *The Neuroscience of Intelligence*. New York: Cambridge University Press.

Haig, David. 2016. "Intracellular Evolution of Mitochondrial DNA (mtDNA) and the Tragedy of the Cytoplasmic Commons." *BioEssays* 38. doi:10.1002/bies.201600003.

Hains, James H., Gary R. Carter, Scott D. Kraus, Charles A. Mayo, and Howard E. Winn. 1982. "Feeding Behavior of the Humpback Whale, *Megaptera novaeangliae*, in the Western North Atlantic." *Fishery Bulletin* 80:259–68.

Haley, Alex. 1972. "My Furthest-Back Person—The African." *New York Times Magazine*, July 16.

Hall, Stephen S. 2006. *Size Matters: How Height Affects the Health, Happiness, and Success of Boys—and the Men They Become*. Boston: Houghton Mifflin.

Hamilton, Garry. 2015. "The Hidden Risks for 'Three-Person' Babies." *Nature* 525:444.

Hammer, Michael F., Doron M. Behar, Tatiana M. Karafet, Fernando L. Mendez, Brian Hallmark, Tamar Erez, Lev A. Zhivotovsky, Saharon Rosset, and Karl Skorecki. 2009. "Extended Y Chromosome Haplotypes Resolve Multiple and Unique Lineages of the Jewish Priesthood." *Human Genetics* 126:707–17.

———, Karl Skorecki, Sara Selig, Shraga Blazer, Bruce Rappaport, Robert Bradman, Neil Bradman, P. J. Waburton, and Monic Ismajlowicz. 1997. "Y Chromosomes of Jewish Priests." *Nature* 385:3.

Hamzelou, Jessica. 2016. "World's First Baby Born with New '3 Parent' Technique." *New Scientist*, September 27.

Handyside, A. H., E. H. Kontogianni, K. Hardy, and R. M. L. Winston. 1990. "Pregnancies from Biopsied Human Preimplantation Embryos Sexed by Y-Specific DNA Amplification." *Nature* 344:768–70.

Haneda, Yata, and Frederick I. Tsuji. 1971. "Light Production in the Luminous Fishes Photoblepharon and Anomalops from the Banda Islands." *Science* 173:143–45.

Happle, Rudolf. 2002. "New Aspects of Cutaneous Mosaicism." *Journal of Dermatology* 29:681–92.

Harding, Cary O. 2017. "Gene and Cell Therapy for Inborn Errors of Metabolism." In *Inherited Metabolic Diseases*. Edited by G. Hoffmann. Berlin: Springer-Verlag.

Harper, Peter S. 1992. "Eugenics, Human Genetics and Human Failings: The Eugenics Society, Its Sources and Its Critics in Britain." *Journal of Medical Genetics* 29:440.

———. 2008. *A Short History of Medical Genetics*. Oxford: Oxford University Press.

———. 2011. "Mary Lyon and the Hypothesis of Random X Chromosome Inactivation." *Human Genetics* 130:169–74.

Harris, Harry. 1974. "Lionel Sharples Penrose (1898–1972)." *Journal of Medical Genetics* 11:1–24.

Harris, Henry. 1999. *The Birth of the Cell*. New Haven: Yale University Press.

Harris, Kelley, and Rasmus Nielsen. 2016. "The Genetic Cost of Neanderthal Introgression." *Genetics* 203:881–91.

Harris, Theodore F. 1969. *Pearl S. Buck: A Biograpahy*. New York: John Day.

Harrison, Ross G. 1937. "Embryology and Its Relations." *Science* 85:369–74.

Hart, Sara A. 2016. "Precision Education Initiative: Moving Toward Personalized Education." *Mind, Brain, and Education* 10:209–11.

Hatfield, A. 2015. "Delineating Cancer Evolution with Single-Cell Sequencing." *Science Translational Medicine* 7:296fs29.

Hatton, T. J. 2014. "How Have Europeans Grown So Tall?" *Oxford Economic Papers* 66:349–72.

Haworth, C. M. A., M. J. Wright, M. Luciano, N. G. Martin, E. J. C. de Geus, C. E. M. van Beijsterveldt, M. Bartels, and others. 2010. "The Heritability of General Cognitive Ability Increases Linearly from Childhood to Young Adulthood." *Molecular Psychiatry* 15:1112–20.

Hayashi, Katsuhiko, Sugako Ogushi, Kazuki Kurimoto, So Shimamoto, Hiroshi Ohta, and Mitinori Saitou. 2012. "Offspring from Oocytes Derived from in Vitro Primordial Germ Cell-like Cells in Mice." *Science* 338:971–97.

Haygood, M. G., B. M. Tebo, and K. H. Nealson. 1984. "Luminous Bacteria of a Monocentrid Fish (*Monocentris japonicus*) and Two Anomalopid Fishes (*Photoblepharon palpebratus* and *Kryptophanaron alfredi*): Population Sizes and Growth within the Light Organs, and Rates of Release into the Seawater." *Marine Biology* 78:249–54.

Hayman, John, Gonzalo Álvarez, Francisco C. Ceballos, and Tim M. Berra. 2017. "The Illnesses of Charles Darwin and His Children: A Lesson in Consanguinity." *Biological Journal of the Linnean Society* 121, no. 2. doi:10.1093/biolinnean/blw041.

Haynes, Stephen R. 2007. *Noah's Curse: The Biblical Justification of American Slavery.* Oxford: Oxford University Press.

Heard, Edith, and Robert A. Martienssen. 2014. "Transgenerational Epigenetic Inheritance: Myths and Mechanisms." *Cell* 157:95–109.

Hearnshaw, L. S. 1979. *Cyril Burt, Psychologist.* Ithaca, NY: Cornell University Press.

Heim, Sverre. 2014. "Boveri at 100: Boveri, Chromosomes and Cancer." *Journal of Pathology* 234:138–41.

Hellenthal, Garrett, George B. J. Busby, Gavin Band, James F. Wilson, Cristian Capelli, Daniel Falush, and Simon Myers. 2014. "A Genetic Atlas of Human Admixture History." *Science* 343:747–51.

Hendriks, Saskia, Eline A. F. Dancet, Ans M. M. van Pelt, Geert Hamer, and Sjoerd Repping. 2015. "Artificial Gametes: A Systematic Review of Biological Progress towards Clinical Application." *Human Reproduction Update* 21:285–96.

Hendry, Tory A., Jeffrey R. de Wet, and Paul V. Dunlap. 2014. "Genomic Signatures of Obligate Host Dependence in the Luminous Bacterial Symbiont of a Vertebrate." *Environmental Microbiology* 16:2611–22.

———, Jeffrey R. de Wet, Katherine E. Dougan, and Paul V. Dunlap. 2016. "Genome Evolution in the Obligate but Environmentally Active Luminous Symbionts of Flashlight Fish." *Genome Biology and Evolution* 8:2203–13.

Henig, Robin Marantz. 2004. *Pandora's Baby: How the First Test Tube Babies Sparked the Reproductive Revolution.* Boston: Houghton Mifflin.

Henikoff, Steven, and John M. Greally. "Epigenetics, Cellular Memory and Gene Regulation." *Current Biology* 26:R644–R648.

Henn, Brenna M., Laura R. Botigué, Carlos D. Bustamante, Andrew G. Clark, and Simon Gravel. 2015. "Estimating the Mutation Load in Human Genomes." *Nature Reviews Genetics* 16:333–43.

Henrich, Joseph. 2016. *The Secret of Our Success: How Culture Is Driving Human Evolution, Domesticating Our Species, and Making Us Smarter.* Princeton: Princeton University Press.

Herzenberg, L. A., D. W. Bianchi, J. Schröder, H. M. Cann, and G. M. Iverson. 1979. "Fetal Cells in the Blood of Pregnant Women: Detection and Enrichment by Fluorescence-Activated Cell Sorting." *Proceedings of the National Academy of Sciences of the United States of America* 76:1453–55.

Hewlett, Barry S., and Casey J. Roulette. 2016. "Teaching in Hunter-Gatherer Infancy." *Royal Society Open Science* 3:150403.

Heyes, Cecilia. 2016. "Born Pupils? Natural Pedagogy and Cultural Pedagogy." *Perspectives on Psychological Science* 11:280–95.

Hikabe, Orie, Nobuhiko Hamazaki, Go Nagamatsu, Yayoi Obata, Yuji Hirao, Norio Hamada, So Shimamoto, and others. 2016. "Reconstitution in Vitro of the Entire Cycle of the Mouse Female Germ Line." *Nature* 539:299–303.

Hill, Helen F., and Henry H. Goddard. 1911. "Delinquent Girls Tested by the Binet Scale." *Training School* 8:50–55.

Hill, Kim R., Brian M. Wood, Jacopo Baggio, A. Magdalena Hurtado, and Robert T. Boyd. 2014. "Hunter-Gatherer Inter-Band Interaction Rates: Implications for Cumulative Culture." *PLOS One* 9:e102806.

Hippocrates. *On Airs, Waters, and Places.* Translated by Francis Adams. MIT Internet Classics Archive. http://classics.mit.edu/Hippocrates/airwatpl.html (accessed August 29, 2016).

Hirschfeld, Ludwik, and Hanka Hirschfeld. 1919. "Serological Differences Between the Blood of Different Races: The Result of Researches on the Macedonian Front." *Lancet* 194:675–79.

Hirschhorn, Joel N., Cecilia M. Lindgren, Mark J. Daly, Andrew Kirby, Stephen F. Schaffner, Noel P. Burtt, David Altshuler, Alex Parker, John D. Rioux, and Jill Platko. 2001. "Genomewide Linkage Analysis of Stature in Multiple Populations Reveals Several Regions with Evidence of Linkage to Adult Height." *American Journal of Human Genetics* 69:106–16.

Hirschhorn, Kurt, Wayne H. Decker, and Herbert L. Cooper. 1960. "Human Intersex with Chromosome Mosaicism of Type XY/XO: Report of a Case." *New England Journal of Medicine* 263:1044–48.

Hirszfeld, Ludwik, and Hanna Hirszfeldowa. 1918. "Essai D'Application des Méthodes Sérologiques au Problème des Races." *L'Anthropologie* 29:505–37.

Hjelm, Brooke E., Brandi Rollins, Firoza Mamdani, Julie C. Lauterborn, George Kirov, Gary Lynch, Christine M. Gall, Adolfo Sequeira, and Marquis P. Vawter. 2015. "Evidence of Mitochondrial Dysfunction within the Complex Genetic Etiology of Schizophrenia." *Molecular Neuropsychiatry* 1:201–19.

Hodge, Gerald P. 1977. "A Medical History of the Spanish Habsburgs: As Traced in Portraits." *JAMA* 238:1169–74.

Hodge, Kathie T., with Bradford Condon. 2010. "A Fungus Walks into a Singles Bar." "Cornell Mushroom Blog," June 2. https://blog.mycology.cornell.edu/2010/06/02/a-fungus-walks -into-a-singles-bar/ (accessed July 25, 2017).

Holt, I. J., A. E. Harding, and J. A. Morgan-Hughes. 1988. "Deletions of Muscle Mitochondrial DNA in Patients with Mitochondrial Myopathies." *Nature* 331:717–19.

Hooper, Paul L., Michael Gurven, Jeffrey Winking, and Hillard S. Kaplan. 2015. "Inclusive Fitness and Differential Productivity across the Life Course Determine Intergenerational Transfers in a Small-Scale Human Society." *Proceedings of the Royal Society B* 282:20142808.

Horvath, Steve. 2013. "DNA Methylation Age of Human Tissues and Cell Types." *Genome Biology* 14:R115.

———, and Andrew J. Levine. 2015. "HIV-1 Infection Accelerates Age according to the Epigenetic Clock." *Journal of Infectious Diseases.* doi:10.1093/infdis/jiv277.

Hotchkiss, R. D. 1965. "Portents for a Genetic Engineering." *Journal of Heredity* 56:197–202.

Howell, Michael, and Peter Ford. 2010. *The True History of the Elephant Man: The Definitive Account of the Tragic and Extraordinary Life of Joseph Carey Merrick.* New York: Skyhorse Publishing.

"How One Sin Perpetuates Itself." 1916. *Evening Star,* March 12.

Hsu, Stephen. 2016. "Super-Intelligent Humans Are Coming." Nautilus, March 3. http://nautil .us/issue/34/adaptation/super_intelligent-humans-are-coming-rp (accessed September 11, 2017).

Huang, Lam Opal, Aurélie Labbe, and Claire Infante-Rivard. 2013. "Transmission Ratio Distortion: Review of Concept and Implications for Genetic Association Studies." *Human Genetics* 132:245–63.

Hubby, J. L., and R. C. Lewontin. 1966. "A Molecular Approach to the Study of Genic Heterozygosity in Natural Populations. I. The Number of Alleles at Different Loci in *Drosophila pseudoobscura.*" *Genetics* 54:577–94.

Hübler, Olaf. 2016. "Height and Wages." In *The Oxford Handbook of Economics and Human Biology.* Edited by John Komlos and Inas R. Kelly and Oxford: Oxford University Press.

Huerta-Sánchez, Emilia, and Fergal P. Casey. 2015. "Archaic Inheritance: Supporting High-Altitude Life in Tibet." *Journal of Applied Physiology* 119:1129–34.

———, Xin Jin, Asan, Zhuoma Bianba, Benjamin M. Peter, Nicolas Vinckenbosch, Yu Liang, and others. 2014. "Altitude Adaptation in Tibetans Caused by Introgression of Denisovan-Like DNA." *Nature* 512:194–97.

Hughes, Langston. 1940. *The Big Sea: An Autobiography.* New York: Knopf.

Hughes, Virginia. 2014. "Epigenetics: The Sins of the Father." *Nature* 507:22.

Hunley, Keith L., Graciela S. Cabana, and Jeffrey C. Long. 2016. "The Apportionment of Human Diversity Revisited." *American Journal of Physical Anthropology* 160. doi:10.1002 /ajpa.22899.

Hunter, John. 1779. "Account of the Free Martin. By Mr. John Hunter, FRS." *Philosophical Transactions of the Royal Society of London* 69:279–93.

Hunter, Marjorie. 1961. "President Lauds Two Poster Girls." *New York Times*, November 14.

Hunter, Neil. 2015. "Meiotic Recombination: The Essence of Heredity." *Cold Spring Harbor Perspectives in Biology* 7:a016618.

Hurst, Gregory D. 2017. "Extended Genomes: Symbiosis and Evolution." *Interface Focus* 7:5, 20170001.

Hurst, Laurence D. 1993. "Evolutionary Genetics: Drunken Walk of the Diploid." *Nature* 365:206–07.

"I.Q. Control." 1938. *Time*, November 7.

Imamura, Masanori, Orie Hikabe, Zachary Yu-Ching Lin, and Hideyuki Okano. 2014. "Generation of Germ Cells in Vitro in the Era of Induced Pluripotent Stem Cells." *Molecular Reproduction and Development* 81:2–19.

Isaacs, Alison T., Fengwu Li, Nijole Jasinskiene, Xiaoguang Chen, Xavier Nirmala, Osvaldo Marinotti, Joseph M. Vinetz, and Anthony A. James. 2011. "Engineered Resistance to *Plasmodium falciparum* Development in Transgenic *Anopheles stephensi*." *PLOS Pathology* 7:e1002017.

Isoda, Takeshi, Anthony M. Ford, Daisuke Tomizawa, Frederik W. van Delft, David Gonzalez de Castro, Norkio Mitsuiki, Joannah Score, and others. 2009. "Immunologically Silent Cancer Clone Transmission from Mother to Offspring." *Proceedings of the National Academy of Sciences of the United States of America* 106:17882–85.

Jablonski, Nina G., and George Chaplin. 2017. "The Colours of Humanity: The Evolution of Pigmentation in the Human Lineage." *Philosophical Transactions of the Royal Society B* 372:1724.

Jackson, John P. 2005. *Science for Segregation: Race, Law and the Case Against* Brown v. Board of Education. New York: New York University Press.

Jacob, François. 1993. *The Logic of Life: A History of Heredity.* Princeton: Princeton University Press.

James, William. 1868. "Review of 'Variation of Animals and Plants under Domestication.'" *North American Review* 107:362–68.

Janick, Jules. 2015. "Luther Burbank: Plant Breeding Artist, Horticulturist, and Legend." *HortScience* 50:153–56.

Janssens, F. A. 2012. "The Chiasmatype Theory: A New Interpretation of the Maturation Divisions, 1909." *Genetics* 191:319–46.

Jeanty, Cerine, S. Christopher Derderian, and Tippi C. Mackenzie. 2014. "Maternal-Fetal Cellular Trafficking: Clinical Implications and Consequences." *Current Opinion in Pediatrics* 26:377–82.

Jégu, Teddy, Eric Aeby, and Jeannie T. Lee. 2017. "The X Chromosome in Space." *Nature Reviews Genetics* 18. doi:10.1038/nrg.2017.17.

Jensen, Arthur R. 1967. "How Much Can We Boost IQ and Scholastic Achievement?" Speech given before the California Advisory Council of Educational Research. San Diego, California.

Johnson, Christopher H. 2013. *Blood & Kinship: Matter for Metaphor from Ancient Rome to the Present.* New York: Berghahn Books.

Johnson, Ronald C., Gerald E. McClearn, Sylvia Yuen, Craig T. Nagoshi, Frank M. Ahern, and Robert E. Cole. 1985. "Galton's Data a Century Later." *American Psychologist* 40:875.

Johnson, Scott C. 2014. "The New Theory That Could Explain Crime and Violence in America." *Matter,* February 17. https://medium.com/matter/the-new-theory-that-could-explain -crime-and-violence-in-america-945462826399#.4jhplsza3 (accessed July 27, 2017).

Johnston, Ian. 2016. "Scientists Break 13-Year Silence to Insist 'Three-Parent Baby' Technique Is Safe." *Independent*, August 11.

Jonkman, Marcel F., Hans Scheffer, Rein Stulp, Hendri H. Pas, Miranda Nijenhuis, Klaas Heeres, Katsushi Owaribe, Leena Pulkkinen, and Jouni Uitto. 1997. "Revertant Mosaicism in Epidermolysis Bullosa Caused by Mitotic Gene Conversion." *Cell* 88:543–51.

Jordan, David Starr, and Vernon Lyman Kellogg. 1909. *The Scientific Aspects of Luther Burbank's Work.* New York: A. M. Robertson.

Jordan, Harvey Ernest. 1913. "The Biological Status and Social Worth of the Mulatto." *Popular Science Monthly*, June.

Jordan, John W., Eyre Whalley, Mary Fisher, Richd. Quinton, Thos. Holme, B. F., Anne Farrow, Hannah Walker, M. Foulger, and Mary Folger, and P. F. 1899. "Franklin as a Genealogist." *Pennsylvania Magazine of History and Biography* 23:1–22.

Jordan, Winthrop D. 2014. "Historical Origins of the One-Drop Racial Rule in the United States." *Journal of Critical Mixed Race Studies* 1:98–132.

Joyce, Gerald F. 2012. "Bit by Bit: The Darwinian Basis of Life." *PLOS Biology* 10:e1001323.

Ju, Young Seok, Inigo Martincorena, Moritz Gerstung, Mia Petljak, Ludmil B. Alexandrov, Raheleh Rahbari, David C. Wedge, and others. 2017. "Somatic Mutations Reveal Asymmetric Cellular Dynamics in the Early Human Embryo." *Nature* 543:714–18.

Juengst, Eric T., Jennifer R. Fishman, Michelle L. McGowan, and Richard A. Settersten. 2014. "Serving Epigenetics Before Its Time." *Trends in Genetics* 30:427–29.

Juric, Ivan, Simon Aeschbacher, and Graham Coop. 2016. "The Strength of Selection Against Neanderthal Introgression." *PLOS Genetics* 12:e1006340.

Kakourou, Georgia, Christina Vrettou, Maria Moutafi, and Joanne Traeger-Synodinos. 2017. "Pre-Implantation HLA Matching: The Production of a Saviour Child." *Best Practice & Research: Clinical Obstetrics & Gynaecology.* doi:10.1016/j.bpobgyn.2017.05.008.

Kalantry, Sundeep, and Jacob L. Mueller. 2015. "Mary Lyon: A Tribute." *American Journal of Human Genetics* 97:507–11.

Kamin, Leon J., and Arthur S. Goldberger. 2002. "Twin Studies in Behavioral Research: A Skeptical View." *Theoretical Population Biology* 61:83–95.

Karp, Robert J., Qutub H. Qazi, Karen A. Moller, Wendy A. Angelo, and Jeffrey M. Davis. 1995. "Fetal Alcohol Syndrome at the Turn of the 20th Century: An Unexpected Explanation of the Kallikak Family." *Archives of Pediatrics & Adolescent Medicine* 149:45–48.

Kaufman, Alan S., Xiaobin Zhou, Matthew R. Reynolds, Nadeen L. Kaufman, Garo P. Green, and Lawrence G. Weiss. 2014. "The Possible Societal Impact of the Decrease in U.S. Blood Lead Levels on Adult IQ." *Environmental Research* 132:413–20.

Kaufman, Jay S. 2014. "Commentary: Race: Ritual, Regression, and Reality." *Epidemiology* 25:485–87.

Kaufman, Seymour. 2004. *Overcoming a Bad Gene: The Story of the Discovery and Successful Treatment of Phenylketonuria, a Genetic Disease That Causes Mental Retardation.* Bloomington, IN: AuthorHouse.

Keevak, Michael. 2001. *Becoming Yellow: A Short History of Racial Thinking.* Princeton: Princeton University Press.

Keith, Arthur. 1911. "An Inquiry into the Nature of the Skeletal Changes in Acromegaly." *Lancet* 177:993–1002.

Kelleher, Erin S. 2016. "Reexamining the P-Element Invasion of *Drosophila melanogaster* Through the Lens of PiRNA Silencing." *Genetics* 203:1513–31.

Kelsoe, John R., Edward I. Ginns, Janice A. Egeland, Daniela S. Gerhard, Alisa M. Goldstein, Sherri J. Bale, David L. Pauls, Robert T. Long, Kenneth K. Kidd, Giovanni Conte, and others. 1989. "Re-evaluation of the Linkage Relationship between Chromosome 11p Loci and the Gene for Bipolar Affective Disorder in the Old Order Amish." *Nature* 342:248.

Kendi, Ibram X. 2016. *Stamped from the Beginning: The Definitive History of Racist Ideas in America.* New York: Nation Books.

Kevles, Daniel J. 1995. *In the Name of Eugenics: Genetics and the Uses of Human Heredity.* Cambridge: Harvard University Press.

Khosrotehrani, Kiarash, and Diana W. Bianchi. 2005. "Multi-Lineage Potential of Fetal Cells in Maternal Tissue: A Legacy in Reverse." *Journal of Cell Science* 118:1559–63.

Khush, Gurdev S. 1995. "Breaking the Yield Frontier of Rice." *GeoJournal* 35:329–32.

Kim, Somi, and Bong-Kiun Kaang. 2017. "Epigenetic Regulation and Chromatin Remodeling in Learning and Memory." *Experimental & Molecular Medicine* 49:e281.

Kingsbury, Noel. 2011. *Hybrid: The History and Science of Plant Breeding.* Chicago: University of Chicago Press.

Kingsford, Charles Lethbridge. 1905. *Chronicles of London.* London: Clarendon Press.

Klapisch-Zuber, Christiane. 1991. "The Genesis of the Family Tree." *I Tatti Studies in the Italian Renaissance* 4:105–29.

Klein, Robert J., Caroline Zeiss, Emily Y. Chew, Jen-Yue Tsai, Richard S. Sackler, Chad Haynes, Alice K. Henning, and others. 2005. "Complement Factor H Polymorphism in Age-Related Macular Degeneration." *Science* 308:385–89.

Knight, Thomas Andrew. 1799. "An Account of Some Experiments on the Fecundation of Vegetables. In a Letter from Thomas Andrew Knight, Esq. To the Right Hon. Sir Joseph Banks, K.B.P.R.S." *Philosophical Transactions of the Royal Society of London* 89:195–204.

Knoblich, Juergen A. 2008. "Mechanisms of Asymmetric Stem Cell Division." *Cell* 132: 583–97.

Knowler, W. C., R. C. Williams, D. J. Pettitt, and A. G. Steinberg. 1988. "Gm3;5,13,14 and Type 2 Diabetes Mellitus: An Association in American Indians with Genetic Admixture." *American Journal of Human Genetics* 43:520–26.

Koltunow, Anna M. G., Susan D. Johnson, and Takashi Okada. 2011. "Apomixis in Hawkweed: Mendel's Experimental Nemesis." *Journal of Experimental Botany* 62:1699–707.

Kondrashov, Alexey S. 2017. *Crumbling Genome: The Impact of Deleterious Mutations on Humans*. Hoboken, NJ: Wiley Blackwell.

Koonin, Eugene V., and Yuri I. Wolf. 2009. "Is Evolution Darwinian or/and Lamarckian?" *Biology Direct* 4:42.

Kornrich, Sabino. 2016. "Inequalities in Parental Spending on Young Children." *AERA Open* 2:1–12.

Koszul, Romain, Matthew Meselson, Karine Van Doninck, Jean Vandenhaute, and Denise Zickler. 2012. "The Centenary of Janssens's Chiasmatype Theory." *Genetics* 191:309–17.

Kouzak, Samara Silva, Marcela Sena Teixeira Mendes, and Izelda Maria Carvalho Costa. 2013. "Cutaneous Mosaicisms: Concepts, Patterns and Classifications." *Anais Brasileiros de Dermatologia* 88:507–17.

Kretzschmar, Kai, and Fiona M. Watt. 2012. "Lineage Tracing." *Cell* 148:33–45.

Kühl, Stefan. 2002. *The Nazi Connection: Eugenics, American Racism, and German National Socialism*. New York: Oxford University Press.

Kumar, Akash, Allison Ryan, Jacob O. Kitzman, Nina Wemmer, Matthew W. Snyder, Styrmir Sigurjonsson, Choli Lee, and others. 2015. "Whole Genome Prediction for Preimplantation Genetic Diagnosis." *Genome Medicine* 7:35.

Kun, Ádám, András Szilágyi, Balázs Könnyű, Gergely Boza, István Zachar, and Eörs Szathmáry. 2015. "The Dynamics of the RNA World: Insights and Challenges." *Annals of the New York Academy of Sciences* 1341:75–95.

Kuzma, Jennifer, and Lindsey Rawls. 2016. "Engineering the Wild: Gene Drives and Intergenerational Equity." *Jurimetrics* 56:279–96.

Kwiatkowski, D. P., G. Busby, G. Band, K. Rockett, C. Spencer, Q. S. Le, M. Jallow, E. Bougama, V. Mangana, and L. Amengo-Etego. 2016. "Admixture into and Within Sub-Saharan Africa." *eLife* 5:e15266.

Lai-Cheong, Joey E., John A. McGrath, and Jouni Uitto. 2011. "Revertant Mosaicism in Skin: Natural Gene Therapy." *Trends in Molecular Medicine* 17:140–48.

Lakritz, Naomi. 1998. "What About the Babies Science Cooks Up in the Lab?" *Calgary Herald*, June 16, p. B1.

Laland, Kevin N. 2017. *Darwin's Unfinished Symphony: How Culture Made the Human Mind*. Princeton: Princeton University Press.

Lam, May P. S., and Bernard M. Y. Cheung. 2012. "The Pharmacogenetics of the Response to Warfarin in Chinese." *British Journal of Clinical Pharmacology* 73:340–47.

Lämke, Jörn, and Isabel Bäurle. 2017. "Epigenetic and Chromatin-based Mechanisms in Environmental Stress Adaptation and Stress Memory in Plants." *Genome Biology* 18:124.

Lander, E. S., and N. J. Schork. 1994. "Genetic Dissection of Complex Traits." *Science* 265:2037–48.

Landolt, A. M., and M. Zachmann. 1980. "The Irish Giant: New Observations Concerning the Nature of His Ailment." *Lancet* 315:1311–12.

Langdon-Davies, John. 1963. *Carlos: The King Who Would Not Die*. Englewood Cliffs: Prentice-Hall.

Lapaire, O., W. Holzgreve, J. C. Oosterwijk, R. Brinkhaus, and D. W. Bianchi. 2007. "Georg Schmorl on Trophoblasts in the Maternal Circulation." *Placenta* 28:1–5.

Laughlin, Harry H. 1920. "Biological Aspects of Immigration." In *Hearings Before the House Committee on Immigration and Naturalization*. Sixty-Sixth Congress, Second Session, April 16–17.

Lavery, Stuart, Dima Abdo, Mara Kotrotsou, Geoff Trew, Michalis Konstantinidis, and Dagan Wells. 2013. "Successful Live Birth Following Preimplantation Genetic Diagnosis for

Phenylketonuria in Day 3 Embryos by Specific Mutation Analysis and Elective Single Embryo Transfer." *JIMD Reports* 7:49–54.

Lawrence, Cera R. 2008. "Preformationism in the Enlightenment." *Embryo Project Encyclopedia.* Senior editor Erica O'Neil. https://embryo.asu.edu/pages/preformationism-enlightenment (accessed August 24, 2017).

Laxova, Renata. 1998. "Lionel Sharples Penrose, 1898–1972: A Personal Memoir in Celebration of the Centenary of His Birth." *Genetics* 150:1333–40.

Lazaridis, Iosif, Dani Nadel, Gary Rollefson, Deborah C. Merrett, Nadin Rohland, Swapan Mallick, Daniel Fernandes, and others. 2016. "Genomic Insights into the Origin of Farming in the Ancient Near East." *Nature* 536, no. 7617. doi:10.1038/nature19310.

——, Nick Patterson, Alissa Mittnik, Gabriel Renaud, Swapan Mallick, Karola Kirsanow, Peter H. Sudmant, and others. 2014. "Ancient Human Genomes Suggest Three Ancestral Populations for Present-Day Europeans." *Nature* 513:409–13.

Lazebnik, Y., and G. E. Parris. 2015. "Comment On: 'Guidelines for the Use of Cell Lines in Biomedical Research': Human-to-Human Cancer Transmission as a Laboratory Safety Concern." *British Journal of Cancer* 112:1976–77.

Lederer, Susan E. 2013. "Bloodlines: Blood Types, Identity, and Association in Twentieth-Century America." In *Blood Will Out: Essays on Liquid Transfers and Flows.* Edited by Janet Carsten. London: John Wiley & Sons.

Ledford, Heidi. 2016. "CRISPR: Gene Editing Is Just the Beginning." *Nature* 531:156.

Lein, Ed, and Mike Hawrylycz. 2014. "The Genetic Geography of the Brain." *Scientific American* 310:70–77.

Lenormand, Thomas, Jan Engelstädter, Susan E. Johnston, Erik Wijnker, and Christoph R. Haag. 2016. "Evolutionary Mysteries in Meiosis." *bioRxiv.* doi:10.1101/050831.

Leontiou, Chrysanthia A., Maria Gueorguiev, Jacqueline van der Spuy, Richard Quinton, Francesca Lolli, Sevda Hassan, Harvinder S. Chahal, and others. 2008. "The Role of the Aryl Hydrocarbon Receptor-Interacting Protein Gene in Familial and Sporadic Pituitary Adenomas." *Journal of Clinical Endocrinology and Metabolism* 93:2390–2401.

Leroi, Armand Marie. 2003. *Mutants: On Genetic Variety and the Human Body.* New York: Viking Penguin.

——. 2014. *The Lagoon: How Aristotle Invented Science.* New York: Viking Penguin.

Lester, Camilla H., Niels Frimodt-Møller, Thomas Lund Sørensen, Dominique L. Monnet, and Anette M. Hammerum. 2006. "In Vivo Transfer of the vanA Resistance Gene from an *Enterococcus faecium* Isolate of Animal Origin to an *E. faecium* Isolate of Human Origin in the Intestines of Human Volunteers." *Antimicrobial Agents and Chemotherapy* 50:596–99.

Lewis, Ronald L. 1974. "Slavery on Chesapeake Iron Plantations Before the American Revolution." *Journal of Negro History* 59:242–54.

Lewis, Simon L., and Mark A. Maslin. 2015. "Defining the Anthropocene." *Nature* 519:171–80.

Lewontin, Richard C. 1970. "Race and Intelligence." *Bulletin of the Atomic Scientists* 26:2–8.

——. 1972. "The Apportionment of Human Diversity." *Evolutionary Biology* 6:381–98.

Li, Mu, and Creswell J. Eastman. 2012. "The Changing Epidemiology of Iodine Deficiency." *Nature Reviews Endocrinology* 8:434–40.

Liang, Puping, Yanwen Xu, Xiya Zhang, Chenhui Ding, Rui Huang, Zhen Zhang, and others. 2015. "CRISPR/Cas9-Mediated Gene Editing in Human Tripronuclear Zygotes." *Protein & Cell* 6:363–72.

Librado, Pablo, Antoine Fages, Charleen Gaunitz, Michela Leonardi, Stefanie Wagner, Naveed Khan, Kristian Hanghøj, and others. 2016. "The Evolutionary Origin and Genetic Makeup of Domestic Horses." *Genetics* 204:423–34.

——, Cristina Gamba, Charleen Gaunitz, Clio Der Sarkissian, Mélanie Pruvost, Anders Albrechtsen, Antoine Fages, Naveed Khan, Mikkel Schubert, and Vidhya Jagannathan. 2017. "Ancient Genomic Changes Associated with Domestication of the Horse." *Science* 356:442–45.

Lifton, Robert Jay. 2000. *The Nazi Doctors: Medical Killing and the Psychology of Genocide.* New York: Basic Books.

Ligon, Azra H., Steven D. P. Moore, Melissa A. Parisi, Matthew E. Mealiffe, David J. Harris, Heather L. Ferguson, Bradley J. Quade, and Cynthia C. Morton. 2005. "Constitutional Rearrangement of the Architectural Factor HMGA2: A Novel Human Phenotype including Overgrowth and Lipomas." *American Journal of Human Genetics* 76:340–48.

Lim, Jana P., and Anne Brunet. 2013. "Bridging the Transgenerational Gap with Epigenetic Memory." *Trends in Genetics* 29:176–86.

Lindholm, Anna K., Kelly A. Dyer, Renée C. Firman, Lila Fishman, Wolfgang Forstmeier, Luke Holman, Hanna Johannesson, and others. 2016. "The Ecology and Evolutionary Dynamics of Meiotic Drive." *Trends in Ecology & Evolution* 31, no. 4. doi:10.1016/j.tree.2016.02.001.

Lindhurst, Marjorie J., Julie C. Sapp, Jamie K. Teer, Jennifer J. Johnston, Erin M. Finn, Kathryn Peters, Joyce Turner, Jennifer L. Cannons, David Bick, and Laurel Blakemore. 2011. "A Mosaic Activating Mutation in AKT1 Associated with the Proteus Syndrome." *New England Journal of Medicine* 365:611–19.

———, Miranda R. Yourick, Yi Yu, Ronald E. Savage, Dora Ferrari, and Leslie G. Biesecker. 2015. "Repression of AKT Signaling by ARQ 092 in Cells and Tissues from Patients with Proteus Syndrome." *Scientific Reports* 5. doi:10.1038/srep17162.

Linnarsson, Sten. 2015. "A Tree of the Human Brain." *Science* 350:37.

Lippman, Walter. 1922. "The Mental Age of Americans." *New Republic,* October 25.

Liu, Yang, Liangliang Zhang, Shuhua Xu, Landian Hu, Laurence D. Hurst, and Xiangyin Kong. 2013. "Identification of Two Maternal Transmission Ratio Distortion Loci in Pedigrees of the Framingham Heart Study." *Scientific Reports* 3. doi:10.1038/srep02147.

Locey, Kenneth J., and Jay T. Lennon. 2016. "Scaling Laws Predict Global Microbial Diversity." *Proceedings of the National Academy of Sciences of the United States of America* 113:5970–75.

Lodato, Michael A., Mollie B. Woodworth, Semin Lee, Gilad D. Evrony, Bhaven K. Mehta, Amir Karger, Soohyun Lee, and others. 2015. "Somatic Mutation in Single Human Neurons Tracks Developmental and Transcriptional History." *Science* 350:94–98.

Loike, John D., and Ruth L. Fischbach. 2013. "New Ethical Horizons in Gestational Surrogacy." *Journal of Fertilization: In Vitro-IVF-Worldwide* 1:2. doi:10.4172/jfiv.1000109.

Long, Edward. 1774. *The History of Jamaica: Or, General Survey of the Ancient and Modern State of That Island.* London: T. Lowndes.

López-Beltrán, Carlos. 1995. "*Les maladies héréditaires*: 18th Century Disputes in France." *Revue d'histoire des sciences* 48:307–50.

———. 2004. "In the Cradle of Heredity: French Physicians and *L'Hérédité Naturelle* in the Early 19th Century." *Journal of the History of Biology* 37:39–72.

Louvish, Simon. 2010. *Chaplin: The Tramp's Odyssey.* London: Faber.

Lucotte, G., and F. Diéterlen. 2014. "Frequencies of M34, the Ultimate Genetic Marker of the Terminal Differenciation of Napoléon the First's Y-Chromosome Haplogroup E1b1b1c1 in Europe, Northern Africa and the Near East." *International Journal of Anthropology* 29:27–41.

Lucotte, Gérard. 2011. "Haplotype of the Y Chromosome of Napoléon the First." *Journal of Molecular Biology Research* 1:12–19.

Lupski, James R. 2013. "Genome Mosaicism—One Human, Multiple Genomes." *Science* 341:358–59.

Lycett, Stephen J., Kerstin Schillinger, Metin I. Eren, Noreen von Cramon-Taubadel, and Alex Mesoudi. 2016. "Factors Affecting Acheulean Handaxe Variation: Experimental Insights, Microevolutionary Processes, and Macroevolutionary Outcomes." *Quaternary International* 411B:386–401.

Lynch, Michael. 2016. "Mutation and Human Exceptionalism: Our Future Genetic Load." *Genetics* 202:869–75.

Lyon, Mary F. 1961. Gene Action in the X-Chromosome of the Mouse (*Mus musculus L.*). *Nature* 190:372–73.

Lyons, Derek E., Andrew G. Young, and Frank C. Keil. 2007. "The Hidden Structure of Overimitation." *Proceedings of the National Academy of Sciences of the United States of America* 104:19751–56.

Ma, Hong, Nuria Marti-Gutierrez, Sang-Wook Park, Jun Wu, Yeonmi Lee, Keiichiro Suzuki, Amy Koski, and others. 2017. "Correction of a Pathogenic Gene Mutation in Human Embryos." *Nature* 548:413–19.

Macdonald, David A., and Nancy N. McAdams. 2001. *The Woolverton Family, 1693–1850 and Beyond: Woolverton and Wolverton Descendants of Charles Woolverton, New Jersey Immigrant.* Albuquerque: Penobscot Press.

Mackintosh, N. J. 1995. *Cyril Burt: Fraud or Framed?* Oxford: Oxford University Press.

Maddox, Brenda. 2002. *Rosalind Franklin: The Dark Lady of DNA.* New York: HarperCollins.

Madrigal, Alexis. 2012. "The Perfect Milk Machine: How Big Data Transformed the Dairy Industry." *Atlantic*, May 1. http://www.theatlantic.com/technology/archive/2012/05/the -perfect-milk-machine-how-big-data-transformed-the-dairy-industry/256423/ (accessed July 30, 2017).

Maher, Brendan. 2008. "Personal Genomes: The Case of the Missing Heritability." *Nature* 456:18–21.

Mahmood, Uzma, and Keelin O'Donoghue. 2014. "Microchimeric Fetal Cells Play a Role in Maternal Wound Healing After Pregnancy." *Chimerism* 5:40–52.

Maienschein, Jane. 1978. "Cell Lineage, Ancestral Reminiscence, and the Biogenetic Law." *Journal of the History of Biology* 11:129–58.

Malan, Valérie, R. Gesny, N. Morichon-Delvallez, M. C. Aubry, A. Benachi, D. Sanlaville, C. Turleau, J. P. Bonnefont, C. Fekete-Nihoul, and M. Vekemans. 2007. "Prenatal Diagnosis and Normal Outcome of a 46,XX/46,XY Chimera: A Case Report." *Human Reproduction* 22:1037–41.

Mandel, Roi. 2014. "Auschwitz Prisoner No. A7733 Finally Finds His Family." *Ynet News*, September 11. https://www.ynetnews.com/articles/0,7340,L-4589762,00.html (accessed September 11, 2017).

Maples, Brian K., Simon Gravel, Eimear E. Kenny, and Carlos D. Bustamante. 2013. "RFMix: A Discriminative Modeling Approach for Rapid and Robust Local-Ancestry Inference." *American Journal of Human Genetics* 93:278–88.

Marchione, Marilynn. 2016. "Three-Parent Kids Grew Up OK." *NBC News*, October 27. http:// www.nbcnews.com/health/health-news/three-parent-kids-grew-ok-n674126 (accessed August 5, 2017).

Marciniak, Stephanie, and George H. Perry. 2017. "Harnessing Ancient Genomes to Study the History of Human Adaptation." *Nature Reviews Genetics* 18:659–74.

Marcotrigiano, Michael. 1997. "Chimeras and Variegation: Patterns of Deceit." *HortScience* 32:773–84.

Maron, Barry J., John R. Lesser, Nelson B. Schiller, Kevin M. Harris, Colleen Brown, and Heidi L. Rehm. 2009. "Implications of Hypertrophic Cardiomyopathy Transmitted by Sperm Donation." *JAMA* 302:1681–84.

Marouli, Eirini, Mariaelisa Graff, Carolina Medina-Gomez, Ken Sin Lo, Andrew R. Wood, Troels R. Kjaer, Rebecca S. Fine, and others. 2017. "Rare and Low-Frequency Coding Variants Alter Human Adult Height." *Nature* 542:186–90.

Marré, Julia, Edward C. Traver, and Antony M. Jose. 2016. "Extracellular RNA Is Transported from One Generation to the Next in *Caenorhabditis elegans*." *Proceedings of the National Academy of Sciences of the United States of America* 113:12496–501.

Martin, Aryn. 2007a. "The Chimera of Liberal Individualism: How Cells Became Selves in Human Clinical Genetics." *Osiris* 22:205–22.

———. 2007b. "'Incongruous Juxtapositions': The Chimaera and Mrs McK." *Endeavour* 31:99–103.

———. 2010. "Microchimerism in the Mother(land): Blurring the Borders of Body and Nation." *Body & Society* 16:23–50.

———. 2015. "Ray Owen and the History of Naturally Acquired Chimerism." *Chimerism* 6:2–7.

Martin, William F., Aloysius G. M. Tielens, Marek Mentel, Sriram G. Garg, and Sven B. Gould. 2017. "The Physiology of Phagocytosis in the Context of Mitochondrial Origin." *Microbiology and Molecular Biology Reviews* 81. doi:10.1128/MMBR.00008–17.

———, Sinje Neukirchen, Verena Zimorski, Sven B. Gould, and Filipa L. Sousa. 2016. "Energy for Two: New Archaeal Lineages and the Origin of Mitochondria." *BioEssays* 38:850–56.

Martínez, María Elena. 2011. *Genealogical Fictions: Limpieza de Sangre, Religion, and Gender in Colonial Mexico*. Stanford: Stanford University Press.

Martiniano, Rui, Lara M. Cassidy, Ros O'Maolduin, Russell McLaughlin, Nuno M. Silva, Licinio Manco, Daniel Fidalgo, and others. 2017. "The Population Genomics of Archaeological Transition in West Iberia." *bioRxiv*. doi:10.1101/134254.

Maryland State Archives. 2007. *A Guide to the History of Slavery in Maryland*. http://msa .maryland.gov/msa/intromsa/pdf/slavery_pamphlet.pdf.

Mather, K., and T. Dobzhansky. 1939. "Morphological Differences Between the 'Races' of *Drosophila pseudoobscura*." *American Naturalist* 73:5–25.

Mathias, Rasika Ann, Margaret A. Taub, Christopher R. Gignoux, Wenqing Fu, Shaila Musharoff, Timothy D. O'Connor, Candelaria Vergara, and others. 2016. "A Continuum of Admixture in the Western Hemisphere Revealed by the African Diaspora Genome." *Nature Communications* 7:12522.

Mathieson, Iain, Iosif Lazaridis, Nadin Rohland, Swapan Mallick, Bastien Llamas, Joseph Pickrell, Harald Meller, Manuel A. Rojo Guerra, Johannes Krause, and David Anthony. 2015. "Genome-Wide Patterns of Selection in 230 Ancient Eurasians." *Nature* 528:499–503.

———, Songül Alpaslan Roodenberg, Cosimo Posth, Anna Szécsényi-Nagy, Nadin Rohland, Swapan Mallick, Iñigo Olalde, and others. 2017. "The Genomic History of Southeastern Europe." *bioRxiv.* doi:10.1101/135616.

Mattson, Sarah N., Nicole Crocker, and Tanya T. Nguyen. 2011. "Fetal Alcohol Spectrum Disorders: Neuropsychological and Behavioral Features." *Neuropsychology Review* 21:81–101.

Mauritsen, Thorsten, and Robert Pincus. 2017. "Committed Warming Inferred from Observations." *Nature Climate Change.* doi:10.1038/nclimate3357.

Maybury-Lewis, David. 1960. "Parallel Descent and the Apinaye Anomaly." *Southwestern Journal of Anthropology* 16:191–216.

Mazumdar, Pauline M. H. 1992. *Eugenics, Human Genetics, and Human Failings: The Eugenics Society, Its Sources and Its Critics in Britain.* London: Routledge.

Mazzarello, Paolo. 1999. "A Unifying Concept: The History of Cell Theory." *Nature Cell Biology* 1:E13–E15.

McCutcheon, John P. 2016. "From Microbiology to Cell Biology: When an Intracellular Bacterium Becomes Part of Its Host Cell." *Current Opinion in Cell Biology* 41:132–36.

McDonald, Michael J., Daniel P. Rice, and Michael M. Desai. 2016. "Sex Speeds Adaptation by Altering the Dynamics of Molecular Evolution." *Nature* 531:233–36.

McGough, Ian John, and Jean-Paul Vincent. 2016. "Exosomes in Developmental Signalling." *Development* 143:2482–93.

McGue, Matt, and Irving I. Gottesman. 2015. "Classical and Molecular Genetic Research on General Cognitive Ability." *Hastings Center Report* 45, Suppl 1, S25–S31.

McGuire, Michelle K., and Mark A. McGuire. 2017. "Got Bacteria? The Astounding, Yet Not-So-Surprising, Microbiome of Human Milk." *Current Opinion in Biotechnology* 44:63–68.

McKim, W. Duncan. 1899. *Heredity and Human Progress.* New York and London: G. P. Putnam's Sons.

McKusick, Victor A. 1985. "Marcella O'Grady Boveri (1865–1950) and the Chromosome Theory of Cancer." *Journal of Medical Genetics* 22:431–40.

Medawar, Peter. 1957. *The Uniqueness of the Individual.* London: Methuen.

Meijer, Gerrit A. 2005. "Chromosomes and Cancer, Boveri Revisited." *Analytical Cellular Pathology* 27:273–75.

"Mendelism Up to Date." 1916. *Journal of Heredity* 7:17–23.

Mercado, Luis, and David F. Musto. 1961. "The William Osler Medal Essay." *Bulletin of the History of Medicine* 35:346.

Mesoudi, Alex. 2016. "Cultural Evolution: Integrating Psychology, Evolution and Culture." *Current Opinion in Psychology* 7:17–22.

———, and Kenichi Aoki, eds. 2015. *Learning Strategies and Cultural Evolution During the Paleolithic.* Replacement of Neanderthals by Modern Humans Series. New York: Springer.

Messner, Donna A. 2012. "On the Scent: The Discovery of PKU." *Distillations,* Spring. Chemical Heritage Foundation. https://www.chemheritage.org/distillations/magazine/on-the-scent -the-discovery-of-pku (accessed September 11, 2017).

Metzger, Michael J., Carol Reinisch, James Sherry, and Stephen P. Goff. 2015. "Horizontal Transmission of Clonal Cancer Cells Causes Leukemia in Soft-Shell Clams." *Cell* 161:255–63.

Meyer, Wynn K., Barbara Arbeithuber, Carole Ober, Thomas Ebner, Irene Tiemann-Boege, Richard R. Hudson, and Molly Przeworski. 2012. "Evaluating the Evidence for Transmission Distortion in Human Pedigrees." *Genetics* 191:215–32.

Meyer-Rochow, V. B. 1976. "Some Observations on Spawning and Fecundity in the Luminescent Fish *Photoblepharon palpebratus.*" *Marine Biology* 37:325–28.

Mézard, Christine, Marina Tagliaro Jahns, and Mathilde Grelon. 2015. "Where to Cross? New Insights into the Location of Meiotic Crossovers." *Trends in Genetics* 31:393–401.

Mikanowski, Jacob. 2012. "Dr. Hirszfeld's War: Tropical Medicine and the Invention of Sero-Anthropology on the Macedonian Front." *Social History of Medicine* 25:103–21.

Mills, Elizabeth Shown, and Gary B. Mills. 1984. "The Genealogist's Assessment of Alex Haley's Roots." *National Genealogical Society Quarterly* 72:35–49.

Mills, Gary B., and Elizabeth Shown Mills. 1981. "Roots and the New 'Faction': A Legitimate Tool for Clio?" *Virginia Magazine of History and Biography* 89:3–26.

Mills, John. 1776. *A Treatise on Cattle.* Dublin: Whitestone, Potts.

Milo, Ron, and Rob Phillips. 2015. *Cell Biology by the Numbers.* New York: Garland Science.

Mirzaghaderi, Ghader, and Elvira Hörandl. 2016. "The Evolution of Meiotic Sex and Its Alternatives." *Proceedings of the Royal Society B* 283:1838.

Mitchell, Kevin. 2016. Twitter post. September 7, 3:07 A.M. https://twitter.com/WiringTheBrain /status/773417464336187392 (accessed August 6, 2017).

Moeller, Andrew H., Alejandro Caro-Quintero, Deus Mjungu, Alexander V. Georgiev, Elizabeth V. Lonsdorf, Martin N. Muller, Anne E. Pusey, Martine Peeters, Beatrice H. Hahn, and Howard Ochman. 2016. "Cospeciation of Gut Microbiota with Hominids." *Science* 353:380–82.

de Montaigne, Michel. 1999. *The Autobiography of Michel de Montaigne.* Edited by Marvin Lowenthal. Boston: David R. Godine, Publisher.

Montialoux, Claire. 2016. "Revisiting the Impact of Head Start." Institute for Research on Labor and Employment, University of California, Berkeley. http://irle.berkeley.edu/revisiting-the -impact-of-head-start/ (accessed August 24, 2017).

Moodley, Yoshan, Bodo Linz, Robert P. Bond, Martin Nieuwoudt, Himla Soodyall, Carina M. Schlebusch, Steffi Bernhöft, James Hale, Sebastian Suerbaum, and Lawrence Mugisha. 2012. "Age of the Association Between *Helicobacter pylori* and Man." *PLOS Pathogens* 8:e1002693.

Moore, David Scott. 2015. *The Developing Genome: An Introduction to Behavioral Epigenetics.* Oxford: Oxford University Press.

Moreno, Inmaculada, Jose Manuel Míguez-Forjan, and Carlos Simón. 2015. "Artificial Gametes from Stem Cells." *Clinical and Experimental Reproductive Medicine* 42:33–44.

Morgan, Francesca. 2010a. "Lineage as Capital: Genealogy in Antebellum New England." *New England Quarterly* 83:250–82.

———. 2010b. "A Noble Pursuit? Bourgeois America's Uses of Lineage." In *The American Bourgeoisie: Distinction and Identity in the Nineteenth Century.* Edited by Julia Rosenbaum and Sven Beckert. New York: Palgrave Macmillan.

Morgan, Thomas Hunt. 1915. *The Mechanism of Mendelian Heredity.* New York: H. Holt.

———. 1925. *Evolution and Genetics.* Princeton: Princeton University Press.

Moris, Naomi, Cristina Pina, and Alfonso Martinez Arias. 2016. "Transition States and Cell Fate Decisions in Epigenetic Landscapes." *Nature Reviews Genetics* 17:693–703.

Morotomi, Masami, Fumiko Nagai, and Yohei Watanabe. 2012. "Description of *Christensenella minuta* Gen. Nov., Sp. Nov., Isolated from Human Faeces, Which Forms a Distinct Branch in the Order Clostridiales, and Proposal of Christensenellaceae Fam. Nov." *International Journal of Systematic and Evolutionary Microbiology* 62:144–49.

Morreale de Escobar, G., María Jesús Obregón, and F. Escobar del Rey. 2004. "Role of Thyroid Hormone During Early Brain Development." *European Journal of Endocrinology* 151, Suppl 3, U25–U37.

Morris, Thomas D. 2004. *Southern Slavery and the Law, 1619–1860.* Chapel Hill and London: University of North Carolina Press.

Moses, A. Dirk, and Dan Stone. 2010. "Eugenics and Genocide." In *The Oxford Handbook of the History of Eugenics.* Edited by Alison Bashford and Philippa Levine. New York: Oxford University Press.

Mroz, Jacqueline. 2012. "In Choosing a Sperm Donor, a Roll of the Genetic Dice." *New York Times,* May 14.

Muehlenbachs, Atis, Julu Bhatnagar, Carlos A. Agudelo, Alicia Hidron, Mark L. Eberhard, Blaine A. Mathison, Michael A. Frace, and others. 2015. "Malignant Transformation of *Hymenolepis nana* in a Human Host." *New England Journal of Medicine* 373:1845–52.

Muinzer, Thomas Louis. 2014. "Bones of Contention: The Medico-Legal Issues Relating to Charles Byrne, 'The Irish Giant.'" *Queen's Political Review* 2:155–66.

Mulchinock, Karen. N.D. "Breaking Our Family's Curse." *Chat* magazine, pp.30–31.

Muller, Hermann J. 1933. "The Dominance of Economics over Eugenics." *Scientific Monthly* 37:40–47.

———. 1949. "Progress and Prospects in Human Genetics." *American Journal of Human Genetics* 1:1–18.

———. 1950. "Our Load of Mutations." *American Journal of Human Genetics* 2:111–76.

———. 1961a. "Human Evolution by Voluntary Choice of Germ Plasm." *Science* 134:643–49.

———. 1961b. "Should We Weaken or Strengthen Our Genetic Heritage?" *Daedalus* 90:432–50.

Müller, Amanda Cecilie, Marianne Antonius Jakobsen, Torben Barington, Allan Arthur Vaag, Louise Groth Grunnet, Sjurdur Frodi Olsen, and Mads Kamper-Jørgensen. 2016. "Microchimerism of Male Origin in a Cohort of Danish Girls." *Chimerism* 6:1–7.

Müller-Wille, Staffan. 2010. "Cell Theory, Specificity, and Reproduction, 1837–1870." *Studies in History and Philosophy of Biological and Biomedical Sciences* 41:225–31.

———, and Hans-Jörg Rheinberger. 2012. *A Cultural History of Heredity.* Chicago: University of Chicago Press.

———, eds. 2007. *Heredity Produced: At the Crossroads of Biology, Politics, and Culture, 1500–1870.* Cambridge: MIT Press.

Murchison, Elizabeth P. 2016. "Cancer in the Wilderness." *Cell* 166:264–68.

———, Cesar Tovar, Arthur Hsu, Hannah S. Bender, Pouya Kheradpour, Clare A. Rebbeck, David Obendorf, Carly Conlan, Melanie Bahlo, and Catherine A. Blizzard. 2010. "The Tasmanian Devil Transcriptome Reveals Schwann Cell Origins of a Clonally Transmissible Cancer." *Science* 327:84–87.

———, Ole B. Schulz-Trieglaff, Zemin Ning, Ludmil B. Alexandrov, Markus J. Bauer, Beiyuan Fu, Matthew Hims, and others. 2012. "Genome Sequencing and Analysis of the Tasmanian Devil and Its Transmissible Cancer." *Cell* 148:780–91.

Murgia, Claudio, Jonathan K. Pritchard, Su Yeon Kim, Ariberto Fassati, and Robin A. Weiss. 2006. "Clonal Origin and Evolution of a Transmissible Cancer." *Cell* 126:477–87.

Musgrave, Stephanie, David Morgan, Elizabeth Lonsdorf, Roger Mundry, and Crickette Sanz. 2016. "Tool Transfers Are a Form of Teaching Among Chimpanzees." *Scientific Reports* 6:34783.

Muthukrishna, Michael, Ben W. Shulman, Vlad Vasilescu, and Joseph Henrich. 2014. "Sociality Influences Cultural Complexity." *Proceedings of the Royal Society B* 281:1774, 20132511.

Myerson, Abraham. 1925. *The Inheritance of Mental Diseases.* Baltimore: Williams & Wilkins Company.

National Academy of Sciences. 2016. *Gene Drives on the Horizon: Advancing Science, Navigating Uncertainty, and Aligning Research with Public Values.* Washington, DC: National Academies Press. doi:10.17226/23405.

National Academy of Sciences. 2017. *Human Genome Editing: Science, Ethics, and Governance.* Washington, DC: National Academies Press. doi:10.17226/24623.

National Human Genome Research Institute. 2000. "Remarks Made by the President, Prime Minister Tony Blair of England (via satellite), Dr. Francis Collins, Director of the National Human Genome Research Institute, and Dr. Craig Venter, President and Chief Scientific Officer, Celera Genomics Corporation, on the Completion of the First Survey of the Entire Human Genome Project." https://www.genome.gov/10001356/ (accessed September 10, 2017).

National Institutes of Health. 2017. "Estimates of Funding for Various Research, Condition, and Disease Categories (RCDC)." https://report.nih.gov/categorical_spending.aspx (accessed July 31, 2017).

National Institutes of Health, Human Gene Therapy Subcommittee. 1990. "The Revised 'Points to Consider' Document." *Human Gene Therapy* 1:93–103.

NCD Risk Factor Collaboration. 2016. "A Century of Trends in Adult Human Height." *eLife* 5:e13410.

———. 2017. "Height: Ranking for People Born from 1896 to 1996." http://www.ncdrisc.org/height-ranking-mean.html (accessed August 12, 2017).

Neimark, Jill. 2016. "The Mitochondrial Minefield of Three-Parent Babies." *Undark*, December 23. https://undark.org/article/three-parent-babies-battle-mitochondria/ (accessed September 11, 2017).

Nelson, J. Lee, Daniel E. Furst, Sean Maloney, Ted Gooley, Paul C. Evans, Anajane Smith, Michael A. Bean, Carole Ober, and Diana W. Bianchi. 1998. "Microchimerism and HLA-Compatible Relationships of Pregnancy in Scleroderma." *Lancet* 351:559–62.

New England Consortium of Metabolic Programs. 2010. "Discovery of the Diet for PKU by Dr. Horst Bickel." YouTube video, January 11. https://www.youtube.com/watch ?v=-rs0iZW0Lb0 (accessed August 24, 2017).

Newnan, Horatio H., Frank N. Freeman, and Karl John Holzinger. 1937. *Twins: A Study of Heredity and Environment.* Chicago: University of Chicago Press.

"New Way to Detect a Dread Disease." 1962. *Life,* January 19, p. 45.

Nielsen, Mark, Ilana Mushin, Keyan Tomaselli, and Andrew Whiten. 2014. "Where Culture Takes Hold: 'Overimitation' and Its Flexible Deployment in Western, Aboriginal, and Bushmen Children." *Child Development* 85:2169–84.

Nielsen, Rasmus, Joshua M. Akey, Mattias Jakobsson, Jonathan K. Pritchard, Sarah Tishkoff, and Eske Willerslev. 2017. "Tracing the Peopling of the World Through Genomics." *Nature* 541:302–10.

Nightingale, Katherine. 2015. "Remembering Mary Lyon and Her Impact on Mouse Genetics." Insight, February 3. http://www.insight.mrc.ac.uk/2015/02/03/remembering-mary-lyon -and-her-impact-on-mouse-genetics/ (accessed August 3, 2017).

Niklas, Karl J., Edward D. Cobb, and Ulrich Kutschera. 2014. "Did Meiosis Evolve Before Sex and the Evolution of Eukaryotic Life Cycles?" *BioEssays* 36:1091–1101.

Nilsson, Eric E., and Michael K. Skinner. 2015. "Environmentally Induced Epigenetic Transgenerational Inheritance of Reproductive Disease." *Biology of Reproduction* 93:145.

Nisbett, Richard E. 2013. "Schooling Makes You Smarter: What Teachers Need to Know About IQ." *American Educator* 37:10.

———, Joshua Aronson, Clancy Blair, William Dickens, James Flynn, Diane F. Halpern, and Eric Turkheimer. 2012. "Group Differences in IQ Are Best Understood as Environmental in Origin." *American Psychologist* 67:503–04.

Nobile, Philip. 1993. "Uncovering Roots." *Village Voice,* February 23, pp. 31–38.

Noble, Charleston, Jason Olejarz, Kevin M. Esvelt, George M. Church, and Martin A. Nowak. 2017. "Evolutionary Dynamics of CRISPR Gene Drives." *Science Advances* 3:1601964.

Noguera-Solano, Ricardo, and Rosaura Ruiz-Gutiérrez. 2009. "Darwin and Inheritance: The Influence of Prosper Lucas." *Journal of the History of Biology* 42:685–714.

Nolte, Ilja M., Peter J. van der Most, Behrooz Z. Alizadeh, Paul I W de Bakker, H. Marike Boezen, Marcel Bruinenberg, Lude Franke, and others. 2017. "Missing Heritability: Is the Gap Closing? An Analysis of 32 Complex Traits in the Lifelines Cohort Study." *European Journal of Human Genetics* 25. doi:10.1038/ejhg.2017.50.

Norrell, Robert J. 2015. *Alex Haley and the Books That Changed a Nation.* New York: St. Martin's Press.

Novembre, John. 2016. "Pritchard, Stephens, and Donnelly on Population Structure." *Genetics* 204:391–93.

Nowell P., and D. Hungerford. 1960. "Chromosome Studies on Normal and Leukemic Human Leukocytes." *Journal of the National Cancer Institute* 25:85–109.

O'Donnell, Karen J., Murdon Abdul Rakeman, Dou Zhi-Hong, Cao Xue-Yi, Zeng Yong Mei, Nancy DeLong, Gerald Brenner, Ma Tai, Wang Dong, and G. Robert DeLong. 2002. "Effects of Iodine Supplementation During Pregnancy on Child Growth and Development at School Age." *Developmental Medicine & Child Neurology* 44:76–81.

Oetting, William S., Marc S. Greenblatt, Anthony J. Brookes, Rachel Karchin, and Sean D. Mooney. 2015. "Germline & Somatic Mosaicism: The 2014 Annual Scientific Meeting of the Human Genome Variation Society." *Human Mutation* 36:390–93.

Oggins, Robin S. 2004. *The Kings and Their Hawks: Falconry in Medieval England.* New Haven: Yale University Press.

Okroi, Mathias, and Leo J. McCarthy. 2010. "The Original Blood Group Pioneers: The Hirszfelds." *Transfusion Medicine Reviews* 24:244–46.

Okuno, Ayako, Ko Hirano, Kenji Asano, Wakana Takase, Reiko Masuda, Yoichi Morinaka, Miyako Ueguchi-Tanaka, Hidemi Kitano, and Makoto Matsuoka. 2014. "New Approach to Increasing Rice Lodging Resistance and Biomass Yield through the Use of High Gibberellin Producing Varieties." *PLOS One* 9:e86870.

Olalde, Iñigo, Morten E. Allentoft, Federico Sánchez-Quinto, Gabriel Santpere, Charleston W. K. Chiang, Michael DeGiorgio, Javier Prado-Martinez, and others. 2014. "Derived Immune and Ancestral Pigmentation Alleles in a 7,000-Year-Old Mesolithic European." *Nature* 507:225–28.

Olalde, Iñigo, Selina Brace, Morten E. Allentoft, Ian Armit, Kristian Kristiansen, Nadin
    Rohland, Swapan Mallick, and others. 2017. "The Beaker Phenomenon and the Genomic
    Transformation of Northwest Europe." *bioRxiv.* doi:10.1101/135962.
Olson, S. 2015. "International Summit on Human Gene Editing: A Global Discussion." National
    Academies of Sciences, Engineering, and Medicine. doi:10.17226/21913.
Order of the Crown of Charlemagne in the United States of America website. http://www
    .charlemagne.org/ (accessed August 12, 2017).
Orel, Vítězslav. 1973. "The Scientific Milieu in Brno During the Era of Mendel's Research."
    Journal of Heredity 64:314–18.
Orgel, L. E. 1997. "Preventive Mitochondrial Replacement." *Chemistry & Biology* 4:167–68.
Osberg, Richard. 1986. "The Jesse Tree in the 1432 London Entry of Henry VI: Messianic
    Kingship and the Rule of Justice." *Journal of Medieval and Renaissance Studies* 16:213–32.
Osborn, Henry Fairfield. 1915. *Men of the Old Stone Age: Their Environment, Life and Art.*
    New York: Charles Scribner's Sons.
———. 1926. "The Evolution of Human Races." *Natural History* 26:3–13.
Osmon, David C., and Rebecca Jackson. 2002. "Inspection Time and IQ: Fluid or Perceptual
    Aspects of Intelligence?" *Intelligence* 30:119–27.
Ostrander, Elaine A., Brian W. Davis, and Gary K. Ostrander. 2016. "Transmissible Tumors:
    Breaking the Cancer Paradigm." *Trends in Genetics* 32:1–15.
Ovid. 2008. *Metamorphoses.* Edited by E. J. Kenney. Translated by A. D. Melville. Oxford:
    Oxford University Press.
Owen, H. Collinson. 1919. *Salonica and After: The Sideshow That Ended the War.* London:
    Hodder and Stoughton.
Owen, Ray D. 1959. "Facts for a Friendly Frankenstein." *Engineering and Science* 22:16–20.
———. 1983. Interview. Oral History Project, California Institute of Technology Archives.
    http://oralhistories.library.caltech.edu/123/ (accessed July 27, 2017).
Pääbo, Svante. 1985. "Molecular Cloning of Ancient Egyptian Mummy DNA." *Nature* 314:
    644–45.
———. 2014. *Neanderthal Man: In Search of Lost Genomes.* New York: Basic Books.
Pagden, Anthony. 1982. *The Fall of Natural Man: The American Indian and the Origins of
    Comparative Ethnology.* Cambridge: Cambridge University Press.
Page, Clarence. 1993. "Alex Haley's Facts Can Be Doubted, But Not His Truths." *Chicago
    Tribune*, March 10.
Paine, Thomas. 1995. "Common Sense." In *Paine: Collected Writings.* Edited by Eric Foner.
    New York: Library of America.
Palladino, Paolo. 1994. "Wizards and Devotees: On the Mendelian Theory of Inheritance and
    the Professionalization of Agricultural Science in Great Britain and the United States,
    1880–1930." *History of Science* 32:409–44.
Pandora, Katherine. 2001. "Knowledge Held in Common: Tales of Luther Burbank and Science
    in the American Vernacular." *Isis* 92:484–516.
Panofsky, Aaron. 2014. *Misbehaving Science: Controversy and the Development of Behavior
    Genetics.* Chicago: University of Chicago Press.
———. 2015. "What Does Behavioral Genetics Offer for Improving Education?" *Hastings Center
    Report* 45, Suppl 1, S43–S49.
Papazyan, Romeo, Yuxiang Zhang, and Mitchell A. Lazar. 2016. "Genetic and Epigenomic
    Mechanisms of Mammalian Circadian Transcription." *Nature Structural & Molecular
    Biology* 23:1045–52.
Parker, Geoffrey. 2014. *Imprudent King: A New Life of Philip II.* New Haven: Yale University
    Press.
Pasmooij, Anna M. G., Marcel F. Jonkman, and Jouni Uitto. 2012. "Revertant Mosaicism in
    Heritable Skin Diseases—Mechanisms of Natural Gene Therapy." *Discovery Medicine*
    14:167–79.
Patin, Etienne, Marie Lopez, Rebecca Grollemund, Paul Verdu, Christine Harmant, Hélène
    Quach, Guillaume Laval, and others. 2017. "Dispersals and Genetic Adaptation of Bantu-
    Speaking Populations in Africa and North America." *Science* 356:543–46.
Paul, Annie Murphy. 2010. *Origins: How the Nine Months before Birth Shape the Rest of Our
    Lives.* New York: Free Press.
Paul, Diane B., and Jeffrey P. Brosco. 2013. *The PKU Paradox: A Short History of a Genetic
    Disease.* Baltimore: Johns Hopkins University Press.

Pauly, Philip J. 1996. "How Did the Effects of Alcohol on Reproduction Become Scientifically Uninteresting?" *Journal of the History of Biology* 29:1–28.

Pawson, Henry Cecil. 1957. *Robert Bakewell: Pioneer Livestock Breeder.* London: Lockwood.

Payer, Bernhard. 2016. "Developmental Regulation of X-Chromosome Inactivation." *Seminars in Cell & Developmental Biology* 56:88–99.

Payne, Brendan, and Patrick F. Chinnery. 2015. "Mitochondrial Dysfunction in Aging: Much Progress but Many Unresolved Questions." *Biochimica et Biophysica Acta* 1847:1347–53.

Peacock, Zachary S., Katherine P. Klein, John B. Mulliken, and Leonard B. Kaban. 2014. "The Habsburg Jaw–Re-examined." *American Journal of Medical Genetics Part A* 164:2263–69.

Pearson, Karl. 1895. "Contributions to the Mathematical Theory of Evolution. III. Regression, Heredity, and Panmixia." *Proceedings of the Royal Society of London* 59:69–71.

———. 1930. *The Life, Letters and Labours of Francis Galton.* Cambridge: Cambridge University Press.

———, and Alice Lee. 1904. "On the Laws of Inheritance in Man." *Biometrika* 3:131–90.

Peiffer, Jason A., Maria C. Romay, Michael A. Gore, Sherry A. Flint-Garcia, Zhiwu Zhang, Mark J. Millard, Candice A. C. Gardner, and others. 2014. "The Genetic Architecture of Maize Height." *Genetics* 196:1337–56.

Pendergrast, W. J., B. K. Milmore, and S. C. Marcus. 1961. "Thyroid Cancer and Thyrotoxicosis in the United States: Their Relation to Endemic Goiter." *Journal of Chronic Diseases* 13:22–38.

Penrose, Lionel S. 1933. *Mental Defect.* London: Sidgwick and Jackson.

———. 1935. "Two Cases of Phenylpyruvic Amentia." *Lancet* 225:23–24.

———. 1946. "Phenylketonuria: A Problem in Eugenics." *Lancet* 247:949–53.

Peters, Brock A., Bahram G. Kermani, Oleg Alferov, Misha R. Agarwal, Mark A. McElwain, Natali Gulbahce, Daniel M. Hayden, and others. 2015. "Detection and Phasing of Single Base de Novo Mutations in Biopsies from Human in Vitro Fertilized Embryos by Advanced Whole-Genome Sequencing." *Genome Research* 25:426–34.

Picard, Martin, Douglas C. Wallace, and Yan Burelle. 2016. "The Rise of Mitochondria in Medicine." *Mitochondrion* 30:105–16.

Pickrell, Joseph K., and David Reich. 2014. "Toward a New History and Geography of Human Genes Informed by Ancient DNA." *Trends in Genetics* 30:377–89.

Pierce, Benjamin A. 2014. *Genetics: A Conceptual Approach.* New York: W. H. Freeman.

Pinhasi, Ron, Daniel Fernandes, Kendra Sirak, Mario Novak, Sarah Connell, Songül Alpaslan-Roodenberg, Fokke Gerritsen, and others. 2015. "Optimal Ancient DNA Yields from the Inner Ear Part of the Human Petrous Bone." *PLOS One* 10:6 e0129102.

Pinsker, Joe. 2015. "America Is Even Less Socially Mobile Than Economists Thought." *Atlantic*, July 23. https://www.theatlantic.com/business/archive/2015/07/america-social-mobility -parents-income/399311/ (accessed September 11, 2017).

Plomin, R., and J. Crabbe. 2000. "DNA." *Psychological Bulletin* 126:806–28.

Plomin, Robert, Joanna K. J. Kennedy, and Ian W. Craig. 2006. "The Quest for Quantitative Trait Loci Associated with Intelligence." *Intelligence* 34:513–26.

Poczai, Péter, Neil Bell, and Jaakko Hyvönen. 2014. "Imre Festetics and the Sheep Breeders' Society of Moravia: Mendel's Forgotten 'Research Network.'" *PLOS Biology* 12:1 e1001772.

Poduri, Annapurna, Gilad D. Evrony, Xuyu Cai, and Christopher A. Walsh. 2013. "Somatic Mutation, Genomic Variation, and Neurological Disease." *Science* 341:6141, 1237758.

———, Gilad D. Evrony, Xuyu Cai, Princess Christina Elhosary, Rameen Beroukhim, Maria K. Lehtinen, L. Benjamin Hills, and others. 2012. "Somatic Activation of AKT3 Causes Hemispheric Developmental Brain Malformations." *Neuron* 74:41–48.

Polderman, Tinca J. C., Beben Benyamin, Christiaan A. de Leeuw, Patrick F. Sullivan, Arjen van Bochoven, Peter M. Visscher, and Danielle Posthuma. 2015. "Meta-Analysis of the Heritability of Human Traits Based on Fifty Years of Twin Studies." *Nature Genetics* 47:702–09.

Poliakov, Léon. 1974. *The Aryan Myth: A History of Racist and Nationalist Ideas in Europe.* New York: Basic Books.

Politi, Yoav, Liron Gal, Yossi Kalifa, Liat Ravid, Zvulun Elazar, and Eli Arama. 2014. "Paternal Mitochondrial Destruction after Fertilization Is Mediated by a Common Endocytic and Autophagic Pathway in *Drosophila*." *Developmental Cell* 29:305–20.

Pollan, Michael. 2001. *The Botany of Desire: A Plant's-Eye View of the World.* New York: Random House.

Porter, Theodore. 2018. *The Unknown History of Human Heredity*. Princeton: Princeton University Press.

Porteus, Stanley. 1937. *Primitive Intelligence and Environment*. New York: Macmillan.

Posth, Cosimo, Christoph Wißing, Keiko Kitagawa, Luca Pagani, Laura van Holstein, Fernando Racimo, Kurt Wehrberger, and others. 2017. "Deeply Divergent Archaic Mitochondrial Genome Provides Lower Time Boundary for African Gene Flow into Neanderthals." *Nature Communications* 8, doi: 10.1038/ncomms16046.

Poznik, G. David, Yali Xue, Fernando L. Mendez, Thomas F. Willems, Andrea Massaia, Melissa A. Wilson Sayres, Qasim Ayub, and others. 2016. "Punctuated Bursts in Human Male Demography Inferred from 1,244 Worldwide Y-Chromosome Sequences." *Nature Genetics* 48:593–99.

Pratt, Catherine. 2007. *Spanish Word Histories and Mysteries: English Words That Come from Spanish*. Boston: Houghton Mifflin.

Pratt, H. P., and A. L. Muggleton-Harris. 1988. "Cycling Cytoplasmic Factors That Promote Mitosis in the Cultured 2-Cell Mouse Embryo." *Development* 104:115–20.

Prescott, William Hickling. 1858. *History of the Reign of Philip the Second, King of Spain*. Boston: Phillips, Sampson and Co.

President's Commission for the Study of Ethical Problems in Medicine and Biomedical and Behavioral Research. 1982. *Splicing Life: A Report on the Social and Ethical Issues of Genetic Engineering with Human Beings*. Washington, DC: President's Commission.

Pressman, Abe, Celia Blanco, and Irene A. Chen. 2015. "The RNA World as a Model System to Study the Origin of Life." *Current Biology* 25:R953–R963.

Price, Alkes L., Chris C. A. Spencer, and Peter Donnelly. 2015. "Progress and Promise in Understanding the Genetic Basis of Common Diseases." *Proceedings of the Royal Society B* 282:1821. doi:10.1098/rspb.2015.1684.

Prichard, James Cowles. 1826. *Researches into the Physical History of Mankind*. London: John and Arthur Arch.

Priest, James Rush, Charles Gawad, Kristopher M. Kahlig, Joseph K. Yu, Thomas O'Hara, Patrick M. Boyle, Sridharan Rajamani, and others. 2016. "Early Somatic Mosaicism Is a Rare Cause of Long-QT Syndrome." *Proceedings of the National Academy of Sciences of the United States of America* 113. doi:10.1073/pnas.1607187113.

Pritchard, Jonathan K. 2017. "An Expanded View of Complex Traits: From Polygenic to Omnigenic." *Cell* 169. doi:10.1016/j.cell.2017.05.038.

———, Matthew Stephens, and Peter Donnelly. 2000. "Inference of Population Structure Using Multilocus Genotype Data." *Genetics* 155:945–59.

———, Matthew Stephens, Noah A. Rosenberg, and Peter Donnelly. 2000. "Association Mapping in Structured Populations." *American Journal of Human Genetics* 67:170–81.

Proctor, Robert N. 1988. *Racial Hygiene: Medicine Under the Nazis*. Cambridge: Harvard University Press.

Provençal, Nadine, and Elisabeth B. Binder. 2015. "The Effects of Early Life Stress on the Epigenome: From the Womb to Adulthood and Even Before." *Experimental Neurology* 268:10–20.

Prowse, Thomas A. A., Phillip Cassey, Joshua V. Ross, Chandran Pfitzner, Talia A. Wittmann, and Paul Thomas. 2017. "Dodging Silver Bullets: Good CRISPR Gene-Drive Design Is Critical for Eradicating Exotic Vertebrates." *Proceedings of the Royal Society B* 284. doi:10.1098/rspb.2017.0799.

Pye, Ruth J., David Pemberton, Cesar Tovar, Jose M. C. Tubio, Karen A. Dun, Samantha Fox, Jocelyn Darby, and others. 2015. "A Second Transmissible Cancer in Tasmanian Devils." *Proceedings of the National Academy of Sciences of the United States of America* 113. doi:10.1073/pnas.1519691113.

Pyne, Stephen Joseph, and William Cronon. 1998. *Burning Bush: A Fire History of Australia*. Seattle: University of Washington Press.

Radian, Serban, Yoan Diekmann, Plamena Gabrovska, Brendan Holland, Lisa Bradley, Helen Wallace, Karen Stals, and others. 2016. "Increased Population Risk of AIP-Related Acromegaly and Gigantism in Ireland." *Human Mutation* 38. doi:10.1002/humu .23121.

Ralph, Peter, and Graham Coop. 2013. "The Geography of Recent Genetic Ancestry Across Europe." *PLOS Biology* 11(5):e1001555.

Raper, Kenneth B. 1946. "The Development of Improved Penicillin-Producing Molds." *Annals of the New York Academy of Sciences* 48:41–56.

Rasmann, Sergio, Martin De Vos, Clare L. Casteel, Donglan Tian, Rayko Halitschke, Joel Y. Sun, Anurag A. Agrawal, Gary W. Felton, and Georg Jander. 2012. "Herbivory in the Previous Generation Primes Plants for Enhanced Insect Resistance." *Plant Physiology* 158:854–63.

Rastan, Sohaila. 2015a. "Mary F. Lyon (1925–2014)." *Nature* 518:36.

———. 2015b. "Obituary: Mary F Lyon (1925–2014)." *Reproductive BioMedicine Online* 30:6, 566–67.

Rayman, Margaret P., and Sarah C. Bath. 2015. "The New Emergence of Iodine Deficiency in the UK: Consequences for Child Neurodevelopment." *Annals of Clinical Biochemistry* 52:705–08.

Reardon, Sara. 2015. "US Congress Moves to Block Human-Embryo Editing." *Nature.* doi:10.1038/nature.2015.17858.

———. 2016. "'Three-parent Baby' Claim Raises Hopes—and Ethical Concerns." *Nature.* doi:10.1038/nature.2016.20698.

———. 2017. "Genetic Details of Controversial 'Three-Parent Baby' Revealed." *Nature* 544:17–18.

Rechavi, Oded, and Itamar Lev. 2017. "Principles of Transgenerational Small RNA Inheritance in *Caenorhabditis elegans*." *Current Biology.* doi:10.1016/j.cub.2017.05.043.

Regal, Brian. 2002. *Henry Fairfield Osborn: Race, and the Search for the Origins of Man.* Burlington: Ashgate.

Regalado, Antonio, and Karby Legget. 2003. "A Global Journal Report: Fertility Breakthrough Raises Questions About Link to Cloning." *Wall Street Journal,* October 13.

Regev, Aviv, Sarah Teichmann, Eric S. Lander, Ido Amit, Christophe Benoist, Ewan Birney, Bernd Bodenmiller, and others. 2017. "The Human Cell Atlas." *bioRxiv.* doi:10.1101/121202.

Reich, David, Richard E. Green, Martin Kircher, Johannes Krause, Nick Patterson, Eric Y. Durand, Bence Viola, and others. 2010. "Genetic History of an Archaic Hominin Group from Denisova Cave in Siberia." *Nature* 468:1053–60.

Reid, James B., and John J. Ross. 2011. "Mendel's Genes: Toward a Full Molecular Characterization." *Genetics* 189:3–10.

Reilly, Philip R. 1991. *The Surgical Solution: A History of Involuntary Sterilization in the United States.* Baltimore: Johns Hopkins University Press.

———. 2015. "Eugenics and Involuntary Sterilization: 1907–2015." *Annual Review of Genomics and Human Genetics* 16:351–68.

Rende, Richard D., Robert Plomin, and Steven G. Vandenberg. 1990. "Who Discovered the Twin Method?" *Behavior Genetics* 20:277–85.

Richards, Martin. 2008. "Artificial Insemination and Eugenics: Celibate Motherhood, Eutelegenesis and Germinal Choice." *Studies in History and Philosophy of Biological and Biomedical Sciences* 39:211–21.

Richards, W. A. 1980. "The Import of Firearms into West Africa in the Eighteenth Century." *Journal of African History* 21:43–59.

Ried, Thomas. 2009. "Homage to Theodor Boveri (1862–1915): Boveri's Theory of Cancer as a Disease of the Chromosomes, and the Landscape of Genomic Imbalances in Human Carcinomas." *Environmental and Molecular Mutagenesis* 50:593–601.

Rietveld, Cornelius A., Tõnu Esko, Gail Davies, Tune H. Pers, and others. 2014. "Common Genetic Variants Associated with Cognitive Performance Identified Using the Proxy-Phenotype Method." *Proceedings of the National Academy of Sciences of the United States of America* 111:13790–94.

———, Sarah E. Medland, Jaime Derringer, Jian Yang, Tõnu Esko, Nicolas W. Martin, Harm-Jan Westra, and others. 2013. "GWAS of 126,559 Individuals Identifies Genetic Variants Associated with Educational Attainment." *Science* 340:1467–71.

Rijnink, Emilie C., Marlies E. Penning, Ron Wolterbeek, Suzanne Wilhelmus, Malu Zandbergen, Sjoerd G. van Duinen, Joke Schutte, Jan A. Bruijn, and Ingeborg M. Bajema. 2015. "Tissue Microchimerism Is Increased During Pregnancy: A Human Autopsy Study." *Molecular Human Reproduction* 21, no. 11. doi:10.1093/molehr/gav047.

Rindermann, Heiner, and Stefan Pichelmann. 2015. "Future Cognitive Ability: US IQ Prediction Until 2060 Based on NAEP." *PLOS One* 10:e0138412.

Riquet, Florentine, Alexis Simon, and Nicolas Bierne. 2017. "Weird Genotypes? Don't Discard Them, Transmissible Cancer Could Be an Explanation." *Evolutionary Applications* 10:140–45.

Risch, Neil, and Kathleen Merikangras. 1996. "The Future of Genetic Studies of Complex Human Diseases." *Science* 273:1516–17.

Ritchie, Stuart. 2015. *Intelligence: All That Matters*. London: Hodder & Stoughton.

Roberts, Dorothy. 2015. "Can Research on the Genetics of Intelligence Be 'Socially Neutral'?" *Hastings Center Report* 45, Suppl 1, S50–S53.

Robinson, Michael F. 2016. *The Lost White Tribe: Explorers, Scientists, and the Theory That Changed a Continent*. Oxford: Oxford University Press.

Robson, K. J., T. Chandra, R. T. MacGillivray, and S. L. Woo. 1982. "Polysome Immunoprecipitation of Phenylalanine Hydroxylase MRNA from Rat Liver and Cloning of Its CDNA." *Proceedings of the National Academy of Sciences of the United States of America* 79:4701–05.

Rodgers, Ali B., and Tracy L. Bale. 2015. "Germ Cell Origins of Posttraumatic Stress Disorder Risk: The Transgenerational Impact of Parental Stress Experience." *Biological Psychiatry* 78:307–14.

——, Christopher P. Morgan, N. Adrian Leu, and Tracy L. Bale. 2015. "Transgenerational Epigenetic Programming Via Sperm MicroRNA Recapitulates Effects of Paternal Stress." *Proceedings of the National Academy of Sciences of the United States of America* 112. doi:10.1073/pnas.1508347112.

Roebroeks, Wil, and Marie Soressi. 2016. "Neandertals Revised." *Proceedings of the National Academy of Sciences of the United States of America* 113:6372–79.

Roewer, Lutz. 2013. "DNA Fingerprinting in Forensics: Past, Present, Future." *Investigative Genetics* 4, no. 1. doi:10.1186/2041-2223-4-22.

Roger, Andrew J., Sergio A. Munoz-Gomez, and Ryoma Kamikawa. 2017. "The Origin and Diversification of Mitochondria." *Current Biology* 27: R1177-R1192.

Rogers, Lois. 1998. "Baby Created from Two Mothers Raises Hopes for Childless." *Sunday Times*, June 14.

"Roots Revisited." 2016. 23andMe blog, May 30. https://blog.23andme.com/ancestry/roots -revisited/ (accessed August 3, 2017).

Rose, Steven. 1972. "Environmental Effects on Brain and Behaviour." In *Race, Culture and Intelligence*. Edited by Ken Richardson, David Spears, and Martin Richards. Harmondsworth, UK: Penguin Books.

Rose, Todd. 2015. *The End of Average: How We Succeed in a World That Values Sameness*. New York: HarperOne.

Rose, Willie Lee. 1976. "An American Family." *New York Review of Books*, November 11.

Rosenberg, Eugene, and Ilana Zilber-Rosenberg. 2016. "Microbes Drive Evolution of Animals and Plants: The Hologenome Concept." *mBio* 7:e01395-15.

Rosenberg, Noah A., and Michael D. Edge. In press. "Genetic Clusters and the Race Debates: A Perspective from Population Genetics." In *Genetic Clusters and the Race Debates: A Perspective from Population Genetics*. Edited by Quayshawn N. Spencer. Oxford: Oxford University Press.

——, Jonathan K. Pritchard, James L. Weber, Howard M. Cann, Kenneth K. Kidd, Lev A. Zhivotovsky, and Marcus W. Feldman. 2002. "Genetic Structure of Human Populations." *Science* 298:2381–85.

Rosenbloom, Arlan L., Jaime Guevara-Aguirre, Ron G. Rosenfeld, and Paul J. Fielder. 1990. "The Little Women of Loja—Growth Hormone-Receptor Deficiency in an Inbred Population of Southern Ecuador." *New England Journal of Medicine* 323:1367–74.

Ross, Cody T., Peter J. Richerson, and Deborah S. Rogers. 2014. "Mechanisms of Cultural Change and the Transition to Sustainability." In *Global Environmental Change*. Edited by Bill Freedman. Springer Netherlands.

Rowley-Conwy, Peter, and Robert Layton. 2011. "Foraging and Farming as Niche Construction: Stable and Unstable Adaptations." *Philosophical Transactions of the Royal Society B* 366:849–62.

Rubin, Beatrix P. 2009. "Changing Brains: The Emergence of the Field of Adult Neurogenesis." *BioSocieties* 4:407–24.

Russell Sage Foundation. 2016. "What We Know About Economic Inequality and Social Mobility in the United States." Blog, July 12. https://www.russellsage.org/what-we-know

-about-economic-inequality-and-social-mobility-united-states (accessed September 11, 2017).

Rutledge, Samuel D., and Daniela Cimini. 2016. "Consequences of Aneuploidy in Sickness and in Health." *Current Opinion in Cell Biology* 40:41–46.

Sabree, Zakee L., Srinivas Kambhampati, and Nancy A. Moran. 2009. "Nitrogen Recycling and Nutritional Provisioning by Blattabacterium, the Cockroach Endosymbiont." *Proceedings of the National Academy of Sciences of the United States of America* 106:19521–26.

Sahoo, Susmita, and Douglas W. Losordo. 2014. "Exosomes and Cardiac Repair After Myocardial Infarction." *Circulation Research* 114:333–44.

Sankararaman, Sriram, Swapan Mallick, Michael Dannemann, Kay Prüfer, Janet Kelso, Svante Pääbo, Nick Patterson, and David Reich. 2014. "The Genomic Landscape of Neanderthal Ancestry in Present-Day Humans." *Nature* 507:354–57.

Sankararaman, Sriram, Swapan Mallick, Nick Patterson, and David Reich. 2016. "The Combined Landscape of Denisovan and Neanderthal Ancestry in Present-Day Humans." *Current Biology* 26:1241–47.

Sawyer, Susanna, Gabriel Renaud, Bence Viola, Jean-Jacques Hublin, Marie-Theres Gansauge, Michael V. Shunkov, Anatoly P. Derevianko, Kay Prüfer, Janet Kelso, and Svante Pääbo. 2015. "Nuclear and Mitochondrial DNA Sequences from Two Denisovan Individuals." *Proceedings of the National Academy of Sciences of the United States of America* 112:15696-700.

Schaefer, Sabine, and Joseph H. Nadeau. 2015. "The Genetics of Epigenetic Inheritance: Modes, Molecules, and Mechanisms." *Quarterly Review of Biology* 90:381–415.

Scheinfeld, Amram. 1944. "The Kallikaks after Thirty Years." *Journal of Heredity* 35:259–64.

Schlebusch, Carina M., Helena Malmström, Torsten Günther, Per Sjödin, Alexandra Coutinho, Hanna Edlund, Arielle R. Munters, and others. 2017. "Ancient Genomes from Southern Africa Pushes Modern Human Divergence Beyond 260,000 Years Ago." *bioRxiv.* doi:10.1101/145409.

Schmerler, Samuel, and Gary M. Wessel. 2011. "Polar Bodies—More a Lack of Understanding Than a Lack of Respect." *Molecular Reproduction and Development* 78:3–8.

Schulman, J. D., and H. J. Stern. 2015. "Low Utilization of Prenatal and Pre-Implantation Genetic Diagnosis in Huntington Disease—Risk Discounting in Preventive Genetics." *Clinical Genetics* 88:220–23.

Schwank, Gerald, Bon-Kyoung Koo, Valentina Sasselli, Johanna F. Dekkers, Inha Heo, Turan Demircan, Nobuo Sasaki, and others. 2013. "Functional Repair of CFTR by CRISPR/Cas9 in Intestinal Stem Cell Organoids of Cystic Fibrosis Patients." *Cell Stem Cell* 13:653–58.

Schwann, Theodor. 1847. *Microscopical Researches into the Accordance in the Structure and Growth of Animals and Plants.* London: Sydenham Society.

Schwartz, James. 2008. *In Pursuit of the Gene: From Darwin to DNA.* Cambridge: Harvard University Press.

Science News Staff. 1997. "Extra Licking Makes for Relaxed Rats." *Science,* September 11.

Scudellari, Megan. 2016. "How IPS Cells Changed the World." *Nature* 534: 310-12. doi:10.1038/534310a.

Secord, James A. 1981. "Nature's Fancy: Charles Darwin and the Breeding of Pigeons." *Isis* 72:162–86.

Segers, Seppe, Heidi Mertes, Guido de Wert, Wybo Dondorp, and Guido Pennings. 2017. "Balancing Ethical Pros and Cons of Stem Cell Derived Gametes." *Annals of Biomedical Engineering* 45. doi:10.1007/s10439-017-1793-9.

Semrau, Stefan, and Alexander van Oudenaarden. 2015. "Studying Lineage Decision-Making in Vitro: Emerging Concepts and Novel Tools." *Annual Review of Cell and Developmental Biology* 31:317–45.

Sender, Ron, Shai Fuchs, and Ron Milo. 2016. "Revised Estimates for the Number of Human and Bacteria Cells in the Body." *PLOS Biology* 14:8, e1002533.

Shabad, L. M., and V. I. Ponomarkov. 1976. "Mstislav Novinsky, Pioneer of Tumour Transplantation." *Cancer Letters* 2:1–3.

Shapiro, E. Donald, Stewart Reifler, and Claudia L. Psome. 1992. "The DNA Paternity Test: Legislating the Future Paternity Action." *Journal of Law and Health* 7:1–47.

Sharpley, Mark S., Christine Marciniak, Kristin Eckel-Mahan, Meagan McManus, Marco Crimi, Katrina Waymire, Chun Shi Lin, and others. 2012. "Heteroplasmy of Mouse MtDNA Is Genetically Unstable and Results in Altered Behavior and Cognition." *Cell* 151:333–43.

Shennan, Stephen. 2011. "Property and Wealth Inequality as Cultural Niche Construction." *Philosophical Transactions of the Royal Society B* 366:918–26.

Shimkin, Michael B. 1955. "M. A. Novinsky: A Note on the History of Transplantation of Tumors." *Cancer* 8:653–55.

Shirley, Matthew D., Hao Tang, Carol J. Gallione, Joseph D. Baugher, Laurence P. Frelin, Bernard Cohen, Paula E. North, Douglas A. Marchuk, Anne M. Comi, and Jonathan Pevsner. 2013. "Sturge-Weber Syndrome and Port-Wine Stains Caused by Somatic Mutation in GNAQ." *New England Journal of Medicine* 368:1971–79.

Shull, George Harrison. 1909. "A Pure-Line Method in Corn Breeding." *Journal of Heredity* 1:51–58.

Siemens, Hermann Werner. 1924. *Die Zwillingspathologie: Ihre Bedeutung, Ihre Methodik, Ihre Bisherigen Ergebnisse.* Berlin: J. Springer.

Silventoinen, Karri, Aline Jelenkovic, Reijo Sund, Yoon-Mi Hur, Yoshie Yokoyama, Chika Honda, Jacob V. B. Hjelmborg, Sören Möller, Syuichi Ooki, and Sari Aaltonen. 2016. "Genetic and Environmental Effects on Body Mass Index from Infancy to the Onset of Adulthood: An Individual-Based Pooled Analysis of 45 Twin Cohorts Participating in the Collaborative Project of Development of Anthropometrical Measures in Twins (CODATwins) Study." *American Journal of Clinical Nutrition* 104:371–79.

———, Sampo Sammalisto, Markus Perola, Dorret I. Boomsma, Belinda K. Cornes, Chayna Davis, Leo Dunkel, Marlies de Lange, Jennifer R. Harris, and Jacob V. B. Hjelmborg. 2003. "Heritability of Adult Body Height: A Comparative Study of Twin Cohorts in Eight Countries." *Twin Research* 6:399–408.

Silver, Ari J., Jessica L. Larson, Maxwell J. Silver, Regine M. Lim, Carlos Borroto, Brett Spurrier, Anne Morriss, and Lee M. Silver. 2016. "Carrier Screening Is a Deficient Strategy for Determining Sperm Donor Eligibility and Reducing Risk of Disease in Recipient Children." *Genetic Testing and Molecular Biomarkers* 20:276–84.

Silvers, Willys. 1979. *The Coat Colors of Mice: A Model for Mammalian Gene Action and Interaction.* New York: Springer-Verlag.

Skerfving, Staffan, Lina Löfmark, Thomas Lundh, Zoli Mikoczy, and Ulf Strömberg. 2015. "Late Effects of Low Blood Lead Concentrations in Children on School Performance and Cognitive Functions." *NeuroToxicology* 49:114–20.

Skinner, Michael K. 2015. "Environmental Epigenetics and a Unified Theory of the Molecular Aspects of Evolution: A Neo-Lamarckian Concept That Facilitates Neo-Darwinian Evolution." *Genome Biology and Evolution* 7:1296–302.

Skoglund, Pontus, Jessica Thompson, Mary Prendergast. 2017. "Reconstructing Prehistoric African Population Structure." *Cell* 171:1–13.

Slack, Jonathan M. W. 2002. "Conrad Hal Waddington: The Last Renaissance Biologist?" *Nature Reviews Genetics* 3:889–95.

Slatkin, Montgomery, and Fernando Racimo. 2016. "Ancient DNA and Human History." *Proceedings of the National Academy of Sciences of the United States of America* 113:6380–87.

Slon, Viviane, Bence Viola, Gabriel Renaud, Marie-Theres Gansauge, Stefano Benazzi, Susanna Sawyer, Jean-Jacques Hublin, and others. 2017. "A Fourth Denisovan Individual." *Science Advances* 3:e1700186.

Smedley, Audrey, and Brian D. Smedley. 2007. *Race in North America: Origin and Evolution of a Worldview.* Boulder: Westview Press.

Smith, Barbara M. D. 1967. "The Galtons of Birmingham: Quaker Gun Merchants and Bankers, 1702–1831." *Business History* 9:132–50.

Smith, Bruce D. 2011. "General Patterns of Niche Construction and the Management of 'Wild' Plant and Animal Resources by Small-Scale Pre-Industrial Societies." *Philosophical Transactions of the Royal Society B* 366:836–48.

Smith, Eric Alden, Kim Hill, Frank Marlowe, David Nolin, Polly Wiessner, Michael Gurven, Samuel Bowles, Monique Borgerhoff Mulder, Tom Hertz, and Adrian Bell. 2010. "Wealth Transmission and Inequality Among Hunter-Gatherers." *Current Anthropology* 51:19–34.

——, Monique Borgerhoff Mulder, Samuel Bowles, Michael Gurven, Tom Hertz, and Mary K. Shenk. 2010. "Production Systems, Inheritance, and Inequality in Premodern Societies." *Current Anthropology* 51:85–94.

Smith, Jane S. 2009. *The Garden of Invention: Luther Burbank and the Business of Breeding Plants.* New York: Penguin Press.

Smith, J. David. 1985. *Minds Made Feeble: The Myth and Legacy of the Kallikaks.* Rockville, MD: Aspen Systems Corp.

——, and Michael L. Wehmeyer. 2012a. *Good Blood, Bad Blood: Science, Nature, and the Myth of the Kallikaks.* Washington, DC: American Association on Intellectual and Developmental Disabilities.

——. 2012b. "Who Was Deborah Kallikak?" *Intellectual and Developmental Disabilities* 50, no. 2. doi:10.1352/1934-9556-50.2.169.

Smythies, John, Lawrence Edelstein, and Vilayanur Ramachandran. 2014. "Molecular Mechanisms for the Inheritance of Acquired Characteristics-Exosomes, MicroRNA Shuttling, Fear and Stress: Lamarck Resurrected?" *Frontiers in Genetics* 5:133.

Sniekers, Suzanne. 2017. "Genome-Wide Association Meta-Analysis of 78,308 Individuals Identifies New Loci and Genes Influencing Human Intelligence." *Nature Genetics* 49:1107–12.

Sojo, Victor, Barry Herschy, Alexandra Whicher, Eloi Camprubí, and Nick Lane. 2016. "The Origin of Life in Alkaline Hydrothermal Vents." *Astrobiology* 16:181–97.

Sonneborn, T. M., ed. 1965. *The Control of Human Heredity and Evolution.* New York: Macmillan.

de Souza, R. A. G. 2012. "Origins of the Elephant Man: Mosaic Somatic Mutations Cause Proteus Syndrome." *Clinical Genetics* 81:123–124.

Soyk, Sebastian, Niels A. Müller, Soon Ju Park, Inga Schmalenbach, Ke Jiang, Ryosuke Hayama, Lei Zhang, Joyce Van Eck, José M. Jiménez-Gómez, and Zachary B. Lippman. 2017. "Variation in the Flowering Gene SELF PRUNING 5G Promotes Day-Neutrality and Early Yield in Tomato." *Nature Genetics* 49:162–68.

Spalding, Kirsty L., Olaf Bergmann, Kanar Alkass, Samuel Bernard, Mehran Salehpour, Hagen B. Huttner, Emil Boström, and others. 2013. "Dynamics of Hippocampal Neurogenesis in Adult Humans." *Cell* 153:1219–27.

——, Ratan D. Bhardwaj, Bruce A. Buchholz, Henrik Druid, and Jonas Frisén. 2005. "Retrospective Birth Dating of Cells in Humans." *Cell* 122:133–43.

Sparrow, Robert. 2012. "Orphaned at Conception: The Uncanny Offspring of Embryos." *Bioethics* 26:173–81.

——. 2015. "Enhancement and Obsolescence: Avoiding an 'Enhanced Rat Race.'" *Kennedy Institute of Ethics Journal* 25:231–60.

Spinner, Nancy B., and Laura K. Conlin. 2014. "Mosaicism and Clinical Genetics." *American Journal of Medical Genetics Part C* 166:397–405.

Springer, Nathan M., and Robert J. Schmitz. 2017. "Exploiting Induced and Natural Epigenetic Variation for Crop Improvement." *Nature Reviews Genetics* 18:563–75.

Spurling, Hilary. 2011. *Pearl Buck in China: Journey to the Good Earth.* New York: Simon & Schuster.

Stansfield, William D. 2006. "Luther Burbank: Honorary Member of the American Breeders' Association." *Journal of Heredity* 97:95–99.

Starr, Douglas P. 1998. *Blood: An Epic History of Medicine and Commerce.* New York: Alfred A. Knopf.

Steckel, Richard H. 2009. "Heights and Human Welfare: Recent Developments and New Directions." *Explorations in Economic History* 46:1–23.

——. 2013. "Biological Measures of Economic History." *Annual Review of Economics* 5:401–23.

——. 2016. "Slave Heights." In *The Oxford Handbook of Economics and Human Biology.* Edited by John Komlos and Inas R. Kelly. Oxford: Oxford University Press.

Stephen Spielberg Film and Video Archive. *Das Erbe.* Video produced in 1935. Accessed at US Holocaust Memorial Museum, courtesy of Bundesarchiv. https://www.ushmm.org/online /film/display/detail.php?file_num=3210 (accessed August 24, 2017).

Stern, Claudio D. 2003. "Conrad H. Waddington's Contributions to Avian and Mammalian Development, 1930–1940." *International Journal of Developmental Biology* 44:15–22.

———, and Scott E. Fraser. 2001. "Tracing the Lineage of Tracing Cell Lineages." *Nature Cell Biology* 3:E216–E218.

Stewart, James B., and Patrick F. Chinnery. 2015. "The Dynamics of Mitochondrial DNA Heteroplasmy: Implications for Human Health and Disease." *Nature Reviews Genetics* 16:530–42.

Stough, C., J. Brebner, T. Nettelbeck, C. J. Cooper, T. Bates, and G. L. Mangan. 1996. "The Relationship Between Intelligence, Personality and Inspection Time." *British Journal of Psychology* 87:255–68.

Strakova, Andrea, Máire Ní Leathlobhair, Guo-Dong Wang, Ting-Ting Yin, Ilona Airikkala-Otter, Janice L. Allen, Karen M. Allum, Leontine Bansse-Issa, Jocelyn L. Bisson, and Artemio Castillo Domracheva. 2016. "Mitochondrial Genetic Diversity, Selection and Recombination in a Canine Transmissible Cancer." *eLife* 5:e14552.

Straney, Shirley G. 1994. "The Kallikak Family: A Genealogical Examination of a 'Classic in Psychology.'" *American Genealogist* 69:65–80.

Stroud, Laura R., George D. Papandonatos, Amy L. Salisbury, Maureen G. Phipps, Marilyn A. Huestis, Raymond Niaura, James F. Padbury, Carmen J. Marsit, and Barry M. Lester. 2016. "Epigenetic Regulation of Placental NR3C1: Mechanism Underlying Prenatal Programming of Infant Neurobehavior by Maternal Smoking?" *Child Development* 87:49–60.

Stubbs, E. L., and J. Furth. 1934. "Experimental Studies on Venereal Sarcoma of the Dog." *American Journal of Pathology* 10:273–86.

Stulp, Gert, and Louise Barrett. 2016. "Evolutionary Perspectives on Human Height Variation." *Biological Reviews of the Cambridge Philosophical Society* 91:206–34.

Sturtevant, A. H., and T. Dobzhansky. 1936. "Inversions in the Third Chromosome of Wild Races of *Drosophila pseudoobscura*, and Their Use in the Study of the History of the Species." *Proceedings of the National Academy of Sciences* 22:448–50.

Sudik, R., S. Jakubiczka, F. Nawroth, E. Gilberg, and P. F. Wieacker. 2001. "Chimerism in a Fertile Woman with 46,XY Karyotype and Female Phenotype: Case Report." *Human Reproduction* 16:56–58.

Sung, Patrick, and Hannah Klein. 2006. "Mechanism of Homologous Recombination: Mediators and Helicases Take on Regulatory Functions." *Nature Reviews Molecular Cell Biology* 7:739–50.

Sweet, James H. 1997. "The Iberian Roots of American Racist Thought." *William and Mary Quarterly* 51:143–66.

Syed, Sana. 2015. "Iodine and the 'Near' Eradication of Cretinism." *Pediatrics* 135:594–96.

Szostak, Natalia, Szymon Wasik, and Jacek Blazewicz. 2016. "Hypercycle." *PLOS Computational Biology* 12:e1004853.

Takatsuka, Hirotomo, and Masaaki Umeda. 2015. "Epigenetic Control of Cell Division and Cell Differentiation in the Root Apex." *Frontiers in Plant Science* 6:1178.

Tan, An S., James W. Baty, Lan-Feng Dong, Ayenachew Bezawork-Geleta, Berwini Endaya, Jacob Goodwin, Martina Bajzikova, and others. 2015. "Mitochondrial Genome Acquisition Restores Respiratory Function and Tumorigenic Potential of Cancer Cells Without Mitochondrial DNA." *Cell Metabolism* 21:81–94.

Tang, Walfred W. C., Toshihiro Kobayashi, Naoko Irie, Sabine Dietmann, and M. Azim Surani. 2016. "Specification and Epigenetic Programming of the Human Germ Line." *Nature Reviews Genetics* 17:585–600.

Tanner, J. M. 1979. "A Concise History of Growth Studies from Buffon to Boas." In *Human Growth*. Edited by Frank Falkner. New York: Plenum Press.

———. 2010. *A History of the Study of Human Growth*. Cambridge: Cambridge University Press.

Taubes, Gary. 2013. "Rare Form of Dwarfism Protects Against Cancer." *Discover* magazine, March 27. http://discovermagazine.com/2013/april/19-double-edged-genes (accessed August 2, 2017).

Tavernise, Sabrina. 2014. "Shoukhrat Mitalipov's Mitochondrial Manipulations." *New York Times*, March 17.

Teich, A. H. 1984. "Heritability of Grain Yield, Plant Height and Test Weight of a Population of Winter Wheat Adapted to Southwestern Ontario." *Theoretical and Applied Genetics* 68:21–23.

Terman, Lewis Madison. 1922. "Were We Born That Way?" *World's Work* 44:655–60.

Teves, Sheila S., Luye An, Anders S. Hansen, Liangqi Xie, Xavier Darzacq, and Robert Tjian. 2016. "A Dynamic Mode of Mitotic Bookmarking by Transcription Factors." *bioRxiv.* doi:10.1101/066464.

Theis, Kevin R., Nolwenn M. Dheilly, Jonathan L. Klassen, Robert M. Brucker, John F. Baines, Thomas C. G. Bosch, John F. Cryan, and others. 2016. "Getting the Hologenome Concept Right: An Eco-Evolutionary Framework for Hosts and Their Microbiomes." *bioRxiv.* doi:10.1101/038596.

Thomas, Mark. 2013. "To Claim Someone Has Viking Ancestors Is No Better than Astrology." *Guardian,* February 25.

Thomas, W. H. 1904. "Medical Treatment of Diabetes." *Journal of the American Medical Association* 42:1451.

Thornton, Alex, and Katherine McAuliffe. 2006. "Teaching in Wild Meerkats." *Science* 313:227–29.

Thurtle, Phillip. 2007. "The Poetics of Life: Luther Burbank, Horticultural Novelties, and the Spaces of Heredity." *Literature and Medicine* 26:1–24.

Tibbles, J. A., and M. M. Cohen. 1986. "The Proteus Syndrome: The Elephant Man Diagnosed." *British Medical Journal* 293:683–85.

Tingley, Kim. 2014. "The Brave New World of Three-Parent I.V.F." *New York Times,* June 27.

Tippett, Patricia. 1983. "Blood Group Chimeras: A Review." *Vox Sanguinis* 44:333–59.

Tissot, Tazzio, Audrey Arnal, Camille Jacqueline, Robert Poulin, Thierry Lefèvre, Frédéric Mery, François Renaud, and others. 2016. "Host Manipulation by Cancer Cells: Expectations, Facts, and Therapeutic Implications." *BioEssays* 38:276–85.

Tollefsbol, Trygve O., ed. 2014. *Transgenerational Epigenetics Evidence and Debate.* London: Academic Press.

Touati, Sandra A., and Katja Wassmann. 2016. "How Oocytes Try to Get It Right: Spindle Checkpoint Control in Meiosis." *Chromosoma* 125:321–35.

Traub, Amy, Laura Sullivan, Tatiana Meschede, Thomas Shapiro. 2017. *The Asset Value of White Privilege: Understanding the Racial Wealth Gap.* New York: Demos.

Treves, Frederick. 1923. *The Elephant Man and Other Reminiscences.* London: Cassell and Company.

Trzaskowski, M., J. Yang, P. M. Visscher, and R. Plomin. 2014a. "DNA Evidence for Strong Genetic Stability and Increasing Heritability of Intelligence from Age 7 to 12." *Molecular Psychiatry* 19, no. 3. doi:10.1038/mp.2012.191.

Trzaskowski, Maciej, Nicole Harlaar, Rosalind Arden, Eva Krapohl, Kaili Rimfeld, Andrew McMillan, Philip S. Dale, and Robert Plomin. 2014b. "Genetic Influence on Family Socioeconomic Status and Children's Intelligence." *Intelligence* 42. doi:10.1016/j.intell.2013.11.002.

Tuchman, Arleen Marcia. 2011. "Diabetes and Race. A Historical Perspective." *American Journal of Public Health* 101, no. 1. doi:10.2105/AJPH.2010.202564.

Tucker, William H. 1994. *The Science and Politics of Racial Research.* Urbana: University of Illinois Press.

———. 2007. "Burt's Separated Twins: The Larger Picture." *Journal of the History of the Behavioral Sciences* 43:81–86.

Tucker-Drob, Elliot M., and Timothy C. Bates. 2015. "Large Cross-National Differences in Gene × Socioeconomic Status Interaction on Intelligence." *Psychological Science* 27:138–49.

Turkheimer, Eric. 2012. "Genome Wide Association Studies of Behavior Are Social Science." In *Philosophy of Behavioral Biology.* Edited by Kathryn S. Plaisance and Thomas Reydon. Springer Netherlands.

———. 2015. "Genetic Prediction." *Hastings Center Report* 45, Suppl 1, S32–S38.

Turkheimer, Eric, Andreana Haley, Mary Waldron, Brian D'Onofrio, and Irving I. Gottesman. 2003. "Socioeconomic Status Modifies Heritability of IQ in Young Children." *Psychological Science* 14:623–28.

Ujvari, Beata, Anne-Maree Pearse, Kate Swift, Pamela Hodson, Bobby Hua, Stephen Pyecroft, Robyn Taylor, and others. 2014. "Anthropogenic Selection Enhances Cancer Evolution in Tasmanian Devil Tumours." *Evolutionary Applications* 7:260–65.

———, Anthony T. Papenfuss, and Katherine Belov. 2016. "Transmissible Cancers in an Evolutionary Context." *BioEssays* 38:S14–S23.

———, Robert A. Gatenby, and Frédéric Thomas. 2016a. "The Evolutionary Ecology of Transmissible Cancers." *Infection, Genetics and Evolution* 39:293–303.
———. 2016b. "Transmissible Cancers, Are They More Common Than Thought?" *Evolutionary Applications* 9:633–34.
Uller, Tobias, and Heikki Helanterä. 2013. "Non-Genetic Inheritance in Evolutionary Theory: A Primer." *Non-Genetic Inheritance*. doi:10.2478/ngi-2013-0003.
Unckless, Robert, and Andrew Clark. 2015. "Driven to Extinction: On the Probability of Evolutionary Rescue from Sex-Ratio Meiotic Drive." *bioRxiv*. doi:10.1101/018820.
Urashima, Tadasu, Sadaki Asakuma, Fiame Leo, Kenji Fukuda, Michael Messer, and Olav T. Oftedal. 2012. "The Predominance of Type I Oligosaccharides Is a Feature Specific to Human Breast Milk." *Advances in Nutrition* 3:473S–482S.
Urban, Tim. 2015. "My Visit with Elon Musk at SpaceX." Business Insider, May 11. http://www.businessinsider.com/my-visit-with-elon-musk-at-spacex-2015-5 (accessed March 22, 2017).
US Department of Health and Human Services, Office of Minority Health. 2017. "Asthma and Hispanic Americans." http://minorityhealth.hhs.gov/omh/browse.aspx?lvl=4&lvlid=60 (accessed August 24, 2017).
US Holocaust Memorial Museum Photo Archives. 1944. "Soviets Exhume a Mass Grave in Zloczow Shortly After the Liberation." Photograph #86588. Courtesy of Herman Lewinter. http://digitalassets.ushmm.org/photoarchives/detail.aspx?id=16276&search=&index=1 (accessed September 8, 2017).
US National Library of Medicine. 2017. "Tay-Sachs Disease." *Genetics Home Reference*, October 10. http://ghr.nlm.nih.gov/condition/tay-sachs-disease.
"U.S. Panel Urges Testing at Birth." 1961. *New York Times*, December 10, p.80.
Vacca, Marcella, Floriana Della Ragione, Francesco Scalabrì, and Maurizio D'Esposito. 2016. "X Inactivation and Reactivation in X-Linked Diseases." *Seminars in Cell & Developmental Biology* 56:78–87.
Valles, Sean A. 2012. "Lionel Penrose and the Concept of Normal Variation in Human Intelligence." *Studies in History and Philosophy of Science Part C: Studies in History and Philosophy of Biological and Biomedical Sciences* 43:281–89.
Vallot, Céline, Jean-François Ouimette, and Claire Rougeulle. 2016. "Establishment of X Chromosome Inactivation and Epigenomic Features of the Inactive X Depend on Cellular Contexts." *BioEssays* 38. doi:10.1002/bies.201600121.
Van der Pas, Peter. 1970. "The Correspondence of Hugo de Vries and Charles Darwin." *Janus* 57:173–213.
Van Dijk, Bob A., Dorret I. Boomsma, and Achile J. M. de Man. 1996. "Blood Group Chimerism in Human Multiple Births Is Not Rare." *American Journal of Medical Genetics* 61:264–68.
Van Dijk, Peter J., and T. H. Noel Ellis. 2016. "The Full Breadth of Mendel's Genetics." *Genetics* 204:1327–36.
Van Eenennaam, Alison L., Kent A. Weigel, Amy E. Young, Matthew A. Cleveland, and Jack C. M. Dekkers. 2014. "Applied Animal Genomics: Results from the Field." *Annual Review of Animal Biosciences* 2:105–39.
Van Lookeren Campagne, Menno, Erich C. Strauss, and Brian L. Yaspan. 2016. "Age-Related Macular Degeneration: Complement in Action." *Immunobiology* 221:733–39.
Vanneste, Evelyne, Thierry Voet, Cédric Le Caignec, Michèle Ampe, Peter Konings, Cindy Melotte, Sophie Debrock, and others. 2009. "Chromosome Instability Is Common in Human Cleavage-Stage Embryos." *Nature Medicine* 15:577–83.
Van Opstal, Edward J., and Seth R. Bordenstein. 2015. "Rethinking Heritability of the Microbiome." *Science* 349:1172–73.
Varney, Robin L., and Mohamed A. F. Noor. 2010. "The Scuttle Fly." *Current Biology* 20:R466–R467.
Vermont Historical Society. "William Jarvis & the Merino Sheep Craze." http://vermonthistory.org/educate/online-resources/an-era-of-great-change/work-changing-markets/william-jarvis-s-merino-sheep (accessed August 6, 2017).
Vernot, Benjamin, Serena Tucci, Janet Kelso, Joshua G. Schraiber, Aaron B. Wolf, Rachel M. Gittelman, Michael Dannemann, and others. 2016. "Excavating Neandertal and Denisovan DNA from the Genomes of Melanesian Individuals." *Science* 352:1172–73.
Villa, Paola, and Wil Roebroeks. 2014. "Neandertal Demise: An Archaeological Analysis of the Modern Human Superiority Complex." *PLOS One* 9:e96424.

The Vineland Training School. 1896. *8th Annual Report*. Vineland, NJ.
———. 1898. *10th Annual Report*. Vineland, NJ.
———. 1899. *11th Annual Report*. Vineland, NJ.
———. 1906. *18th Annual Report*. Vineland, NJ.
———. 1907. *19th Annual Report*. Vineland, NJ.
———. 1909. *21st Annual Report*. Vineland, NJ.
———. 1910. *22nd Annual Report*. Vineland, NJ.
Vines, Gail. 1997. "Mary Lyon: Quiet Battler." *Current Biology* 7:R269.
Vinkhuyzen, Anna A. E., Naomi R. Wray, Jian Yang, Michael E. Goddard, and Peter M. Visscher. 2013. "Estimation and Partition of Heritability in Human Populations Using Whole-Genome Analysis Methods." *Annual Review of Genetics* 47:75–95.
Vinovskis, Maris. 2008. *The Birth of Head Start: Preschool Education Policies in the Kennedy and Johnson Administrations*. Chicago: University of Chicago Press.
Visscher, Peter M., Brian McEvoy, and Jian Yang. 2010. "From Galton to GWAS: Quantitative Genetics of Human Height." *Genetics Research* 92:371–79.
Visscher, Peter M., Matthew A. Brown, Mark I. McCarthy, and Jian Yang. 2012. "Five Years of GWAS Discovery." *American Journal of Human Genetics* 90:7–24.
———, Sarah E. Medland, Manuel A. R. Ferreira, Katherine I. Morley, Gu Zhu, Belinda K. Cornes, Grant W. Montgomery, and Nicholas G. Martin. 2006. "Assumption-Free Estimation of Heritability from Genome-Wide Identity-by-Descent Sharing between Full Siblings." *PLOS Genetics* 2:e41.
———, Stuart Macgregor, Beben Benyamin, Gu Zhu, Scott Gordon, Sarah Medland, William G. Hill, and others. 2007. "Genome Partitioning of Genetic Variation for Height from 11,214 Sibling Pairs." *American Journal of Human Genetics* 81:1104–10.
de Vries, Hugo. 1904. "The Aim of Experimental Evolution." *Carnegie Institution of Washington Yearbook* 3:39–49.
———. 1905. "A Visit to Luther Burbank." *Popular Science Monthly*, August.
Waddington, C. H. 1957. *The Strategy of the Genes: A Discussion of Some Aspects of Theoretical Biology*. London: George Allen & Unwin.
Wade, Nicholas. 1980. "UCLA Gene Therapy Racked by Friendly Fire." *Science* 210:509.
———. 1981a. "Gene Therapy Caught in More Entanglements." *Science* 212:24–25.
———. 1981b. "Gene Therapy Pioneer Draws Mikadoesque Rap." *Science* 212:1253.
———. 2002. "Scientist Reveals Secret of Genome: It's His." *New York Times*, April 27.
Walfred W. C. Tang, Sabine Dietmann, Naoko Irie, Harry Leitch, Vasileios Floros, and others. 2015. "A Unique Gene Regulatory Network Resets the Human Germline Epigenome for Development." *Cell* 161:1453–67.
Walker, Richard F., Jia Sophie Liu, Brock A. Peters, Beate R. Ritz, Timothy Wu, Roel A. Ophoff, and Steve Horvath. 2015. "Epigenetic Age Analysis of Children Who Seem to Evade Aging." *Aging* 7:334–39.
Wang, Haoyi, Hui Yang, Chikdu S. Shivalila, Meelad M. Dawlaty, Albert W. Cheng, Feng Zhang, and Rudolf Jaenisch. 2013. "One-Step Generation of Mice Carrying Mutations in Multiple Genes by CRISPR/Cas-Mediated Genome Engineering." *Cell* 153:910–18.
Wang, Michael. 2012. "Heavy Breeding." *Cabinet* magazine, Issue 45.
Wang, Zhang, and Martin Wu. 2015. "An Integrated Phylogenomic Approach Toward Pinpointing the Origin of Mitochondria." *Scientific Reports* 5:7949.
Warner, Rebecca H., and Henry L. Rosett. 1975. "The Effects of Drinking on Offspring: An Historical Survey of the American and British Literature." *Journal of Studies on Alcohol* 36:1395–1420.
Warren, Wendy. 2016. *New England Bound: Slavery and Colonization in Early America*. New York: Liveright Publishing.
Wasse, Mr. 1724. "Part of a Letter from the Reverend Mr. Wasse, Rector of Aynho in Northamptonshire, to Dr. Mead, Concerning the Difference in the Height of a Human Body." *Philosophical Transactions of the Royal Society* 33:87–88.
Wasser, Solomon P., ed. 1999. *Evolutionary Theory and Processes: Modern Perspectives. Papers in Honour of Eviatar Nevo*. Dordrecht: Kluwer Academic.
Waxman, Sorrell H., Stanley M. Gartler, and Vincent C. Kelley. 1962. "Apparent Masculinization of the Female Fetus Diagnosed as True Hermaphrodism by Chromosomal Studies." *Journal of Pediatrics* 60:540–44.

Weedon, Michael N., Guillaume Lettre, Rachel M. Freathy, Cecilia M. Lindgren, Benjamin F. Voight, John R. B. Perry, Katherine S. Elliott, and others. 2007. "A Common Variant of HMGA2 Is Associated with Adult and Childhood Height in the General Population." *Nature Genetics* 39:1245–50.

Weil, François. 2013. *Family Trees: A History of Genealogy in America.* Cambridge: Harvard University Press.

Weintraub, Karen. 2013. "Three Biological Parents and a Baby." *New York Times*, December 16.

Weismann, August. 1889. *Essays upon Heredity and Kindred Biological Problems.* Edited by Selmar Schönland, Arthur Everett Shipley, and Edward Bagnall Poulton. Oxford: Clarendon Press.

———. 1893. *The Germ-plasm: A Theory of Heredity.* New York: Scribner's.

Weiss, Sheila Faith. 2010. *The Nazi Symbiosis: Human Genetics and Politics in the Third Reich.* Chicago: University of Chicago Press.

Wellcome Library. "The Lionel Penrose Papers." Digital Collections. Codebreakers: Makers of Modern Genetics. https://wellcomelibrary.org/collections/digital-collections/makers-of -modern-genetics/digitised-archives/lionel-penrose/ (accessed August 24, 2017).

Wellcome Trust Case Control Consortium. 2007. "Genome-Wide Association Study of 14,000 Cases of Seven Common Diseases and 3,000 Shared Controls." *Nature* 447:661–78.

White, William Allen. 1922. "What's the Matter with America?" *Collier's*, July 1, pp.3–4.

White House Photographs. 1961. "John F. Kennedy Meets the McGrath Family." November 14. Kennedy Presidential Library and Museum, Boston. Digital Identifier: JFKWHP-1961-11-14-A.

Whitelaw, Emma. 2015. "Disputing Lamarckian Epigenetic Inheritance in Mammals." *Genome Biology* 16:60.

Whiten, Andrew, Christine A. Caldwell, and Alex Mesoudi. 2016. "Cultural Diffusion in Humans and Other Animals." *Current Opinion in Psychology* 8:15–21.

Wiedemann, H. R., G. R. Burgio, P. Aldenhoff, J. Kunze, H. J. Kaufmann, and E. Schirg. 1983. "The Proteus Syndrome: Partial Gigantism of the Hands and/or Feet, Nevi, Hemihypertrophy, Subcutaneous Tumors, Macrocephaly or Other Skull Anomalies and Possible Accelerated Growth and Visceral Affections. *European Journal of Pediatrics* 140:5–12.

Wilde, Jonathan J., Juliette R. Petersen, and Lee Niswander. 2014. "Genetic, Epigenetic, and Environmental Contributions to Neural Tube Closure." *Annual Review of Genetics* 48:583–611.

Wilkins, Adam S., and Robin Holliday. 2009. "The Evolution of Meiosis from Mitosis." *Genetics* 181:3–12.

Williams, Michael. 2003. *Deforesting the Earth: From Prehistory to Global Crisis.* Chicago: University of Chicago Press.

Williams, R. C., W. C. Knowler, D. J. Pettitt, J. C. Long, D. A. Rokala, H. F. Polesky, R. A. Hackenberg, A. G. Steinberg, and P. H. Bennett. 1992. "The Magnitude and Origin of European-American Admixture in the Gila River Indian Community of Arizona: A Union of Genetics and Demography." *American Journal of Human Genetics* 51:101–10.

Wilschut, Rutger A., Carla Oplaat, L. Basten Snoek, Jan Kirschner, and Koen J. F. Verhoeven. 2015. "Natural Epigenetic Variation Contributes to Heritable Flowering Divergence in a Widespread Asexual Dandelion Lineage." *Molecular Ecology* 25:1759–68.

Winston, Andrew S. 1998. "Science in the Service of the Far Right: Henry E. Garrett, the IAAEE, and the Liberty Lobby." *Journal of Social Issues* 54:179–210.

Witkowski, Jan. 2015. *The Road to Discovery: A Short History of Cold Spring Harbor Laboratory.* Cold Spring Harbor: Cold Spring Harbor Laboratory Press.

Wolinsky, Howard. 2007. "A Mythical Beast: Increased Attention Highlights the Hidden Wonders of Chimeras." *EMBO Reports* 8:212–14.

Wolynn, Mark. Mark Wolynn, The Family Constellation Institute website. http://www .markwolynn.com/ (accessed August 24, 2017).

Woo, S. L., A. S. Lidsky, F. Güttler, T. Chandra, and K. J. Robson. 1983. "Cloned Human Phenylalanine Hydroxylase Gene Allows Prenatal Diagnosis and Carrier Detection of Classical Phenylketonuria." *Nature* 306:151–55.

Wood, Andrew R., Tonu Esko, Jian Yang, Sailaja Vedantam, Tune H. Pers, Stefan Gustafsson, Audrey Y. Chu, and others. 2014. "Defining the Role of Common Variation in the Genomic and Biological Architecture of Adult Human Height." *Nature Genetics* 46:1173–86.

Wood, Edward J. 1868. *Giants and Dwarfs*. London: R. Bentley.

Wood, Roger J. 1973. "Robert Bakewell (1725–1795), Pioneer Animal Breeder, and His Influence on Charles Darwin." *Folia Mendeliana* 58:231.

———, and Vítězslav Orel. 2001. *Genetic Prehistory in Selective Breeding: A Prelude to Mendel*. Oxford: Oxford University Press.

Wright, Donald R. 1981. "Uprooting Kunta Kinte: On the Perils of Relying on Encyclopedic Informants." *History in Africa* 8:205–17.

Wright, Nicholas A. 2014. "Boveri at 100: Cancer Evolution, from Preneoplasia to Malignancy." *Journal of Pathology* 234:146–51.

Wright, Robert. 1995. "TRB: Dumb Bell." *New Republic*, January 2.

Wu, Hao, Junjie Luo, Huimin Yu, Amir Rattner, Alisa Mo, Yanshu Wang, Philip M. Smallwood, Bracha Erlanger, Sarah J. Wheelan, and Jeremy Nathans. 2014. "Cellular Resolution Maps of X Chromosome Inactivation: Implications for Neural Development, Function, and Disease." *Neuron* 81:103–19.

Wu, Yuxuan, Dan Liang, Yinghua Wang, Meizhu Bai, Wei Tang, Shiming Bao, Zhiqiang Yan, Dangsheng Li, and Jinsong Li. 2013. "Correction of a Genetic Disease in Mouse Via Use of CRISPR-Cas9." *Cell Stem Cell* 13:659–62.

Wykes, David L. 2004. "Robert Bakewell (1725–1795) of Dishley: Farmer and Livestock Improver." *Agricultural History Review* 52:38–55.

Xu, Na, Chia-Lun Tsai, and Jeannie T. Lee. 2006. "Transient Homologous Chromosome Pairing Marks the Onset of X Inactivation." *Science* 311:377–89.

Xue, James, Todd Lencz, Ariel Darvasi, Itsik Pe'er, and Shai Carmi. 2016. "The Time and Place of European Admixture in the Ashkenazi Jewish History." *PLOS Genetics* 13:336–45.

Yablonka-Reuveni, Zipora. 2011. "The Skeletal Muscle Satellite Cell: Still Young and Fascinating at 50." *Journal of Histochemistry & Cytochemistry* 59:1041–59.

Yamanaka, Shinya. 2012. "The Winding Road to Pluripotency." Nobel Lecture, December 7. https://www.nobelprize.org/nobel_prizes/medicine/laureates/2012/yamanaka-lecture.html (accessed August 3, 2017).

Yang, Jian, Andrew Bakshi, Zhihong Zhu, Gibran Hemani, Anna A. E. Vinkhuyzen, Sang Hong Lee, Matthew R. Robinson, and others. 2015. "Genetic Variance Estimation with Imputed Variants Finds Negligible Missing Heritability for Human Height and Body Mass Index." *Nature Genetics* 41:1114.

Yin, Kangquan, Caixia Gao, and Jin-Long Qiu. 2017. "Progress and Prospects in Plant Genome Editing." *Nature Plants* 3:17107.

Yong, Ed. 2013. "Chinese Project Probes the Genetics of Genius." *Nature* 497:297.

———. 2016a. "A Google Maps for the Human Body." *Atlantic*, October 14. https://www.theatlantic.com/science/archive/2016/10/a-google-maps-for-the-human-body/504002/ (accessed September 10, 2017).

———. 2016b. *I Contain Multitudes: The Microbes Within Us and a Grander View of Life*. New York: Ecco.

Young, Arthur. 1771. *The Farmer's Tour Through the East of England*. London: W. Strahan.

Yu, Neng, Margot S. Kruskall, Juan J. Yunis, Joan H. M. Knoll, Lynne Uhl, Sharon Alosco, Marina Ohashi, Olga Clavijo, Zaheed Husain, and Emilio J. Yunis. 2002. "Disputed Maternity Leading to Identification of Tetragametic Chimerism." *New England Journal of Medicine* 346:1545–52.

Yudell, Michael. 2014. *Race Unmasked: Biology and Race in the Twentieth Century*. New York: Columbia University Press.

Yunis, Edmond J., Joaquin Zuniga, Viviana Romero, and Emilio J. Yunis. 2007. "Chimerism and Tetragametic Chimerism in Humans: Implications in Autoimmunity, Allorecognition and Tolerance." *Immunologic Research* 38:213–36.

Zenderland, Leila. 1998. *Measuring Minds: Henry Herbert Goddard and the Origins of American Intelligence Testing*. Cambridge: Cambridge University Press.

Zeng, Xiao Xia, Kian Hwa Tan, Ailing Yeo, Piriya Sasajala, Xiaowei Tan, Zhi Cheng Xiao, Gavin Dawe, and Gerald Udolph. 2010. "Pregnancy-Associated Progenitor Cells Differentiate and Mature into Neurons in the Maternal Brain." *Stem Cells and Development* 19:1819–30.

Zerubavel, Eviatar. 2012. *Ancestors and Relatives: Genealogy, Identity and Community*. Oxford: Oxford University Press.

Zhang, John, Guanglun Zhuang, Yong Zeng, Jamie Grifo, Carlo Acosta, Yimin Shu, and Hui Liu. 2016. "Pregnancy Derived from Human Zygote Pronuclear Transfer in a Patient Who Had Arrested Embryos After IVF." *Reproductive BioMedicine Online* 33:529–33.

———, Hui Liu, Shiyu Luo, Zhuo Lu, Alejandro Chávez-Badiola, Zitao Liu, Mingxue Yang, and others. 2017. "Live Birth Derived from Oocyte Spindle Transfer to Prevent Mitochondrial Disease." *Reproductive BioMedicine Online* 34:361–68.

Zhivotovsky, L. A. 1999. "Recognition of the Remains of Tsar Nicholas II and His Family: A Case of Premature Identification?" *Annals of Human Biology* 26:569–77.

Zhou, Qinghua, Haimin Li, Hanzeng Li, Akihisa Nakagawa, Jason L. J. Lin, Eui-Seung Lee, Brian L. Harry, Riley Robert Skeen-Gaar, Yuji Suehiro, and Donna William. 2016. "Mitochondrial Endonuclease G Mediates Breakdown of Paternal Mitochondria upon Fertilization." *Science* 353:394–99.

Zhu, Xiaofeng, J. H. Young, Ervin Fox, Brendan J. Keating, Nora Franceschini, Sunjung Kang, Bamidele Tayo, and others. 2011. "Combined Admixture Mapping and Association Analysis Identifies a Novel Blood Pressure Genetic Locus on 5p13: Contributions from the CARe Consortium." *Human Molecular Genetics* 20:2285–95.

Zickler, Denise, and Nancy Kleckner. 2015. "Recombination, Pairing, and Synapsis of Homologs during Meiosis." *Cold Spring Harbor Perspectives in Biology* 7:a016626.

———. 2016. "A Few of Our Favorite Things: Pairing, the Bouquet, Crossover Interference and Evolution of Meiosis." *Seminars in Cell & Developmental Biology* 54. doi:10.1016 /j.semcdb.2016.02.024.

Zimmer, Carl. 2005. "Children Learn by Monkey See, Monkey Do. Chimps Don't." *New York Times*, December 13.

———. 2008a. *Microcosm: E. Coli and the New Science of Life*. New York: Pantheon Books.

———. 2008b. "The Search for Intelligence." *Scientific American* 299:68–75.

———. 2010. *Brain Cuttings: Fifteen Journeys through the Mind*. New York: Scott & Nix.

———. 2011. "Discovering My Microbiome: 'You, My Friend, Are a Wonderland.'" *Loom*, June 27. http://phenomena.nationalgeographic.com/2011/06/27/discovering-my-microbiome -you-my-friend-are-a-wonderland/ (accessed August 24, 2017).

———. 2014a. "Why Do We Have Blood Types?" *Mosaic*, July 15.

———. 2014b. "White? Black? A Murky Distinction Grows Still Murkier." *New York Times*, December 26, A20.

———. 2015a. "Breakthrough DNA Editor Born of Bacteria." *Quanta*, February 5.

———. 2015b. "The Cords That Aren't Cut." *New York Times*, September 15, D3.

———. 2017. "A Speedier Way to Catalog Human Cells (All 37 Trillion of Them)." *New York Times*, August 22, D6.

Zimmermann, Michael B. 2008. "Research on Iodine Deficiency and Goiter in the 19th and Early 20th Centuries." *Journal of Nutrition* 138:2060–63.

———, Pieter L. Jooste, and Chandrakant S. Pandav. 2008. "Iodine-Deficiency Disorders." *Lancet* 372:1251–62.

Zirkle, Conway. 1946. "The Early History of the Idea of the Inheritance of Acquired Characters and of Pangenesis." *Transactions of the American Philosophical Society* 35:91–151.

ACKNOWLEDGMENTS

For a book about heredity, I must start my thanks with my family. My daughters, Charlotte and Veronica, have been the textbook definition of good sports over the past couple of years. They've put up with their father's endless musings about how heredity has or hasn't influenced who they are. Thanks also go to my brother, Ben, and parents, Marfy Goodspeed and Richard Zimmer, for our long conversations about our family history and for the light they have cast into the shadowy corners of my memory. Thanks most of all go to my beloved wife, Grace. This book would not have existed if not for her. She guided me along as I tried to tame a swirl of thoughts into an idea for a book, sustaining me through the low months when the writing seemed like it would never end, reading every page of the manuscript carefully to let me know what was unclear or unnecessary, and always reminding me of the life we shared beyond my computer monitor.

Deep gratitude goes to Stephen Morrow, my editor at Dutton, for the boundless energy and attention to detail he brought to this book, his vigor undiminished since he edited my earliest books, *At the Water's Edge*, *Parasite Rex*, and *Soul Made Flesh*. I would also like to thank my agent, Eric Simonoff, for always reserving his enthusiasm for the ideas that he finds truly compelling and then giving his all to help me turn them into books.

I'm grateful to the Alfred P. Sloan Foundation for awarding me a fellowship for this book, allowing me to carry out more of the research necessary for such a wide-ranging subject. I'd also like to thank the editors at newspapers and magazines who enabled me to explore some of the topics that made their way into this book, including Michael Mason and Celia Dugger at the *New York Times*, and Michael Moyer and Thomas Lin at *Quanta*. My experience with getting my genome sequenced came about while I worked on a series of articles published by *Stat*, called "Game of Genomes." I'm grateful to Jason Ukman, Jeff DelViscio, and Rick Berke for allowing me to head off into uncharted terrain for those pieces.

For assistance in research, transcribing, and fact-checking, I'd like to thank Helen Bellison, Nakeirah Christie, Asu Erden, Kevin Hwang, Jeremiah Johnston, Saatchi Kalsi, Haleigh Larson, Lauren McNeel, Neal Ravindra, Kevin Wang, and Maddy Zoltek. Thanks to Alice Colwell and Erika Richards for help in German translation. I'm very grateful to Jay Shendure at the University of Washington, who read through the entire manuscript for scientific accuracy, as did Graham Coop and members of his lab at the University of California, Davis: Doc Edge, Erin Calfee, Vince Buffalo, Nancy Chen, Emily Josephs, Sivan Yair, Kristin Lee, and Anita To.

Many people helped me get hold of information that went into this book. I'm particularly grateful to people who hosted me on research visits, especially Patricia Martinelli at the Vineland Historical and Antiquarian Society, as well as the staff at the Luther Burbank Home and Gardens in Santa Rosa, California.

I communicated with many people about the topics I explored in this book—either in person, over the phone, or by e-mail. Some of them also read over portions of my book and shared their responses. I'd like to thank them all for being so generous with their time and knowledge: Erol Akcay, Joshua Akey, Tracy Bale, Tracey Beck, Ethan Bier, Catherine Bliss, Russell Bonduriansky, Christine Brown, Tony Capra, Francisco Ceballos, Christopher Chabris, George Church, Declan Clarke, Nathaniel Comfort, Graham Coop, Ian Deary, Jack Dekkers, Brian Dias, Jill Doerfler, Joseph Ecker, Erle Ellis, Yaniv Erlich, Kevin Esvelt, William Foulkes, Keolo Fox, Valentino Gantz, Mark Gerstein, Simon Gravel, John Greally, Robert Green, Hank Greeley, Sean Harper, Joe Heinrich, Joel Hirschhorn, Greg Hurst, Insoo Hyun, Amiyaal Ilany, Anthony James, Anthony Jose, Fred Kaplan, Eimear Kenny, Johannes Krause, Leonid Kruglyak, Sushant Kumar, Amanda Larracuente, Iosif Lazairidis, Zachary Lippman, Isabelle Mansuy, Robert Martienssen, Christopher Mason, Iain Mathieson, John McCutcheon, Maurizio Meloni, Elizabeth Murchison, Alondra Nelson, Faranza Parshankar, Diane Paul, Nathan Pearson, Joseph Pickrell, Ron Pinhasi, Danielle Posthuma, James Priest, Jonathan Pritchard, Erik Puffenberger, Jennifer Raff, David Reich, Stuart Ritchie, Noah Rosenberg, Beth Shapiro, Adam Siepel, Robert Sparrow, Kevin Strauss, Sonia Sutti, Kim TallBear, Ali Torkamani, Sici Tsoi, Tobias Uller, Peter Visscher, Christopher Walsh, Eske Willerslev, and Melinda Zeder.

INDEX

Aboriginal Australians, 448–51, 454, 466–68
Ache people, 465
acquired traits, 56–57, 426–27. *See also* Lamarckism
acromegaly, 274–75, 277
Adams, James Truslow, 567–68
Adygei people, 223
African Americans, 166–67, 179–80, 206, 212, 315–16, 567
age-related macular degeneration (AMD), 277
Agius, Emmanuel, 534, 538
Agricultural Revolution, 31–33, 468–69, 476–77, 492, 495, 563
AIP gene, 275
Akan people, 180, 461
Akey, Joshua, 245
Alexandra, Princess of Wales, 352
alkaptonuria, 60, 117
alleles, 126, 149, 151, 279–80, 475, 539
Allen, Elizabeth, 92, 105
alphaproteobacteria, 418–19
Altmann, Richard, 417–19
Alzheimer's disease, 529, 538–39
American Association for the Advancement of Science, 512–13
American Indians, 19, 198–99
American Museum of Natural History, 235
American Psychological Association, 102, 314
American Society of Human Genetics, 500
amino acids, 116, 134, 138, 139, 410. *See also* proteins
aneuploidy, 367–68
ankylosing spondylitis, 535–36
anthropology
    and cultural inheritance, 448, 451–52
    and cumulative culture, 460–61, 465
    and Du Bois's research, 203
    and paleogenetics, 225–26
    and racial classifications, 196
    and scientific racism, 206–7
    and skin color, 201
    and tracing lineages, 178
    and wealth inequality, 470
    See also paleoanthropology
Anthropometric Laboratory, 290
antibiotic resistance, 141–42
antibodies, 143, 217, 345, 377, 401, 556–58
Apinayé people, 13
apple trees, 41–42
"The Apportionment of Human Diversity" (Lewontin), 208
Arabidopsis thaliana, 443
archaea, 139, 144
archaeology, 178, 225–27
Archebaud, John, 159
Aristotle, 14–16, 19, 24, 219, 324–25, 331–32, 484
artificial insemination, 501–4. *See also* in vitro fertilization
Aryans, 123, 498

Asbury, Kathryn, 317
Ashkenazi Jews, 180, 212, 220, 222–24
Atabrine, 385–86
Augustus Caesar, 253
Aurignacian culture, 226–27
aurochs, 498
Australian Aboriginals, 448–51, 467
Austro-Hungarian Empire, 221
autoimmune diseases, 389, 536
Avdonin, Alexander, 175–76
Avery, Oswald, 140–41, 508
Ayash, Chen Aida, 541

bacteria
    and cell division, 323–24
    and cell theory, 326
    and CRISPR system, 143–44, 488–89
    and discovery of genes, 124
    environmental influences of, 564
    and evolution of DNA-based life, 138–42
    and genetic engineering, 508–9, 511
    and immune response, 345, 401, 487–89
    and light organs, 406
    and microbiomes, 407–17, 466
    and mitochondria, 144–45, 417–21, 466
    and origin of eukaryotes, 144–45
    and paleogenetics, 225, 247
Bailey, Harriet, 197
Bakewell, Robert, 33–36, 44
bald eagle evidence, 171, 173–74, 499
Bale, Tracy, 441
Baltimore, David, 523–24, 527, 534, 565
Banfield, Jill, 488–89
Bantu people, 233
Barnes, Jennifer, 447
Barrett, Helen, 313
Barrett, Louise, 270
Barry, Joan, 171–74
Bartels, Friedrich, 312
Bateson, William, 60, 137
Bath, Sarah, 306–7
Beck, Tracy, 131
Begley, Sharon, 527
bell curve, 259–60, 263, 290–91
Belov, Kathy, 393
Berg, Paul, 523–24
beta-thalassemia, 510, 512, 525
BHH gene, 525
Bianchi, Diana, 387, 388, 390
Bickel, Horst, 126–29
Bier, Ethan, 551–55, 558, 560–61, 572
Biesecker, Leslie, 356–57
bifidobacteria, 416
Binet, Alfred, 77, 96
Blaine, Delabere, 395
Blashko, Alfred, 350–51

ABOUT THE AUTHOR

CARL ZIMMER writes the Matter column for the *New York Times* and has contributed to *The Atlantic, National Geographic, Time*, and *Scientific American*. He has won the Stephen Jay Gould Prize, among many other honors, for his journalism. Zimmer teaches science writing at Yale University. His previous books include *Parasite Rex, Evolution: The Triumph of an Idea*, and *Microcosm*.